Mechanical Engineering Series

Frederick F. Ling
Editor-in-Chief

Mechanical Engineering Series

J. Angeles, **Fundamentals of Robotic Mechanical Systems: Theory, Methods, and Algorithms, 2nd ed.**

P. Basu, C. Kefa, and L. Jestin, **Boilers and Burners: Design and Theory**

J.M. Berthelot, **Composite Materials: Mechanical Behavior and Structural Analysis**

I.J. Busch-Vishniac, **Electromechanical Sensors and Actuators**

J. Chakrabarty, **Applied Plasticity**

K.K. Choi and N.H. Kim, **Structural Sensitivity Analysis and Optimization 1: Linear Systems**

K.K. Choi and N.H. Kim, **Structural Sensitivity Analysis and Optimization 2: Nonlinear Systems and Applications**

G. Chryssolouris, **Manufacturing Systems**

G. Chryssolouris, **Laser Machining: Theory and Practice**

V.N. Constantinescu, **Laminar Viscous Flow**

G.A. Costello, **Theory of Wire Rope, 2nd Ed.**

K. Czolczynski, **Rotordynamics of Gas-Lubricated Journal Bearing Systems**

M.S. Darlow, **Balancing of High-Speed Machinery**

W. R. DeVries, **Analysis of Material Removal Processes**

J.F. Doyle, **Nonlinear Analysis of Thin-Walled Structures: Statics, Dynamics, and Stability**

J.F. Doyle, **Wave Propagation in Structures: Spectral Analysis Using Fast Discrete Fourier Transforms, 2nd ed.**

P.A. Engel, **Structural Analysis of Printed Circuit Board Systems**

A.C. Fischer-Cripps, **Introduction to Contact Mechanics**

A.C. Fischer-Cripps, **Nanoindentations, 2nd ed.**

J. García de Jalón and E. Bayo, **Kinematic and Dynamic Simulation of Multibody Systems: The Real-Time Challenge**

W.K. Gawronski, **Advanced Structural Dynamics and Active Control of**

(continued after index)

Yakov M. Tseytlin

Structural Synthesis in Precision Elasticity

Springer

Yakov M. Tseytlin
Instrument Society of America
Providence, RI, USA

Editor-in-Chief
Frederick F. Ling
Earnest F. Gloyna Regents Chair Emeritus in Engineering
Department of Mechanical Engineering
The University of Texas at Austin
Austin, TX 78712-1063, USA
and
Distinguished William Howard Hart
Professor Emeritus
Department of Mechanical Engineering,
Aeronautical Engineering and Mechanics
Rensselaer Polytechnic Institute
Troy, NY 12180-3590, USA

Structural Synthesis in Precision Elasticity

Library of Congress Control Number: 2005938551

ISBN 10: 0-387-25156-1
ISBN 13: 978-0-387-25165-1
ISBN 10: 0-387-25157-X (ebook)

Printed in the United States of America.

9 8 7 6 5 4 3 2 1 SPIN 10949818

springeronline.com

DEDICATION

This book is dedicated with love to Mark, Franklin, Lev, and to bright memory of Adel, Ivan, Michael, and Meyna.

Mechanical Engineering Series

Frederick F. Ling
Editor-in-Chief

The Mechanical Engineering Series features graduate texts and research monographs to address the need for information in contemporary mechanical engineering, including areas of concentration of applied mechanics, biomechanics, computational mechanics, dynamical systems and control, energetics, mechanics of materials, processing, production systems, thermal science, and tribology.

Series Preface

Mechanical engineering, and engineering discipline born of the needs of the industrial revolution, is once again asked to do its substantial share in the call for industrial renewal. The general call is urgent as we face profound issues of productivity and competitiveness that require engineering solutions, among others. The Mechanical Engineering Series is a series featuring graduate texts and research monographs intended to address the need for information in contemporary areas of mechanical engineering.

The series is conceived as a comprehensive one that covers a broad range of concentrations important to mechanical engineering graduate education and research. We are fortunate to have a distinguished roster of consulting editors, each an expert in one of the areas of concentration. The names of the consulting editors are listed on page vi of this volume. The areas of concentration are applied mechanics, biomechanics, computational mechanics, dynamic systems and control, energetics, mechanics of materials, processing, thermal science, and tribology.

PREFACE

Elasticity has been utilized in many scientific and industrial instruments and precision machines. Elastic mechanisms, transducers, flexures, and cantilevers in various forms have found application in a variety of high precision measuring heads, automatic sorting and converting machines, mass balances, torsion pendulums, scanning force and tunnel microscopy, X-ray interferometers, and also in consumer computer products, galvanometers, gyroscopes, and micro- and nano-mechanical systems. These systems provide smooth, external friction free, and wear free motion. Pneumatic measurement and control systems on the basis of precision elasticity are involved in many industrial applications.

Elastic systems have a special structure because the elastic deformation of the solid body, when the body is capable of returning to its undeformed configuration upon removal of external effects, for most materials is very limited. The synthesis of a precision elastic structure requires a thorough knowledge of theoretical basics and special approach in the direction of form stability, physical and geometric nonlinearity limitation, parameters optimization, stress-strain uncertainties and fatigue lifetime evaluation at reference and operating conditions, and effective correctors application.

Let us recall that the structural synthesis is an iteration process, the first steps of which are partly heuristic, based on the requests, collective experience, and digestible approach. The application of more specialized computerized methods is usually effective in the following steps of the structure creation after choosing the comprehensive model. We also remind that the terms precision and accurate ("exact") by definition have different meanings. If the first term is referred to repeatability and small variance of the specified functions and parameters in the synthesized structure, the second one means closeness of the agreement between a calculated value and "exact" theoretical and/or experimental measurement result. In this book we care about the synthesis of precision structures, but in many cases this synthesis is based on approximate methods which have certain theoretical support. The necessary accuracy in many cases then is provided by adjustable correctors in straight and feedback chains, which may be mechanical or have other physical nature. However, the satisfactory accuracy cannot be achieved if the system does not have necessary precision. This does not mean that accuracy is not important for structural synthesis in precision elasticity. It is very important because we *a priori* need to know effective functional relationships between structural constituents and their parameters. Besides that, the correctors' options are limited and some structures do not require or allow them. Another way to achieve necessary accuracy is the adequate calibration and mapping. Therefore, we have developed and discuss in the book the informational principle of acceptable uncertainty, which provides us with effective tool for structural synthesis in precision elasticity. This criterion constrains uncertainty on the basis of necessary and satisfactory requirements.

The book reflects the summary of theoretical and experimental studies whose conclusions are effective for optimized structural synthesis in precision elasticity, as well as demonstrate a large experience and options in the synthesis, production, and application of precision elastic guides, mechanisms, correctors, transducers, instruments, and machines. The main focus of this book is in the possible simplification of the corresponding analytical apparatus by using kinematical equivalents, matrix methods, appropriate quasi-linear approximations, conformal mapping of approximate contours, and function expansion with enough accurate minimal polynomials. This approach allows us to substitute some known unwieldy formulae and methods that are not convenient for digestible and tractable synthesis. In many cases, satisfactory approximations are more effective for structural synthesis in precision elasticity than the accurate ("exact") theoretical solutions, which in some applications are indeed not adequate to a real object. In addition, the latter cannot be implemented in real elastic structures without a loss of accuracy as a result of an uncertainty in shapes, dimensions of elastic elements, and elastic parameters of materials especially in micro- and nano-scale structures.

The strain and stress states of elastic systems are usually described by tensor and matrix apparatuses with coordinate related constituents. These analytical methods are perfectly developed but in some cases they expressed through unwieldy equations, which are not digestible and tractable, require numerical integration and special software application. They are especially complicated for a nonlinear system. Therefore, for the structural synthesis in precision elasticity, an accurate enough approximation and linearization are very useful.

It may be shown that the comprehensive combination of tension, torsion, and flexure can be found in the elastic helicoids. Helicoidal motion is the most general in kinematics. Indeed, the helicoid type stresses-strain combination is present in many real structural systems, including turbine blades, propellers, and typical flexures especially at the spatial system of applied forces and constraints.

The helicoidal elastic transducers can be the most sensitive and easily integrated in modern intelligent systems. Helical forms are also found in natural structures such as DNA. Therefore, we pay significant attention in the book to evaluation of the main features of elastic helicoids. Precision elastic structures have many other elastic elements: flat and corrugated diaphragms, punched plates, rings, bellows, elastic shells, band, leaf, and notch spring mechanisms whose related features are also discussed in detail.

The book consists of two main parts:

A. The elastic systems functional analysis and structural synthesis methods, including effective approximations and references to the history of their development;

B. The application and development of precision functional elastic systems at reference and operating conditions, including the observation of archives with effective synthesized structures.

Each part provides theoretical basics and a large variety of examples and recommendations. This book gives theoretical and practical tools to researchers, precision machines, instruments and miniature systems designers, engineers, metrologists, and also to students of certain engineering courses. It is a handbook as well as a textbook.

Structural synthesis in precision elasticity usually includes four necessary steps:

i) Specification of working range, required precision, level of expected automation and speed of operation, level of effective space constrains and/or miniaturization, reference and operating conditions of application;

ii) Synthesis of elastic systems' schemes with optimization of their structure in the direction of elements' features and their links' uncertainty minimization, application of typical and verified solutions, and original combination of effective structures or specification parameters of original elastic elements;
iii) Quasi-static and dynamic parameters of applied elastic elements and their systems optimization with the digestible and tractable methods for evaluation of materials' elastic features and limitation of predictable non-elastic options;
iv) Application effective correctors, an adjustment resolution and compensatory range estimation for specified and predictable conditions of traceability.

These problems are discussed in part A and part B of the book. Despite that this book is dedicated to the general problems of the structural synthesis in precision elasticity, most of the practical examples and applications are concerned with the measuring systems as the precision is their main goal.

The author intended to show close connection between the elastic precision structures developed during the 20th century and even before and the new elastic systems for atomic force microscopy and other recently created advanced structures in precision elasticity. In the last decade, the elastic precision structures have found the most impressive application in the micro-electromechanical and opto-mechanical systems (MEMS and MOMS) with easy miniaturization and manufacturing, small energy consumption, and independent functioning without grease or any other special supply. The effective application of miniature precision elastic structures has started in the nanotechnology branches including nonoelectromechanical systems (NEMS), imaging and atomic force microscopes' cantilevers and cantilever and string arrays with different features and shapes, atom, molecular, subattogram, and subattonewton sensitivity, microtweezers for the manipulation of bacteria and small particles.

This book reflects the author's experience and represents his ideas and opinions. Results of many other researchers in this field are also discussed in the book. However, we do not discuss in this book ordinary extension and compression springs well known from the general mechanics as well as the spring motors with spiral springs and other similar elements, which do not have special precision requirements.

Although every effort has been made to eliminate errors, experience dictates their inevitable presence in such comprehensive work. When detected, the author would be grateful for guidance on errors or omissions. Extensive references are included to enable the reader to verify discussed problems and solutions. We should also note that in the preparation of this book reasonable efforts have been made to publish reliable data and information, but the author cannot assume responsibility for the validity of all referenced materials or for the consequences of their inappropriate use.

The author hopes that this book will help to develop new and more effective structures in precision elasticity.

Acknowledgements

The author would like to express his gratitude to the Russian and American manufacturing companies, academic and research Institutions who supported author's efforts in research and development of effective precision systems. Thanks to the Plenum Publishing Corporation and Allerton Press who translated many of my articles dedicated to the problems of precision mechanics and metrology from Russian to English and published them in US. Thanks also to the Instrumentation Society of America and especially to E.W. Taylor who supported author's efforts in presentation research results at the International Instrumentation Symposiums. Author expresses great gratitude to the publisher's editors whose help in publication of this book and preparation of the index is essential. The author is grateful to professor of MIT A.H. Slocum and Ph.D., P.E. A.J. Rafanelli who supported author's initial intension to write the book on American soil. Special appreciation goes to the Brown University Writing Center associates T.R. Sheehan and M.L. Richmond who helped with discussion and development of the book's text structures.

Contents

A.

Synthesis of Elastic Systems

In part A we discuss how to obtain accurate and reliable kinematics and sensitive transducers in elastic systems with tension, torsion, and flexure; combination of straight and feedback elastic links for increasing the working range and minimization of bias and random uncertainties in the transmission chains; the general and specific relations including elastic form stability and its possible bifurcation, functional bifurcating output, the greatest transmission ratios in elastic mechanism, and limitations of the physical and geometric nonlinearity.

1.

Problems and Methods of Synthesis

In this chapter we discuss classification of the main functional constituents in precision elastic systems, general problems of precision kinematics and dynamics, and specifics of elastic kinematics and dynamics. Let us recall that the main difference between analysis and synthesis of precision structures is in the specified features and goals.

1-1. Schematics of Typical Elastic Systems

Here we present the functional classification of rotary and translational guides, planar and spatial elastic mechanisms, elastic sensitive transducers with amplitude and frequency output, including the bifurcating types, flexible links, typical elastic kinematics, force, and influence correctors.

General classification. Elastic systems (ELS) in the precision elasticity are usually used as the elastic kinematic systems (ELKS for plural and single versions) and as the transducers (ELT) with amplitude (AM) and frequency (FR) output (Fig. 1-1). ELT include also amplitude and frequency sensors of different kind, which may be connected with the transducer's probe and have microscopic dimensions.

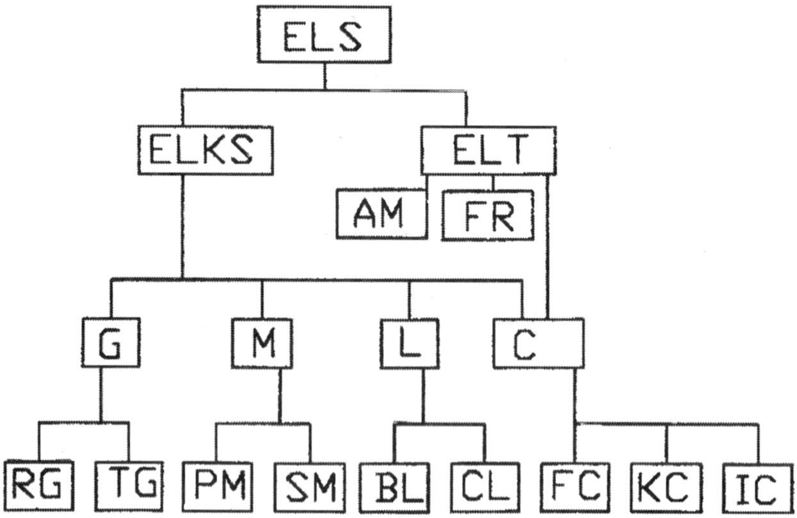

Fig. 1-1. Classification of the typical elastic systems.

The ELKS can be divided in accordance with their functional application into guides (G) for rotary (RG) and translational (TG) motion, mechanisms (M) with planar (PM) and spatial (SM) kinematics, and elastic links (L) with flexible bands (BL) and elastic couplings (CL). Correctors (C) with compensatory options for force (FC), kinematics (KC), and influence (IC) quantities from working, transportation, and storage conditions represent a special group that can be used in ELKS and ELT as well. Influence quantities include scalar, vector, and space field static and dynamic factors such as: temperature, pressure, gravitational, electrostatic and electromagnetic, radiation fields, forced vibration parameters; humidity, velocity, and dust content of ambient air, etc.

Conditional images. Let us use for the schematic unification the conditional images of strip springs, bands, and diaphragms in their typical deformed state. These conditional images, however, do not directly correspond to a real form of the elastic line (elastica) that depends on the acting load. For example, the elastic line for the flat strip without load is a straight line, at the pure bending it is a circular arc, and at the parallel displacement of the ends it is a smooth line with a point *IP* of inflection (Fig. 1-2a). The conditional image of the initially flat flexible beam (thin strip, band) in the schematics will be as shown in Fig. 1-2b and independent from the type of load and the real form of the elastic line. In this case L_{BC} is the length of strip in the schematic scale, $r_{con} \geq 0.1\, L_{BC}$; $s_{con} \leq 0.3\, L_{BC}$ (but not less than 2.5mm). An arc with radius r_{con} corresponds to a quarter or slightly less portion of a circle. In the latter case, the arcs should be connected by a straight line whose length does not exceed $2r_{con.}$ The place for clamping a rigid body 1 (lever) to the strip may be marked by a black point (solid semicircle). Both ends of the strip's conditional image are straight and parallel to each other. One end can be rotated to angle θ, which is less than 20° (shown by dashed line).

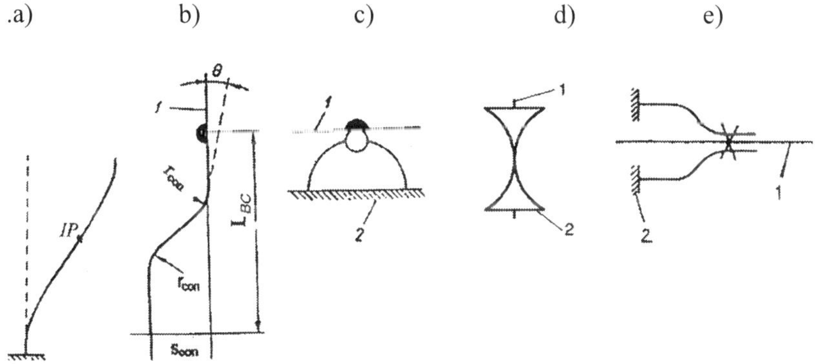

Fig. 1-2. Schematics of the typical elastic elements.

The conditional images for many typical ELKS are presented in Tables 1-1 through 1-6 of this section. Let us discuss some typical schematics of wide used ELKS (their structures' elements and calculation methods are shown in Chaps. 3-5, 7, and 8). The *RG* (a conditional image is shown in Fig. 1-2c) provide rotary motion to the movable, usually rigid body 1 (a crank, connecting rod, rocker) relative to stand 2 (base, fixed element). For elastic RG we use flat strip springs (leaf spring with conditional image in Fig.1-2 b and set of leaf springs), notch hinges (a conditional image in Fig. 1-2d), angle hinges, torsion wires and strips. The main type of deformation in the leaf springs and notch hinges is a planar flexure; in the angle hinges there is a constrained torsion similar to that in the thin-walled open sections, and the torsion-tension suspensions have a torsion deformation with quasi-constant longitudinal tension.

The elastic rotary guides.

RG-I. The *leaf spring hinge* (Table1-1, ELKS of type RG-I) has stand 1 (fixed element), flexible spring strip 2, and movable element 3. The strip is flat in the unloaded state, but in some cases a strip with a preliminary flexed form is used. The movable element can rotate to the angle θ around the axis of rotation, which is parallel to the longer side of the strip's cross section. However, the position of this axis along the strip, as will be shown in Chaps. 3 and 4, is not constant and depends on the applied loads. The leaf spring hinges are usually used for the pendulums and levers where there are no requirements for high accuracy of motion kinematics.

The *double leaf spring hinge* has a larger limit for rotation of movable element 3 (ELKS of type RG-I*d*) relative to stand 1. Intermediate moving element 5 also rotates to an angle $\theta_1 \approx \frac{1}{2}\theta$ at the deformation of the flexible elastic strips 2 and 4.

RG-II. The universal *cross-strip hinge* (ELKS of type RG-II) has stand 1, flexible strips 2 and 4, and movable guided rigid element 3. Lucas-Troeger elastic cross-strip pivots RG-II*s*, *st* with standardized dimensions (*see* Chap. 4) were built and manufactured in the USA on this basis in 1960-1990s.

Table 1-1. Rotary elastic guides' structural schematics

Name, type, range	Schematic	Name, type, range	Schematic
Single*leaf spring ELKS RG-I θ≤ 15°, for pendulums & levers	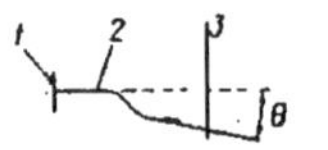	*Notch hinge ELKS RG-IV*c*-circular, IV*e*-elliptic, IV*f*-filleted, IV*co*-concave, for AFM etc.	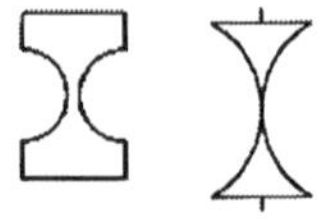
Double leaf spring RG-I*d* θ≤ 30°	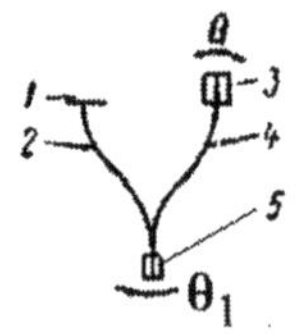	Double and multi-motion hinge with strip RG-V*s* θ≤ 10°; with notch hinges RG-V*n*; with " quasi-toroidal" notch hinge RG-V *nqt*	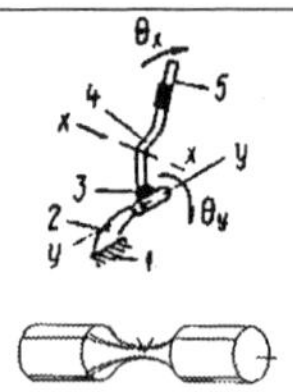
*Cross-strip ELKS RG-II θ≤ 30°, Lucas-Troeger pivots symmetrical ELKS RG-II*s*, and with standardized dimensions RG-II*st*	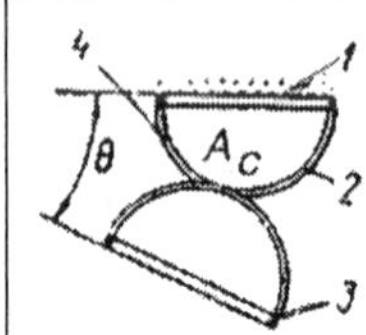	Profiled hinges ELKS RG-VI*s*-single, -VI*d*-double, -VI*c*-cruciform θ≤ 3°,for measuring instruments	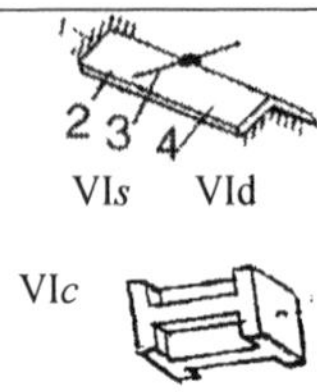
Asymmetric cross-strip ELKS RG-II*a* Imaginary-crossing-strip ELKS RG-II*i* Two curved-strip ELKS RG-II*cu*	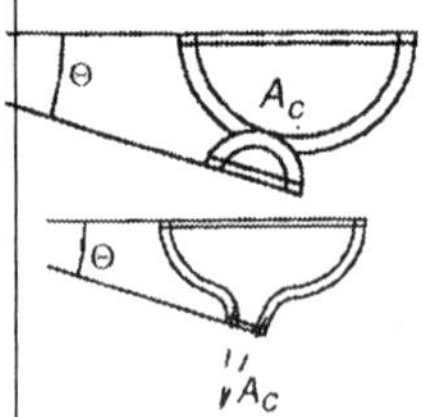	Cartwheel hinge ELKS RG-VII*c* Hexfoil hinge ELKS RG-VII*h* monolithic structure, for miniature systems	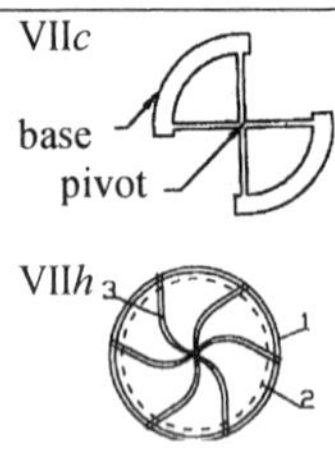
Double cross-strip ELKS RG-II*d* θ≤ 60°	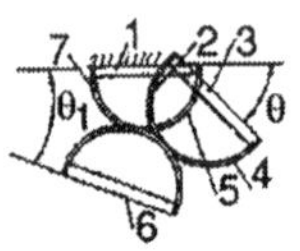	Torsion filar hinge ELKS RG-VIII For gyroscope, galvanometer	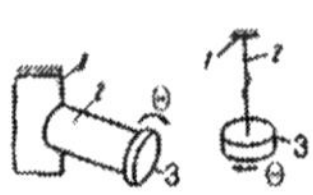
Three cross- strip ELKS RG-III for gyroscopes θ≤ 3° Three strip "fork-like" RG-III*af* and "lance-like" (spade) RG-III*as* θ≤ 10°	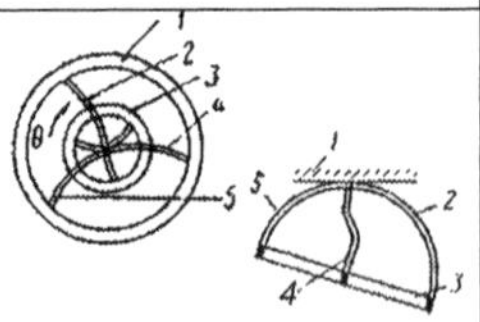 RG-III*af*	Multi-torsion hinge ELKS RG-VIII*m* for gyroscope	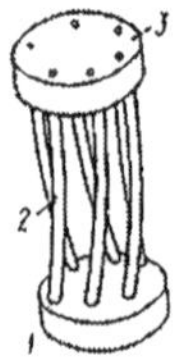
Multi-strip hinge ELKS RG-III*m* for instruments	*See* Fig. 4-3f	Double filar suspension ELKS RG-IX for electromechanical instruments, AFM	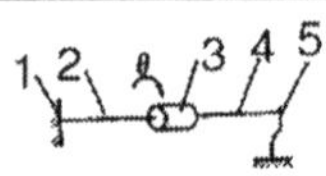

Note: * marks the guides whose features are discussed in this book in detail.

Point A_c of strips crossing can be in asymmetric position (RG-II*a*) and even imaginary in the case of the so-called non-crossing or imaginary (protracted)-crossing-strip hinge (RG-II*i*). Cross-strip hinge with initially curved strips (RG-II*cu*) found some applications as well.

Double cross - strip hinges (RG-II*d*) have stand 1, flexible strips 2, 4, 5, 7, movable guided rigid element 3, and intermediate moving element 6. Thus, this ELKS consists of two consecutively acting cross-strip hinges. Double cross -strip hinges allow element 3 to rotate at a relatively large angle θ (approximately $\pm$ 60° from the initial position) with comparatively small deflection of the axis of rotation. Intermediate element 6 also rotates to the angle $\theta_1 \approx ½\ \theta$.

RG-III. *Three cross-strip hinges* can have different structures. Three cross-strip hinges with three equal crossing strips (RG-III) have stand 1, three equal flexible strips 2, 4, 5, and movable guided element 3 which can rotate to some angle θ. The strips are flat in the initial non-deformed position and mutually positioned at the angle 120°. The disadvantages of this device are due to its relatively complicated structure and a small angle of rotation for movable element 3 (θ in the order of 1-3°). The advantage of such a hinge is in the more stable position of an axis of rotation. Another " fork-like" or "lace-like"(spades-like) three strip hinge (RG-III*af*, RG-III*as*) with initially straight basic 4 and two additional curved flexible strips 2 and 5 also provides the rotational motion of element 3 in relation to stand 1. In this case the geometric and elastic parameters of the basic and additional strips should be different with the definite optimal relationship for the axis of rotation position better stability (*see* Chap.4).

Multi-strip hinges (RG-III*m*) have the number of strips that exceeds three. In particular, four-strip hinges are used (*see* Fig. 4-3f) in the suspension of pendulums. They are similar to the three-strip hinges in their structure and functions, but their strips are mutually positioned under a different angle in the initial state. The angle of rotation with this hinges is usually within $\theta = \pm$ 3°.

RG-IV. *Notch hinges* with a filleted (RG-IV*f*), circular (RG-IV*c*), elliptic (RG-IV*e*), or other concave (RG-IV*co*) profile find application in a large variety of precision, micro- and nano-scale systems: piezoelectric motors, pickup head actuators, flexure hinges for an optical disk precision positioning, scanning atomic force (AFM) and tunnel (STM) microscopes, X-ray interferometers, and measuring systems for precision plunger pairs in the diesel fuel apparatus.

RG-V. A *double motion strip or double-notch hinge* (RG-V*s*,*n*) are also known which provide the rotation of the movable element around the two mutually perpendicular axes. This hinge may be considered a strip or notch hinge of the second kind. It consists of the sequence of two leaf spring hinges (1, 2, 3, and 3, 4, 5) that are mutually unfolded at 90°, or of two notch hinges unfolded at 90°. In some applications the single notch hinge finds use as a hinge with rotation options in two mutually perpendicular directions, one of which has a flexural stress and another has torsional stress. There are some applications for notch hinges with "quasi-toroidal" (circle of rotation, universal circular flexure) shape (RG-V*nqt*), which provide equal flexural stiffness in any direction within 360°.

RG-VI. The *profiled torsion hinges* can have different schematics and structures. A single-angle spring torsion hinge (RG-VI*s*) has stand 1, elastic angle 2, and movable element 3 that can rotate around the angle's shear center which coincides with twist center. Because here we have a constrained type of torsion, as for thin-walled beams, element 3 can have simultaneously a rotation and displacement along the axis and in the plain that is perpendicular to the axis of rotation. Besides that, the cantilever application of alternative load can cause unstable work of this hinge. Profiled torsion hinges can have elastic bodies with a so-called *cruciform shape* (RG-VI*c*) that provides better resistance to the flexure in the two planes. The *double-angle hinges* (RG-VI*d*) have more stable kinematics under changeable load; they have two parts of elastic angles 2 and 4 that are twisted in opposite directions at the rotation of element 3.

RG-VII. The *cartwheel hinge* (RG-VII*c*) is similar to a cross-strip pivot. Both consist of a pair of leaf-type flexures, usually at right angles to each other, with the point of intersection being considered the pivot point. Cartwheel hinges differ from cross-strip pivots in that the flexure spokes coincide at the hinge axis and are joined. Such geometry is both symmetric and amenable to manufacture from a monolithic piece of material. For some structures, the cartwheel hinge may be integrated within complete monolithic mechanisms, therefore reducing the problems associated with such an assembly. Another version of the near axis connected strip-hinges is the *hexfoil* with the doubled three mutually connected strips 3 (VII*h*). The *hexfoil rotary flexure* has been proposed (*see* NASA TechBriefs, 2002) as pivots suitable for use in precise optical instruments. In the application that inspired the hexfoil concept, there is a requirement for limited-rotation mirror gimbal 2 that would maintain coincidence among the axis of rotation and a fiducial mark on the mirror, over the entire range of rotation on the base 1. Its monolithic miniature version is similar to the cartwheel, but with two axially displaced rigid cylindrical bodies 1and 2 and six (three pairs) parallelogram-like spokes.

RG-VIII. The *torsion filar hinges* have cantilever thin elastic rod 2 with a round or rectangular cross section. If the flexural rigidity of the rod is small (conditionally marked by sinusoid on schematic), it should have longitudinal tension that may be provided by the mass of rotating element 3 at the vertical position of the rod clamped to stand 1. This device is called a hanger. The torsion hinge was already known in 1750 and used then in the torsion balances. It has a new application in the special torsion balances (Quinn *et al.*, 1992) for the gravitational constant estimation, and as fused silica suspension (first proposed by V.B. Braginskii) for the VIRGO gravitational waves detector. Initially the torsion hinge was made from wire with a round cross section, and only since 1826 mechanics have been using the rectangular strip with lesser torsion rigidity at an equal cross-sectional area. The *multi-torsion* hinges with several twisting thin rods 2 are also used (RG-VIII*m*).

RG-IX. The *double tension-torsion suspensions* have constant longitudinal tension by clamping one of strip ends to spring 5. Movable element 3 is positioned between strips 2 and 4, which by the tension-torsion suspension provide stability to its position in axial and radial directions. The tension-torsion suspension incorporates thin strips with an elongated rectangular or oval cross section. The double tension-torsion suspensions were widely used in the precision electromechanical instruments (galvanometers, micro-ampermeters, oscillographs, etc.) and for some models of AFM. The structures with strip suspension may be compared with the suspension structures based on the two pairs of the round filaments. The strip suspension has a larger load capability at equal torsion rigidity and a lesser torsion rigidity at the equal cross-sectional area.

The *elastic translational guides.*

These guides (the conditional image is shown in Fig. 1-2e) provide translational motion for guided movable rigid element 1(slider) relative to stand 2. For this purpose different strip and notch devices, plates, diaphragms, and bellows are used. Elastic parallelogram (with elastic strips) and quasi-parallelogram (with notch hinges, plates, diaphragms, and bellows) systems are typical (*see* Table 1-2).

TG-I. An *ordinary spring parallelogram* is a two-strip device. It has stand 1, flat strips 2 and 4, and movable element 3 (carriage-slider) which translationally moves in relation to the stand. The elastic strips are parallel in the initial unloaded position, and the distances between the points of their clamping AB and CD are equal. The basic direction of the motion of a carriage is shown by arrow s_1 and the transverse translational motion is shown by arrow s_2. The displacement s_1 is usually perpendicular to the initial axis of strips, and s_2 is parallel to the strip direction. The displacements s_1 and s_2 are functionally related.

TG-II. The *spring parallelogram with rigid cover plates* 5 clamped to flexible strips 2 and 4 is more stable in the presence of changes in the load conditions on movable carriage 3. However, it has increased structural friction.

TG-III. The springs 2 and 4 parallelogram with additional curved strips 5 and 6 on both sides is also more stable in the presence of changeable load conditions, but this more complicated structure does not provide rectilinear motion and has fewer industrial applications than ELKS of type TG-II.

TG-IV. The *rectilinear spring compound movement* provides a slider 7with motion that is close to the straight line. In this case, the parasitic displacement of movable element (slider) 7 in the direction s_2 can be made small enough at the significant working motion in the direction s_1. ELKS TG-IV consists of stand 1, flexible strips 2, 3, 5, 6, main slider 7, and intermediate slider 4. Thus, this ELKS has two consequently acting ordinary spring parallelograms. The rectilinear spring compound movements are practically used for guiding the movable element at displacement s_1 up to 100 mm with a trajectory that is close to the straight line. It is because its absolute motion in direction s_2 is a sum of the transverse motion s_3 of intermediate element 4 and relative motion s_{74} of carriage 7 towards element 4 in the opposite direction, i.e. $| s_2 | = | s_3|-|s_{74}| < | s_3 |$. The modules of values s_3 and s_{74} are not exactly equal as a result of the different conditions in longitudinal loading for corresponding elastic elements. Intermediate element 4 has a displacement that is translational but is not so close to the rectilinear one as that of basic element 7. ELKS TG-IV are used to guide the scanning reference mirrors in interferometers and spectrometers. The analogous system can be built on the basis of notch hinges (RG-IV*n*) and rigid levers as well. Note that the notch hinges may have very small dimensions, accurate, and simple in assembly. Therefore they effective in micro-nano-systems , where the external friction should be eliminated. However, their strain-stress conditions are much more complicated than in the hinges with the flexible strips. We discuss this problem in more detail in Chap. 3.

Table 1-2. Translational elastic guides' structural schematics

Name , type, and application limit	Schematic	Name, type, and application limit	Schematic
*Spring parallelogram ELKS TG-I $S_1/L < 0.1$	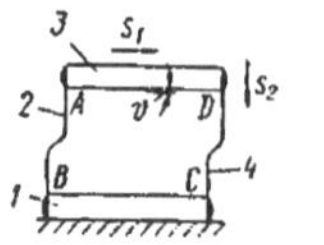	Spring parallelogram with enslaving ELKS TG-V	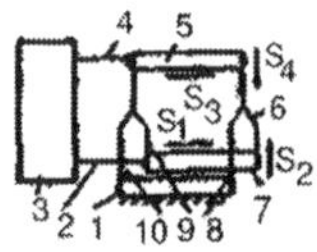
Spring parallelogram with cover plates ELKS TG-II	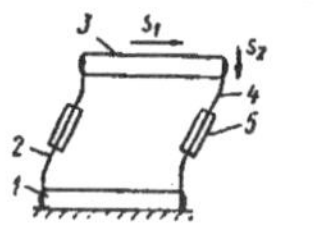	Two-strip guide ELKS TG-VI	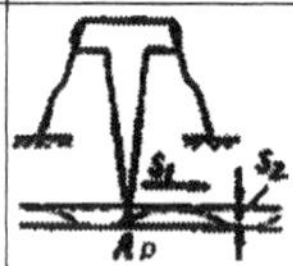
*Spring parallelogram with additional strips ELKS TG-III	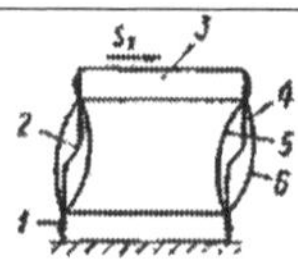	*Spring quasi-parallelogram with notch hinges ELKS TG-VII	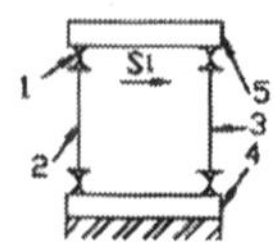
Rectilinear spring compound movement ELKS TG-IV $S_1/L < 0.25$ Rectilinear compound movement with notch hinges TG-IV*n*	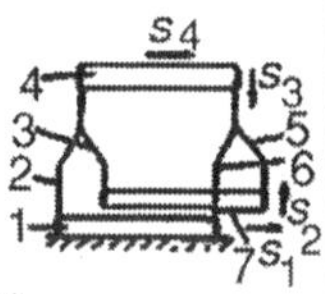 ** 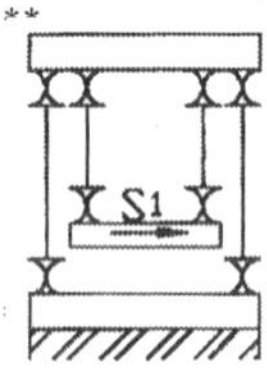	Spring quasi-parallelogram with plates, diaphragms ELKS TG-VIII or bellows TG-VIII*b*	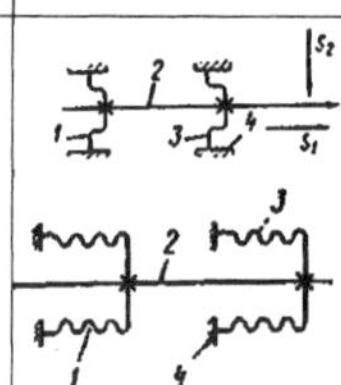
Symmetrical rectilinear spring movement ELKS TG-IV*s* Symmetrical rectilinear movement with notch hinges TG-IV*sn*	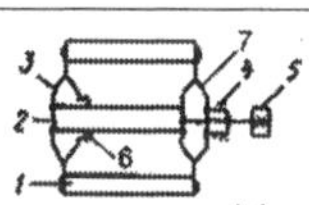 ** 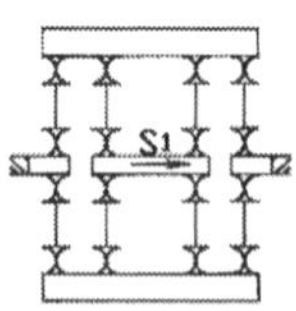	Double-ring guide ELKS TG-IX	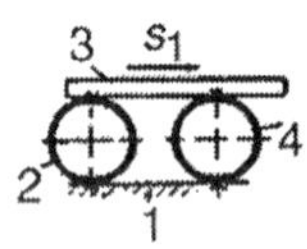

Note: * marks the guides whose features are specially discussed in this book in detail.
** Notch-hinge analog

The *symmetrical spring movement system* (TG-IV*s*) that consists of two symmetrically positioned (relative to the common movable carriage) rectilinear compound movements is more reliable for providing a rectilinear trajectory. This structure has stand 6 and 4, driver 5, two double spring parallelograms 3 and 7, main movable carriage 2 (slider), and two symmetrical auxiliary carriages 1. The analogous notch-hinge version (TG-IV*sn*) has similar elements.

TG-V. A *spring parallelogram with enslaving* provides a trajectory that is also close to the straight line. This structure was proposed by J. E. Plainevaux in the 1950s and has auxiliary feedback kinematic link 3 between displacement s_1 of movable carriage 7 and displacement s_3 of moving intermediate element 5. This kinematic link forcibly compensates for the difference in the loading conditions of the double parallelogram's elastic elements. ELKS TG-V consists of stand 1, flexible strips 6, 8, 9, and 10, driver 2 for movable element 7, and driver 4 for intermediate element 5. Drivers 2 and 4 are kinematically linked by transmission 3 to provide a definite relationship between displacements s_1 and s_3 in order to get the condition $|s_2| = |s_4| - | s_{75}| \approx 0$ at the specified range of loading on the movable carriage. The structure of feedback link 3 between drivers 2 and 4 may be different: a gear transmission with micrometric lead screws, lever transmission, etc.

TG-VI. A two-strip system of the *trapezoid type* can be used as a translational guide for motion of point A_p close to rectilinear, i.e., $|s_2| << |s_1|$. It is natural that a rectilinear motion will be achieved only at certain geometric relationships in the elements of this ELKS and with some degree of approximation.

TG-VII. The s*pring quasi-parallelogram with notch hinges* is based on four notch hinges 1, rigid links 2, 3, stand 4, and carriage 5 whose motion is a translational one. This structure is used in the miniature devices, micro- and nano-electromechanical systems, X-ray interferometers, micro-lithography, and as submicron positioning systems in other applications.

TG-VIII. The *spring quasi-parallelograms on the diaphragms* are also useful for application in the translational guides. These structures usually have two flexible diaphragms 1 and 3, and movable rigid element 2 (slider) that translationally (close to rectilinear) moves in relation to stand 4 and yields the condition, $|s_2| << |s_1|$. For decreasing the guide's rigidity in direction s_1 for element 2's working motion, the solid flat diaphragms can be changed to plates with the shaped holes or to the corrugated diaphragms (*see* Chap. 3). Besides that, one of the plates (diaphragms) can be substituted by the equivalent leaf spring if this structure is more convenient for synthesis purposes. In cases where accuracy requirements for the guiding motion are not high, hermetic options are necessary, and the range of motion for a movable element is large, the *elastic guides with two bellows* are used (TG-VIII*b*). They consist of bellows 1 and 3 that clamped to movable element 2, which has translational motion in relation to stand 4.

TG-IX. The double-*ring translational guide* is also known. This guide has stand 1, two rings 2 and 4 clamped to the stand, and movable slider 3. The rings in this structure can be substituted by the semi-rings that yield *C* version of ELKS TG-IX*c*.

The *elastic transmission mechanisms.* These systems in the functional application are similar to the mechanisms with rigid elements. They serve for the transmission and transformation of motion (displacement). The mandatory coordinate is specified for a point of the driven element as a function of the coordinate for the driving element's point. The kinematic transformation function (ratio in the linear system) may be specified as a derivative from the function of the driven element's linear s_2 or angular θ_2 displacement with respect to displacement of the driving element s_1 or θ_1. It may be expressed as a dimensionless value $i = ds_2 / ds_1$

or $i = d\theta_2 / d\theta_1$ with $i > 1$ in an amplifying mechanism or $i < 1$ in an attenuation mechanism. The transformation ratio may have an expression in certain dimensional units if the motion of output and input in the mechanism are of a different nature, e.g., $i_\theta = d\theta / ds$ where the output represents an angle of rotation θ and the input represents a linear displacement s, or vice versa. This type of transformation ratio is often specified as sensitivity. Besides that, the rigidity or stiffness of driving is specified in relation to the input j or output j_θ, respectively. The relative uncertainty of the transmission mechanism is equal to a differential from the logarithm of the transmission function: $Unr = d \log i\,(s)$.

The mechanisms may be divided into the planar and the spatial ones in accordance with the relative motion of their elements. The planar mechanism has a trajectory of all moving points in one plane or in parallel planes. For elastic systems this definition may be applied with respect to the motion of their real or conditional rigid movable elements. The *conditional rigid element* of ELKS is such an additional element that can be clamped tangentially to an elastic element without distortion of its features. The mass of this clamped conditional element is assumed to be equal to zero, and its contact with the elastic element should be through a tangential line or in a point.

Planar mechanisms. The structure of the *elastic-lever* and *elastic-hinge-lever's* mechanisms for amplification and motion direction's conversion may have many different versions (*see* Table 1-3).

PM-I. The specifics of the *elastic-lever* mechanism (PM-I*el*) include the elastic flexible nature of its levers. In this respect, it differs from the *elastic hinge-lever mechanisms* (PM-I*hl*) where the rigid levers are installed on the elastic hinges. The *elastic* hinge-lever mechanism (PM-I*hl*), for example, has four notch hinges and two rigid levers one of which has arm u. The elastic hinges may have a strip or notch structure. The latter yields a miniaturization and spatial option (SM-I*hl*) with flexural and torsional hinge motion by force F_2 and F_1, respectively.

PM-I*el.* The elastic "*crank-slider*" *mechanism* is a widely spread in the measuring heads version of the elastic-lever mechanism. It consists of rigid slider 2 installed on elastic translational guides 1, elastic connecting rod 3, and flexible crank 4. The displacement of the slider in direction s_1 transfers through connecting rod 3 to the motion of elastic crank 4. If a conditional rigid body (shown by a dashed lines) is clamped to point A_c of crank 4, the body's motion will be planar and close to the rotation on angle θ.

PM-II. The *two-strip reed mechanism* transforms a small linear motion s_1 to the rotation of the driven element for an output angle θ. The mechanism consists of two flexible strips 2 and 4, stand 1, driving element 5 on the elastic translational guide, and driven element 3. The strips are parallel in the initial position. One of the strips is fixed by its end to stand 1, and the end of another strip is driven by the driving element for displacement s_1. The main advantage of the mechanism is in its simple structure and technology of manufacturing. The defects of the mechanism ELKS of type PM-II are in the small transmission ratio ($i_\theta = d\theta / dS_1 < 0.1$ °/μm) and small working range of the driven element's rotation ($\theta < 5°$). There are known attempts to increase the transformation ratio of ELKS PM-II by increasing the strips' length, but in this case special restricting elements are required (*see* Chap.4).

Table 1-3. Elastic transmission mechanisms' structural schematics

Name, type, limits of mechanism	Schematic examples	Application
Elastic-lever ELKS PM-I*el* i=1/4-5	Crank-slider 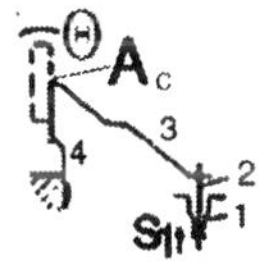	Spring measuring heads
Planar elastic hinge-lever ELKS PM-I*hl*	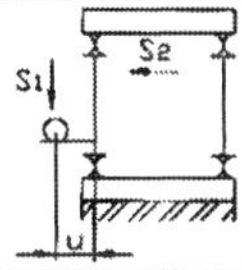	Actuators
Spatial elastic hinge-lever ELKS SM-I*hl*	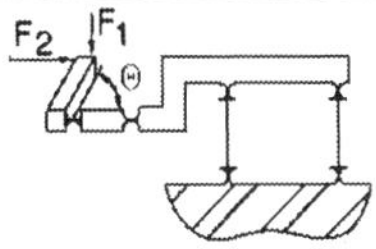	Actuators
* Two-strip reed ELKS PM-II $\theta < 5°$; $i_\theta < 0.1$ °/ μm	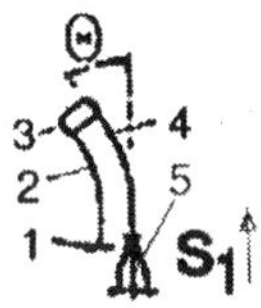	Sheffield, NPL comparators
*Buckled-strip ELKS PM-III $\theta < 20°$ Double buckled bowed strip PM-III*d* Relay–strip ELKS PM-III*r* (Fig 1-3d) Bistable ELKS PM-IIIb	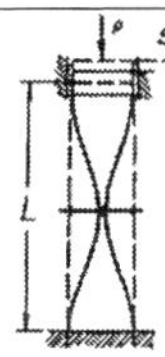 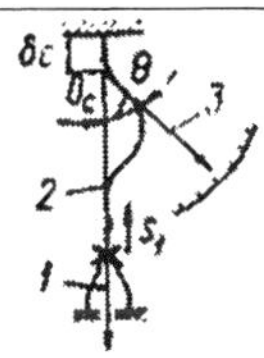 PM-III*d* PM-III	Null-indicators Contact Amplifier, see *Appl.Phys.Lett* **87**, 021904 (2005) MEMS
*Planar cable mechanism ELKS PM-IV $i < 3.5$ Bridge-notch analog ELKS PM-IV*br*	PM-IV P M-IV*br* 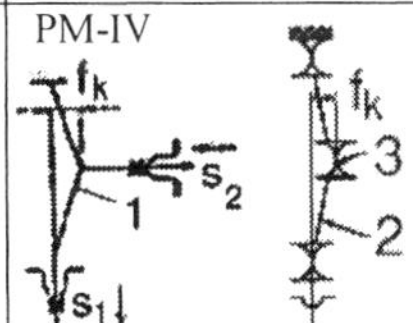	Spring measuring heads, *see* also Fig. 4-12d for PM-IV*br*+ TG-IV*sn*
Attenuation strip-spring ELKS PM-V*s* Attenuation notch-lever ELKS PM-V*n*; ELKS-SM-V*n* Spring-ratchet ELKS PM-V*sr*	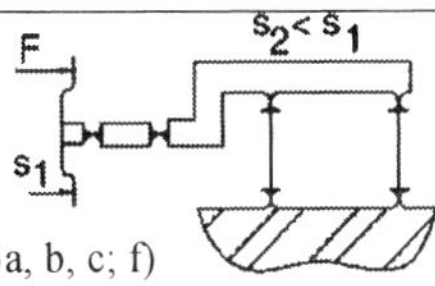 Fig. 1-3a, b, c; f)	Servo-drivers; measuring heads
*Pretwisted strip in double-helicoid ELKS SM-VI Practical $\theta < 3$ turns $0.8 < i_\theta < 10$°/μm;	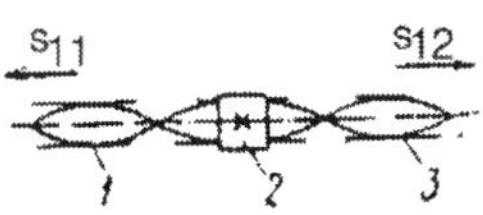	Precision instruments

Table 1-3. Cont'd.

Name, type, limits of mechanism	Schematic examples	Application
* Coiled strip helicoid ELKS SM-VII $\theta < 10$ turns; $0.1 < i_\theta < 1°/\mu m$	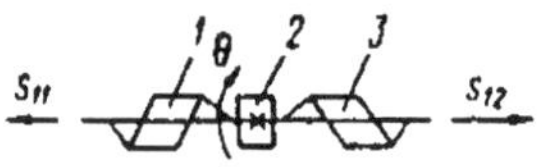	Instruments
* Spatial cable ELKS SM-VIII $\theta < 90°$; $0.02 < i_\theta < 3°/\mu m$	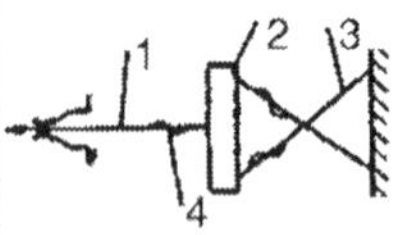 SM-VIII	Instruments
Belt-like ELKS SM-IX $i_\theta < 1°/\mu m$	SM-IX 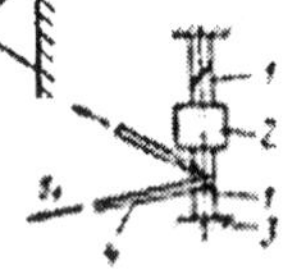	Infra-red detector

Note: * marks the mechanisms whose features are discussed in this book in detail.

PM-III. The *buckled-strip mechanism* (Table 1-3, ELKS of type PM-III) is based on a buckled thin beam 2's motion under compression. The beam is connected to driving slider 1 on the elastic translational guide, and its driven pointer 3 rotates to angle θ at the translational motion of the slider. The function of motion in the ELKS of type PM-III is nonlinear, but the position of center O_c for the rotation of the pointer at a small angle θ is practically constant (the change of δ_c is close to zero). Double buckled bowed strip device (PM-III*d*) is effective in adjustment systems. A bowed beam flexure with adjusting screw and a limiter to prevent flexure snap jump has found application in precision machines. The *relay-like elastic mechanism* PM=III*r* has two stable positions. This mechanism is used in some micro-switches. It consists of *E*-shaped strip 1 (Fig. 1-3d) whose side parts are compressed beyond the limit of stability (in a buckling state) and the central part is stretched. The practically instantaneous switch from contact group 3 to contact group 4 occurs when we press knob 2. Similar bistable elastic mechanisms with parallel buckled beams (ELKS PM-III*b*) are used in MEMS (*see* M.P. Brenner *et al.*, 2003, *PNAS*, **100**, 9663-9667; US Patents 6,215,081; 6,748,818).

PM-IV. The *planar cable mechanism* has flexible strips or flexible wires 1 that form a flexural sag f_k. The motion s_1 at the end of the wire transforms in this mechanism into increased motion s_2 of the middle point, or vise versa, motion s_2 reduces into motion s_1. Similar *bridge- notch mechanism* (PM-IV*br*) has two rigid segments 2 connected by the flexible notch hinges 3 to a driving element, stand and to each other. The latter mechanism can have miniature dimensions and is usually used for the amplification of the input motion of microactuator.

PM-V. The structure of the *attenuation transmission* may be executed on the basis of the *strip-spring mechanism* (Table 1-3, ELKS of type PM-V*s*, Fig. 1-3a,b) with the sequence of a soft and hard spring strips, or on the basis of the *notch-lever* planar or spatial mechanism (Table 1-3, ELKS of type PM-V*n* and ELKS of type SM-V*n*). The latter mechanism in the planar (PM) or spatial (SM) versions indeed provides an attenuation function $s_2 < s_1$ (F) with the same sequence for a soft (driving force F) and hard spring, but by means of notch hinges incorporated in the hard monolithic structures. The attenuation mechanism can have a structure with the ordinary spring parallelogram (see Table 1-2, ELKS of type TG-I) where $|s_2| << |s_1|$, or the spring parallelogram driven by strip spring 1 (Fig.1-3a) that is flexible and softer than the parallelogram's own driving rigidity (Patent SU 167308), a soft driving cantilever (*see* Fig.4-9b with version B), and in the form of the inverted planar amplifying mechanisms. The bowed flexure is a kind of the reverse buckled bridge mechanism with small sag (L.W. Nickols, 1955). This device works as a strut with adjustable length. It has two flexible strips 1and 7 (Fig. 1-3f), two plates 4 and 5 clamped to the middle of each strip, compressed spring 2 between these plates, bolt 3, and nut 6 (shown conditionally). It was assumed that each part of the strips could be represented as two consequent cantilevers which transverse flexure with sag $2y$ causes mutual small longitudinal motion of strip's ends within $4x$.

The spring parallelogram in principle may be used as an "attenuator" with transmission ratio $i = s_2 / s_1 < 1$, but this transmission is not a linear one. The attenuation mechanism with rigid driven strip 2 (Fig. 1-3b) and softer driving strip 3 is a less stable one. For attenuation purposes ELKS of type PM-II with the driving connecting element 1 (reversed PM-II) may be successfully used. The modification of such a mechanism with three strips 2, 3, and 6 (Fig. 1-3c) simultaneously provides two opposite displacements $s_2 = - s_3$ for driven elements 4 and 5 when driving element 1 moves only in one direction s_1. This structure was used in R. V. Jones's interferometer for the symmetrical opening of the slot.

Unfolded double-reed version. The combination of two sequential reed mechanisms I and II positioned in the mutually perpendicular planes (test sample was built by the author in 1990's) provides the conversion of element 1's (Fig. 1-3e) displacement s_1 to a quasi-linear displacement s_2 of transducer's anchor 3 (slider) on strip 2. This structure does not require an additional translational guide for the slider and is useful when case 4 is a tube with small available diameter D_T.

Hydraulics and pneumatic bellows devices can also be used as amplification and attenuation mechanisms (*see* Chap. 3).

The *spring-ratchet mechanism* (PM-V*sr*) provides motion only in one direction from driver 1 (Fig. 1-3g) , but this mechanism has an external friction between the coils of the cylindrical spring 2 and the central driven shaft 3 in the direction of the idle motion. This feature differs from other "pure" elastic mechanisms.

Spatial mechanisms.

SM-VI. The *pretwisted strip in the double-helicoid mechanism* has special features. This is a spatial mechanism with a helical motion that transforms a small mutual displacement of the ends in the pre-tensioned strip $s = s_{11} - s_{12}$ (elongation) into significant rotation θ of driven element 2 around the strip's axis. ELKS of type *SM*-VI has the greatest transmission ratio ($i_\theta \leq 10°/\mu m$) among all known transmission mechanisms with the quasi-linear kinematic function (*see* Chap. 2). Such a characteristic at a small gradient and small working forces F_Q (the rigidity $j_\theta = \partial F_Q / \partial\theta$ can be in the range of a fracture of centinewton per one degree of a driven element's rotation) does not belong to any other elastic transmission mechanism with quasi-linear transformation. Besides that, the increase of the transmission ratio at the rational parameters of the pretwisted strip and negligible driven load yields a possible decrease of the necessary driving force. The pretwisted strip in a double-helicoid has two pretwisted segments 1 and 3 with opposite directions of helix that are connected to each other through a straight crosspiece with clamped driven element 2. The pretwisted segments are subjected to untwist under elongation of the helicoid strip that transforms the linear motion of its ends to the rotation of its crosspiece.

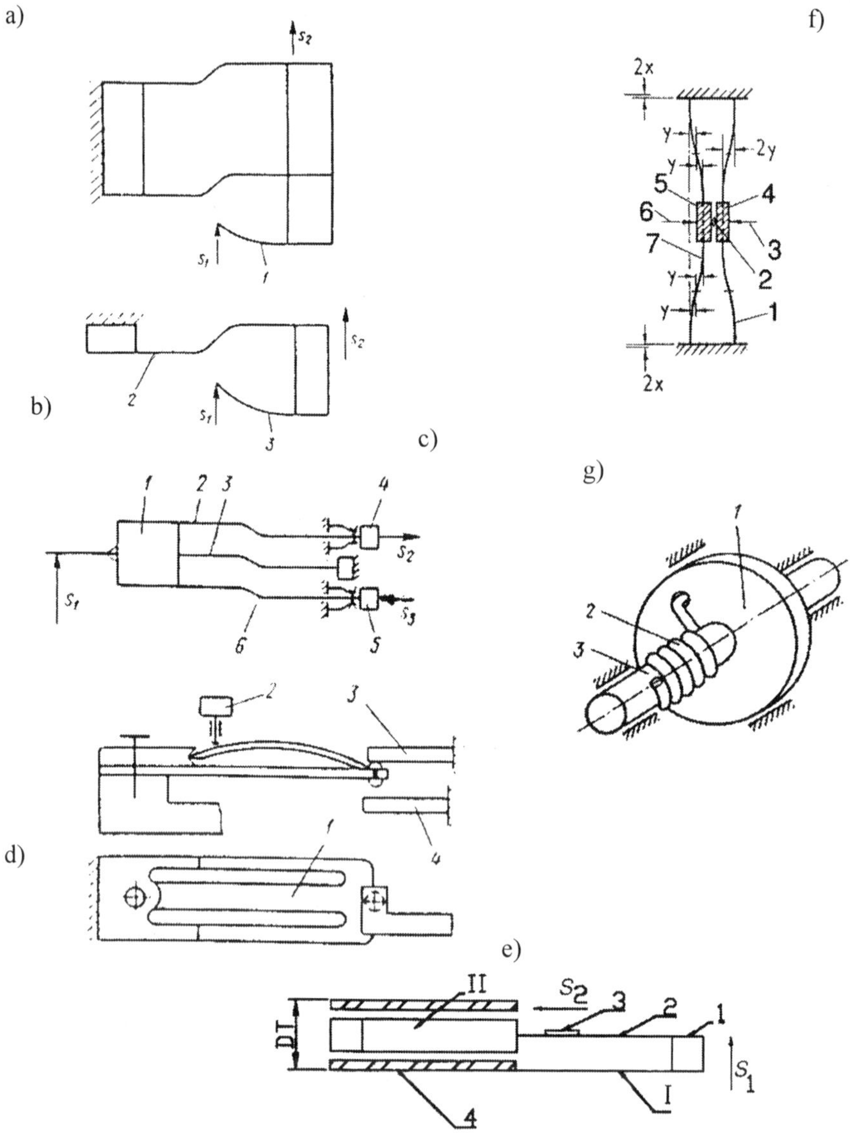

Fig. 1-3. Typical attenuation kinematic systems (panel f is after L.W. Nickols, 1955).

SM-VII. The *coiled strip mechanism* provides a better transformation of the longitudinal forces into torque than ELKS of type SM-VI. The double-coiled strip has two segments 1 and 3 that are pre-coiled with opposite helix directions and connected by the straight crosspiece with fixed driven element 2. With elongation $s_1 = s_{11} - s_{12}$, the coiled segments of the strip as helical shells get untwisting and the driven element 2 rotates of an angle θ. The rational value for the transformation ratio of such mechanisms is in the range of 0.1 – 1 °/μm. The function of motion $\theta = F(s)$ in this case is not a linear one. Versions of this mechanism have asymmetric double-coiled strips with a different number of coils on each coiled segment.

SM-VIII. The *spatial cable mechanism* with thin twisted wires transforms the small translational motion s of slider 1on a translational guide into the rotational displacement θ of element 2 with the rotation of crossed wires 3 at their tension. The versions of this mechanism differ if the tension-torsion suspension 4 changes to a second pair of the crossed or parallel wires. It is evident that the ultimate angle of rotation for element 3 is less than 90°. The transmission ratio in the initial quasi-linear part of the elongation is up to 3 °/μm, after which it decreases. The main defect of the mechanism is in the nonlinearity of its transmission function.

SM-IX. The *belt-like mechanism* transforms a small longitudinal displacement s_1 of the strip's end 4 into the angular rotation θ of element 2, which is clamped to the torsion suspensions 1on base 3. This structure has a small damping and a stable initial position with uncertainty that is less than 5 x 10^{-10} mm. The heat of infrared radiation in R. V. Jones' infrared detector was enough to change the length of strip 4 by the detectable value s_1 (*see* R.V. Jones, 1988).

Band transmissions.

BL-I-III. The flexible thin metallic or plastic bands are used in the band transmissions. Uncertainties of a band transmission and driven roll rotation angle ($\Delta\varphi_{d1}$) depend on errors in the diameters of rolls, their surface geometry, and eccentricity, as well as the band deformation from its tension Δs_{et}, bending Δs_b, sag, change of temperature, and catenary's options. Methods of band clamping to the rolls and carriages are also important. The typical structures of the band transmissions are shown in Table 1-4.

Table 1-4. Band transmissions' typical structural schematics

Name and type of ELKS	Schematic	Closing
Single-band with one working roll, ELKS BL-I	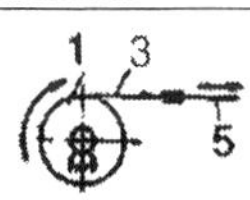 	Force snap
Double-band with one working roll, ELKS BL-I*a*	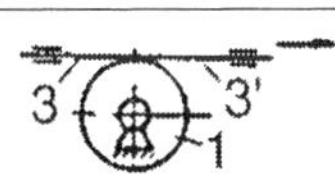 	Kinematical
Single-band with two working rolls, ELKS BL-II	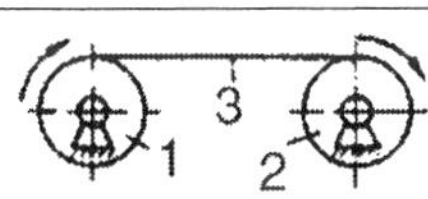 $i_{12}=d_1/d_2$; $\Delta s_{et}=\Delta F_p L_{3c}/EA$; $\Delta s_b \propto \frac{H^{3/2}}{3d_2^2(F_{p0}\mp\Delta F_p)^{3/2}}(1+i_{12}^{-2})$ $\Delta\varphi_{dl}=2(\Delta s_{et}+\Delta_{sb})/d_1+\ldots$	Force snap

Table 1-4 Cont'd.

Name and type of ELKS	Schematic	Closing
Closed band with two working rolls, ELKS BL-II*a*	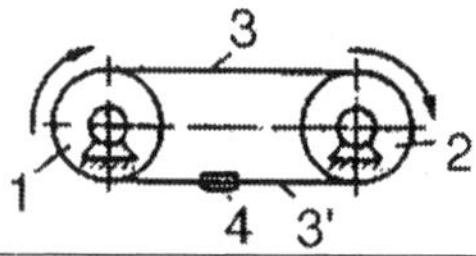	Kinematical
Single-band reversible, ELKS BL-III	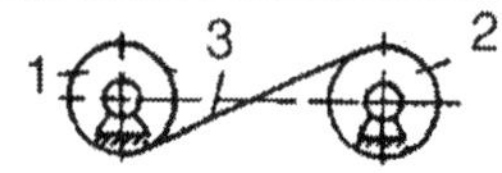	Force snap
Double-band reversible, ELKS BL-III*a*	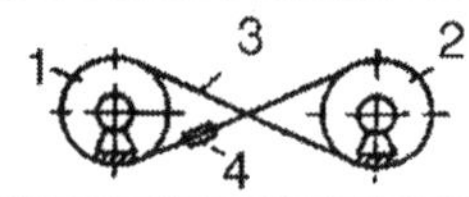	Kinematical
*Rolamite, ELKS BL-IV	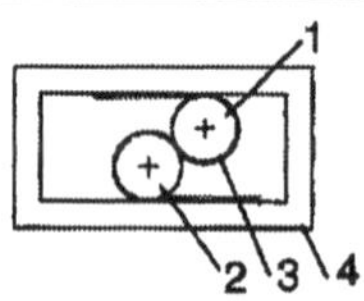	Force-Kinematical
Rolamite with two bands, ELKS BL-IV*a*	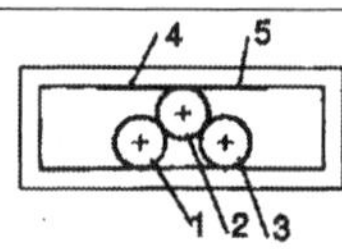	Kinematical

Note: * marks the transmission whose features are discussed in this book in detail. L_{3c} is the band 3's free length, E is its material modulus of elasticity, H is section flexural rigidity, and A is cross-sectional area; F_{p0} is the band's tension and ΔF_p its uncertainty.

Numbers 1 and 2 denote the working rolls (diameters d_1 and d_2) that define the transmission's kinematics, number 3 denotes the band (3 and 3' are the branches of a band with section area A and flexural rigidity H), element 4 represents the connecting rods in the free sections of the band, element 5 is a slider. The band may have one branch (single-band ELKS of type BL-I, BL-II, BL-III) or two branches (double-band ELKS of type BL-I*a*, BL-II*a*, and BL-III*a*). The latter in ELKS of type BL-III*a* do not intersect each other because they have mutual displacement along the axis of rolls (double-row position). The single-band transmissions require a closing fastener by force F_p (forcible), and the closed or double-band transmissions may have a kinematical closing snap. The connection between the band and the rolls can be executed by one of three methods: i) rigid connection by clamping the band to the roll, ii) frictional connection with the smooth rolls, and iii) forcible connection between a perforated band with toothed rolls. Transmissions with the clamped band are used more often because they provide precision and can work under vibration without sliding between the band and the roll. Such transmission (BL-I*a*), for example, was successfully used in the precision comparator of firm "Sigma". The defect of this transmission is in a small angle of rotation for working rolls (less than 360°). The "Sigmatic comparators" (*see* F. T. Farago, 1968) with ribbon transmission have overall amplification in the range from $300^{\times}$ to $5000^{\times}$. Effective application of a steel belt positioning mechanism for nanometer motion control presented in A. Shimokohbe (2002) publication. Let us recall that friction in the belt system is proportional to $e^{\mu_f \theta_r}$, where μ_f is the coefficient of friction and θ_r is the arc of contact of smaller roll (pulley). Transmissions with a free band (friction connection) cannot eliminate possible sliding between a metallic band and rolls, and therefore, can reliably work only in the absence of external strikes and forced vibrations. Another option is to use magnetic rotation stabilizer or flexible rubber bands with initial tension up to 4% of their length. The advantage of the latter transmissions is in the unlimited angle of the rolls' rotation but these transmissions have limited precision. Transmissions with perforated bands and toothed rolls are rarely used in precision applications because they are more complicated to manufacture and less accurate. These transmissions have indeed limited kinematic options. Only one type of transmission ELKS of type BL-II*a* is used with the smooth or perforated closed bands. The known *flexible gears* are more effective than the perforated bands with toothed rolls.

Band transmissions usually have driving and driven rolls with equal diameters $d_{r1} = d_{r2}$ (the transmission ratio is equal one). However, band transmissions can successfully work with increasing transmission ratio up to $i = d_{r1} / d_{r2} = 3$ as well as with decreasing transmission ratio which is equal to $i = 1/3$.

BL-IV. The *rolamite* precision band mechanism has a wide working range, minimal external friction in the same order as for ball bearings, negligible sliding, and pulling force that depends on the elastic and geometric parameters of the band. The mechanism consists of two rolls 1 and 2, and flexible band 3 that wraps the rolls, and whose ends are clamped to cluster 4. This device may be used as a translational or rotational guide as well as a transmission system with continuous (round rolls) motion or motion that switches (Chap. 5) to the sub-ranges (polygon rolls). There are known structures of the rolamite with weak tension in the band, where its free segments are not straight, and also versions with two bands 4, 5 (ELKS of type BL-IV*a*) and three or more rolls 1, 2, and 3.

Couplings and correctors.

Couplings. The precision kinematic chains may have *elastic coupling* links that prevent a backlash at the reverse motion. The different versions of elastic couplings with metallic and non-metallic elastic elements are used in mechanical engineering and instrumentation (*see* examples in Table 1-5, ELKS of type CL-I- IV). The flexible shafts are useful for the torque transmission. However, the usage of the elastic couplings for transmission of precision measuring information is very rare.

Correctors. In this book we will mainly discuss mechanical correctors and compensators. Let us remind that correctors and compensators in feedback chains may also have non-mechanical nature: electrical, optical, etc.

KC. The *kinematical correctors* may be divided into geometric (KC-I, KC-III) and elastic (KC-II) types. The geometric correctors are used to change the length and initial position of the levers and other elements of the elastic transmission mechanisms. Their structure is similar to that used in regular mechanisms with rigid bodies: adjustable eccentrics, telescopic clampers, and movable plates. The elastic kinematical correctors usually have the form of cantilever flat springs with adjustable length, which allows us to change their resilience smoothly. In addition, the stiffness of the elastic elements can be changed with adjustable cover plates and stops by which any part of the elastic lever can be constrained (enslaved). The precision adjustment of the initial position for elastic elements in the kinematic chain can be provided by a forcible deformation of the clamping elements.

FC-I. The *driving force correctors* usually have a special structure. It is known that variations in forces from all elastic kinematic elements as well as from all force-generating elements have been algebraically summed in the point of the resultant force application. The main problem for the force correctors is to decrease the change ΔQ in forces applied to the driving element in its working range. This can be achieved by the installation of an elastic corrector with negative rigidity. A negative rigidity in the *force push-spring* lever with balance corrector (FC-I*pu*) is based on the lever 1 profile that reduces arm *a* of the force pushing down spindle 2 at its higher position even if the force in the spring 3 increases. Balancing body 4 on the lever compensates for the changes in the spindle force due to a different orientation of the structure in the gravity field. A negative rigidity in the *force pull- spring lever* corrector (FC-I*pl*) is based on the increase of arm *a* of spring 1's action on lever 2 at the higher position of spindle 4. This compensates for the increase in the force of spring 3, which pushes spindle 4 down at its higher position. Balancing body 5 on the lever compensates the changes in the spindle force due to a different orientation of the structure in the gravity field.

FC-II-III. *Force spring buckling* correctors (FC-II) and *force thin beam-buckling* correctors (FC-III) are based on a decrease in the longitudinal force and changing arm *a* of its action. The latter may be unstable at the change of its orientation.

FC-IV. The *spiral spring* on the mandrel 1 has an approximately constant within 2-5% pulling force F_Q. The spring has preliminary tension between its coils, which may cause an action of external friction.

Table 1-5. Typical elastic couplings and correctors' structural schematics

Name, Type, application	Schematic examples	Name, Type, application	Schematic examples
Plate-ring coupling ELKS CL-I Shaft cross < 3° Pins5; rings 2,3; shaft 1,4	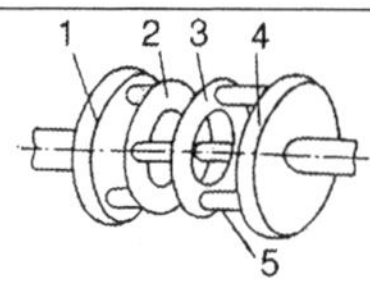	Force-spring-buckling ELKS FC-II Negative rigidity at the compress-spring buckling	
Strip (*s*) or notch (*n*) coupling ELKS CL-II*s*, *n* Shaft cross <3° Strips 2, 3, 5,6; joint 4; shafts 1,7	II*s* II*n* 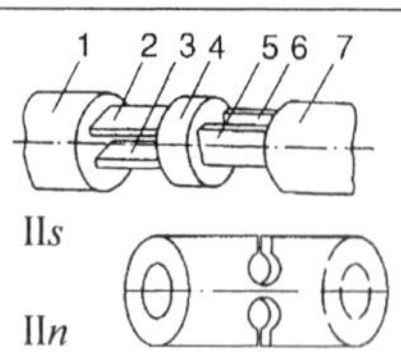	Force thin beam-buckling ELKS FC-III Negative rigidity	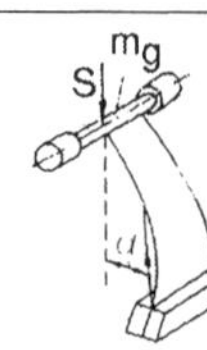
Diaphragm coupling ELKS CL-III 1,3- shafts, 2-diaphragm (disk)	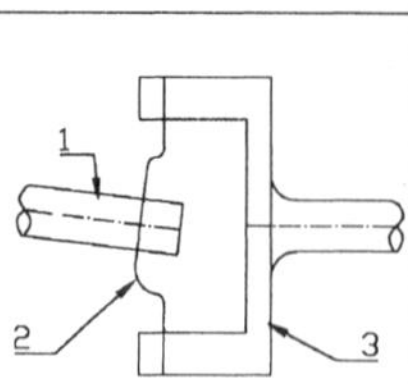	Force constant-spiral ELKS FC-IV Small change of pulling force @ the motion large range	$F_Q=M/R=EJ/R^2\approx$const 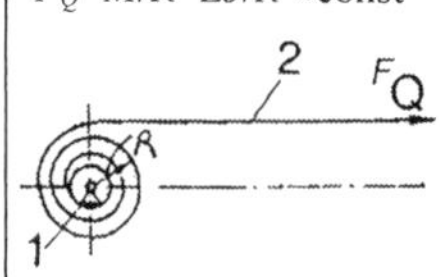
Bellows coupling ELKS CL-IV 1,3-shafts 2-bellows	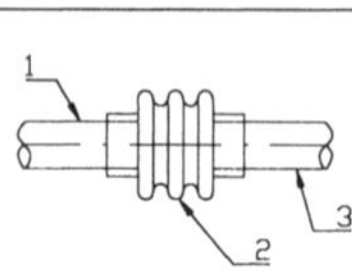	Force sine-strip spring, ELKS FC-V Negative rigidity	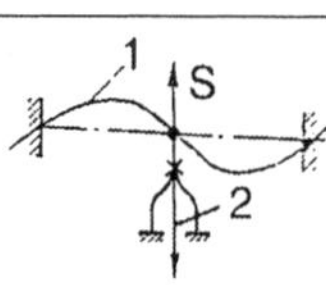

Table 1-5 Cont'd.

Name, Type, application	Schematic examples	Name, Type, application	Schematic examples
Kinematical corrector, change geometry ELKS KC-I, change in the elements' length or position angle		*Force sine-strip with balances, ELKS FC-V*b* Negative rigidity plus balances for orientation change	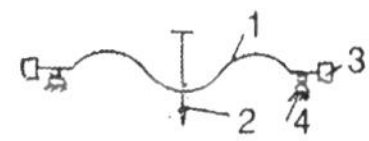
Kinematical elastic corrector, change rigidity of elastic elements ELKS KC-II		*Influence damper ELKS IC-I Air, liquid, dynamic factors damper for: strike, vibration, etc.	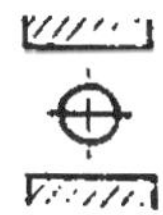
Kinematical eccentric corrector, ELKS KC-III Adjustment of the elements' initial position	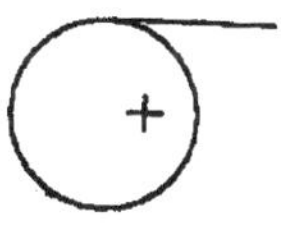	*Influence-speed retraction ELKS IC-II Stabilization in the speed of elastic deformation	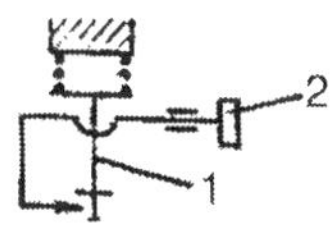
Force push-spring –lever, balance ELKS FC-I*pu* Negative rigidity in pushing spring lever shoulder	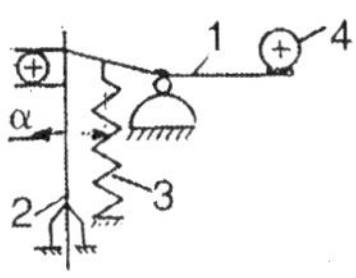	Influence temperature with bimetallic elastic elements ELKS IC-III *bi*	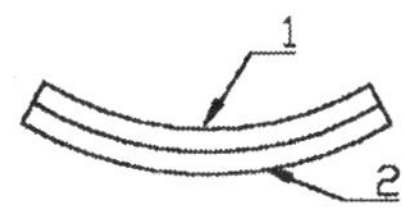
*Force pull-spring -lever, balance ELKS FC-I*pl* Negative rigidity in pulling lever spring shoulder	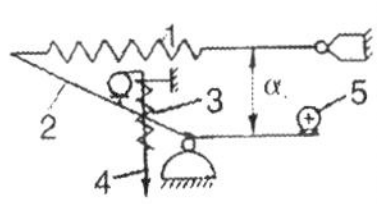	Influence orientation with balance masses ELKS IC-IV*b*	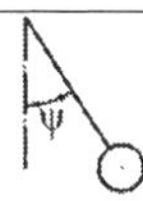
		Influence restrictors from overloading at work, transportation and in stock, IC-V	See Chaps. 3, 4, 8

Note: asterisk * marks the correctors whose features are discussed in this book in detail.

FC-V. The *sine form* of spring strip 1 produces a negative rigidity in its action on spindle 2 (GB Patent 617076, 1949). The ends of this sine spring are fixed on the base of the structure under specific adjustable angles. The same effect of a negative rigidity has a three half-waves sine spring 1 in ELKS FC-V*b* whose ends are clamped to the small levers with elastic pivots 4 and balancing bodies 3 (SU patent 262448, 1970). A change in spindle 2 and the whole structure's orientation in the gravity field causes rotation of the levers with balancing bodies 3 on their pivots 4 by which the sine-spring ends accept a new angular position. As a result, the acting on the spindle force changes accordingly, compensating for the influence of gravity.

IC-I, II. The *influence-damper corrector* (IC-I) can be used as a liquid or air damper in the vibration field or in the dynamic motion of the elastic structure's elements. The *influence-speed retraction's corrector* (IC-II) is useful when the speed of spindle 1's motion released after its retraction should be stable. In this case, the

speed of spindle 1's motion after its release by pushing stop 2 is practically independent from the speed of stop 2's displacement.
IC-III*bi*. Influence *temperature compensators* in many cases are built from bimetal elastic elements where layers 1 and 2 have different coefficients of linear expansion. As a result, change of temperature causes change in the element's curvature.
IC-IV*b*. Influence *orientation compensators* are often based on the action of balance masses examples of which shown in the schematics of systems FC-I*pu* , FC-I*pl*, and FC-IV*b* (Table 1-5).
IC-V. ELKS may also have *influence restrictors* and stops for safety purposes to protect elastic elements from overloading under the working conditions impact, during transportation, and while in stock (*see* Chaps. 3, 4, and 8).

Actuators, sensors, and transducers.

The automation in all possible applications needs precision actuators, sensors, and transducers. The most advanced actuators in precision elasticity are built on the basis of flexures, torsions, and piezoelectric effects. Use of other actuators' principles is more restricted as a result of limited range, not smooth enough (intermittent, stick-slip) motion, and low precision (*see* Fig. 1-8b).
Sensor may be considered a specific part of transducer, which accepts the quantity to be transformed in the measurable by the transducer's system. The precision elastic transducer (ELT) fulfills the most accurate transformation from one type of physical quantity to another.

Table 1-6. Typical elastic transducers' structural schematics

Name and type of transducer	Schematic examples	Application and notes
Amplitude diaphragm-pneumatic ELKS of type AM-I	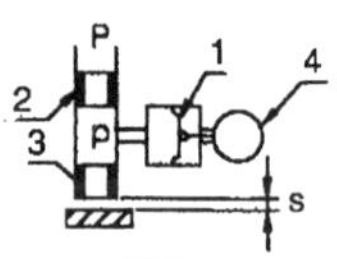	Measurement and detection of pressure p, ordinary and differential w/flat and corrugated diaphragm
Amplitude bellows-pneumatic ELKS of type AM-II	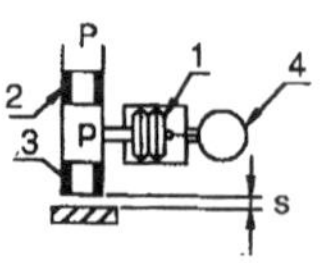	Measurement of pressure p and displacement s; ordinary and differential
*Amplitude pretwisted-tube ELKS of type AM-III	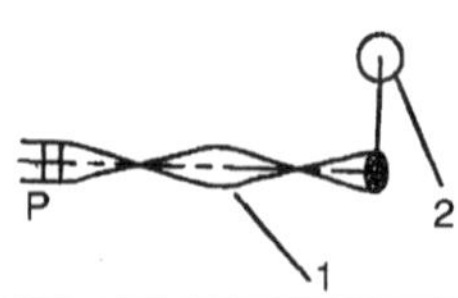	Measurement of pressure p with internal kinematical transmission options
Amplitude cantilever detector: tenso, piezo, optical, capacity, tunneling. ELKS of type AM-IV		Measurement of force with high sensitivity and low damping, less than 10^{-9}
*Amplitude-quantized (sorting) photoelectric with double helicoid, ELKS of type AM-V	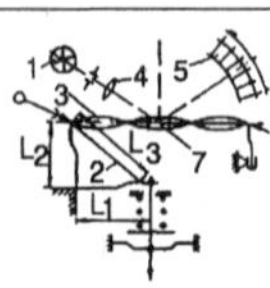	Sorting by dimensional groups with high accuracy and rate up to 10 measurement per second

Table 1-6. Cont'd.

Name and type of transducer	Schematic examples	Application and notes
*Amplitude photoelectric with impulse output AM-VI, continuous analog AM-VI*a*, or end signals AM-VI*es*	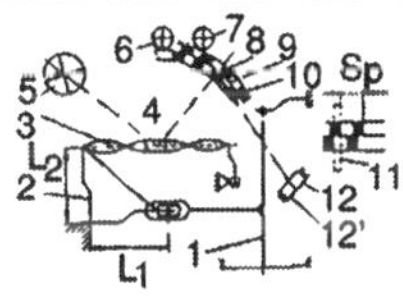	Measurement and sorting by dimensional groups at minimal number of photoelectric elements
Frequency round or flat string ELKS of type FR-I	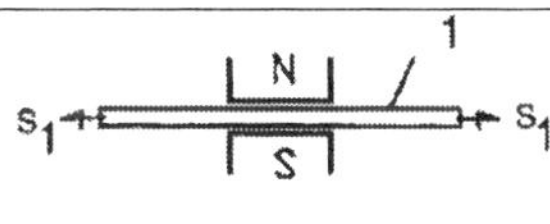	Strain, stress, and displacement measurement
Frequency twisting tapes ELKS of type FR-II	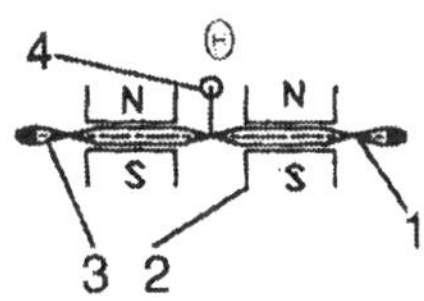	Instrumentation, space guidance components, telecommunication
*Frequency helicoidal multivibrator with pretwisted strip ELKS of type FR-III	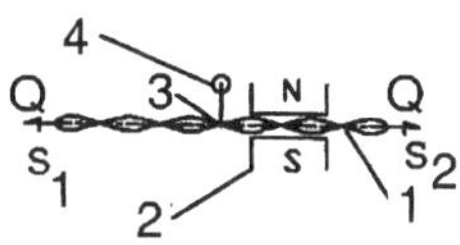	Precision geometrical measurement, effective frequency range: 2500 –3600 Hz
*Frequency quasi-helicoidal multivibrator with coiled strip ELKS of type FR-IIIqco, pretwisted tube FR-IIIqtu, spatial cable mechanism ELKS of type FR-IIIqca	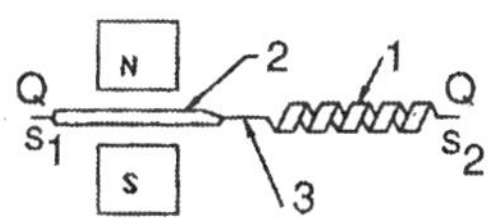	Options in measurement techniques
*Frequency modulated cantilever sensor ELKS of type FR-IV		Atomic force microscopy, piconewton , subnanometer attogram resolution, fine bio/chemical sensor
Frequency monocrystal-membrane, ELKS of type FR-V	-	Measurement of pressure with frequency change in tens kHz

* marks the transducers whose features are discussed in this book in detail.

It has the advantages of a low threshold, great sensitivity, absence of friction and backlash, and long lifetime expectancy. The important feature of the contemporary transducer is its ability to be linked to an electrical, preferably digital, system with low energy consumption or to an effective pneumatic system in industrial applications. Table 1-6 shows several types of ELTs that have good potentials.

AM-I-II. Pneumatic amplitude transducers can be based on an elastic diaphragm sensor 1 (AM-I) or elastic bellow sensor 1 (AM-II). ELKS of type AM-I and AM-II in the double-orifice version may have two separate orifices: restrictor orifice 2 with inner diameter d_2 and measuring nozzle orifice 3 with inner diameter d_3. Intermediate pressure p in the chamber between these orifices depends on the escape clearance s at measuring nozzle orifice 3: $p = P / \{1+[\mu_3 A_3 / (\mu_2 A_2)]^2\}$, where μ_2 and μ_3 are the coefficients of air flow through the corresponding orifice, $A_3 = \pi d_3 s$ is the escape area for a flow of air through nozzle orifice 3; $A_2 = \pi d_2^2/4$ is the area for flow of air through restrictor orifice 2; and P is a constant pressure from the air source. The diaphragm and bellows displacement under pressure p can be read on the dial of indicator 4 or be used by automatic output system. Single-orifice version is used in rotameters which based on the measuring of air flow speed with flouting pointer and do not require elastic sensors. Specifics of ejector system see in Chap. 8.

AM-III. ELKS of type AM-III are based on the *pretwisted elastic tube* features (*see* Chap. 3). This ELS 1 has internal kinematic amplification of its elongation under high inside pressure p to pointer 2's rotation.

AM-IV. *Amplitude cantilevers.* ELKS of type AM-IV is a cantilever with bending amplitude detection. The advantage of such an amplitude transducer (sensor) is the small size and large sensitivity. It may be built from small piezoelectric or tensoresistive crystal. Its deflection from the initial position may be detected by optical (bouncing optical ray, interference) and electrical means. Moreover, this may be an array of cantilevers or membranes as a part of the monolithic structure with a small possible creep, low temperature sensitivity, and a negligible damping.

AM-V. The amplitude-quantized *photoelectric transducer* for precision dimensional sorting has an elastic double-helicoid 3 with mirror 7 on its crosspiece. An angular position of the latter defines the photo-resistor in block 5 lightened by a ray that passes from the source of light 1 through lens 4 to mirror 7 and further reflects to the block 5 of photoelectric sensors. The angular position of mirror 7 with high precision depends on the longitudinal extension of the double-helicoid that unambiguously depends on elastic lever 2 with arms L_1 and L_2 and the measuring spindle connected to it positions.

AM-VI. *Amplitude-photo-impulse* transducer consists of spindle 1 on the translational guide, elastic lever 2 with arms L_1 and L_2, elastic double-helicoid 3 with mirror 4, source of lighting rays 5, dial 8, and reflecting annular rings with rasters 9 and 10 whose angular position can be adjusted by gears 6 and 7. Light image 11 reflects from the rasters and passes to photoreceivers 12 and 12*'* where electrical pulses are generated during the light image motion. The main advantage of the impulse structure in comparison with the positioning structure is that it is better protected from instability of the light's intensity and electrical voltage applied to the photoresivers. A similar amplitude-analog photoelectric transducer (AM-VI*a*) has a continuous output or relay-like end outputs (AM-VI*es*) which functionally depend on the position of the ray on the surface of reflecting annular rings and photoelectric sensor (sensors) whose sensing surface is shifted from the axis of the annular rings. The reflecting annular rings in this case do not have raster. The disadvantage of these latter transducers is the dependence of their uncertainty on the instability of light intensity and voltage supply.

FR-I. The ordinary *string transducer* with flat or round string 1 in the magnetic field with poles N, S transforms a change in string longitudinal tension or stretch s_1 into the corresponding change of its natural flexural frequency. The string is connected to a self-exciting oscillator with a positive feedback that keeps the vibration of string 1 permanent.

FR-II. Frequency output of *twisting-tapes* system is based on the change in the ratio of the natural flexural frequencies of two tapes 1 and 3 that have twist in opposite directions with the input rotation along their axis by lever 4. Both tapes are positioned in the magnetic field of permanent magnets 2.

FR-III. The *helicoidal multivibrator* is a bifurcating transducer with an amplitude output at the pretwisted strip's crosspiece 3 rotation in response to the change in double-helicoid 1's longitudinal tension $|s_1\text{-}s_2|$, and with a simultaneous change in its natural flexural frequency output. The latter is supported by the self-exciting oscillator with a positive feedback that is connected to the elastic helicoid positioned in the magnetic field of permanent or electrical magnet 2 and to the frequency meter. The angle of rotation for the elastic double-helicoid's crosspiece 3 may be detected by a pointer 4, the mirror clamped to it , through photoelectrical or capacitance sensor, Hall's effect sensor, and also by laser beam diffraction patterns (*see* Chap. 8). The advantages of this structure (introduced by Y.M. Tseytlin in 1980, Patents SU 706,677; 721, 658) are the highest sensitivity among all quasi-linear mechanical systems, long lifetime expectancy, low energy consumption, and a wide application with integration into contemporary intelligent computerized systems.

FR-III*q*. The *quasi-helicoidal multivibrator* with frequency and amplitude outputs can be made on the basis of coiled strip that is connected to the initially flat string 2 positioned in the magnetic field. The amplitude output is produced by the rotation of joint 3 at the change of coiled strip 1's longitudinal tension $|s_1\text{-}s_2|$. The joint 3's angular position and string 2's natural flexural frequency may be detected by the same means as in the case of the helicoidal multivibrator (ELKS of type FR-III). Another quasi-helicoidal multivibrator (FR-IIIqtu, *qca*) can be made on the basis of pretwisted tube (Chap. 3) and the spatial cable mechanism SM-VIII as well.

FR-IV. The *cantilever vibrating sensors* are the main part in the AFM with atomic resolution of surface structures, molecules detection, and options of piconewton (10^{-12} N) forces and femtogram (10-18 g) mass evaluation (*see* Chaps 3, 5, 8). They have small sizes with length L within hundreds of micrometers (microcantilevers) and usually have a probe slightly shifted from the moving end ($\delta p/L \ll 1$). Cantilevers with the smaller sizes in nanometer range (nanocantilevers) are also available.

FR-V. ELKS of type FR-V based on the change of frequency and speed of the acoustical waves' propagation through elastic diaphragm surface by pressure.

1-2. General Problems of Precision Kinematics

It is known that at each instant, the motion of any rigid body can be thought of as the superposition of both a translational and rotational motion. The motion of the rigid body is a combination of the translational motion with a linear speed **V** and the rotational motion with angular speed **Ω**, which in a given instant corresponds to the certain origin O in Fig. 1-4a. Both translational and rotational velocities are the vector quantities. Hence in this case, we have bivector (motor) of two superposed vectors, **V** and **Ω**. Their mutual position may be estimated by the so-called complex (dual) angle $\overset{\circ}{\Phi}$: a contour between lines of vectors and their binormal b^k, such as

$$\overset{\circ}{\Phi} = \gamma^{\circ} + \varpi b^k , \tag{1-1}$$

where γ° is the actual angle of the vectors' crossing, b^k is the shortest distance between the lines of their action, called the axis of a complex angle, and ϖ is the Clifford's symbol (dual unit , $\varpi^2 = 0$).

Typically, we have the following six cases of rigid body motion:

1) $\mathbf{V} = 0$, $\mathbf{\Omega} = 0$ corresponds to the rest or the instant stop;

2) $\mathbf{V} \neq 0$, $\mathbf{\Omega} = 0$ corresponds to the instant translational motion (with respect to the velocity);
3) $\mathbf{V} = 0$, $\mathbf{\Omega} \neq 0$ corresponds to the rotation around an instant axis;
4) $\mathbf{V} \neq 0$, $\mathbf{\Omega} \neq 0$, and $\gamma^{\circ} = 90^{\circ}$ corresponds to the plane-parallel motion;
5) $\mathbf{V} \neq 0$, $\mathbf{\Omega} \neq 0$, and $\gamma^{\circ} = 0$ or 180° corresponds to the instant helical motion (kinematic helix);
6) $\mathbf{V} \neq 0$, $\mathbf{\Omega} \neq 0$, and γ° is any possible angle.

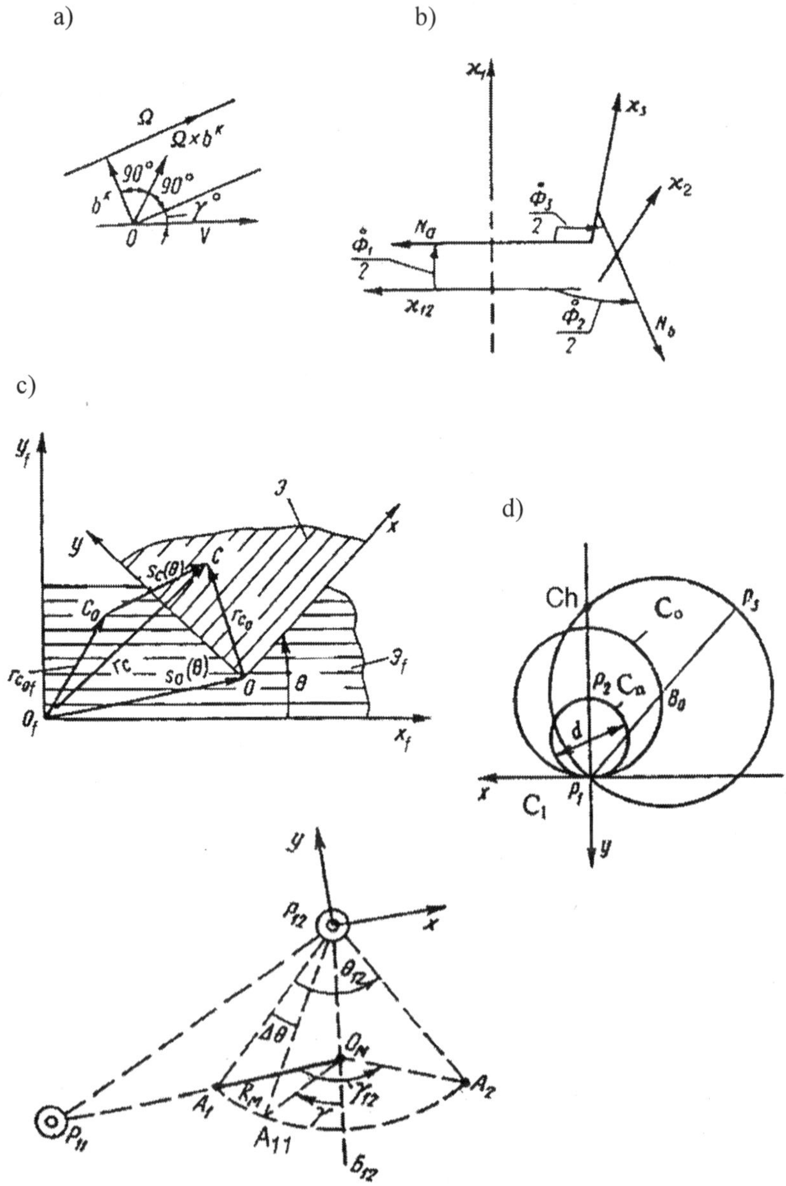

Fig. 1-4. Kinetostatic of motion

Case 6 may be converted to case 5. Therefore, the most general case of rigid body motion is the kinematical helix or the helical motion.

Instant motion. The instant helical motion may be defined by the ratio of the translational and rotational velocities. This ratio is called the helix parameter, κ_p. If vectors **V** and **Ω** are defined through corresponding projections in Cartesian coordinates as Ω_x , Ω_y , Ω_z , and V_x, V_y, V_z , then the helix parameter is equal to :

$$\kappa_p = (\Omega_x V_x + \Omega_y V_y + \Omega_z V_z) / \Omega^2. \tag{1-2}$$

The coordinates of point O on the axis of the kinematic helix are equal to:

$$X_0 = (\Omega_y V_z - \Omega_z V_y) / \Omega^2, \tag{1-3}$$

$$Y_0 = (\Omega_z V_x - \Omega_x V_z) / \Omega^2, \tag{1-4}$$

$$Z_0 = \kappa_p, \tag{1-5}$$

where $$\Omega^2 = \Omega_x^2 + \Omega_y^2 + \Omega_z^2 \tag{1-6}$$

and the axis of the helix is parallel to vector **Ω**. Vector **Ω** of the angular velocity is called the main vector of the motor (**V**, **Ω**), vector of translational velocity **V** is called the moment of the motor (bivector), and point O is called the center of conversion. Hence, the helix represents the particular case of the motor at which axes of its main vector and moment coincide.

Finite spatial motion. The general relations for the finite displacement of the rigid body are also known. Two sequential helical displacements with the dual angles $\overset{\circ}{\Phi}_1$ and $\overset{\circ}{\Phi}_2$ in relation to any arbitrary axes of the space with unit helixes $\kappa_{p1}=\chi_1$ and $\kappa_{p2}=\chi_2$ may be replaced by one resulting equivalent helical motion. The axis with unit helix $\kappa_{p3} = \chi_3$and the dual angle $\overset{\circ}{\Phi}_3$ of the resulting helical motion may be found by the following method (Fig. 1-4b). Draw the axis $\kappa_{p12}=\chi_{12}$ of the complex angle $\angle\chi_1\chi_2 = \overset{\circ}{\Phi}_\kappa$ and the straight line N_a crossing the axis of helix χ_1 at right angle and having dual angle $\overset{\circ}{\Phi}_1 / 2$ with the axis of the dual angle $\overset{\circ}{\Phi}_\kappa$. After that draw the straight line N_b crossing the axis of the helix χ_2 at right angle and having the complex angle $\overset{\circ}{\Phi}_2 / 2$ with the same axis of the angle $\overset{\circ}{\Phi}_\kappa$. The double dual angle between the lines N_a and N_b is equal to the complex angle $\overset{\circ}{\Phi}_3$ of the resulting helical motion, and the angle axis (N_a, N_b) with unit helix χ_3 is the axis of the kinematic helix of the resulting displacement (Halphen's theorem, 1873; *see* F.M. Dimentberg, 1978).

A motor $\mathbf{S}_\kappa$ of the finite displacements may be described by three angles of rotation with respect to the coordinate axes (ϑ_x, ϑ_y, ϑ_z as the main vector projections) and by three translational displacements along the coordinate axes (S_x, S_y, S_z as the main moment projection). The motor of forces $\mathbf{Q}_\kappa$ may also bc described by three projections of the forces on the coordinate axes (Q_x, Q_y, Q_z as the main vector projections) and three couple moment projections (M_x, M_y, M_z as the main couple moment projections). To find the displacement of a point of the moving rigid body, apply unit force (or unit couple moment) vector to the point in corresponding direction. Then the displacements $S_{\kappa i}$ will be equal to the scalar product of finite displacement motor $\mathbf{S}_\kappa$ of the rigid body and the unit motor $\mathbf{Q}_{\kappa e}$ of imaginary (virtual) force, $S_{\kappa i} = \mathbf{Q}_{\kappa e} \bullet \mathbf{S}_{\kappa}$. This is the so-called dummy-unit-load method.

To find a scalar product of the motors we need to multiply their coordinate constituents. In this case, the first motor is represented as a row matrix and the second one as a column (shown below as transposed ^{T} row) matrix:

$$\mathbf{Q}_{\kappa e} \bullet \mathbf{S}_{\kappa} = \left[M_x\ M_y\ M_z\ Q_x Q_y Q_z\right][\vartheta_x\ \vartheta_y \vartheta_z\ s_x\ s_y s_z]^T$$
$$= \vartheta_x M_x + \vartheta_y M_y + \vartheta_z M_z + s_x Q_x + s_y Q_y + s_z Q_z\ . \qquad (1\text{-}7)$$

This scalar product of the motors is equal to the work of the force motor on the displacement motor. The transformation matrix A_{mn} is used to transfer from one coordinate system to another (motor's tensor).

Plane motion. The most typical type of rigid body's motion on a translational and rotational guide, as well as in a planar mechanism, is the plane-parallel one (case 4). Translation (case 2) and rotation around a fixed axis (case 3) represent case 4's idealized particular cases. In kinematics, the plane-parallel motion is considered the sum of the displacements of translational $s_0(\theta)$ together with point O and rotational θ around this point of the moving plane $\mathrm{Э}$ with reference to the fixed plane $\mathrm{Э}_f$ (Fig. 1-4c). If there is a functional connection between translational and rotational motion of the plane $\mathrm{Э}$, the vector-radius $r_c\ (\theta)$ of any point C on the moving plane can be expressed by the equation in complex form with argument θ as follows,

$$r_c\ (\theta) = s_0(\theta) + r_{c0}\ e^{i\theta}, \qquad (1\text{-}8)$$

where r_{c0} is the vector-radius of point C at the initial position of the moving plane $\mathrm{Э}$ when moving xoy and fixed $x_f o_f y_f$ coordinates coincide. Then small displacements $S_c(\theta)$ of point C in the fixed coordinates may be described by the complex equation

$$s_c(\theta) = s_0(\theta) + r_{c0}\ e^{i\theta} - r_{c0}\ . \qquad (1\text{-}9)$$

The function $e^{i\theta}$ can be expanded by converging series: $e^{i\theta} = \sum\limits_{n=0}^{\infty} \dfrac{i^n \theta^n}{n!}$. (1-10)

If function $s_0(\theta)$ is an analytical one, it may be represented by converging Taylor's series. In this case, we finally have an equation for the displacement $s_c(\theta)$:

$$s_c(\theta) = \sum_{n=0}^{\infty} \frac{d^n s_0(\theta)}{n!\, d\theta^n}\Big|_{\theta=0} \theta^n + r_{c0} \sum_{n=0}^{\infty} \frac{i^n \theta^n}{n!} - r_{c0} = \sum_{n=1}^{\infty} i^n (r_{c_0} - P_{n0}) \frac{\theta^n}{n!}, \qquad (1\text{-}11)$$

where P_{n0} are the so-called initial poles of motion,

$$P_{n0} = -\frac{1}{i^n} \frac{d^n s_0(\theta)}{d\theta^n}\Big|_{\theta=0}\ . \qquad (1\text{-}12)$$

The pole of the n-order is a point with vector-radius $r_c = P_{n0}$ on the moving plane, which at the position under consideration coincides with the trajectory point whose n-order derivative is equal to zero. It is known that any plane non-translational displacement of the rigid body from one position to another close position may be produced by one rotation on a small angle around the certain point of the fixed plane, called the instantaneous center of rotation (*ICR*), which coincides in the moment with the instantaneous center of velocities (*ICV*) on the body. This point corresponds to the pole P_1 of the first order (n=1) which coordinates are

$$x_{P_1} = x_{ICR} = x_o - dy_o / d\,\theta; \qquad (1\text{-}13)$$

$$y_{P_1} = y_{ICR} = y_o + dx_o / d\,\theta. \qquad (1\text{-}14)$$

Coordinates of the point on the immovable plane that corresponds to the pole of the second order are equal to:

$$x_{P_2} = x_o + d^2x_o / d\,\theta^2; \qquad (1\text{-}15)$$

$$y_{P_2} = y_o + d^2y_o / d\,\theta^2. \qquad (1\text{-}16)$$

As a reminder, the locus of *ICR*s is called the immovable centrode C_e, and the locus of *ICV*s is called the movable centrode C_{e1}. It is clear that when the plane body moves, the above mentioned centrodes roll one upon another without sliding, and their point of contiguity corresponds to *ICR* and *ICV*. The curvature of the trajectory of a body point's motion depends on the difference between the centrode curvatures $(K - K_1)$, but not on the curvature of each of the centrodes. If the difference between the curvatures is constant, the centrodes may be substituted by a circle C_0 (Fig. 1-4d) with diameter $2d$ and the straight line C_1 at the same curvature difference:

$$K\text{-}K_1 = K_0 - K_{10} = K_0 = 1/\,d, \qquad (1\text{-}17)$$

where $K_0 = 1/\,d$ is a curvature of the circle with the radius d, and the curvature of straight line $K_{10} = 0$. On the basis of (13, 14) and (15, 16) we obtain the equation for estimating radius d as $d = \sqrt{[y_0'(\theta) + x_0''(\theta)]^2 + [y_0''(\theta) - x_0'(\theta)]^2}$. (1-18)

Circle C_0 with diameter $2d$ is called a circle of rolling, and circle C_n with diameter d is called a circle of rotation or rotational circle. The straight line C_1 is the pole tangent and the perpendicular to it in point P_1 (*ICR*) straight line is the pole normal. The trajectory of the points of the circle rolling on a straight line is known as a trochoid. The point B_o on the circle of rolling, for which the curvature of the trajectory K=0 and its derivative K'=0, is called the Ball's point. The trajectory of the Ball's point in the limited path is close to the straight tangent line with the order that is not less than three. Chebyshev's point (*Ch*) has a trajectory even closer to rectilinear one, witch contiguity to the tangent line is not less than the fifth order. The trajectory of the displacement of point P_c of the moving plane that corresponds at the first moment to the *ICR* (P_1) may be described by the following equations in x and y coordinates with the origin at P_{10};

$$x_{P_c} = \frac{2K_1 - K}{(K_1 - K)^2}\frac{\theta^3}{3!} + \frac{4K_1K' - KK' - 3K_1K'}{(K_1 - K)^3}\frac{\theta^4}{4!}; \qquad (1\text{-}19a)$$

$$\begin{aligned} y_{P_c} = & -\frac{1}{K_1 - K}\frac{\theta^2}{2!} + \frac{K_1' - K'}{K_1 - K}\frac{\theta^3}{3!} \\ & + \frac{(K_1 - K)(K_1'' - K'') - 2(K_1' - K')^2 + 3K_1^2 - 3KK_1 + K^2}{(K_1 - K)^3}\frac{\theta^4}{4!}. \end{aligned} \qquad (1\text{-}19b)$$

At a small rotation angle θ of the circle rolling on the straight line, we have

$$x_{P_c} = (\theta\text{-}\sin\theta) / (K_0 - K_{10}) \approx -\,\theta^3\,(K_{10}\text{-}K_0) / [3!\,(K_{10} - K_0)^2\,]; \qquad (1\text{-}20a)$$

$$y_{P_c} = -\,(1\text{-}\cos\theta) / (K_{10} - K_0) \approx -\,\theta^2 / [2!\,(K_{10} - K_0)]. \qquad (1\text{-}20b)$$

Hence, at a small rotation angle θ of the circle rolling on the straight line, the cycloid trajectory of point P_c is close to the semi-cubic parabola. Comparing the equations (12)-(14), (15), (16), and (20), we can define the geometric meaning of the kinematic poles. Pole P_1 of the first order corresponds to the *ICR*, pole P_2 of the second order (pole of the rotation) corresponds to the center of the rolling circle, pole P_3 of the third order (second pole of the rotation) has coordinates that depend not only on the difference of centrodes' curvatures, but on the curvature of each centrode and its derivative

$$x_{P_3} = (2K_1\text{-}K) / (K_1\text{-}K)^2, \quad (1\text{-}21)$$

$$y_{P_3} = (K'_1 - K') / (K_1 - K). \quad (1\text{-}22)$$

Note that at the rotation, point P_c gets a small linear displacement in comparison with that for other points of the moving plane. Therefore the pole of the first order or the point close to it at a small angle of rotation should be chosen as a conditional center of the moving plane's rotation (fulcrum). This is especially important for an elastic rotational guide that does not have a real guiding surface. Closeness of the plane motion to the rotation may be characterized by a stable position of the chosen fulcrum. If the two close finite positions of the moving body at its plane-parallel motion are known, and the position of two respective *ICR* are also known, then the problem of finding the trajectory of the moving body's points becomes more definite. Suppose we have two finite positions A_1 and A_2 (Fig. 1-4e) for point A and one very close to them position A_{11} or A_{22} of the same point. In this case we can always provide one and only one definite circle with radius R_M through the points A_1, A_2, A_{11} or the points A_1, A_2, A_{22}. Let us specify the two close finite positions $\Delta\theta$ of the body by pole P_{12} for the finite rotation and angle of rotation θ_{12}. We denote by P_{11} the *ICR* of the body in the first position. Then the center O_M of the sought circle lies on the intersection of the bisector $Б_{12}$ of angle θ_{12} and the straight line connecting P_{11} with point A_1. The angles of radius R_M corresponding rotation are denoted as γ and γ_{12}. Providing the circle through four points A_1, A_2, A_{11}, and A_{22} is only possible if point A is positioned on the so-called cyclic Burmester's curve of the third order. The trajectory of such points is close to the circle with a contiguity of the third order. In general, there are two known methods of functions (trajectories of motion) approximation: by optimal interpolation and with standard deviation minimization.

Precision interpolation. The approximation of the trajectories of motion with interpolation can be achieved by several single-multiple knots or by one high-multiple knot. The mixed versions of approximation also can be used. The possible accuracy of approximation is usually estimated by the summation of all of the knots' multiples. Let us, for example, compare the accuracy of approximation of a function with a circle using different methods. The equation for the segment of a circle with its center at point O (Fig. 1-5a) and central angle $\upsilon_{ok} << 1$ may be represented as

$$\Delta x_{ck} = (y^2 + 2yr\, \upsilon_{ok}) / 2r, \quad (1\text{-}23)$$

where $(\Delta x_{ck} / r) << 1$.

An expansion of the approximating function $\Delta x_{ck} = f(y)$ by Taylor's series is presented as $\Delta x_{ck} = f(y) = f(0) + f'(0)\, y + f''(0)\, y^2/2 + f'''(0)\, y^3/6$

$$+ f^{IV}(0)\, y^4/24 + \ldots = a_0 + a_I y + a_{II} y^2 + a_{III} y^3 + a_{IV} y^4 + \ldots \text{and } f(0) = 0. \quad (1\text{-}24)$$

If we do an approximation of the second order with the three-multiple knot in point A, then, from the condition of an equality of (23) and (24), their first and second derivatives at y=0, function Δx_{ck} is equal to: $\Delta x_{ck}^{(A)} = a_I y + a_{II} y^2$. (1-25)

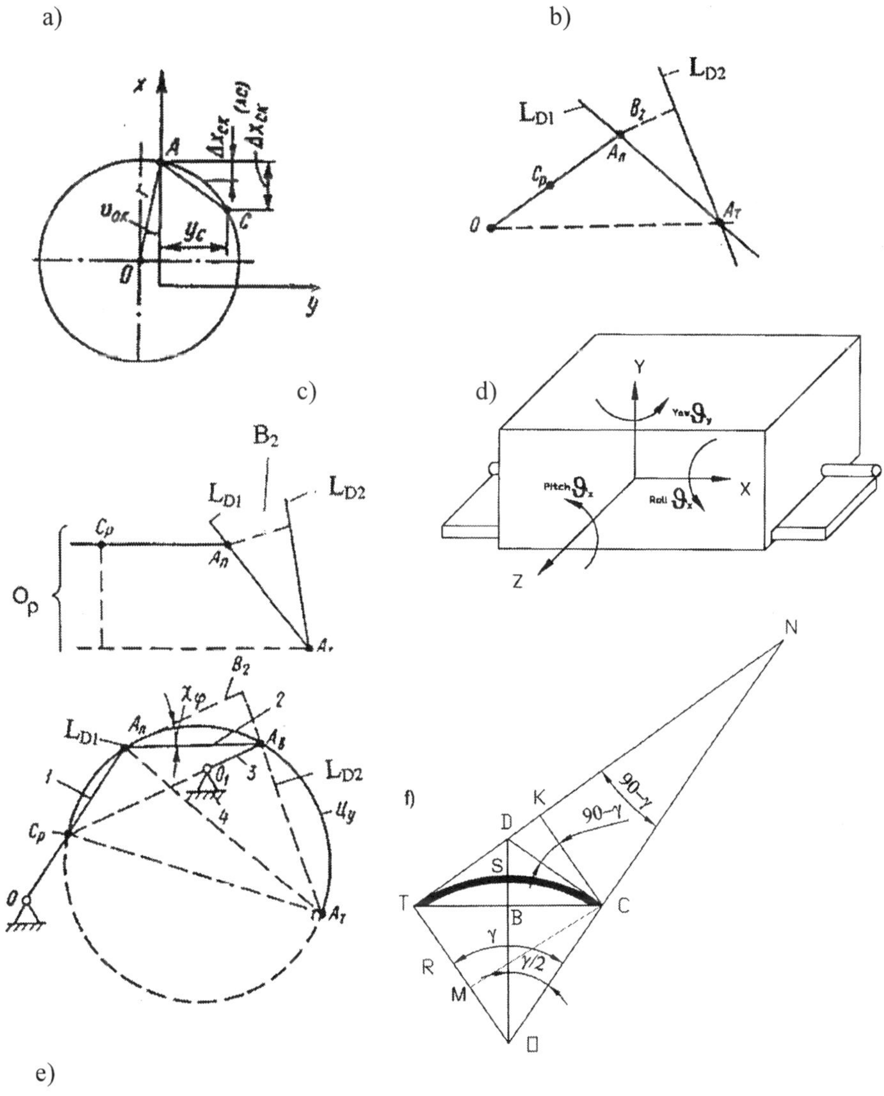

Fig. 1-5. Trajectory approximation and its uncertainty estimation

For an approximation with one two-multiple knot in point A and an one-multiple knot in point C, a segment of the circle passing through these points may be presented in the form $\Delta x_{ck}{}^{(AC)} = a_1 y + y^2 \left(\Lambda_{XC} - y_C a_1\right) / y^2{}_C$. (1-26)

Now to simplify the analysis, we will lower the degree of the polynomial (24) by Chebyshev's polynomials with zero point, which have minimum deflection from zero (*see* Table 1-8). Consider that y argument is within the limits $0 < y < y_C$ and we have the option to omit the member of (24) with coefficient a_{IV}. In this case, we obtain the following values of maximum deflections at $y = y_C$ and $y = ½\, y_C$, respectively

$$|\Delta x - \Delta x^{(A)}_{ck}|_{y=y_C} = |-0.4308\, a_{III} y_C^3 + 1.3923\, a_{III} y_C^3| \approx |\,0.96 y_C^3 a_{III}|; \qquad (1\text{-}27)$$

$$|\,\Delta x - \Delta x^{(AC)}_{ck}\,|_{y=(1/2)y_C} = |\,-0.2154\ a_{III} y_C^3 + 0.1074 a_{III} y_C^3| \approx |\,0.108 y_C^3 a_{III}| \qquad (1\text{-}28)$$

$$|\,\Delta x - \Delta x^{(AC)}_{ck}|_{y=(1/2)y_C} < |\,\Delta x - \Delta x^{(A)}_{ck}|_{y=y_C}\,. \qquad (1\text{-}29)$$

Derived results show the advantage of the mixed knots method of approximation at accepted allowances. It is clear that in other specific cases the analysis of approximation accuracy may be conducted in a similar way.
Minimization of standard deviation. Another more spread and easy method of approximation is the method of minimum standard deviation δ_σ, where

$$\delta_\sigma = \sqrt{(\beta-\alpha)^{-1} \int_\alpha^\beta [\Phi_0(\varphi) - \Phi(\varphi, a, b, c \ldots)]^2 d\varphi}\ , \qquad (1\text{-}30)$$

$\Phi_0(\varphi)$ is a specified function of the argument φ in the range $[\alpha;\beta]$ and $\Phi(\varphi, a, b, c\ldots)$ is the approximate trajectory's function with sought transformation mechanism's parameters $a, b, c\ldots$The differentiation of this expression by all of sought parameters and equalizing the derivatives to zero will give the necessary number of equations for the parameters optimization.
Chain's uncertainty. Elastic kinematic chains may be with (i) amplification, (ii) reduction, or (iii) parallel transmission tasks. Let us recall that an error Δx_m of m-element's transformation feature multiplies sequentially by the structural diagram elements' influence coefficients K_m such that the chain's output error $\Delta y = K_m \Delta x_m$. The error's influence coefficient K_m is less in the amplifying chain if element m is further from the chain's input. However, if the element's error depends on the input value from preceding elements in the linear manner, its position in the transmission chain may have no influence. If the element's error depends on the input value in the second or larger power, its deposit in the chain's output error may be less at its position closer to the chain's input. The effective distribution of the transmission ratios between the chain's elements depends on the error factor, strain, stress, and the force factor. This is one of the important problem in structural synthesis.
Motion errors. Let us recall that the rotational circle is the locus where curvature K of the trajectories is equal to zero. The ideal rectilinear motion has trajectory with $K = 0,\ K'_\vartheta = 0,\ K''_\vartheta = 0\ldots,\ K^{(n)}_\vartheta = 0$, where $n \to \infty$. An ideal rotational motion has a trajectory with constant curvature $K \neq 0$=constant. The real motion of the body may be considered the motion with the almost specified trajectory but with some deflections from it. As an ideal trajectory for rotational motion, we choose one with an immovable axis, and for a translational motion we choose rectilinear motion as its idealization. It is known from the kinematics that *ICR* for the adding transfer and relative motions, as well as the resulting (absolute) motion, in any instant moment are on one straight line and the distances between them are in reverse proportion to the angle velocities of these motions. Simplified schematics for differential of actual error Δ_{2a} calculations were proposed (A.T. Draudin, 1967) on this basis for the rotational guides (Fig. 1-5b) and rectilinear guides (Fig. 1-5c). This differential is equal to:

$d\,\Delta_{2a} = B_2\, d\,\vartheta$, (1-31)

where B_2 is the arm of the line of action L_{D2} for the driving point in relation to the instantaneous center A_T (point of intersection of the lines of action) of the erroneous motion on the line of action L_{D1} for the driven point; $d\vartheta$ is a differential of the angle of rotation in the erroneous motion. Note that the position of the center O_p of rotation for the rectilinear guide is in the infinity. Typical angular errors in the rectilinear motion (Fig. 1-5d) include roll angle ϑ_x around the longitudinal axis of motion X, yaw angle ϑ_y out of plane of motion around the Y transverse axis, and pitch angle ϑ_z in the plane of motion around the Z transverse axis. The shown notations of axes may be changed, but functional connections between the named slope error angles and the plane and direction of motion are supposed to be the same. We can show that the problem of finding the uncertainty of the guides is similar to the problem of finding the parameters of the connecting rod motion in an articulated mechanism with the rigid elements (EKS). Such a four-link mechanism (Fig. 1-5e) $OA_nA_BO_1$ has a rocker 1, connecting rod 2, crank 3, and stand 4. We can consider that the connecting rod corresponds to a guided element. In this case point A_n of the connecting rod is a driven one and is positioned on the line of action L_{D1}; point A_B is a driving one for the sequential kinematic chain and is positioned on the line of action L_{D2}. Hinge O is the center of the curvature of point A_n's motion trajectory and can be considered the center of the ideal motion. Hinge A_n is the center of the relative motion of the connecting rod and can be considered a center of the erroneous motion. Finally, point C_P corresponds to pole P_1 of the first order (*ICR*), or to the center of the resulting (absolute) motion of the connecting rod. Hence this point is the center of real motion. Point A_r of the intersection of the lines of action is positioned on

circle $Ц_y$ that passes through points A_n, A_B, C_P and has a diameter equal to $\overline{C_pA_T}$. In the case of a reversal in the direction of motion of rocker 1 (angular velocity of the rocker will be zero $\Omega_1 = 0$, but the angular acceleration $\Omega'_1 \neq 0$), an instantaneous center of acceleration (*ICA*)is also positioned on circle $Ц_y$. The differential of the relative motion may be described through the relative velocity of point A_B that is

equal to: $V_{A_BA_n} = \overline{A_nA_B}\,\Omega_2$, (1-32)

where Ω_2 is the angular velocity of the connecting rod in the relative motion. Then the projection of point A_B's erroneous relative motion differential on the direction of line of action L_{D2} may be expressed by the formula

$$d\,\Delta_{2a} = \overline{A_nA_B}\,\cos\chi_\varphi\,\Omega_2\,dt = B_2\,d\vartheta, \tag{1-33}$$

where $B_2 = \overline{A_nA_B}\,\cos\chi_\varphi$; $d\vartheta$ is the differential of link 2's rotation angle in the relative (erroneous) motion. Expressions (32) and (33) are analogous in form and content that is convenient for further description of the problem. Note that $\Delta_{2a} = 0$ at the identical lines of action for the driven and driving points. This conclusion corresponds to the known Abbe's principle in the linear measurement that shows the absence of the guides' influence on the uncertainty of measurement at identical lines of action for comparing dimensions. Hence, the important feature here is the mutual position of lines of action, but not the type or design specifics of the guides. Let us recall the specifics of Grashof's criterion for the four-bar mechanisms with at least one link making full revolution: $\overline{OA}_B+\overline{OO_1} \le \overline{OA_n} + \overline{A_nA_B}$.

Circle segment's curvature and its ends shift. Let us recall that curvature of a circle segment is equal to $\aleph=R^{-1}$ (Fig. 1-5f). In some problems of kinematics, the curvature $1/R=\aleph$ and length $L=R\gamma$ of the circle segment should be linked with the relative shift $\overline{KC}$ of its ends. Similar problems often appear in the pure bending of beams and their thermal deformations (*see* Chap. 7). We can find this link from triangles *ODC*, *OTN*, *OMC*, and *CDK*, where $\overline{DC} = R\tan(\gamma/2)$, $\overset{\frown}{TC} = L$, $\gamma = L/R$, and

$\overline{KC} = \overline{TM} = R\tan(L/2R)\sin(L/R) = \aleph^{-1}\tan(L\aleph/2)\sin(L\aleph)=L(1-\cos\gamma)/\gamma$. One can approximate these trigonometric functions at a small angle γ by series expansion. If for example $(L/R) \le 0.1$, then $\overline{KC} = L^2/2R$. For larger angles γ, we can use other approximations presented in Table 1-9. Let us also recall that sag $\overline{BS}$ =R(1-cosγ/2).

Therefore this sag at $\gamma \le 0.1$ equals $\overline{BS} \approx R\gamma^2/8$.

Transmission ratio. The main kinematic characteristics of the transmission mechanisms are the displacement function $F(s)$ or $F(\theta)$, with the linear s or angular θ argument, and kinematic transmission function $i(s)$ or $i(\theta)$ which at the discrete points may be characterized by the dimensionless transmission ratios $i = ds_{exit} / ds_{enter}$, $i = d\theta_{exit} / d\theta_{enter}$ or by the dimensional transmission ratios $i_\theta = d\theta_{exit} / ds_{enter}$, $i_s = ds_{exit}/ d\theta_{enter}$. In precision mechanisms, specified displacement function and specified transmission ratios should be conducted with required accuracy. For that reason, the accurate formulae, or their accurate enough approximations, have to be used at the mechanisms' synthesis. The most effective approximations are (i) the least-squares or weight-estimated deflections and (ii) the uniform one with minimum deflection from zero. The used least–maximum–error uniform approximation for the mechanism's functions based on the Chebyshev's polynomials, shown in Tables 1-7, 1-8.

Table 1-7. Chebyshev's polynomials P_n (Z) with minimum deflection from the zero in the range [-1; +1]

n-degree	P_n (Z)	Roots	Maximum deflection points	Maximum deflection
2	$Z^2 - 1/2$	± 0.7071	0; ± 1	1/2
3	Z^3 - (3/4) Z	0; ± 0.8660	± 0.5; ± 1	1/4
4	$Z^4 - Z^2 + 1/8$	± 0.3827; ± 0.9239	0; ± 0.7071; ± 1	1/8
5	Z^5 –(5/4) Z^3 + (5/16) Z	0; ± 0.5878; ±0.9511	± 0.3090; ± 0.8090; ± 1	1/16

After M.M. Kempinskii, 1972; $P_{40}(Z)=Z^4-0.8284Z^2$ has zero point at Z=0.

Table 1-8. Chebyshev's polynomials R_n (Z) with zero point that have minimum deflection from the zero in the range [0; +1]

n-degree	R_n (Z)	Roots	Maximum deflection points	Maximum deflection
2	$Z^2 - 0.8284Z$	0; 0.8284	0.4142; 1	0.1716
3	$Z^3 - 1.3923\ Z^2 + 0.4308\ Z$	0; 0.4641; 0.9282	0.1961; 0.7320; 1	0.0385
4	$Z^4 - 1.9209Z^3 + 1.1135Z^2 - 0.1835\ Z$	0; 0.2813; 0.6791; 0.9604	0.1127; 0.4802; 0.8478; 1	0.0091
5	$Z^5 - 2.4373Z^4 + 2.0477Z^3 - 0.6781Z^2 + 0.0698Z$	0; 0.1862; 0.4875; 0.7887; 0.9749	0.0728; 0.3291; 0.6458; 0.9021; 1	0.0022

After M.M. Kempinskii, 1972.

Chebyshev's polynomials and theorems. If the function has argument ξ with range limits $[a, b]$ other than one, there should be a substitution $\xi=(a-b)\ Z/\ b$ or $\xi=bZ$ when the limits are symmetrical to zero. At the symmetrical limits of the function arguments, it is more likely to use $P_n(Z)$ polynomials (Table 1-7) for function approximation; for functions with limits in one direction from zero (positive or negative), it is more likely to use $R_n(\ Z)$ polynomials (Table 1-8). It should be mentioned that Chebyshev's polynomials with minimum relative deflection from zero $P^o{}_n\,(Z)$, $R^o{}_n(Z)$ are also in use: $P^o{}_n\,(Z)= Z\,P_{n-1}(Z)$, $R^o{}_n(Z) = Z\,R_{n-1}(Z)$. Let us recall the content of Chebyshev's theorem which states that the Chebysev's polynomial $P_n(x)$ will least deflect from a continuous function $F(x)$in the segment $[a,b]$ if the difference $\Delta(x)= P_n(x)$ - $F(x)$ approaches in this segment its ultimate values $(n+2)$ times consequently changing its sign. If during the approximation, we yield a formula with some coefficient that is not equal to one at the series member with the highest power, then the maximum deflection of the approximated function will be equal to the product of the used polynomial maximum deflection and that coefficient. Let us recall the content of another Cebyshev's theorem which states that if the integer function of q degree has the coefficient T_q at the highest degree member, the values of this function within the segment from $x=a$ to $x=b$ cannot be in more tight limits than $\pm 2T_q\left(\frac{b-a}{4}\right)^q$. Chebyshev's polynomials are very effective for lowering the power of series members, with a small loss of accuracy. It is also useful to approximate trigonometric functions by power series expansion with a maximum possible economy (minimal polynomials), as shown in Table 1-9. The roots of Chebyshev's polynomials are effectively used in precision mechanics for accuracy adjustment purposes and correctors' range calculations. Other possible approximations of the trigonometric functions by power series expansion (*see* G. A. Korn, T.M. Korn, 1968) are widely known and will be used accordingly. Practical examples of the effective approximation of specific functions will be shown in the later sections.

Table 1-9. Trigonometric functions series expansion with maximum economization (minimal polynomials)*

Range of X	sin X	1-cos$X$	tan X	$\sin^{-1}X$	$\tan^{-1}X$
$-0.1 \le X \le +0.1$	0.999X ± 0.0001	$X^2/2$ ± 0.0001	1.002X ± 0.0001	1.001X ± 0.0001	0.997X ±0.0001
$-0.3 \le X \le +0.3$	0.989X ±0.0012	$X^2/2$ ± 0.0003	1.024X ± 0.0025	1.011X ± 0.0012	0.978X ±0.002
$-0.5 \le X \le +0.5$	0..969X ± 0.005	0.489X^2 ± 0.0005	1.068X ± 0.012	1.035X ± 0.006	0.945X ±0.009
$-0.8 \le X \le +0.8$	0.922X ± 0.02	0.478X^2 ± 0.003	1.20X ± 0.07	1.11X ± 0.04	0.88X ± 0.03
$-\pi/2 \le X \le +\pi/2$	0.72X ±0.14	0.421X^2 ±0.04	-	-	-

* After M.M. Kempinskii, 1972.

Note: $\sin^{-1}X$= arcsin X; $\tan^{-1}X$ = arctan X; involute function tan$X$-X = inv X= ϑ_{in}; inv $X = 0.3333X^3$ in range $-0.1 \le X \le +0.1$;inv X=0.345X^3 in range $-0.3 \le X \le +0.3$; inv$X$ = 0.380X^3-0.003X in range $-0.5 \le X \le +0.5$; and inv$X$ = 0.003X + 0.300X^3 in range $-0.8 \le X \le +0.8$. The most accurate expansion of the inverse involute function $X(\vartheta_{in})$ may be found in (F. L. Litvin, 1994) as:

$$X(\vartheta_{in}) = (3\vartheta_{in})^{1/3} - \frac{2}{5}\vartheta_{in} + \frac{9}{175}3^{2/3}\vartheta_{in}^{5/3} - \frac{2}{175}3^{1/3}\vartheta_{in}^{7/3} + \ldots \quad \text{for } \vartheta_{in} < 1.8.$$

Spline functions. Polynomials are a very powerful tool in digestible and tractable analysis and synthesis by interpolation. However, for rapidly changing values of the functions to be approximated the degree of polynomial has to be increased that usually requires application of some computational techniques. More effective way is to divide the basic interval into subintervals each of which is approximated by piecewise polynomials lower degree. The polynomials are joined together in splines at the intervals endpoints (knots) with guarantee of differentiability (smoothness) of the resulting function at these points. Let us recall that the word spline means a thin flexible strip that provides a minimal curvature. A practical guide to splines is carefully discussed in C. de Boor monograph, 2001, where are shown limitations of polynomial approximations (the distance from polynomial), and in many other published sources. Splines are very helpful with CAD.

Sine and *tangent mechanism.* The precision rigid lever mechanisms' schematics are based on the trigonometric functions application. Their kinematic pairs usually have point-like contact between balls and plains. Elementary lever mechanisms for transformation of translational motion into rotational, and vice versa, may be divided on this basis into two groups: (i) "sine" mechanisms with a ball on the lever and the plain on the slider, (ii) "tangent" mechanisms with the ball on the slider and plain on the lever. This classification is helpful for the optimized synthesis of the mechanisms with the rigid lever elements and in some cases for the structural synthesis of the precision elastic kinematic systems as well (*see* Chap. 4).

1-3. Specifics of Elastic Kinematics and Dynamics

The *generalized functional diagram* of ELKS may be presented in the form shown in Fig. 1-6a, where the functional links between input $s(L)$driving member 1 and exit member 3 are depicted by the solid lines. An ELKS typically has a so-called internal feedback from compliance (dashed line) and in some cases it has link 6 of the external feedback. The shown diagram is a typical one for an elastic transfer mechanism with elastic element(s) 2. For elastic guides 4 the schematic becomes simpler because transformation member 2 is absent and members 1 and 3 have a rigid connection. Element 3 may be a real or virtual (conditional) rigid body. Compensators and correctors 5 may be connected to any of the links. The output of the ELKS transformation can be in amplitude 7 or/and the frequency 8 forms. The working space 9 in the surrounding media depends on the influence functions f(V_ψ) for the specific influence quantities V_ψ (*see* Chap. 7). It is easy to show that ELKS are similar both in functional features and structure to the equivalent kinematic systems (*EKS*) with rigid links that are connected by regular kinematic couples (hinges, slides, etc.). Let us recall the similarity of expressions for the behavior of a pendulum in motion and the static behavior of the elastic tape with weight on the movable end (*see* M. Born, 1956). Both expressions are based on the minimal principle of total energy, which comprises the potential and kinematics one. The analogies between different objects in precision mechanics may be found in many other cases such as membrane analogy for beam's twisting, where the *ICR* is equivalent with the center of twist and shear center, elastic-plastic deformation and static parameters of dry powder layers on the plane, etc.

Functional formula. The ELKS functional formula at the input displacement s may be presented in the following form

$$F(s) \supset F_k(s) + F_j(Q) + \delta F_j(Q) + \delta F_{kc}, \tag{1-34}$$

where $F(s)$ is the exit displacement of the ELKS, $F_k(s)$ is the exit displacement of the EKS or in the ELKS straight chain without resistance to motion from sequential elastic elements, $F_j(Q)$ and $\delta F_j(Q)$ are the linear and non-linear feedback of ELKS compliances, and δF_{kc} represents the environmental conditions that influence the ELKS kinematics. The feedback compliance shows the influence of the load opposite to motion direction applied to the driven link of the ELKS on its kinematics. Hence the ELKS' kinematics depends on their structure, input motion, stresses from direct and backward loads, and the conditions of work. But it is required that at some normal ranges of work conditions, ELKS behavior should be unambiguous or the uncertainty of its work should be less than a permissible limit. It is clear from the functional formula that for any real ELKS its kinematics may be represented by ideal motion as well as systematic and random errors (uncertainties) whose sources, however, are specific. An estimation of the accuracy of systems with feedback may be done by methods of structural analysis. The output coordinate s_{exit} in ELKS usually functionally depends on the resisting force (moment) F_Q (s_{exit}) of the subsequent driven mechanism or transducer, whose action usually causes a decrease of the exit coordinate s_{exit} in ELKS, or involves the internal compliance feedback.

The *kinematic chain* with different elastic transfer mechanisms at straight and feedback links may be calculated with the matrix apparatus:

$$s_1 = a_{11}s_2 + a_{12}F_{Q\,2}, \quad (1\text{-}35)$$

$$F_{Q1} = a_{21}s_2 + a_{22}\,F_{Q\,2}, \quad (1\text{-}36)$$

where s_1 and s_2 are the displacements at the input and exit of the elastic system, respectively; F_{Q1} and $F_{Q\,2}$ are the driving force at the input and resisting force at the output of the system, respectively. The first members on right-hand side of (35) and (36) correspond to the straight link (F_{Q2} =0) and the second members correspond to the internal feedback. These equations may be presented in the matrix form as follows:

$$\chi_1 = \begin{bmatrix} s_1 \\ F_{Q\,1} \end{bmatrix} = \begin{bmatrix} a_{11} & a_{12} \\ a_{21} & a_{22} \end{bmatrix} \begin{bmatrix} s_2 \\ F_{Q\,2} \end{bmatrix} = a_{ij}\,\chi_2 \,. \quad (1\text{-}37)$$

Hence the transmission ratio of the straight chain at $F_{Q2} = 0$ is equal to

$$i_{sc} = s_2 / s_1 \big|_{F_{Q2}=0} = 1/\,a_{11}. \quad (1\text{-}38)$$

The transmission ratio of the internal feedback with resistance rigidity j_{sr2} of the subsequent system at $F_{Q2} = j_{sr2}s_2 + F_{Q02}$ is equal to:

$$i_{ifb} = a_{12}\,j_{sr2}\,, \quad (1\text{-}39)$$

where F_{Q02} is an initial resisting force.

At outputs 7 (mechanism) and 8 (transducer) in Fig. 1-6a, we have $\chi_2 = \chi_{\text{out}}$ and $j_{sr2} = j_{\,res.\text{out}}$. The compliance of links 2 and 4 in the straight transformation chain occurs when the resisting force is applied to body 3. The environmental conditions are represented by the border 9 (Fig. 1-6a) with influence functions and quantities, which act on the system 1-4, its shields, correctors 5, 6, and the measurand $s(L)$.

a) b)

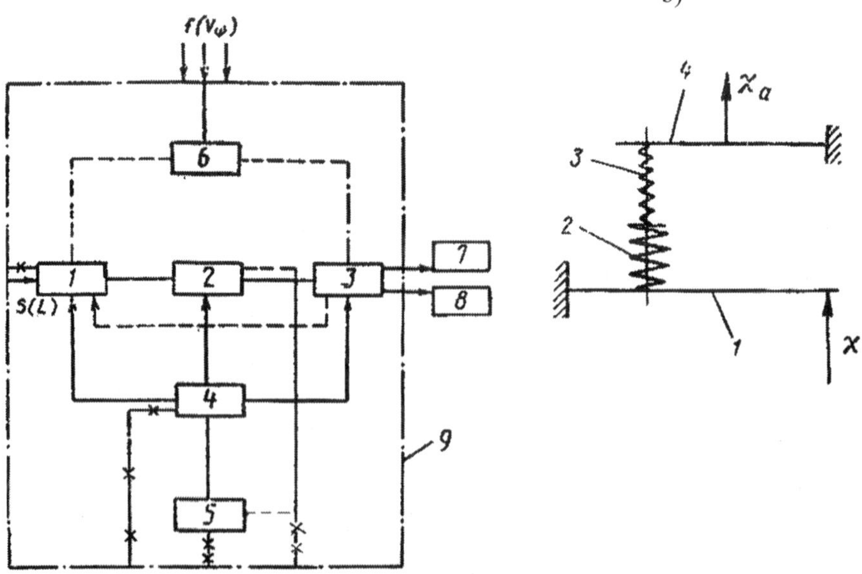

Fig. 1-6. ELKS functional block diagram and schematic for ELKS matrix calculation

If the transformation chain has several elastic elements, the resulting matrix $a_{ij}^{\Sigma,n}$ with internal feedback from the resisting action of the following n elastic elements will be defined as the product of partial matrices $a_{ij}^{(n)}$. For example, at four elastic elements 1-4 (Fig. 1-6b), among which are shown two leaf springs 1, 4 and two compression springs 2 and 3, we have

$$a_{ij}^{(1)}\chi = a_{ij}^{(1)} a_{ij}^{(2)} a_{ij}^{(3)} a_{ij}^{(4)} \chi_a = \begin{bmatrix} a_{11}^{\Sigma,4} & a_{12}^{\Sigma,4} \\ a_{21}^{\Sigma,4} & a_{22}^{\Sigma,4} \end{bmatrix} \chi_a = a_{ij}^{\Sigma,4} \chi_a \,. \tag{1-40}$$

$$\text{If } \chi_a = \chi_{out} = \begin{bmatrix} s_{out} \\ j_{res.out} s_{out} + F_{Q0out} \end{bmatrix}_{n=4}, \tag{1-40a}$$

then the resulting matrix for all of the foregoing elements is $a_{ij}^{\Sigma,4}$ and the transformation ratio of the straight transformation chain equals

$$i_{sc} = [a_{11}^{\Sigma,4}]^{-1}. \tag{1-41}$$

The transformation ratio of the internal feedback is equal to:

$$i_{ifb} = a_{12}^{\Sigma,4} j_{res.out}. \tag{1-42}$$

Thus, the whole transformation ratio of the elastic system equals

$$i_{ELKS} = [a_{11}^{\Sigma,4} + a_{12}^{\Sigma,4} j_{res.out}]^{-1}. \tag{1-43}$$

Matrices of the elastic bonds have systematic and random uncertainties as a result of the errors in the real elastic element functions and parameters, imperfection of their materials, defects of assembly, and the systematic and random action f (V_ψ) of the environmental influence quantities V_ψ (*see* Chaps. 6, 7).

Schematic of ELKS calculations.

It is typical for ELKS to have statical indeterminacy as a result of "excessive bonds" that may be eliminated by superposition force and displacement methods if the system is linear. Superposition method does not work properly for nonlinear systems. Vector geometric bonds are effective for analysis, while kinematic characteristics are more effective in solving the ELKS synthesis problems. In the latter case, it can be convenient to use the kinematic poles, inharmonic ratios of the perspective, etc.

Analytical and numerical solutions.

The theory of elasticity and the study of elastic element's behavior are based on the solution of ordinary and partial differential equations in Cartesian (x, y, z), polar (r, ϑ), or cylindrical (r, ϑ, z) coordinates, as well as surface curvilinear coordinates, with specified boundary-value conditions. Table 1-10 shows typical differential equations used in precision elasticity which have analytical and/or numerical solutions. The analytical solution (*An*) can be exact for simple problems, but in many cases these solutions are an approximate with some degree of precision. The numerical solution and integration (*Nu*) of differential equations with the addition of boundary-value conditions is more accurate than approximate one but not as digestible and tractable as the analytical solution may be. Numerical solution may also be used for special functions and integrals: elliptic, probability, and others. The results of the numerical solution are usually represented in table, nomogram, or chart's form, which limits the amount of stored information.

Table 1-10. Typical differential and integral equations in precision elasticity

№	Equation	Solution used in Chapter	Application examples
1	$y''(x) = M(x) / EJ$	*An* 3	Beam flexure linear relations
2	$\left(\frac{1}{\rho}-\frac{1}{\rho_0}\right) \approx y''_x [1+(y'_x)^2]^{-3/2} = M(x) / EJ$	*An, Nu* 3	Beam flexure nonlinear relations
3	$\frac{\partial s_u^{(2)}}{\partial y} - \frac{\partial s_u^{(1)}}{\partial z} = 2\theta_\gamma\, x$	*An* 2, 3	Beam torsion
4	$\frac{\partial s_u^{(2)}}{\partial y} - \frac{\partial s_u^{(1)}}{\partial z} = 2\theta_\gamma\, x + F\,\vartheta(z,y)$	*An* 2	Pretwisted beam torsion and tension
5	$\frac{\partial^4 w}{\partial x^4} + 2\frac{\partial^4 w}{\partial x^2 \partial y^2} + \frac{\partial^4 w}{\partial y^4} = \nabla^4 w = \frac{p}{D}$	*An, Nu* 3	Plates flexure, rectangular, pressure *p*
6	$\nabla^4 w = (\frac{\partial^2}{\partial r^2} + \frac{1}{r}\frac{\partial}{\partial r} + \frac{1}{r^2}\frac{\partial^2}{\partial \vartheta^2})(\frac{\partial^2 w}{\partial r^2} + \frac{1}{r}\frac{\partial w}{\partial r} + \frac{1}{r^2}\frac{\partial^2 w}{\partial \vartheta^2}) = \frac{p}{D}$	*An, Nu* 3	Plates flexure, round shape, pressure *p*
7	$\rho\psi_a'' + \psi_a' - \psi_a/\rho_r = \frac{1}{2}\,\vartheta_n^2;\quad \psi_a\vartheta_n = pR\rho_r^2/\,2Eh,$ $\psi_a = -\,H_r\rho_r\,/\,Eh$	*An* 3	Absolutely ** flexible diaphragm
8	$\psi'' + d_1\psi' + d_2\psi + d_3\theta + d_4\theta\psi + d_5\theta^2 + d_6\psi\theta' = f_\psi(ql_i);\ i=1,2,3$ $\theta'' + c_1\theta' + c_2\theta + c_3\psi + c_4\psi\theta + c_5\theta^2 + c_6\theta\theta' = f_\theta(qm_j);\ j=1,2$	*Nu* 3	Shells $\Phi_0<2\pi$, non-closed, nonlinear**
9	$m\ddot{x} + C_x\dot{x} + jx = F(t)$, Lagrange equation of the first kind (D'Alembert principle); $\ddot{y} + 2C_\gamma\dot{y} + \omega_0^2\, y = \frac{F_0}{m}e^{i\omega t};\ C_\gamma = \frac{C_y}{2m}$	*An, Nu* 4, 5, 8	Linear dynamics and vibration with damping
10	$(EJ_y\, u'' + EJ_{xy}\, v'')'' + \rho_\gamma A\,\ddot{u} = 0$ $(EJ_{xy} u'' + EJ_x\, v'')'' + \rho_\gamma A\,\ddot{v} = 0$	*An, Nu* 5	Pretwisted beam vibration
11	$T_k = \frac{1}{2}\int_0^L \rho_\gamma(\dot{u}^2 + \dot{v}^2)A dz$	*An, Nu* 5, 8	Kinetic energy
12	$\ddot{z} - C\theta\dot{z} + \beta z^3 + \omega^2 z = 0$, where β is a force parameter	*An* 7	Nonlinear self-excited vibration
13	$\frac{d}{dt}\frac{\partial T_k}{\partial \dot{q}_i} - \frac{\partial T_k}{\partial q_i} = F_{Qi}$, where q_i are the virtual coordinates of motion, F_{Qi} are the generalized resistance forces	*An, Nu* 4, 5, 8	Dynamnic Lagrange equation of the second kind
14	$P_k = \frac{1}{2}\int_0^L (Qw' + M_x\, v'' + M_y u'' + M_z\theta')dz$,	*An* 2, 3	Beam potential energy ***

* After J.T. Oden and E.A. Ripperger, 1981, ** after L.E. Andreeva, 1981; *** ½ corresponds to Clapeyron's theorem; №11 and 13 are used with Hamilton's principle of least actions.

Table 1-10 Cont'd.

N^o	Equation	Solution used in Chapter	Application examples
15	$d^4\phi/dz^4 + c_1^2 d^2\phi/dz^2 = m_T/E\Gamma^\upsilon$; ϕ is torsion angle; $m_T = dM_T/dz$; M_T is torque; $c_\Gamma = \sqrt{GJ_T/E\Gamma^\upsilon}$; J_T is torsional constant of section; Γ^υ is the warping constant	*An* 4	Strained thin-walled beam structure*
16	$\nabla^4 w = \omega^2 \sqrt{\rho_\gamma h/D}$	*An, Nu* 5, 8	Plates vibration

Notations in Table 1-10 are as follows: J, J_x, J_y, J_z, Jxy are the corresponding moments of inertia of sectional area with respect to coordinate axes x, y, z; u, v, w are the displacements in x, y, z directions, respectively; w is the transverse displacement in thin plates and diaphragms, EJ is the flexural stiffness (rigidity) of beam section , ρ_0 and ρ are the initial and deformed system's radii of curvature, r and ϑ are the polar coordinate system radius and angle, respectively; ρ_γ is the density of material, $\rho_{\gamma L} = \rho_\gamma A$ is the beam mass per unit length, A is the cross-sectional area, h is the thickness of a beam, plate, or shell, b is the width of cross section; $M(x)$ is the bending moment; E is Young's modulus; $s_u^{(1)}, s_u^{(2)}$ are the displacements in the direction of coordinates 1 and 2 in a cross-sectional plane, respectively; θ_γ is an angle of rotation per unit length of the twisted body; $F_\vartheta(x,y)$ is a function which reflects local rotations (hypothesis of V. M. Marchenko) of the elements of the pretwisted (initially, "naturally" twisted) beam's cross section; p is a transverse load (pressure) per unit area of the diaphragms' middle plane; $D = Eh^3/12(1-\nu_p^2)$ is the flexural (cylindrical) rigidity of a plate with thickness h and of material with Young's modulus E and Poisson's ratio ν_p, θ is the angle of rotation, $\psi = H_r R \rho_r / [E h_a^2 (1+\chi_\Phi)]$ is the function of radial tension H_r of the shell or diaphragm with thickness h_a and working radius R of the diaphragm or the radius of the central axis of tubular shell; $f_\psi(ql_i), f_\theta(qm_j)$ are the functions with the unitless (dimensionless) parameters of load ql_i, qm_j (L.E. Andreeva, 1981), $\rho_r = r/R$ is a current dimensionless radius; ϑ_n are the angles of rotation of the normal to the diaphragm or shell surface at arbitrary point; d_1-d_6, c_1-c_6 are the constant factors, $\nabla^4 = \nabla^2\nabla^2$, $\nabla^2 = \nabla \bullet \nabla = \frac{\partial^2}{\partial x^2} + \frac{\partial^2}{\partial y^2} + \frac{\partial^2}{\partial z^2}$ is the Laplace operator; $F_0 e^{i\omega t}$ is the exciting force with angular frequency ω and the amplitude F_0, $F(t)$ is an exciting variable force in time, $\omega_0 = 2\pi f_0$ is the natural angular frequency proportional to linear (f_0) frequency of the vibrating system; C_y , C_x , and C_θ are the damping factors in the directions denoted by subscripts, j represents stiffness or spring constant, and m is a moving mass, which at rotation corresponds to mass moment of inertia, I_g . Expressions $(\,)'_\zeta = d(\,)/d\zeta$ or $(\,)'_\zeta = \partial(\,)/\partial(\zeta)$ are the ordinary and partial derivatives, respectively. Apostrophes mark the derivatives along the coordinate axis denoted as subscripts. Dots above letters denote the derivatives with respect to time t (e.g., $\dot{x}$ is the speed and $\ddot{x}$ is the acceleration along the x coordinate). The geometric form of the thin wall shell of rotation may be closed or non-closed. The latter is a more general case with the possibility of changing the central angle Φ_o by $\Delta\Phi$ where $\Delta\Phi / \Phi_o = \chi_\Phi$. The closed shell has angle $\Phi_o = 2\pi$ and $\Delta\Phi = 0$. Working elastic displacement of the shell and beam can be large. This corresponds to the nonlinear relationship between forces and deformations. The system of differential equations for non-closed shell of rotation in the large displacements may be represented in dimensionless values. The solution for the system of simultaneous differential equations (8, Table 1-10) with the constant coefficients d_i and c_i is obtained through the numerical integration with the addition of boundary-value conditions in certain elastic elements: diaphragms, bellows, and manometric tubes. The stress and strain of elastic shells is defined on the basis of found functions ψ and θ as a result of this simultaneous system numerical solution (*see* L.E. Andreeva, 1981) with the Newton-Kantorovich method (*see* S.G. Mikhlin and K.L. Smolitskiy, 1967).

The effective solutions of the corresponding differential equations shown in Table 1-10 for typical elastic elements of precision elasticity are discussed in detail in later chapters of this book.

Singular functions. The analytical solution of differential equations is based on their consequent integration as shown in the following expression,

$\int_n dx \int_{(n-1)} dx ... \int_1 y^{(n)}(x)\, dx = F(x) + \sum_n C_n f_n(x)$, where the number n of integrals is equal to a degree of the differential equation, $F(x)$ represents the input function, and C_n represents the scope of the integration's constants, which values are defined through the specified boundary-value conditions. Straightforward integration of the differential equations is typical for elastic beams' bending and other similar cases with distributed forces and moments. However, in many instances, the elastic structure is subjected to the locally applied discrete forces, moments, or masses. In these cases, the singular step and impulse functions are helpful to present discrete variables as a quasi-distributed. The Dirac delta function $\delta(x)$ is one of the most powerful singular functions (I.M. Gel'fand *et al.*, 1964) for solution of many differential equations in applied physics, including precision elasticity. The main feature of this function is shown in the following expression,

$$\int_a^b \delta(x - x_0)\varphi(x)dx = \varphi(x_0)\,, \qquad (1\text{-}44)$$

where $\varphi(x_0)$ is the discrete value of the function $\varphi(x)$ at the coordinate x_0. Delta function $\delta(x\text{-}x_0)$ is equal to zero everywhere except at x_0, where it is infinite, and its integral equals one. The result of such integration is equal to zero if $x_0 < a$ or $x_0 > b$. The delta function is one of the so-called generalized functions, among which are symmetrical and asymmetrical unit-step functions $U(x)$. All these functions are symbolic, but they can be approximated by the limits of continuously differentiable functions shown in following Table 1-11.

Laplace transform. The solution of many linear differential equations and their systems is easier to derive with their intermediate Laplace transformation. Let us recall that the direct Laplace transformation $\mathfrak{L}$ of a function $f(t,a)$ is represented by the following integral, $\mathfrak{L}\, f(t,a) = \int_0^\infty f(t,a)e^{-\lambda t} d\lambda = F(\lambda)$, and the inverse of Laplace transformation is provided by the integral $\mathfrak{L}^{-1} F(\lambda) = \frac{1}{2\pi j}\int_{c-j\infty}^{c+j\infty} F(\lambda)e^{t\lambda}d\lambda = f(t)$.

After Laplace transformation, the differential equation becomes a corresponding algebraic one that simplifies the solution process. In the Laplace field, the differentiation corresponds to the multiplication by the complex variable λ , and the integration corresponds to the division by this variable. Some Laplace transforms of the singular functions are shown in Table 1-12.

Table 1-11. Approximation of impulse $\delta(x)$, its derivative double impulse $\delta'(x)$, and unit-step $U(x)$functions

Singular function $S(x)$	Approximation	Limit
$\delta(x)$	$\frac{1}{\pi}\frac{\varepsilon}{x^2+\varepsilon^2}$	$\varepsilon\to 0$
	$\frac{1}{2\sqrt{\pi t}}\exp(-\frac{x^2}{4t})$	$t\to 0$
	$\frac{\alpha}{\pi(\alpha^2x^2+1)}$	$\alpha\to\infty$
	$\frac{\alpha}{\sqrt{\pi}}\exp(-\alpha^2x^2)$	$\alpha\to\infty$
	$\frac{\alpha}{\pi}\frac{\sin\alpha x}{\alpha x}$	$\alpha\to\infty$
$\delta'(x)$	$-\frac{2}{\pi}\frac{\varepsilon x}{(x^2+\varepsilon^2)^2}$	$\varepsilon\to 0$
$U(x)$	$\frac{1}{2}+\frac{1}{\pi}\tan^{-1}(\alpha x)$	$\alpha\to\infty$
	$\frac{1}{2}[erf(\alpha x)+1]$	$\alpha\to\infty$

Table 1-12. The Laplace transform $F(\lambda)$of the singular functions $S(x)$

No	$F(\lambda)$	$S(x)$	Function type
1	$\frac{1}{\lambda}e^{-a\lambda}$	$U(x-a)=0$ at $x<a$; $U(x-a)=1$ at $x\geq a$	Unit function at and after point a;
2	$\frac{1}{\lambda^2}e^{-a\lambda}$	$(x-a)U(x-a)$	Unit integral at point a
3	$e^{-a\lambda}$	$\delta(x-a)$	Unit impulse at point a
4	$\lambda e^{-a\lambda}$	$\delta'(x-a)$	Double impulse at point a

Let us consider, for example, the simply supported beam of Fig. 1-7a with an applied bending moment M_b at the midpoint $a_M L=L/2$ and a uniformly distributed load of q intensity.

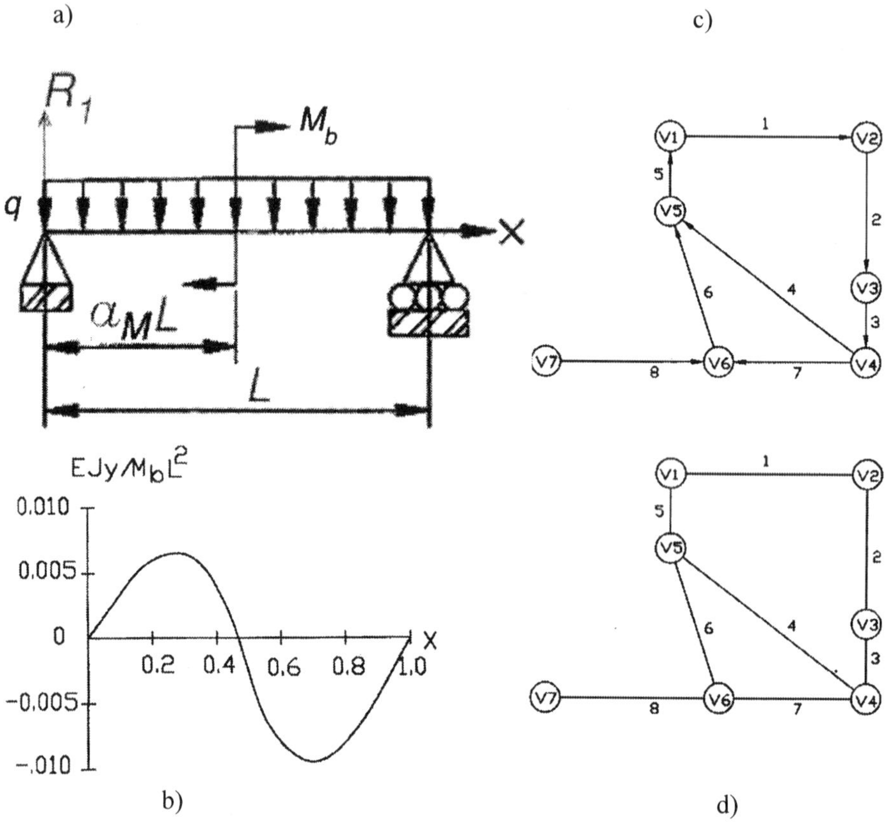

Fig. 1-7. Diagrams of: beam load schematic and its relative deflection; directed and undirected graphs.

We can find the beam's deflection by solving the differential equation of the fourth order $EJ\frac{d^4y}{dx^4} = f(x)$, for which the right-hand side corresponds to the shear reaction force R_1 on the left support of the beam, $R_1 = \frac{qL}{2} - \frac{M_b}{L}$, the uniformly distributed load with q intensity, and the bending moment M_b should be multiplied by the double impulse unit function $\delta'(x-L/2)$ at the middle point ($a_M = ½$). Therefore, we have in relative form with $X=x/L$ the following expression

$f(x) = R_1\delta(x) - qU(x) + M_b\delta'(X-1/2)$ and the $\mathfrak{L}$ transform of $f(x)$ is

$F(\lambda) = R_1 - \frac{q}{\lambda} + M\lambda e^{-1/2\lambda}$. Letting $Y(\lambda)$ be the $\mathfrak{L}$ transform of $y(x)$, the Laplace transform equation for the bending beam is

$$EJY(\lambda) = \frac{F(\lambda)}{\lambda^4} = \frac{R_1}{\lambda^4} - \frac{q}{\lambda^5} + \frac{M_b \exp(-\lambda/2)}{\lambda^3} + \frac{C}{\lambda^2}. \quad (1\text{-}45)$$

Here C is a constant of integration to be determined later from the boundary-value conditions.

The reverse $\mathfrak{L}^{-1}$ transform of (45) yields

$$EJy(x) = \frac{R_1 x^3}{6} - \frac{qx^4}{24} + \frac{M_b(x-1/2L)^2}{2} U(x-1/2L) + Cx\,, \tag{1-46}$$

where the unit function $U(x)=0$ at $x< \frac{1}{2}L$ and $U(x)=1$ at $x\geq \frac{1}{2}L$.

Using the boundary condition at $x=L$ in equation (46), where $y(L)=0$, replacing R_1 by its value, and solving for C, we yield the result

$$EJy(x)= (\frac{qL}{2} - \frac{M_b}{L})\frac{x^3}{6} - \frac{qx^4}{24} + \frac{M_b(x-\frac{1}{2}L)^2}{2} U(x-\frac{1}{2}L) + (\frac{M_bL^2}{24} - \frac{qL^4}{12})\frac{x}{L}\,.$$

The use of space impulse functions has made it possible to express this problem in one differential equation (*see* M.F. Gardner *et al.*, 1942). If we now eliminate the uniformly distributed load by assuming that $q=0$, then we obtain the solution with the relative expression of the longitudinal beam's coordinate X=x/L as

$$\frac{EJy(x)}{M_bL^2} = \frac{(X-1/2)^2}{2} U(X-1/2) - \frac{X^3}{6} + \frac{X}{24}\,, \tag{1-47}$$

which has exactly the same magnitude of the members for similar loading of the beam as derived by S.T. Smith (2000) with the application of the special functions, which he calls "Macauley functions".

Another solution, without singular functions for the simply supported beam with applied in midpoint moment, is presented by J.M. Gere (2001) which corresponds to

$$y(x) = \frac{M_bx}{24LEJ}(L^2 - 4x^2) \text{ for the range } 0 \leq x \leq L/2. \tag{1-48}$$

Note that both solutions have similar results if the applied moment M_b has the same direction. If moment M_b is applied in some point with coordinate $a_M L \neq L/2$, the method of solution for the deflection problem by singular functions remains the same. However, the value of the constant C is different in this case. Let us rewrite equation (46) as follows

$$EJy(x) = -\frac{M_bx^3}{6L} + \frac{M_b(x-a_M L)^2}{2} U(x-a_M L) + C_a x\,. \tag{1-49}$$

Using the boundary condition at $x=L$ in this equation, where $y(L)=0$, and solving for C_a, we yield $C_a = \frac{1}{L}[\frac{M_bL^2}{6} - \frac{M_b(L-a_ML)^2}{2}]$. We can see that asymmetrical application of the moment M_b causes the asymmetrical deflection of the simply supported beam even if the shift of the moment application from the midpoint is a small one. For example, if $a_M=0.49$, then $C_a = \frac{M_bL^2}{27.31}$ and (49) in the relative representation

corresponds to $$\frac{EJy(x)}{M_bL^2} = -\frac{X^3}{6} + \frac{(1-a_M)^2}{2} U(1-a_M) + \frac{X}{27.31}\,. \tag{1-50}$$

The chart of the beam's dimensionless deflection for this case is shown in Fig. 1-7b.

The singular functions application is effective not only in the kinetostatics of beams, but also in the dynamic systems with the mix of distributed and concentrated masses and forces (*see* Chaps. 4, 5, 8).

Graphs application. It is convenient to use graphs' diagrams for the first stages of the structural synthesis in static and dynamics as well. We can represent members of the structure or the targeted operations in the vertices (nodes) of a graph whose edges reflect the functional connections or the operation time. The graph may be directed, or digraph, (Fig. 1-7c) with ordered by arrows edges and undirected one (Fig. 1-7d). The functional graph has trees, circuits, and cycles. The planar graph does not have intersected edges. Graph's matrix presentation is based on the incidence (rectangular between vertices and corresponding edges) and adjacency (square between connected vertices) matrices. They usually have binary presentation.

The incidence M(Gr) and adjacency $A(Gr)$ matrices of a graph are related by the following theorem: $A(Gr) = M(Gr)M(Gr)^T - D_n$, where $M(Gr)^T$ is the transposed incidence matrix and D_n is the diagonal matrix whose members are equal to the degree $d(V_i)$ of a corresponding vertex, $D_{ii}=d(V_i)$. The number of edges incident from (or exiting) a node represents the out degree of this node in a digraph. The number of edges incident to (or entering) a node represents the in degree of this node in a digraph. The edges of one column in a digraph's incidence matrix have different signs (+ and -). We will not discuss graph's theory relations in more detail because graphs options are well known (B. Bollobás, 1998). The reader can find examples of effective application of graphs for design of precision and complex systems in publications by D.M. Buede (2000), and many other sources.

The finite element method. Finite element analysis and modeling is a very powerful method for stress-strain research in the elastic body. This method allows us to perform calculations on the basis of matrix representation for the elements with different geometric, non-uniform elastic properties and loading. In recent years there have been many powerful computer's software packages developed for finite element analysis (*FEA*) and modeling (*FEM*) that are capable of solving very complicated problems in the field of elasticity including vibrations, thermal elasticity, and other (*see* A.A. Becker, 2004).

Two types of bifurcation.

We suppose that all relationships in the precision elasticity should be unambiguous. However, the bifurcation problem is well known and essential in the nonlinear elastic systems.

Buckling. The simplest example of bifurcation is the buckling effect. An important aspect is that there are multiple equilibrium states in the buckled area for the same beam with different orientation (*see* L. Euler , M. Born, 1956; S. Antman, 1995). Elastic instability of nanomechanical beams has specific features. They buckle due to residual strain and asymmetry.

Split output signal. Another type of bifurcation that deals with the split of the output signal into two physical types can be found in precision elastic transducers. And the input quantity in this system is the same for both outputs (*see* Chaps. 5 and 8) that helps to avoid the Abbe's error and to compensate for some influence quantities.

Dynamic features.

The nonlinear systems' functions are more complicated and required dynamic differential equations or combined amplitude-frequency approach. The general problems of linear and nonlinear dynamics in elastic mechanisms are discussed in the publication edited by A. G. Erdman, 1993. Dynamic behavior of precision linear systems is usually represented by the differential equations of the first order for the first-order system, and the second order differential equation for the second-order system. The dynamic properties of a linear system can be described by one of the following characteristics: a) the transfer function $W(\lambda)$ which represents a ratio between the Laplace transform of the input function to that of the output function; b)the amplitude-phase frequency response $W(i\omega) = A_{\omega}(\omega)\, e^{i\varphi(\omega)}$;c)the combination of amplitude-frequency response and phase-frequency response $A_{\omega}(\omega)$, $\varphi(\omega)$;d)the real part of frequency response $R(\omega) = \text{Re } W(i\omega)$, e)the imaginary part of frequency response $J(\omega) = \text{Im } W(i\omega)$;f) impulse transient response $\theta_{ie}(t)$ which corresponds to the input unit impulse; g)transient response $\theta_e(t)$ to the step input. All the above mentioned characteristics for a linear function may be represented in terms of one another. Let us recall that the amplitude-frequency response $A_{\omega}(\omega)$ and phase-frequency $\varphi(\omega)$ response (used separately) are not enough for the calculation of the dynamic response of the linear system in a general case. But for minimal-phase systems, $A_{\omega}(\omega)$ or $\varphi(\omega)$ represent the full dynamic characteristic (*see* M.G. Boguslavskii, Y.M. Tseytlin, 1976).

Friction and damping.

Let us recall that we recognize different types of damping: i) hysteretic, ii) anelastic, iii) the external friction in sliding, rolling, and iv) structural in rigid joints.

Dry external friction. It is important that external friction is practically excluded from the elastic mechanisms. As a result, they have smooth motion even at a very low speed. They do not require any lubrication with grease or oil. Dry external friction is a source of an "intermittent (stick-slip) movement" with possible significant sporadic jumps, especially in the systems with low rigidity. The latter is typical in mechanical ELKS. Fig. 1-8a schematically shows a balance drive used for evaluating of measuring efforts. In the figure, 1 is the balance (drive), 2 is the instrument spindle, 3 is the elastic link equivalent to the stiffness j_{dr} of the drive, and 4 is the elastic link that characterizes the effective stiffness j_M of the instrument mechanism's elastic components. This method is a standard procedure for estimating contact instrument's measuring efforts. However, the method does not provide stable results if the precision instrument's mechanism has kinematic pairs with external friction that is sufficiently high. The balance indicates in such a case a pronounced non-uniformity in motion and instability of readings for the same marks on the instrument scale. This phenomenon has an intrinsic nature and occurs even with very smooth motion and uniform displacement of the instrument casing at low speed with respect to the balance in a single direction. The motion differential equation of such a system can be written in the form (*see* I.S. Amosov, Y.M. Tseytlin, 1969)

$$m_g \ddot{x} - j_{dr}(1 + j_m/j_{dr})(Vt - x) + (Q_{fr} - \beta_{fr}\dot{x}) = 0, \tag{1-51}$$

where m_g is a moving mass of slider 2; $\dot{x}$ is the speed of slider 2's motion; $\ddot{x}$ is the slider's acceleration , t is time of motion; $(Q_{fr} - \beta_{fr}\dot{x})$ is a friction force in the presence of a constant component Q_{fr} and a falling speed characteristic with a coefficient of proportionality β_{fr}; V is the speed of driver 1's displacement (considered to be constant).

The stick-slip motion is characterized by the linear size δS_k of a jump, its duration, and the duration t_1 of the halt. The jumps of the driven member with the discrete displacement of the driver are determined in such a system from (51) and may attain the value of the order

$$\delta S_k = 2\ \delta Q_{fr} / [j_{dr}\ (1 + j_m/j_{dr})], \qquad (1\text{-}52)$$

where j_M is the gradient in N/m of the instrument's measuring effort; j_{dr} can be called the stiffness of the balance drive (the reciprocal of the sensitivity related to the displacement of the balance's pan), δQ_{fr} is the difference between the static and kinematic friction forces. The value δQ_{fr} depends to a certain extent on the duration of the halt and it increases with a rising t_1 (recall the Coulomb's natural logarithm relations). For the direction of the motion shown in the schematic, the value of t_1 and therefore of δQ_{fr} increases with a reduction of the drive (balance) stiffness j_{dr}.

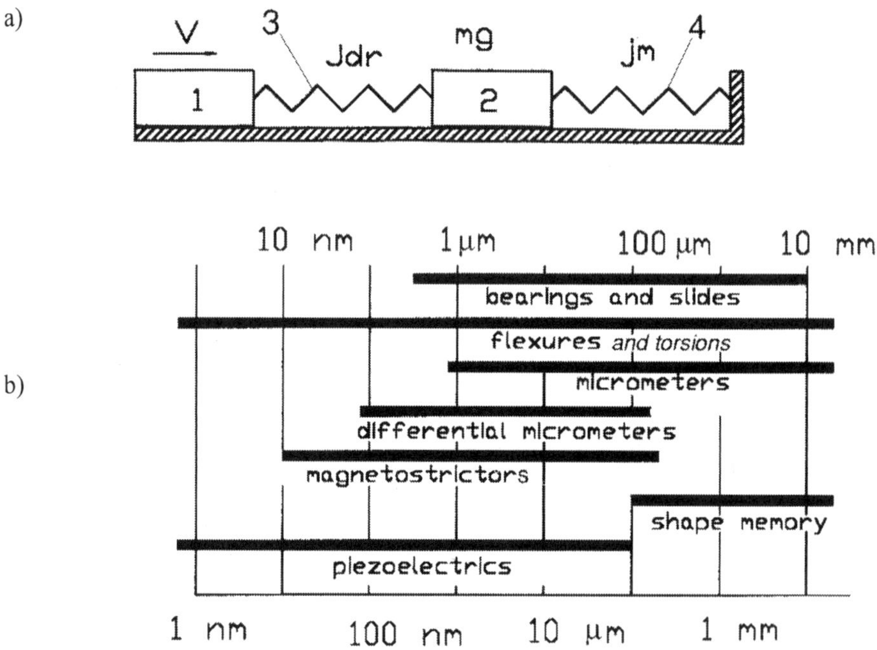

Fig. 1-8. Limitations of smooth motion in linear actuators.

Factor δQ_{fr} may be replaced in (52) for tentative computations by $\delta Q_{fr\,\infty}$, which denotes the difference between the static and kinematic friction forces for an infinitely large duration of contact in the spindle sliding guides. For steel and steel materials in the kinematic pairs of the tested instrument, $\delta Q_{fr\,\infty}$ was, for instance, equal to 0.2 N. Let us recall that in general the kinematic sliding friction is about 80% of the static friction (B.N.J. Persson, 1998). Significant influence on this type of friction has chemical nature, hardness of material surfaces and their surface finish.

Equation (52) explains why it is very difficult to achieve a low threshold using mechanical actuators and translational stages with rigid guides, especially if the lubrication of the kinematic pairs is not feasible (*see* Fig. 1-8b).

The elimination of a stick-slip motion is important for the miniature drives of AFM, where the relations between the static friction force, mass of slider, its lateral stick-slip displacements, and the "critical velocity" above which the motion becomes quasi-steady has been found. It is supposed by many authors that elastic deformation with internal friction does not have stick-slip motion. However, it was found that plastic deformation with the motion of dislocations in metals might have stick-slip motion (*see* B.N.G. Persson, 1998), which causes appearance of upper σ_a (similar to "static friction") and lower σ_b (similar to "kinematic friction") yield points (*see* Fig. 1-9a) . Moreover, we assume (Iu. V. Tarbeev, Y. M. Tseytlin *et al.*, 1983a) that elastic deformation itself might have some kind of stick-slip motion as a result of quantum friction, which occurs at low thermal occupation number $N_{th}=K_B T/(2\pi\hbar f) \leq 1$(*see* A. Gaidarzhy *et al.*, 2005), and microplasticity in the elastic materials which causes limitations in precision elasticity. Let us recall, however, that the stick-slip step scanners based on combination of fast and slow motion of slider's base are successfully used in scanning probe microscopy with providing large range of motion and precise steps after short system training.

Effective actuators. There are three basic types of actuators used with precision stages: manual drives, stepper or dc motors, and piezoelectric transducers. Manual and motor drives can move components over long distances, constrained only by the size of the manual drive or, in the case of the motor, by the length of the lead screw. These drives have restrictions (Fig. 1-8b) in the feasible precision in the order of 1 μm if they have bearings, slides, gears in their kinematic chains. Piezoelectric actuators can move components over very short distances with nanometric precision. Usually they have 0.1% extension of their size at a normal temperature (*see* Chap. 8) after polarization under an electric field at temperature 300 °C and have been used in the kinematic chains with mechanical amplifiers. But these actuators may have significant nonlinearity in voltage-displacement characteristic. Magnetostrictors (*see* Chap.7) are used in the systems with temperatures from cryogenic up to 250 °C. Estimations for bearing and slides, micrometers, and piezoelectric actuators shown in Fig. 1-8b are similar to that by Melles Griot company (presented on web site http://alignment.mellesgriot.com/nano, 11/14/02).

The shape memory features in some metal alloys (nitinol *Ti-55Ni*, *Ti-Pd*, *Cu-Al*, *Fe-Mn*) are effective for the servo-systems in medical and other applications, but they are usually not accurate enough for precision elasticity. Nitinol's work is based on the transformation from martensite to austenite structure under action of temperature. Shape memory effect may be in ceramics (zirconia etc.) as well. Ultraviolet light-induced shape memory polymers are also known.

Structural compensation of the errors (uncertainties) also has some specifics in ELKS. The function of errors in the output of ELKS is naturally expressed as the relationship between the change of displacement in the output Δs_{out} and in the input Δs_{inp}:

$$\Delta s_{out} = [i_{sc} / (1 + i_{sc}\, i_{ifbe})]\, \Delta s_{inp}, \qquad (1\text{-}53)$$

where Δs_{inp} is the uncertainty in the input of the system as a result of the errors in the straight chain Δs_{sc} of the transformation and equivalent feedback Δs_{ifbe}, i.e.,

$$\Delta s_{inp} = \Delta s_{sc} - \Delta s_{ifbe} = (\Delta s_{out.p} / i_{sc}) - i_{ifbe}\, \Delta s_{out.ifbe}, \qquad (1\text{-}54)$$

$$\Delta s_{out} = \frac{\Delta s_{out.p} - i_{sc}\, i_{ifbe}\, \Delta s_{out.ifbe}}{1 + i_{sc}\, i_{ifbe}}, \qquad (1\text{-}54a)$$

where index p denotes an uncertainty in coordinate of quantity in s_{out}'s dimension. Here i_{sc} is the transformation ratio of the straight chain of the displacement's transformation when there is an absence of resistance (counteracting) forces, i_{ifbe} is the transformation ratio of the equivalent (internal and external) feedback chain, and the internal one realizes under the resistance forces; Δs_{sc}, Δs_{ifbe} are the uncertainties of the straight and equivalent feedback chains of transformation in the units of the input displacement; $\Delta s_{out.p}$, $\Delta s_{out.\text{ifbe}}$ are the uncertainties of the same links but in the units of the output coordinate Δs_{out}. The condition of the systematic uncertainty (bias) compensation in accordance with (54a) is as follows ξ_{sc} - $i_{sc}\, i_{ifbe}\, \xi_{sifbe} = 0$,(1-55) where $\xi_{sc} = \Delta s_{out.p} / s_{out}$; $\xi_{sifbe} = \Delta s_{out.ifbe} / s_{out}$. Hence from (55), a systematic uncertainty of the two links connected in parallel will be mutually compensated if they have the same signs. If relative error ξ_{sc} in the straight chain and relative uncertainty ξ_{sifbe} in the equivalent feedback chain have random values, then the combined uncertainty also has a random nature with mean value ξ_s and standard uncertainty σ_ξ. Therefore, with an algebraic summation of the mean values and a square summation of the standard deviations, we have:

$$\sigma_\xi = \frac{\sqrt{\sigma_{\xi sc}^2 + (i_{sc} i_{ifbe})^2 \sigma_{ifbe}^2}}{1 + i_{sc} i_{ifbe}}. \tag{1-56}$$

The condition of minimum random uncertainties is defined from analysis of the random uncertainties' variance D_ξ

$$D_\xi = \frac{D_{\xi sc} + i_0^2 D_{\xi ifbe}}{(1 + i_0)^2}, \tag{1-56a}$$

where $i_0 = i_{sc} i_{ifbe}$; $D_\xi = \sigma_\xi^2$; $D_{\xi sc} = \sigma_{\xi sc}^2$; $D_{\xi ifbe} = \sigma_{\xi ifbe}^2$

If $dD_\xi / di_0 = 0$, then we obtain the condition for the optimal adjustment of the transmission ratios, $i_0 = D_{\xi sc} / D_{\xi ifbe}$. (1-56b)
The minimal value of the variance is equal to:

$$D_{\xi\,\min} = \frac{D_{\xi sc} D_{\xi ifbe}}{D_{\xi sc} + D_{\xi ifbe}}. \tag{1-56c}$$

It is seen from (56c) that the variance of uncertainty in ELKS with negative feedback is smaller than the variance of uncertainty in each element. This result is impossible in ELKS without feedback because the consecutive connection of the elements always provides addition of their variances. If the random uncertainties of these chains are functionally connected, then the standard uncertainty in ELKS is equal to:

$$\sigma_\xi = \left[\left(\frac{\partial \xi_s}{\partial \xi_{sc}} \right)^2 \sigma_{\xi sc}^2 + \left(\frac{\partial \xi_s}{\partial \xi_{sifbe}} \right)^2 \sigma_{\xi ifbe}^2 + 2\upsilon_\xi \frac{\partial \xi_s}{\partial \xi_{sc}} \frac{\partial \xi_s}{\partial \xi_{sifbe}} \sigma_{\xi sc} \sigma_{\xi ifbe} \right]^{1/2}$$

,where υ_ξ is the correlation coefficient. If the relationship's regression is a linear one and $\upsilon_\xi = +1$, then $\sigma_\xi = (\partial \xi_s / \partial \xi_{sc})\, \sigma_{\xi sc} + (\partial \xi_s / \partial \xi_{ifbe})\, \sigma_{\xi ifbe}$.
The condition for compensation of the random errors (uncertainties) is an analogous to the condition for compensation of the systematic errors,

$$\sigma_{\xi c} - i_{sc}\, i_{ifbe}\, \sigma_{\xi ifbe} = 0, \tag{1-56d}$$

where $\sigma_{\xi c}$, $\sigma_{\xi ifbe}$ are the standard uncertainties of the straight and equivalent feedback links, respectively. This specifically shows that the errors (uncertainties) from damping may be compensated in the system with a feedback.
Stability of the elastic kinematic links with change in their loading and conditions of application may be characterized by the coefficient of stability for change in loading,

$$k_{Lc} = [F_j(Q) + \delta F_j(Q)]\, F^{-1}_{k}(s); \quad F(s) = (1+k_{Lc})\, F_k(s), \tag{1-57}$$

and by the coefficient of stability with change in the application conditions:

$$k_{Lcc} = \delta F_{kc}(1+k_{Lc})^{-1} F^{-1}_{k}(s). \tag{1-58}$$

The environmental conditions' influence δF_{kc} can be caused by different factors and ultimately depends on the value of an influence quantity (*see* Part B), the direction, amplitude, frequency of its action, its change in time (velocity) and space, and the sequence of the different factors' superposition. The latter is very important for nonlinear systems. The uncertainty (errors) of the elastic mechanisms usually have an amplification coefficient i which corresponds to the mechanism's transformation ratio. As a result of this, the following limitation is required:

$$\delta F_{kc}\, i \leq \eta_{Uv}\{\Delta s_{\delta}\}, \tag{1-59}$$

where $\{\Delta s_{\delta}\}$ is the allowed uncertainty (error) and η_{Uv} is the criterion for error's constituents (uncertainty) negligibility (*see* Part B). At the same time, for any linear kinematic chain there should be a limitation in δF_{kc} value:

$$\delta F_{kc} < F_k(S)\,(1 + k_{Lc}). \tag{1-60}$$

In this case, $k_{Lcc} \leq \eta_{Uv}$.

1-4. Features of Elastic Materials

Internal mechanical damping.

Table 1-13. Internal damping in materials

Material Profile	Designation dimensions, mm	Load (Temp. °C)	*S.D.* %	Stress MPa (strain)	$\zeta_A \times 10^{-4}$ (*f*, Hz)	Q_{fac} (*f*, Hz) max at 4 K, vac.	Δ_τ, % (*f*, Hz)
Micro-strips	Br4Sn-3Zn	torsion	0.1	(90°)	...	30 air (2k)	...
	Br Be 2	torsion	0.2	(90°)	<0.7	...	4.3×10^{-3}
	PtAg20	torsion	0.05	(90°)	...	...	...
Wire ϕ 0.25	*CuBe2*	bend (27)		0.65 σ_e	...	714 (1 k)	...
Steel	4340, mild	torsion bend	≥ 0.15	34.5 17.5 38.5	 4.5 7.0	 1110 714	...
Stainless steel strips	420	torsion	0.09	34.5	6	...	5×10^{-2}
Steel wire	0.20 mm dia	bend	...	0.65 σ_e	...	830 (1 k)	...
Invar	*Fe-Ni*36	bend (27)	...	0.65 σ_e	...	5.5×10^3 (1 k)	...
Copper Alloys	90Cu-10Zn, Brass	torsion bend	0.4	34.5	 15(50) 30 (600)	 333(50) 167 (600)	...
Brass-strip	12*Zn-Cu*	torsion	0.3	34.5	...	...	1.8×10^{-2}
Aluminum alloys	Duralumin 6063-T6	torsion bend	0.25	34.5 7 42	 2.5 25	 2000 200	...
Titanium alloy	Ti-6Al-4V 0.25 dia	torsion (7) bend (27)	0.5	 0.65 σ_e	...	 660 (1 k)	0.4 (1)
Nitinol	*Ti-55Ni*	torsion	26	69	...		>1

Table 1-13 Cont'd.

Material Profile	Designation dimensions, mm	Load (Temp. °C)	*S.D.* %	Stress MPa (strain)	$\zeta_A \times 10^{-4}$ (*f*, Hz)	Q_{fac} (*f*, Hz) max at 4 K, vac.	Δ_τ, % (*f*, Hz)
Piezo** ceramics	PZT-5A PZT-4	unspecified	2.0 0.2	…	…	85 500	…
Silicon	*Si* (111)(110) Fiber ϕ0.69	indent.^ bend (27)	0.035	11.8×10^3		2.8×10^3 10^4-10^5	0.002 (100 k)
Quartz ground piezo fused fiber	SiO_2, ϕ2.3-5 μm	unspecified	0.001	…	0.05 (65k)	10^5 (65k) ##10^6(1k)	…
Silica all-fused annealed fiber	SiO_2 0.25 dia 0.2 dia	tension (780) (-200) bend (27) bend/tors.	…	600	0.5 0.005	10^4 10^6 10^6(1 k) ^^10^8	…
Silicon carbide	SiC 0.1 mm dia	bend (27)	…	…	…	3.3×10^3 (1 k)	…
Silicon nitride	Si_3N_4	unspecified	0.002	…	0.125	4×10^4	…
Tungsten wire fiber	*W* ϕ0.5 ϕ0.0013-0.002 mm	bend (27) bend violin mode	0.28*	$0.65\,\sigma_e$ 500	…	1.4×10^3 (1 k) #3.4×10^5 (0.8) #5.8×10^4 (273) ##5×10^3 (0.1k)	10^{-1} (15k)

S.D. is the specific damping capacity, $\Delta_\tau=\Delta E/E\approx\Delta G/G$ denotes modulus defect, *f* is the test frequency; * at *f*=165kHz.Damping data without specified temperature correspond to "room" temperature. ** *see* data in Table 8-11. ^ data are after E.R. Weppelmann *et al.*, 1993; ^^ data for very pure all-fused silica (level of impurities < 3×10^{-5}%) obtained by V.B. Braginsky *et al.*, 1996, # data from C. I. González et al., 1994. Sign … shows that this parameter is not applicable or the author does not have reliable specific information and recommends to verify data in the CRC Handbook by J.M. Shackelford and W. Alexander, 2001; Smithells Metals Reference Book, 2004; publications by W.N. Sharpe, 2002; B. Bhushan, ed., 2004; andA.H. Slocum, 1992.

Damping. A detailed discussion of the many different microstructural mechanisms that generate hysteretic and anelastic damping in materials is beyond scope of this book (on this subject *see* B.J. Lazan, 1968; B.N.J. Persson, 1998). However, some specifics for the materials with low damping will be outlined below. There are several interrelated damping quantifiers that are used to describe energy dissipation in a structure. They are related in the following manner:

$Q_{fac} = 1/\eta_Q = \pi/\delta_\tau = \phi^{-1} = \dfrac{2\pi W}{\Delta W} = \dfrac{\omega_{k0}}{\Delta\omega_k} = \pi\tau_d f_0$, where η_Q is the loss (damping) factor of material, Q_{fac} is the quality Q-factor, ϕ is the phase angle between stress and strain (hysteretic factor, loss angle), $\delta_\tau = \ln \dfrac{A(t)}{A(t+T_t)}$ is a logarithmic decrement of amplitude ratio $A(t)$ in consequent oscillation cycles with period T_t, ΔW is the part of the energy W dissipated during one cycle of vibrations, τ_d is the time constant of the exponentially decay ring-down amplitude, ω_{k0} is the angular frequency associated with k-th eigenmode,$\Delta\omega_k$ is the full width at half maximum of that mode, and f_0 is the fundamental frequency of the structure. At anelastic damping, logarithmic decrement may be represented by the following expression $\delta_\tau = \pi\Delta_\tau \frac{\omega\tau_r}{1+\omega^2\tau_r^2}$, where ω is the angular frequency, Δ_τ is the modulus defect, and $\tau_r=\tau_{r0}\exp(W_A/K_BT)$ is the relaxation time which depends on activation energy W_A and absolute temperature T. We remind that τ_{r0} is a constant and K_B is the Boltzmann's constant. Note that modulus defect equals $\Delta_\tau=2\delta_{\tau\max}/\pi=2Q_{fac\ \max}^{-1}$. The damping factor (ratio) associated with second order systems $\zeta_A = \frac{\delta_\tau}{\sqrt{4\pi^2+\delta_\tau^2}}$, Q-factor is equal to $Q_{fac}=(2\zeta_A)^{-1}$, and phase angle between response and the forcing function (ω) is defined as $\tan\phi = \frac{2\zeta_A\omega/\omega_0}{(1-\omega^2/\omega_0^2)}$, where ω_0 is the resonant frequency. Table 1-13 shows damping factors for some commonly used materials. A more specific damping factor (*see* A.H.Slocum, 1992) is the loss factor of material η_s that is geometry and load dependent This factor may be expressed by the following relationship, $\eta_s = \alpha_\eta\eta_Q/\beta_\eta$, where α_η and β_η are the damping energy and strain energy integrals, respectively, as the functions of load, stress type, and element's geometry. Data in Table 1-13 for damping factor ζ_A correspond to loading with $\alpha_\eta/\beta_\eta=1.57$ at certain stress σ and frequency f of deformation. Experimental data of Q-factor within 1- 1000 Hz for vibrating in pendulum suspension metal wires are fit the bell-like curve, except for invar which does not show a functional dependence on frequency in specified range (*see* G. Cagnoli *et al.*, 1999). Q-factor in fused silica is of two orders higher than for metals.

Hysteretic damping. In the case of hysteretic damping, specific damping capacity is normally measured in % on solid cylinders stressed by torsion and defined as the ratio of vibrational energy dissipated during one cycle to strain energy of vibration at the beginning of the cycle. Specific damping capacity (*S.D.*) may also be defined as the difference in % between twist angle at quasi-static loading and relaxation twist angle after unloading. In Chap.5 we discuss some specifics about possible practical mixture of hysteretic and anelastic damping.

Damping in elastic body. The nonelastic stress-strain behavior of a linear elastic solid body in time is described in reology with different macroscopic models. Zener's (Poynting, Thomson-Kelvin's) model, for example, has the following representation

$\sigma + \tau^*_\xi\dot{\sigma} = E_R(\xi + \tau_\sigma^*\dot{\xi})$, where τ_σ^* and τ^*_ξ is the time of deformation's relaxation at constant stress and constant strain, respectively; E_R is a relaxation modulus of elasticity that defines the relationship between stress σ and strain ξ after the finish of relaxation process. The losses of energy in the vibrating "linear solid body" may be frequency dependent. In the theory of elasticity it is assumed that the viscose constituent of the dissipate function for the internal friction ψ_{fr} depends on the derivative of a deformation tensor with respect to time and is an essentially positive quadratic form: $\psi_{fr} = 0.5\ \hat{\eta}_{iklm}\dot{u}_{ik}\dot{u}_{lm}$, where $\hat{\eta}_{iklm}$ is a tensor of viscosity of the fourth rank which has evident symmetric features; $\dot{u}_{ik}$ and $\dot{u}_{lm}$ are the derivatives of the tensor of deformation with respect to time. Thus, it is assumed that the viscose constituent of the internal friction is at least symmetric and independent on the direction of deformation. In the theory of internal friction for vibrating systems there are several known stress (F)-strain (θ) hysteretic loops: elliptical (Fig. 1-9b), double concave (Fig. 1-9c), and straight-sided (Fig. 1-9d). Only the area (the work of inelastic forces) of this loop with certain coefficient of energy absorption ψ (the ratio of the loop area A_{hyst} to full energy of deformation) and the damping coefficient C_θ(the proportion coefficient between velocity of deformation and forces of friction) are considered important.

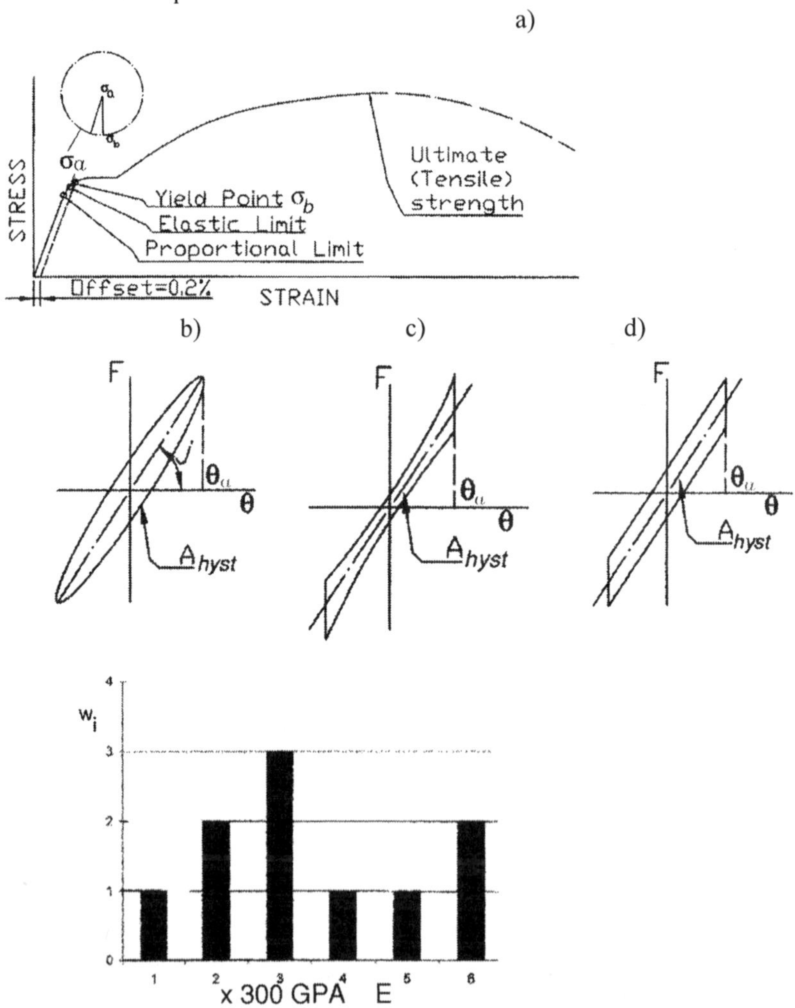

Fig. 1-9. Typical features of elastic materials.

At viscose friction, for example, the equivalent damping coefficient, the stiffness of the elastic element *j,* and absorption coefficient have the following relationship:

$C_\theta = \frac{\psi j}{2\pi\omega}$, where ω is the angular frequency of vibrations (*see* M.Z. Kolovsky *et al.*, 2000). In this case, the hysteretic loop is considered an elliptic (Fig.1-9b) one with the area, which is equal to: $A_{hyst} = \pi C_\theta \omega \theta_a^2$, where θ_a is the amplitude of deformation. However, the form of a hysteretic loop may not be an indifferent one in precision elasticity, especially at the alternative speed of a driving motion in the elastic mechanism and transducer.

Small dissipation. The main advantage of the elastic structures is a small level of energy losses at the deformation. This feature occurs in the full degree for the effective structures and right materials of their elements. There are classical and quantum constraints of the accessible threshold and minimal dissipation of the energy, or the largest achievable *Q*-factor (largest product $Q_{fac}f$).

The main dissipative processes in the solid deformable bodies are as follows: thermoelastic losses, losses from the phonon-phonon and phonon-electron interactions, and the defects of the lattice in crystals. In addition to these "fundamental" sources of the energy losses, there are losses that are linked to the elastic system's structure and environment of application: losses from gas friction, surface losses (better with less surface/volume ratio R_{sv}), losses from the links between different types of vibration in the elastic body, and between elastic elements and the rigid bodies' joints. Energy losses can also be caused by mutual influence of different type transducers and the interaction of environmental influence fields.

For the synthesis of systems with a small dissipation, it is necessary to choose materials of elastic elements with small damping (internal friction). Some coating monolayers suppress mechanical energy dissipation. For example, alkyl and methyl monolayers on the silicon surface suppress energy dissipation in megahertz-range micromechanical silicon oscillators as compared to the more common silicon oxide coating .

Metrology and uncertainty in parameters' estimation.

Dimensional stability. It is known that practically all of precision parts change their dimensions in time. Even such stable parts as gage blocks from special chrome steel and length standards from platinum-iridium in stabilized keeping conditions and without use change their length in fraction of μm during two-three year period and should be periodically recalibrated. The reason of this change is in instability of the phase and structural state of material, adsorption, and the relaxation of residual stresses. The dimensional stability of materials may be characterized by precision elastic limit-*PEL*, microyield strength-*MYS* (*see* Table 1-14) at $PEL=10^{-5}$-10^{-8}, and microstress relaxation-*MRS.* In other cases, oxidation and adsorption can play certain role in dimensional instability. Hence, the choice of material is a very important problem in the structural synthesis of precision systems. Certain recommendations will be discussed in the following sections and chapters. *See* also E.G. Wolff, 2004.

Parameters stability and anisotropy. Let us recall that the properties of elastic materials with the same chemical compound in precision elasticity applications depend on the heat and mechanical treatment as well as the main sizes, forms, and types of working stress. Data for Young's moduli E, rigidity (shear) moduli G, Poisson's ratios ν_p, and the temperature expansion coefficients presented in the technical literature differ for the same materials within 15% or more. These properties are different on the near-surface layers and for thin films (A. G. Every, 2002). In crystals they depend on the direction of lattice planes (M. J. Madou, 2002). For example, Young's moduli are different for aluminum crystal in directions [100] and [111] by 20%; for copper by 285%; for iron by 218%. In the general representation of Hook's law, there are up to 21 independent constituents in the tensor E_{ijkl} with reduced indices and without symmetrical elements for anisotropic linear elastic body. " Due to symmetry of crystals, several more of these may vanish (M.J. Madou, 2002) until, for an isotropic medium, they number two only (E and ν_p)". The latter couple may be represented as E and G as well. The anisotropy coefficient in crystals is defined as $\alpha_{an} = \frac{2E_{44}}{E_{11} - E_{12}}$. $\alpha_{an} = 1$ for an isotropic crystal, it equals 1.57 for single-crystal silicon, and larger than 8 for very anisotropic crystals.

Fig. 1-9e shows distribution for carbon nanotubes' Young's modulus values, E, with different w_i frequency of their publication in contemporary scientific literature. Let us recall that carbon nanofibers may have even more uncertainty in elastic properties than single wall carbon nanotubes (*SWNT*) and multi-wall carbon nanotubes (*MWNT*). Thus, all data on the elastic properties of materials in any reference sources (including this book) are conditional and in most cases are not accurate enough for the direct synthesis of the precision system. It is also true for geometric features, which in many cases cannot be manufactured and assembled to the close tolerances required for the precision structure.

Contact deformation. For example, it is necessary specially estimate the contact deformation for materials with low elastic moduli (Tseytlin Y. M., 1994). The elastic contact deformation w_c due to the action of forces F_Q can be computed for any surfaces with an elliptic shape of the contact region according to the known formula based on Hertz (1881) theory

$$w_c = \varepsilon_\varepsilon \sqrt[k]{F_Q^z / \mu_\varepsilon^z \Gamma_\varepsilon} = A_\varepsilon F_Q^\chi, \tag{1-61}$$

where ε_ε is a coefficient of mutual orientation of the surfaces in contact, μ_ε is a characteristic of the elasticity of materials in contact, and Γ_ε is the geometric characteristic of the shape of the bodies in contact. Ratio z/k is approximately 2/3 for metals in accordance with the Hertz problem. The proposed computational - experimental method (patent SU 1601503) allows one to evaluate local deformations of measuring tip and the measured object under the influence of the measuring force without the necessity of *a priori* evaluation of the actual values of geometric and elastic properties of the bodies in contact (*see* A.I. Sergeeva *et al.*, 1981, 1986, 1990). Equation (61) implies that the experimental deformation in points loaded with force F_Q is equal to: $w_c = \delta w_{12} F_Q^\chi / (F_{Q2}^\chi - F_{Q1}^\chi) = A_\varepsilon F_Q^\chi$, where $\delta w_{12} = w_c(F_{Q2}) - w_c(F_{Q1})$. Thus, in order to evaluate the local deformation under the force F_Q, it suffices to determine the difference δw_{12} between local deformations under two values of the measuring force $F_{Q1,2}$. In order to find the more accurate characteristics of the power χ in dependence $w_c = A_\varepsilon F_Q^\chi$, we evaluate the values of coefficient A_ε and exponent χ with the aid of the least squares method applied to N pairs of experimental data w_i, F_{Qi} and minimization of correlation ρ_χ

$$\chi = \frac{N\sum_i(\ln F_{Qi}\cdot\ln w_i) - \sum_i \ln F_{Qi}\sum_i \ln w_i}{N\sum_i(\ln F_{Qi})^2 - (\sum_i \ln F_{Qi})^2}; \quad \min\ \rho_\chi = \frac{\chi\sqrt{[N\sum_i(\ln F_{Qi})^2 - (\sum_i \ln F_{Qi})^2]}}{\sqrt{N\sum_i(\ln F_{Qi}^{\chi}/\Omega_{ij})^2 - (\sum_i \ln F_{Qi}^{\chi}/\Omega_{ij})^2}},$$

where $\Omega_{ij} = (F_{Qi}^{\chi} - F_{Qj}^{\chi})/\delta w_{ij}$.

We can see that the correlation coefficient ρ_χ increases as the denominator in this expression decreases. However, the minimum of the denominator as a second central moment corresponds to centralizing it within the most probable value of the contact deformations and the exponent χ. Coefficient A_ε can be found according to a well known algorithm $\ln A_\varepsilon = (1/N)\sum_i \ln w_i = (1/N)\sum_i \chi \ln F_{Qi}$.

Adhesion. The Hertz theory is applicable to the bodies, which do not have the adhesion surface energy. One of the first rigorous theoretical treatment of the adhesion of elastic spheres with their attraction is due to the research work of Johnson, Kendal, and Roberts (1971). Their results are called *JKR* theory (*see* J. Israelachvili, 1992). The *JKR* theory is applicable to soft compliant materials with high surface energy and large radii of indenter. Another *DMT* (Derjaguin-Muller-Toporov, 1975) theory applies to harder, less compliant materials with low surface energy and small radii of indenter. This theory assumes that contact deformation profile remains Hertzian. The transition solution M-D (Maugis-Dugdale, 1992, *see* D. Maugis, 2000) is for materials between JKR (λ_{MD}>5) and DMT(λ_{MD}<0.1) approach, where $\lambda_{MD}=2\sigma_0/(\pi\gamma_A j_K^2/R)^{1/3}$, σ_0 is a constant maximum adhesive traction, γ_A is the work of adhesion, R is the effective radius of contacting spheres, j_K is the effective elastic modulus. We can show, for example, that in the DTM case, exponent $\chi = \frac{2}{3}\frac{\ln(F_Q+F_{ad})}{\ln F_Q}$ because in these conditions (*see* Piétrement and Troyon, 2000) the indentation depth may be presented as $w_c=\mu_\varepsilon^{-2/3}R^{-1/3}(F_{adh}+F_Q)^{2/3}=A_\varepsilon F_Q^{\chi}$, where F_{adh} is the acting adhesive force and R is the radius of indenter. Therefore, χ= 2/3 if F_{adh} is equal to zero. It can be shown with precision of three percent or less that adhesian parameters may be presented by lower degree polynomials of the second and first degree respectively if we substitute Chebyshev's polynomials $R_n(z)$ of fourth , third , and second degree (Table 1-8) in Piétrement and Troyon (2000)expressions.

To prevent adhesion is important for mechanisms, measurement techniques, MEMS, and NEMS where it can induce friction and wear even in elastic structures at mutual unintended contacts of their elements (B. Bhushan, 2004).

Non-contact measurement. Surface roughness, material-dependent phase shifts, and other non-uniform features of an elastic body can have a large influence on the uncertainty in non-contact measurements as well, especially in the measuring of thin coating thickness (Y. M. Tseytlin *et al.*, 1984a, 1994). In some cases, the uncertainty in coating thickness measurements caused by surface roughness can be decreased by calibration with the special transformable standard (Patent SU 1603185, 1990). Material-dependent phase shift causes uncertainties in precision measurement by interferometer of nano-steps between thin film and substrate. To avoid this influence in thin films thickness measurement traceability, we proposed to use uniform primary film step standards with coating them above non-uniform ones for calibration by interferometer and transfer nanometer unit by contact styles measurement to non-uniform nano-step standards directly (*see* Y.M. Tseytlin, 1980a; Y.M. Tseytlin *et al.*, 1982a, 1984a, 1984c). Recently the phase shift problem has been discussed by L.P. Yatsenko *et al.*, 2004.

Quantum effects. In accordance with quantum aspects (L.D. Landau and E.M. Lifshitz, 1958), a solid body model is "microscopic" if its smallest size is less than $L_{mic} \leq \hbar v_s /(K_B T)$, where v_s is the velocity of sound in the body; $\hbar$ is the reduced Planck's constant; K_B is Boltzmann constant, and T is the absolute temperature. L_{mic} at $T \leq 293$ K is comparable to an atom's diameter, thermal phonon wavelength, and to the thickness of very thin films or coatings on the solid body surfaces. It has been shown with quantum mechanics methods (Iu.V. Tarbeev ,Y. M. Tseytlin *et al.*, 1983a) that the limit for the smallest measurable displacement is equal to $\Delta L \geq (\hbar \ \tau / m)^{1/2}$, where m is the levitation body's mass and τ is the measurement impulse duration. For example, at $\tau = 10^{-3}$ s, and $m = 10^4$ grams $\Delta L \geq 10^{-19}$ m.

Fundamental length. The remaining question concerns length measurement in extremely small space-time intervals. It is Planck's length $L_g = (\hbar \ \gamma_N / c^3)^{1/2} = 10^{-35}$ m, where γ_N is Newton's gravitational constant, and c is the speed of light (299,752,458 m /s). Hence, in quantum mechanics the principal limit of length uncertainty is very small. However, this believed theoretical limit does not have any metrological support. Let us recall the Heisenberg's uncertainty principle in accordance with which it is impossible to specify or determine simultaneously both the position Δx and velocity ΔV of particle with "full accuracy" because $\Delta x \Delta V = \hbar/2m$ and the standard quantum limit $\Delta x_{SQL} = \sqrt{\frac{\hbar}{2m\omega_0}}$, where ω_0 is angular resonant frequency, and m is the mass of a simple harmonic oscillator. But the recent research on the basis of Casimir force measurements proves some change in the Newtonian constant on the distances that less than 300 nm (*see* R.S. Decca *et al.*, 2003).

Information data of typical Materials

Strength. The relative data for different materials for ELKS applications are shown in Table 1-14. As a reference in this table is accepted data for the micro-strips from bronze *Br*4*Sn*-3*Zn* applied in the most accurate mechanical transmission mechanisms with elastic double helicoids, thin helicoidal multivibrators, and torsion suspensions in oscillographs and galvanometers. The main mechanical properties for these strips were carefully studied and practically applied by this author in synthesis of many precision elastic mechanisms and transducers.

Stress-strain. Fig. 1-9a recalls a typical stress-strain diagram of the elastic plastic materials which features are reflected in Table 1-14. Let us note that similar curves for brittle materials do not have horizontal plateau. However, some slip in the crystal plane is possible even in these kinds of materials. We remind also that stress expressed in Pa = 1 N/m^2 represents the force value applied to the surface unit, and strain represents the unitless ratio of deformation value to the initial dimension of the elastic body.

More specific recommendations for material behavior are presented in the following chapters.

Table 1-14. Mechanical properties of elastic materials

Material		σ_t° (σ_f°)	σ_e°	$\upsilon_{en}^\cap$ max	Modulus of elasticity		F.T.	ρ_γ°	σ_r^0, MYS
Profile	Designation Dim. ,t_T, load				E° (E/ρ_γ)	G° (ν_p°)			
Micro-strips	Br4Sn-3Zn **	1.0	1.0	1.0	1.0 (1.0)	1.0 (1.00)	...	1	1
	Br Be 2(320)*	1.39	1.25	2.0	1.2(1.2)	1.1 (1.26)	...	0.96	1
	PtAg20	1.74	2.00	0.5	1.5(0.7)	1.5 (0.9)	...	2.17	...
Steel	4340(538)* 4340(204)* 4340(427)*	1.11 1.72 1.35	1.49 2.09 1.76	2.58 1.85 1.81	1.84	1.8 (1.12)	...	0.89	>1
Stainless steel strips	420 (315)*	1.41	1.72	1.44	1.78	1.73 (0.97)	...	0.91	
Steel wire	0.30 mm Dia	2.34 3.4	1.24	...	1.06-1.8	...	...	0.89	
Invar	*Fe-Ni*36	0.40	0.35	...	1.33	(1.08)	...	0.91	...
Copper Alloys, Hard	70*Cu*-30*Zn* 90*Cu*-10*Zn* -	0.46 0.37 0.53	0.55 0.47 0.51	0.56 0.62 0.48	1.1 0.96	1 0.96	...	0.96 1	0.045- 0.3
Brass-strip	12*Zn-Cu*	0.46	0.47	0.48	0.93	(1.22)	...	0.98	0.045
Aluminum alloys	7075-T6	0.45	0.64	0.61	0.61	0.64 (1.18)	...	0.32	0.4- 0.8
Titanium alloy	Ti-6Al-4V	0.9- 1.2	1.13	1.98	0.96- 2.14	1 (1.3)	...	0.5	1-2
Nitinol	*Ti*-55*Ni*	0.76	0.26	1.85	0.90	0.97	...	0.73	...
Carbon Whisckers Nanotubes	d<10 nm S(M)WNT	3.8$^{\$}$- 9.2 26^ 39.8	...	...	1.8$^{\$}$ 6.0- 9.17^^ 10^(43)	9.75 (0.32- 0.79) 9(0.5)	9.7	0.15 - 0.29	...
Boron carbide	B_4C	0.31 t/c$^{\$}$ 2.58	...	...	4.1 (11.4)	(0.75)	4.0	0.29	...
Silicon crystal	Si (111) (110) (100) bend	0.17 (0.4- 10) (3.5)	2.7- 8.9	0.06	1.68 1.15	1.3- 1.4 (0.79- 1.00)	0.96- 1.65	0.27	...
Fused silica quartz oriented	SiO_2, including microfibers	0.05 -5 t/c$^{\$}$ 0.97	0.07 11.4	0.02	0.62- 0.95	0.70 (0.50- 0.64)	0.9	0.26 - 0.30	...
Silicon carbide	SiC SiC *Whisker* reinforced *Al* alloy 0.2 vol.	0.18 -3.5 t/c$^{\$}$ 1.25 12^ (5.0)	 26.7 0.37	...	3.55- 3.97 11.7^	4.06 (0.58- 0.90)	3-5.7	0.36	...
Silicon nitride	Si_3N_4 Hot pressed Deposited whisker	0.87 0.33 (5.3) 12.4	17.8	...	1.33 up to 2.8 *** 3.4	2.4 (0.72- 0.97)	1.5-8	0.35	...

Table 1-14 Cont'd.

Material		σ_t°	σ_e°	σ_{en}°	Modulus of elasticity		F.T.	ρ_γ°	σ_r^0,
Profile	Designation Dim. ,t_T, load	(σ_f°)		max	E° (E/ρ_γ)	G° (ν_p°)			MYS
Poly-silicon	Si, tension bending	1.38 - 2.6 (2.5)	2.29 - 4.28	2.88 - 3.27	1.42	(0.79)	0.85-1.9	0.26	…
Tungsten Carbide	*W* *WC*	1.3-3.5	3.5-5.0	…	3.6 6.4	3.6 (0.99)	…	2.19 1.78	…

* Quenching temperature t_T is shown in parenthesis with an asterisk in °C; all properties at 20°C; **Young's modulus E =112.8 GPa; shear modulus G = 44.1 GPa; Poisson's ratio ν_p = 0.279; tensile (UTS)strength σ_t = 1.13 GPa; yield stress σ_e= 0.785 GPa; endurance limit σ_{en} =0.260 GPa (approximately, on the basis of 10 million cycles); elasticity temperature coefficient α_E =- 4.5×10^{-4} K^{-1} (approximately); temperature coefficient of linear expansion α_L = 15.5×10^{-6} K^{-1},E/ρ_γ=12.84 MN m/kg=12.84×10^6 m^2/s^2; specific gravity ρ_γ= 8.8 g/cm^3 (8,800 kg/m^3), and *MYS* σ_r=300 MPa. *F.T.* represents fracture toughness in, MPa m$^{1/2}$, and σ_f° is the relative fracture strength σ_f / σ_t for brittle material. Thermal properties of elastic materials are shown in Table 7-6. Superscript ° indicates relative values; recommended factor of safety is not less than 1.3. *** Data (A. Khan *et al.*, 2004; D. Munz *et al.*, 1999) for amorphous silicon nitride films and cantilevers; data ^ after E. Wong *et al.*, 1997, P. Poncharal *et al.*, 1999, V.N. Popov, 2004; data^^for diamond-like after X.Blase *et al.*(2004), data $ of Thornel fiber ∅ 11μm; for AlBeMet (62%*Be*, 38%*AL*) alloy MYS is equal σ_r=17.3 MPa; t/c$ data belong to tensile (above) and compression (below) test. Sign … shows that this parameter is not applicable or the author does not have reliable information and recommends to verify data in Handbook by J.M. Shackelford and W. Alexander, 2001; Smithells Metals Reference Book, 2004; publications by W.N. Sharpe, 2002; M.J. Madou, 2002; and B. Bhushan, ed., 2004.

2.

Strain-Stress and Transformability Evaluation

Accurate strain-stress transformation is one of the most important problems of structural synthesis in precision elasticity. This chapter is dedicated to these problems and the greatest transformability in quasi-linear elastic systems.

2-1. Continuum Mechanics: Energy Similarity, Greatest Transformability

Strain and stress tensors, matrix relations, straight and feedback links' approximation; principle of energy similarity in nonlinear systems are the main aspects in this section. Continuum mechanics can be effective down to several of lattice constants. In the calculation of solid body's deformation, the problems that exist may be linear or nonlinear. The nonlinearity may be geometric at large angles of material element rotation or physical when exceeding the material limit of proportionality. These kinds of nonlinearity supposed to be independent. It is clear that in ELKS, in principle, there can be four combinations: 1) linear physically and geometrically, 2) nonlinear physically and linear geometrically, 3) linear physically and nonlinear geometrically, and 4) nonlinear physically and geometrically.

Strain tensor. A deformation of the elastic body is expressed as the change in distances between its points. And the square of the elemental distance dl between close points before deformation can be defined through its projections on the coordinate axis x_i (i = 1, 2, 3) as $dl^2 = dx_1^2 + dx_2^2 + dx_3^2$.

Let u_i, u_k, u_l be the components of the displacement of the body's points. Then after deformation, the distance between the same points will change and become equal to dl_1, which is usually expressed with the tensor $\hat{\varepsilon}$ of deformation:

$$dl_1^2 = dl^2 + 2\hat{\varepsilon}\ dx_i\, dx_k, \text{ where } \quad \hat{\varepsilon}_{ik} = \frac{1}{2}\left(\frac{\partial u_i}{\partial x_k} + \frac{\partial u_k}{\partial x_i} + \frac{\partial u_l}{\partial x_i}\frac{\partial u_l}{\partial x_k}\right). \tag{2-1}$$

Indices i, = 1,2,3 and k – 1,2,3 correspond to certain coordinate axes. The tensor $\hat{\varepsilon}_{ik}$ depends on nine vector quantities and can be expressed in matrix form as

$$\hat{\varepsilon}_{ik} = \begin{bmatrix} \varepsilon_{11} & \varepsilon_{12} & \varepsilon_{13} \\ \varepsilon_{21} & \varepsilon_{22} & \varepsilon_{23} \\ \varepsilon_{31} & \varepsilon_{32} & \varepsilon_{33} \end{bmatrix} \quad \text{or} \quad \hat{\varepsilon}_{ik} = \begin{bmatrix} \varepsilon_x & \gamma_{xy} & \gamma_{xz} \\ \gamma_{yx} & \varepsilon_y & \gamma_{yz} \\ \gamma_{zx} & \gamma_{zy} & \varepsilon_z \end{bmatrix} \tag{2-2}$$

Each type of absolute or relative tensor is defined by a linear and homogeneously induced transformation rules relating the tensor components in different reference coordinate systems. The tensor has a symmetric form and as a result $\varepsilon_{ik} = \varepsilon_{ki}$. The tensor's components ε_{ik} are "quasi-vectors" which are parallel to the coordinate axis that corresponds to the second index k and lies on the plane, which is perpendicular to the coordinate axis that corresponds to the first index i. In the Cartesian coordinate system $x,,y,z$, the tensor of deformation at small strains is shown in (2-2),where ε_x , ε_y , ε_z are the relative elongation (stretches) strain in the direction of axes x, y, and z, respectively; γ_{xy} , γ_{xy} , γ_{xy} are the relative shifts (changes in the initial angles between the directions of axes x, y, and z).

Stress tensor. The stress status of the deformed body in any point is described by the tensor of stresses which in Cartesian coordinates is shown as

$$\hat{\sigma}_{ik} = \begin{bmatrix} \sigma_x & \tau_{xy} & \tau_{xz} \\ \tau_{yx} & \sigma_y & \tau_{yz} \\ \tau_{zx} & \tau_{zy} & \sigma_z \end{bmatrix} . \tag{2-3}$$

The tensor's components correspond to the stresses on the planes which are perpendicular to the x,y, and z axes (see the cubical element in Fig. 2-1a). Stresses σ_x, σ_y, σ_z are the normal , and τ_{xy} , τ_{zy}, τ_{zx} are the tangent (shear) ones.

The *potential energy* function of a three-dimensional linear elastic isotropic body may be represented by the components of the stress and strain tensors in Cartesian coordinates as follows

$$U_{v0} = ½ \, (\, \sigma_x\varepsilon_x + \sigma_y\varepsilon_y + \sigma_z\varepsilon_z + \tau_{xy}\gamma_{xy} + \tau_{yz}\gamma_{yz} + \tau_{xz}\gamma_{xz} \,). \tag{2-4}$$

By nature, the stress stable components and external forces must satisfy the conditions of equilibrium, and the elastic deformations must satisfy the conditions of compatibility and continuity of the material. Generalized Hooke's law supposes linear dependence between the components of stress and deformation tensors. The elastic characteristics of the material include modulus of the normal elasticity E (Young's modulus) and rigidity (shear) modulus G that are related to each other through Poisson's ratio ν_p as $E = 2\,(1 + \nu_p)$ G. Young's modulus for the bending of wide strips and plates is usually assumed equal to $E' = E/\,(1-\nu_p^2)$.

Mohr circle. A.I. Lurie (1970) showed that three Mohr's semicircles (Fig. 2-1a) can be used for the analysis of any symmetric tensor of second rank and especially stress tensor. In this case we have the following semicircles:

(*i*) C_1 with center's abscissa coordinate $(\sigma_2+\sigma_3)/2$ and radius $(\sigma_2-\sigma_3)/2$;

(*ii*) C_2 with center's abscissa coordinate $(\sigma_3+\sigma_1)/2$ and radius $(\sigma_1-\sigma_3)/2$;

(iii) C_3 with center's abscissa coordinate $(\sigma_1+\sigma_2)/2$ and radius $(\sigma_1-\sigma_2)/2$.

Note that σ_1, σ_2, and σ_3, positioned on the abscissa axis, are the principal stresses on the element's faces (*see* the image of the elemental cube on the upper right insert) with the normal in the principal directions, which are free from the shear (tangential) stresses. The shear stresses on this diagram are graphically reflected on the ordinate axis. The hatched area of feasible stresses σ_N and τ_N is inside circle C_2 and beyond circles C_1 and C_3. Maximum shear stress $\tau_{\max}$ corresponds to the top point of circle C_2.

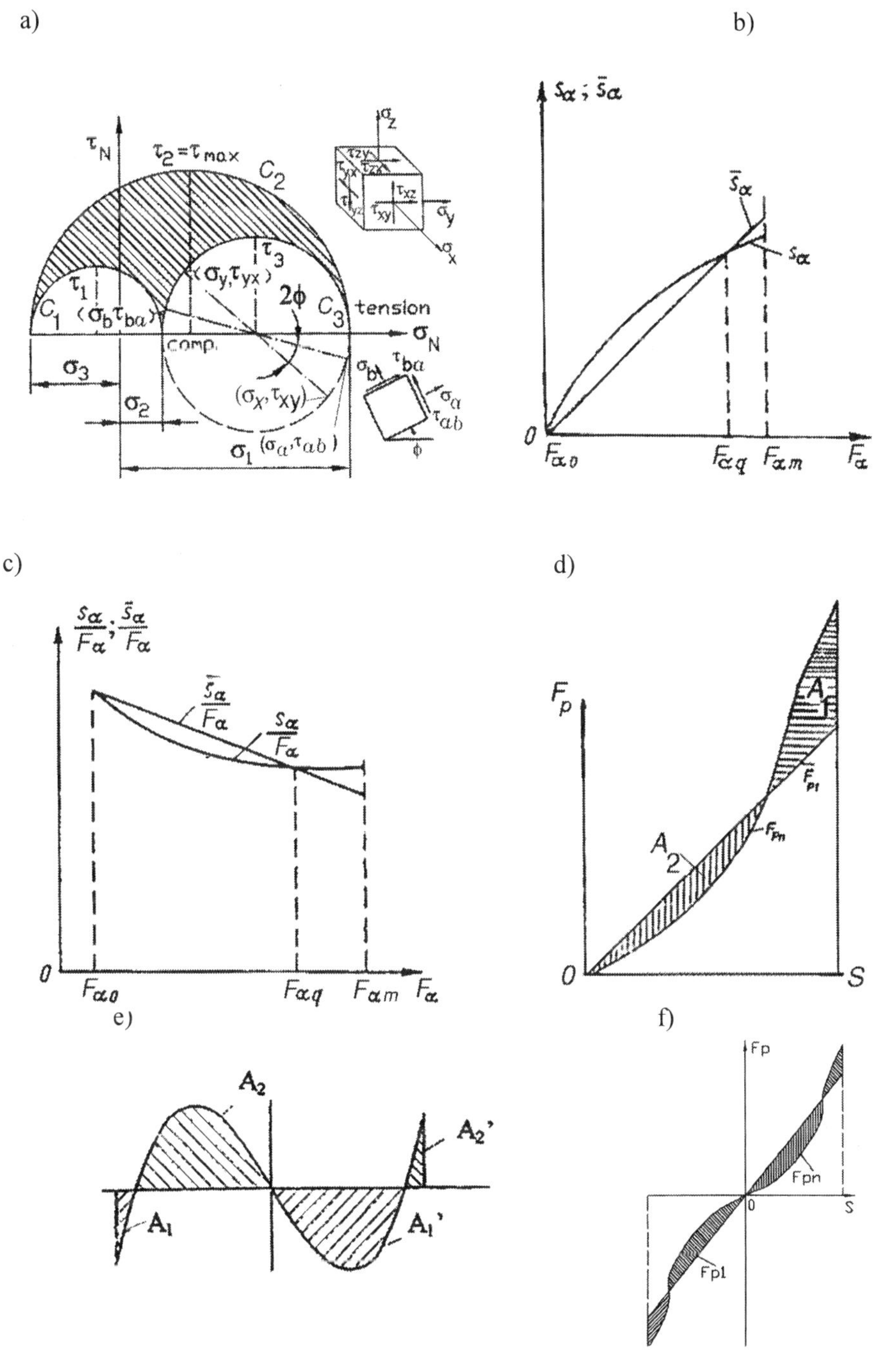

Fig 2-1. Representation of the stress tensor constituents by Mohr's circles and effective approximation of nonlinear strain-stress functions.

This stress is applied to the faces of an elemental cube with normal N orientation at cosines $\sqrt{0.5}$; 0; - $\sqrt{0.5}$. The shear stresses, which correspond to the tops of circles C_3 and C_1, are equal to: $\tau_3=(\sigma_1-\sigma_2)/2$ and $\tau_3=(\sigma_2-\sigma_3)/2$, respectively.

If $\sigma_3=0$ and $\sigma_2=\sigma_1$, circle C_3 becomes a point and circles C_2 and C_1 with radius $\sigma_1/2=\sigma_2/2$ coincide. In this case, the maximum shear stress in the element equals to the radius $\sigma_1/2$ of these circles. There are the following stresses on the face, which has equal angles with the principal axes: the normal stress

$\sigma_N=(\sigma_1+\sigma_2+\sigma_3)/3=\frac{1}{3}I_1\left(\hat{\sigma}\right)$, where $I_1\left(\hat{\sigma}\right)$ is the first invariant of stress tensor $\hat{\sigma}$; the full shear stress on the same face equals $\tau_N=\frac{2}{3}\sqrt{\tau_1^2+\tau_2^2+\tau_3^2}$. We remind that the stress tensor may be divided into two parts (spherical or hydrostatic and deviator) such as $\hat{\sigma}=\frac{1}{3}(\sigma_1+\sigma_2+\sigma_3)\hat{e}+Dev\hat{\sigma}$, where $\frac{1}{3}(\sigma_1+\sigma_2+\sigma_3)=\frac{1}{3}\sigma$ is the mean value of three normal stresses on the mutually perpendicular faces, $\hat{e}$ is the unit tensor, and Dev $\hat{\sigma}$ is the stress tensor deviator (shows the deviation from hydrostatic stress state). It follows from this presentation that intensity of the shear stresses is equal to:

$\tau_N=\sqrt{-\frac{2}{3}I_2\left(Dev\hat{\sigma}\right)}=\sqrt{\frac{2}{3}(\tau_1^2+\tau_2^2+\tau_3^2)}$, where $I_2\left(\hat{\sigma}\right)$ is the second invariant of stress tensor $\hat{\sigma}$.

Let us recall that Otto Mohr in 1882 graphically represented by circle the relationships between normal and shear stresses on elemental faces at an arbitrary angle ϕ (see Fig. 2-1a). Circular representation of plain stresses was described by C. Culmann even earlier in 1866. We use full circle C_3 to visualize this very effective method in the case of the plane stress, where points (σ_x, τ_{xy}) and (σ_y, τ_{yx}) on the circle correspond to the normal and shear stresses on the faces which are perpendicular to the axes x and y, respectively. Tension corresponds to the right side and compression corresponds to the left side of the circle. Points (σ_a, τ_{ab}) and (σ_b, τ_{ba}) on the circle correspond to the normal and shear stresses on the faces that are tilted of angle ϕ to the previous ones (*see* down right insert). It is clear from discussed relationships that the principal stress occurs when the shear stress is zero. Therefore, the principal stresses for the plane stress problem are given by

$$\sigma_{1,2}=\frac{\sigma_x+\sigma_y}{2}\pm\sqrt{(\sigma_x-\sigma_y)^2+4\tau_{xy}^2},$$

the equivalent stress is equal to $\sigma_{eq}=\sqrt{\sigma_1+\sigma_2-\sigma_1\sigma_2}$, and the maximum shear stress in the plane under consideration is $\tau_{max}=\pm\frac{1}{2}\sqrt{(\sigma_x-\sigma_y)^2+4\tau_{xy}^2}$.

The reader can find more detail about using of Mohr's circle for stress, strain, and moment of inertia evaluation in the publication by J.M. Gere, 2001.

Mohr integral. The initial calculation dependencies represent the system of simultaneous differential equations in the partial derivatives with given boundary-value conditions (constraints) at the places of clamping of elastic elements and at other typical points on the elastic line. In many cases, it is convenient to use calculus of variations, the principle of potential energy's minimum, Castigliano's theorems, Mohr's integral, etc.

Mohr's integral for the definition of elastic displacements may be presented in matrix form that yields the unification of intermediate analytical transformations. The integrand in this case should be represented as the product of the square and column matrices with the subsequent integration of the square matrix members. As a result, the square matrix members become the unit Mohr's integrals $\int_0^L M_{ij}^e dl$, and the column matrix members represent the external acting forces F_{Qi}. The integrand of the unit Mohr's integral has the products of the moments of all external force's unit-vectors and the unit-moment. Members of the yielded column matrix correspond to the components of the generalized displacements $s_k^{(i)}$ in k direction and are further algebraically summarized. The structure of the above described calculations can be illustrated by the third rank matrices:

$$[s_k^{(i)}] = \int_0^L \begin{bmatrix} M_{11}^e & M_{12}^e & M_{13}^e \\ M_{21}^e & M_{22}^e & M_{23}^e \\ M_{31}^e & M_{32}^e & M_{33}^e \end{bmatrix} \begin{bmatrix} F_{Q1} \\ F_{Q2} \\ F_{Q3} \end{bmatrix} dl = \begin{bmatrix} \int_0^L M_{11}^e dl & \int_0^L M_{12}^e dl & \int_0^L M_{13}^e dl \\ \int_0^L M_{21}^e dl & \int_0^L M_{22}^e dl & \int_0^L M_{23}^e dl \\ \int_0^L M_{31}^e dl & \int_0^L M_{32}^e dl & \int_0^L M_{33}^e dl \end{bmatrix} \begin{bmatrix} F_{Q1} \\ F_{Q2} \\ F_{Q3} \end{bmatrix},$$

$$s_k = \sum_i s_k^{(i)}, \text{ where } i = 1, 2, 3. \qquad (2\text{-}5)$$

Multiplication of such matrices is conducted in accordance with the common rule, " the row by the column".

Variational methods.

Potential energy and complementary work. Castigliano's theorem for linear elastic structures states the rule for the acting force calculation through the derivative of the potential energy U_v with respect to the displacement in specific direction:

$$\partial U_v / \partial s_i = F_{Q\,i}\,. \qquad (2\text{-}6)$$

Generalized Castigliano's theorem is a satisfactory one for linear as well as for nonlinear structures at the displacement calculation with the formula

$$\partial R_{vC} / \partial F_{Qi} = s_i\,, \qquad (2\text{-}7)$$

where $R_{vC} = (\Sigma F_{Qi} s_i - U_v)$ is a complementary work, F_{Qi} represent the generalized forces and s_i represent the generalized displacements. We should remember that for practical calculations the complementary work needs to be expressed through the generalized forces, but that expression is complicated enough for nonlinear systems.

Linearization.

Effective approximations. The calculations may be simplified if the deflections from linear relationships between forces and displacements are small. This can be done with a method for the best approximation of the nonlinear displacement function s_α by linear $\bar{s}_\alpha$ one. Let us suppose that some linear functions exist as follows:

$$\bar{s}_\alpha = f_1(\eta_\alpha; F_{Q\alpha}); \quad \frac{\bar{s}_\alpha}{F_{Q\alpha}} = f_2(\eta_k; F_{Q\alpha}); \quad \bar{s}_\alpha = f_3[\eta_\varsigma; F_k(s)],$$

for which the following expressions are fair (to simplify further formulae, index Q in force presentation is omitted):

$$s_\alpha - \bar{s}_\alpha = \tilde{\gamma}(F_\alpha) = \eta_k R_n(x)\,; \tag{2-8}$$

$$\frac{s_\alpha}{F_\alpha} - \frac{\bar{s}_\alpha}{F_\alpha} = \tilde{\gamma}(F_\alpha) = \varsigma_k R_n(x)\,; \tag{2-9}$$

$$s_\alpha - \bar{s}_\alpha = s_\alpha - f_3[\eta_\varsigma; F_k(s)] = \tilde{\gamma}(F_\alpha), \tag{2-10}$$

where $\tilde{\gamma}(F_\alpha)$ is a function with the least deflection from zero in the range of force change from $F_{\alpha 0}$ to $F_{\alpha m}$; $R_n(x)$ is Chebyshev's polynomial of the n-th degree and with minimum deflection from zero in the range [0; +1] or [-1;+1]; η_α, η_k, η_ς, and ς_k are the coefficients. Coefficients η_α and η_k should be found in the solution process. Conditions (8) and (9) allow us to approximate the nonlinear function with linear one by using the respective interpolation at Chebyshev's polynomial roots

$$s_\alpha(F_{\alpha q}) - \bar{s}_\alpha(\eta_\alpha; F_{\alpha q}) = 0\,, \tag{2-11}$$

or
$$\frac{s_\alpha(F_{\alpha q}) - \bar{s}_\alpha(F_{\alpha q})}{F_{\alpha q}} = 0, \tag{2-12}$$

at condition that $F_{\alpha 0} = 0$, $F_{\alpha q} = x_q F_{\alpha m}$, where x_q are the roots of Chebyshev's polynomial $R_n(x)$. Condition (10) allows us to correct the preliminary kinematics calculations. Equation (11) is effective when the absolute errors $|s_\alpha - \bar{s}_\alpha|$ of calculation should be minimized (Fig. 2-1b). Equation (12) is a typical one when the least deflections from zero should have the uncertainties of calculations related (Fig. 2-1c) to the acting forces $|(s_\alpha - \bar{s}_\alpha)/F_\alpha|$. In the simplest case of approximation, we can use Chebyshev's polynomial of the second degree with zero point, i.e., $R_2(X) = X^2 - 0.8284\,X$ with roots $X_o = 0$ and $X_q = 0.8284$. The following condition in accordance with the Chebyshev's theorem (*see* Chap.1) should be satisfied to obtain a uniform and accurate approximation with the chosen polynomial

$$\tfrac{1}{2}\,[\,|\tilde{\gamma}(F_{\alpha r})| - |\tilde{\gamma}(F_{\alpha c})|\,] < \{\Delta s\}, \tag{2-13}$$

where $F_{\alpha r} = F_{\alpha m} x_r$; $F_{\alpha c} = F_{\alpha m} x_c$; $x_r = 0.4142$ and $x_c = 1$ are the points of the maximum deflection of Chebyshev's polynomial $R_2(X)$; and $\{\Delta s\}$ is the allowed uncertainty of calculation. The signs of the deflections $\tilde{\gamma}(F_{\alpha r})$ and $\tilde{\gamma}(F_{\alpha c})$ should be different.

If condition (13) is not satisfied, we should in general choose Chebyshev's polynomial of the higher degree or decrease the quantity $F_{\alpha m}$. As a result of the latter change, the approximation range becomes narrower and the nonlinearity of the monotonic function $s_\alpha = f_4 (F_\alpha)$ or $s_\alpha / F_\alpha = f_5 (F_\alpha)$ decreases.

Energy similarity. We can show that the best approximation of a nonlinear function by a linear one can be achieved if the integral of certain Chebyshev's polynomials in the range [0;1] or [-1; +1] is equal to zero. This condition is effective for ELKS synthesis and may be called **the principle of energy similarity.** Let us look at the detail of this method. An output function $F_{Pn}(s) = F_{Pn}(x)$ of the nonlinear kinematic system in the deformation-force coordinates may be compared with a linear output function (Fig. 2-1d) which is a straight line $F_{p1}(s) = P_1(x)$ from the coordinate origin. This straight line should have the greatest possible number n of intersections with the given nonlinear function, and the difference

$$\varphi_n(x) = F_{Pn}(x) - P_1(x) \qquad (2\text{-}14)$$

represents the nonlinearity. Indeed the number of intersections n corresponds to the degree of the possible approximating algebraic function and the number of its interpolation knots. Therefore, the best least absolute error uniform approximation in this case can be done by Chebyshev's polynomial P_n(x) for which degree n is equal to the number of above mentioned intersections, $\varphi_n(x) = T_n P_n(x)$, where T_n is the coefficient at the expanded nonlinear function member with the largest exponent. But if the integral of this polynomial in the range of its arguments is equal to zero, then we have a condition that simultaneously leads us to the least squares approximation as well. Note that the integral of the function $F_{Pn}(s)$ represents the energy (area A_{e1}) of the nonlinear system deformation, and the integral of the function $P_1(s)$ represents the energy (area A_{e2}) of the linear deformation. Hence, we have the condition of the energy similarity of two systems under consideration if $A_{e1}\text{-}A_{e2} = |A_1|\text{-}|A_2| = \int_{x_0}^{x_m} \varphi_n(x)dx = \int_{x_0}^{x_m} T_n P_n(x)dx = 0$. At the same time, the least squares approximation requires minimization of the following expression

$$q_{\min} = [I_{\min} / (x_m - x_0)]^{0.5}, \quad \text{where} \qquad (2\text{-}15)$$

$$I = \int_{X_0}^{X_m} [(F_{Pn}(x) - P_1(x)]^2\, dx = \int_{X_0}^{X_m} [F_{pn}(x) - \Sigma P_i f_i(x)]^2\, dx. \qquad (2\text{-}16)$$

Hence, the minimum of value $I = I_{\min}$ will be under the condition

$$\sum_{i=0}^{n} \partial I / \partial P_i = 2 \int_{X_0}^{X_m} [F_{Pn}(x) - \sum_i^n P_i f_i(x)]\, [\sum_i^n f_i(x)]\, dX = 0. \qquad (2\text{-}17)$$

The requirement $\int_{X_0}^{X_m} g(x)dx = 0$ should be satisfied for optimization of the function $P_1(x)$ and for yielding $I_{\min}$, which is necessary for the least squares approximation of the function $F_{Pn}(x)$ by the straight line. So if the $g(x) = \varphi_n(x) = A_n P_n(x)$, the integral of Chebyshev's polynomial should be equal to zero for the simultaneous approximation with the uniform least absolute error and least squares error requirements satisfaction. As a result, the **principle of energy similarity** may be described as follows:

If the area under the elastic nonlinear function F_{pn} of ELKS is equal to the area which is limited by the straight line F_{p1} passing through the origin of coordinates and intersecting this nonlinear function the greatest number of times or $|A_1|=|A_2|$ (Fig. 2-1d), and at the same time the algebraic sum of areas limited by Chebyshev's polynomial with the same degree as the number of above mentioned intersections, is equal to zero (equivalent to the condition in Fig. 2-1f under which the polynomial $P_3(x)$integral is equal to zero and $|A_1+A'_1|=|A_2+A'_2|$ in Fig. 2-1e), then the elastic nonlinear function approximated by that polynomial is an optimal one and has the least absolute and least squares evaluated nonlinearity.

Hence, this principle of energy similarity requires that the integral of Chebysev's polynomial used for the approximation is equal to zero. It can be shown that some of Chebyshev's polynomials satisfy this requirement and others do not. For Chebyshev's polynomials $P_n(X)$ with the range [-1;+1], the integral

$\int_{-1}^{1} P_n(X)\ dX = 0$ if n is an odd number because

$$P_n(x) = 2^{(1-n)} \cos [n \cos^{-1}(x)] = 2^{-n} \{[x + (x^2-1)^{0.5}]^n + [x - (x^2-1)^{0.5}]^n \}. \quad (2\text{-}18)$$

For the least relative deflection Chebyshev's polynomial $P^{o}_n(X)=X\,P_{n-1}(X)$ with the range [0; 1], the integral $\int_0^1 P^{o}_n(X)\,dX = 0$ at $n=3$ because $P^{o}_3(X) = X^3 - 0.5\,X$.

The integrals for polynomials $R_n(X)$ and $R^{o}_n(X)$ are not equal to zero at any n. This energy similarity method was successfully used (A.N. Grishin, 1970) to minimize the relative uncertainty of the spring-pneumatic transducer for measuring eye pressure p whose readout $W(p)$ function was expressed in the form $p = A_w W + B_w W^3$. The approximating polynomial was the least relative one ($P^{o}_3 = X^3 - 0.5X$) whose integral in the X range [0; +1] is equal to zero. When a straight line passed through the origin of the coordinates and the root point of the polynomial (W_e=0.707W_{max}), then the area limited by the above mentioned transducer's function was halved. Therefore, the algebraic sum of this area is equal to zero.

Let us also recall that the least square approximation of the integral as in (16) can be performed by the function expansion with the help of Jacobi and Legendre polynomials.

The Greatest Linear Transformation.

The research of elastic body's shape for the largest transformation ratio of motion shows that the greatest transformation of translational s to rotational θ motion may be achieved if the elastic body transformable chain represents a very thin flexible helicoid. It can be theoretically proved on the basis of Castigliano's theorem. The kinematic transformation ratio i_{ij} at the deformation of elastic body is equal in accordance with this theorem to the following expression:

$$i_{ij} = (\partial\theta_i / \partial s_j) = (\partial U_v / \partial M_i^{\Phi}) : (\partial U_v / \partial F_j)\,, \quad (2\text{-}19)$$

where $U_v = \iiint_V a(\varepsilon) dV$ is the potential energy of deformed body with volume V,

$a(\varepsilon)$ is a work of elastic deformation per unit volume, M_i^{Φ} is the virtual counteracting (resisting) moment on the output of transformation chain, and F_j is the force on the input of transformation chain.

It is possible to prove by application of the fundamental lemma of the variation calculus and the kinematic analogies that the largest sensitivity in the straight chain of transformation ($M_i^{(\psi)} \to 0$) for linear displacements into angular ones has a body with the shape similar to a thin elastic helicoid. The analytical demonstration of this conclusion is an unwieldy one. Therefore, we do not discuss it in this book. However, it has strong support in the precision instrumentation practice (*see* Chap. 4).

2-2. Kinematics of an Elastic Helicoidal Beam

This section is dedicated to the general and specific strain-stress relations in an elastic helicoid, or so-called initially pretwisted beam. These relations can be linear at a small deformation or the nonlinear at a large deformation. They also can be stable or unstable under certain load conditions.

Elastic helicoid.

The strain-stress state of the helicoidal (pretwisted) beam differs in several ways from the corresponding state of the prismatic beams. First, pretwisted beams have torsion deformation (untwist) under the stretching load applied along their axis. Second, their length changes under the applied torque; normal stresses occur with twisting, and tangent stresses with stretching. These special features are significant at larger ratios $\Lambda_\lambda=(b/h)^2 = \lambda^{-1}$ between the width b=2$b_1$and the thickness h=2h_1 of the beam's cross section.

Helix parameter. The translational motion in pretwisted helicoids has a functional relation to the rotation through the helix parameter κ_p. Let the cross-section profile correspond to the following equation

$$r_a = [r(\upsilon)\cos\upsilon]\,\mathbf{i_a} + [r(\upsilon)\sin\upsilon]\,\mathbf{j_a}, \tag{2-20}$$

where $r(\upsilon)$ is the equation for the polar curve of the cross-section profile's contour. Let this profile conduct the helical motion with parameter κ_p, which corresponds to the certain translational motion along the axis at the profile rotation for one radian. Parameter κ_p has a positive sign if the helical motion is a right-handed, and a negative sign if the helical motion is a left-handed one. The equation of the helicoid side surface in matrix form is as follows

$$r_1 = M_{1a}\, r_a, \tag{2-21}$$

$$\text{where } M_{1a}= \begin{bmatrix} \cos\vartheta & \sin\vartheta & 0 & 0 \\ -\sin\vartheta & \cos\vartheta & 0 & 0 \\ 0 & 0 & 1 & \kappa_p^{-1}\vartheta \\ 0 & 0 & 0 & 1 \end{bmatrix}, \tag{2-22}$$

and ϑ is the angle of the cross section's rotation (Fig. 2-2a) around the helicoids longitudinal axis ζ.

On the basis of (20), (21), and (22), we finally obtain the equation of the helicoids' side surface $r_1 = [r\cos(\upsilon+\vartheta)]\,\mathbf{i} + [r\sin(\upsilon+\vartheta)]\,\mathbf{j} + \kappa_p^{-1}\vartheta\,\mathbf{k}$,

where $\mathbf{i_a}$, $\mathbf{j_a}$, $\mathbf{i}$, $\mathbf{j}$, $\mathbf{k}$ are the unit-vectors of the corresponding coordinate axes.

The important feature of the helicoidal surface may be presented by the equation $Y n_x - X n_y = \kappa_p n_z$, where n_x, n_y, n_z are the normal vector's projections on the

Cartesian coordinate axes X, Y, Z. The initial helix parameter is equal to the relative twist of the unloaded pretwisted beam: $\kappa_{p0} = d\,\vartheta_0 / d\varsigma$. (2-23)

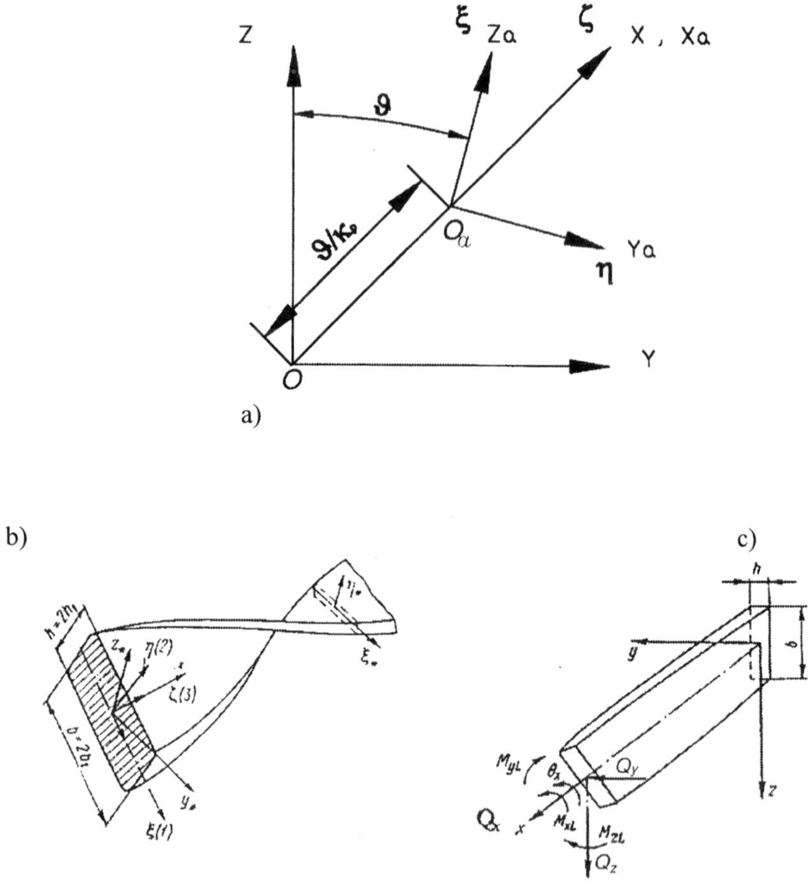

Fig. 2-2. Fixed and following coordinate systems for the helix motion and pretwisted helicoid loading.

The beam is pretwisted uniformly if κ_{p0}= constant. Then $\kappa_{p0} = \vartheta_L / L$, where ϑ_L is the change in the angle position of the cross section in the range of the beam's length (L). The corresponding points of different cross sections are positioned on the helix lines with helix angles which are equal to: $\beta_0 = \tan^{-1}(\kappa_{p0}R)$, (2-24)
where $R = [(\xi-\xi_j)^2 + (\eta-\eta_j)^2]^{0.5}$ is the distance of a point to axis ς_j, and ξ_j and η_j, are the coordinate axes moving with the cross section. If $\beta_0 << 1$, we can assume with necessary accuracy that $\beta_0 = \kappa_{p0}R$. (2-24a)

Internal force Q_ζ, and moment M_ξ, M_η, M_ζ factors related to the axes shown in the subscripts for the pretwisted beam are expressed (B.F. Shorr, 1968) through the coefficients of rigidity in the following matrix form:

$$\begin{bmatrix} Q\zeta \\ M\xi \\ M\eta \\ M\zeta \end{bmatrix} = \begin{bmatrix} j\zeta & j\varepsilon\xi & j\varepsilon\eta & j\varepsilon\theta \\ j\xi\varepsilon & j\xi & j\xi\eta & j\xi\theta \\ j\eta\varepsilon & j\eta\xi & j\eta & j\eta\theta \\ j\theta\varepsilon & j\theta\xi & j\theta\eta & j\theta \end{bmatrix} \begin{bmatrix} \zeta' \\ \aleph\xi \\ \aleph\eta \\ \theta' \end{bmatrix}, \tag{2-25}$$

where $j_{kl}=j_{lk}$ $(k,l=\varepsilon,\ \xi,\ \eta,\ \theta)$; $\aleph_\xi$, $\aleph_\eta$ are the curvatures; $\zeta'=\varepsilon$ is the elongation strain and θ' is the torsion per beam's unit length. The coefficients in matrix of rigidity for moderately pretwisted beams are defined through formulae, which are shown in Table 2-1.

Table 2-1. The coefficients in matrix of rigidity for moderately pretwisted beams

Coordinate	$\zeta'=\varepsilon$	ξ	η	θ
$\zeta'=\varepsilon$	$j_\zeta=j_\varepsilon=EA$	$j_{\varepsilon\xi}=0$	$j_{\varepsilon\eta}=0$	$j_{\varepsilon\theta}=\kappa_{p0}E(J_p-T_o)$
ξ	$j_{\xi\varepsilon}=0$	$j_\xi=EJ_\xi$	$j_{\xi\eta}=0$	$j_{\xi\theta}=\kappa_{p0}E(J_{p\xi}-T_\xi)$
η	$j_{\eta\varepsilon}=0$	$j_{\eta\xi}=0$	$j_\eta=EJ_\eta$	$j_{\eta\theta}=-k_{p0}E(J_{p\eta}-T_\eta)$
θ	$j_{\theta\varepsilon}=\kappa_{p0}E(J_p-T_o)$	$j_{\theta\xi}=\kappa_{p0}E(J_{p\xi}-T_\xi)$	$j_{\theta\eta}=-k_{p0}E(J_{p\eta}-T_\eta)$	$j_\theta=GT$

*After B.F. Shorr, 1968. Some indices may be interchangeable in different publications, e.g., $\varepsilon\leftrightarrow\zeta$, because $\varepsilon=\zeta'$.

The notations in Table 2-1 are as follows: $T=T_o+2(1+\nu_p)\,\kappa_{p0}{}^2(J_r^0-T_r^0)$, A is the area, $J_\xi=\int_A\eta^2\,dA$; $J_\eta=\int_A\xi^2\,dA$ are the axial moments, $J_p=\int_A R^2\,dA$ is the polar moment, $J_{p\xi}=\int_A R^2\eta\,dA$, $J_{p\eta}=\int_A R^2\xi\,dA$ are the polar-axial moments of inertia; $J_r^0=\int_A R^4\,dA$, $T_o=\int_A(R^2+\varphi'_\psi)\,dA$; $T_\xi=\int_A(R^2+\varphi'_\psi)\eta dA$, $T_\eta=\int_A(R^2+\varphi'_\psi)\xi dA$, $T_r^0=\int_A[R^4+(\varphi'_\psi)^2]dA$; $\varphi'_\psi=\varphi'_\eta(\xi-\xi_j)-\varphi'_\xi(\eta-\eta_j)$, and $\varphi'_\eta=\partial\psi^0/\partial\eta$; $\varphi'_\xi=\partial\psi^0/\partial\xi$, where ψ^0 and T_o are the function of rotation and torsional constant for non-twisted beam with the same cross section, respectively. Let us recall that the torsional rigidity for the solid beam with the complicated contour of solid cross section may approximately be estimated by the Saint-Venant's formula $T_o\approx A^4/(40\,J_p)$, where A is the area and J_p is the polar moment of inertia for the corresponding cross section. However, this formula may be not accurate for cross sections with the concave contours.

Table 2-2. The coefficients in matrix of compliance for moderately pretwisted beams

Coord	$\zeta'=\varepsilon$	ξ	η	θ
$\zeta'=\varepsilon$	$v_\varepsilon=\dfrac{1+v_\theta\, j_{\varepsilon\theta}^2/j_\varepsilon}{j_\varepsilon}$	$v_{\varepsilon\xi}=\dfrac{v_\theta j_{\varepsilon\theta}\, j_{\xi\theta}}{j_\varepsilon j_\xi}$	$v_{\varepsilon\eta}=\dfrac{v_\theta\, j_{\varepsilon\theta}\, j_{\eta\theta}}{j_\varepsilon j_\eta}$	$v_{\varepsilon\theta}=-\dfrac{v_\theta\, j_{\varepsilon\theta}}{j_\varepsilon}$
ξ	$v_{\xi\varepsilon}=\dfrac{v_\theta j_{\varepsilon\theta}\, j_{\xi\theta}}{j_\varepsilon j_\xi}$	$v_\xi=\dfrac{1+v_\theta\, j_{\xi\theta}^2/j_\xi}{j_\xi}$	$v_{\xi\eta}=\dfrac{v_\theta\, j_{\eta\theta}\, j_{\xi\theta}}{j_\eta j_\xi}$	$v_{\xi\theta}=-\dfrac{v_\theta\, j_{\xi\theta}}{j_\xi}$
η	$v_{\eta\varepsilon}=\dfrac{v_\theta\, j_{\varepsilon\theta}\, j_{\eta\theta}}{j_\varepsilon j_\eta}$	$v_{\eta\xi}=\dfrac{v_\theta\, j_{\eta\theta}\, j_{\xi\theta}}{j_\eta j_\xi}$	$v_\eta=\dfrac{1+v_\theta\, j_{\eta\theta}^2/j_\eta}{j_\eta}$	$v_{\eta\theta}=-\dfrac{v_\theta\, j_{\eta\theta}}{j_\eta}$
θ	$v_{\theta\varepsilon}=-\dfrac{v_\theta\, j_{\varepsilon\theta}}{j_\varepsilon}$	$v_{\theta\xi}=-\dfrac{v_\theta\, j_{\xi\theta}}{j_\xi}$	$v_{\theta\eta}=-\dfrac{v_\theta\, j_{\eta\theta}}{j_\eta}$	*

* $v_\theta=[j_\theta-(j_{\varepsilon\theta}^2/j_\varepsilon)-(j_{\xi\theta}^2/j_\xi)-j_{\eta\theta}^2/j_\eta]^{-1}$. After B.F. Shorr, 1968. Some indices may be interchangeable in different publications, e.g., $\varepsilon\leftrightarrow\zeta$ because $\varepsilon=\zeta'$.

The coefficients in the reciprocal to (25) matrix of compliance for moderately pretwisted beams are defined through the formulae, which are shown in Table 2-2, where $\nu_\theta=[GT_0(1+\beta_\rho^2)]^{-1}$; $\beta_\rho^2=2(1+\nu_p)k_{p0}^2(J_r-T_r)/T_o$; $J_r-T_r=J_r^0-T_r^0-[(J_p-T_o)^2/A]-[(J_{p\xi}-T_\xi)^2/J_\xi]-[(J_{p\eta}-T_\eta)^2/J_\eta]$.

Matrix of strain. The components of the beam deformation in the X, Y, Z system may be expressed in matrix form as the product of the matrix of compliance ν_{ij} by the column matrix of loads $F_i(Q_x, M_x, M_y, M_z)$ in the fixed Cartesian coordinates :

$$\begin{bmatrix} S_x' \\ -S_y'' \\ S_z'' \\ \theta_x' \end{bmatrix} = \begin{bmatrix} \nu_x & \nu_{xz} & \nu_{xy} & \nu_{x\theta} \\ \nu_{zx} & \nu_z & \nu_{zy} & \nu_{z\theta} \\ \nu_{yx} & \nu_{yz} & \nu_y & \nu_{y\theta} \\ \nu_{\theta x} & \nu_{\theta z} & \nu_{\theta y} & \nu_\theta \end{bmatrix} \begin{bmatrix} Q_x \\ M_z \\ M_y \\ M_x \end{bmatrix} = [\nu_{ij}]\ [Q_i], \tag{2-26}$$

where ν_{ij} are the coefficients of compliance in the corresponding directions X, Y, Z, θ_x (Fig.2-2a, c); S_x, S_y, S_z, θ_x are the elastic displacements in the directions of the coordinate axes X, Y, Z and the angle of rotation around axis X; S'_x, S''_y, S''_z, θ'_x are the derivatives of the displacements with respect to the longitudinal coordinate x. The values of compliance coefficients ν_{kl} in the "following" coordinates ξ_*, η_*, ζ_* with directions ξ, η, ζ, $\vartheta=\theta$ are related to coefficients ν_{ij} in the Cartesian coordinates X, Y, Z, and θ by the equations presented in Table 2-3.

Table 2-3. The compliance coefficients in the Cartesian coordinates

Coordinate	x	z	y	θ
x	$\nu_x=\nu_\varepsilon$	$\nu_{xz}=\nu_{\xi\varepsilon}\cos\vartheta_0 x-\nu_{\eta\varepsilon}\sin\vartheta_0 x$	$\nu_{xy}=\nu_{\xi\varepsilon}\sin\vartheta_0 x+\nu_{\eta\varepsilon}\cos\vartheta_0 x$	$\nu_{x\theta}=\nu_{\varepsilon\theta}$
z	$\nu_{zx}=\nu_{\xi\varepsilon}\cos\vartheta_0 x-\nu_{\eta\varepsilon}\sin\vartheta_0 x$	$\nu_z=\nu_\xi\cos^2\vartheta_0 x+\nu_\eta\sin^2\vartheta_0 x-\nu_{\xi\eta}\sin 2\vartheta_0 x$	*	$\nu_{z\theta}=\nu_{\xi\theta}\cos\vartheta_0 x-\nu_{\eta\theta}\sin\vartheta_0 x$
y	$\nu_{yx}=\nu_{\xi\varepsilon}\sin\vartheta_0 x+\nu_{\eta\varepsilon}\cos\vartheta_0 x$	*	$\nu_y=\nu_\xi\sin^2\vartheta_0 x+\nu_\eta\cos^2\vartheta_0 x+\nu_{\xi\eta}\sin 2\vartheta_0 x$	$\nu_{y\theta}=\nu_{\xi\theta}\sin\vartheta_0 x+\nu_{\eta\theta}\cos\vartheta_0 x$
θ	$\nu_{\theta x}=\nu_{\varepsilon\theta}$	$\nu_{\theta z}=\nu_{\xi\theta}\cos\vartheta_0 x-\nu_{\eta\theta}\sin\vartheta_0 x$	$\nu_{\theta y}=\nu_{\xi\theta}\sin\vartheta_0 x+\nu_{\eta\theta}\cos\vartheta_0 x$	ν_θ

* $\nu_{zy}=\nu_{yz}=(\nu_\xi-\nu_\eta)\sin\vartheta_0 x\cos\vartheta_0 x+\nu_{\xi\eta}\cos 2\vartheta_0 x$; $\vartheta_0=k_{p0}$. After B.F. Shorr. Some indices may be interchangeable in different publications, e.g., $\varepsilon\leftrightarrow\zeta$ because $\varepsilon=\zeta'$.

The state of symmetry corresponds to the equality of the certain compliance's components: $\nu_{kl}=\nu_{lk}$; $\nu_{ij}=\nu_{ji}$. There is a simplification for the thin beams with a profile that is longer in the ξ direction because the shift constituent is a very small one in the η direction. As a result, the derivative of the function of rotation for such beams is equal to $\varphi'_\psi=R^2$, where $R^2=\xi^2+\eta^2$. Let us open matrix (26) and conduct the necessary integration to find the elastic displacement components at the bending of a cantilever pretwisted beam

$$S_x=\int_0^L(\nu_x Q_x+\nu_{xz}M_z+\nu_{xy}M_y+\nu_{x\theta}M_x)\,dx, \tag{2-27}$$

$$S_y = - \int_0^L\int_0^x (\nu_{zx}Q_x + \nu_z M_z + \nu_{zy} M_y + \nu_{z\theta}M_x) \, dx^2; \tag{2-28}$$

$$S_z = \int_0^L\int_0^x (\nu_{yx}Q_x + \nu_{yz} M_z + \nu_y M_y + \nu_{y\theta}M_x) \, dx^2; \tag{2-29}$$

$$\theta_x = \int_0^L (\nu_{\theta x}Q_x + \nu_{\theta z} M_z + \nu_{\theta y} M_y + \nu_{\theta}M_x) \, dx. \tag{2-30}$$

There should be preliminary substitution into (27) to (30) of the expressions for the coefficients of compliance ν_{ij} in the fixed system of coordinates through the corresponding coefficients ν_{kl} in the "following" system of coordinates. The bending moments in the beam section (Fig. 2-2c) are equal to:

$$M_z = M_{zL} - Q_y(L\text{-}X) - Q_xS_y, \tag{2-31}$$

$$M_y = M_{yL} - Q_z(L\text{-}X), \tag{2-32}$$

$$M_x = M_{xL} + Q_zS_y, \tag{2-32a}$$

where moments M_{xL}, M_{yL}, M_{zL} and forces Q_x, Q_y, and Q_z are applied to the end of the beam . The moment Q_xS_y may be neglected if displacement S_y and longitudinal force Q_x are small. Then the solution of integrals in (27) – (30) is simplified.

Motion stability.

If displacement S_y is significant and the moment Q_xS_y cannot be neglected, a differential equation of the second order is obtained after the opening of matrix (26):

$$-S_y'' = [M_{zL} - Q_y(L\text{-}x)\text{-}Q_xS_y](\nu_\xi \cos^2 \vartheta_0 x + \nu_\eta \sin^2 \vartheta_0 x)$$
$$+ (\nu_\xi - \nu_\eta) \sin \vartheta_0 x \cos \vartheta_0 x[M_{yL} + Q_z(L \;\text{-}x)] + \nu_{\xi\theta} \cos \vartheta_0 x[M_{xL} + Q_zS_y],$$

which transforms in the corresponding nonhomogeneous differential equation

$$S_y'' - Q_xS_y(\nu_\xi \cos^2 \vartheta_0 x + \nu_\eta \sin^2 \vartheta_0 x) + Q_zS_y \nu_{\xi\theta}\cos \vartheta_0 x =$$
$$-(M_{zL} - Q_yL) \nu_\xi - [Q_y \nu_\xi + (M_{yL} + Q_zL) (\nu_\xi - \nu_\eta) \vartheta_0]x$$
$$+ [(M_{zL} - Q_yL) (\nu_\xi - \nu_\eta) \vartheta_0^2 + Q_z (\nu_\xi - \nu_\eta) \vartheta_0] x^2 + [Q_y(\nu_\xi - \nu_\eta) \vartheta_0^2$$
$$+ (2/3) (M_{yL} + Q_zL) (\nu_\xi - \nu_\eta) \vartheta_0^3] x^3 - (2/3) Q_z (\nu_\xi - \nu_\eta) \vartheta_0^3 x^4$$
$$- M_{xL} \nu_{\xi\theta}\cos \vartheta_0 x. \tag{2-33}$$

The right-hand side of (33) can be approximated by a polynomial of the fourth degree of x. The left-hand side of this equation after corresponding transformations can be presented in the form of Mathieu's equation (N.W. MacLachlan, 1964). The final equation at negative force Q_x and Q_z =0 may be presented as follows

$$S_y'' + S_y Q_x \frac{\nu_\xi + \nu_\eta}{2} \left[1 - \frac{\nu_\eta - \nu_\xi}{\nu_\xi + \nu_\eta} \cos \vartheta_0 x \right] = 0 , \tag{2-34}$$

where Q_x, ν_ξ , ν_η are the parameters; cos $\vartheta_0 x$ is the periodic function. It is known, however, that Mathieu's equations have fields with both stable and unstable solutions. The physical essence of these specifics is important for the beam shape stability at bending. It means that the pretwisted thin beam may lose stability at bending if the beam is loaded by a longitudinal compressing load (Table 2-4). This loss of stability is followed by large changes in deformation at a small change in the load on the boundary of the critical field. A similar effect is also possible for initially plane straight beams, which are loaded with a compressing force and twisting moment. Critical compressing force in these cases is much smaller than the one for compressed straight beams in accordance with the known Euler's equations (*see* Chap. 3).

Table 2-4. Loss of beam's stability under compression and torsion

Type of load	Schematic of loading	The critical value of load
Compression and torsion of flat beam		$\theta_{tr}=\frac{4}{3}\frac{n_2-n_1}{(1-n_1)(1-n_2)}$ * Elliptical cross section: $n_1=1-\frac{2}{(1+\nu_p)(1+h_1^2 b_1^{-2})}$; $n_2=1-\frac{2}{(1+\nu_p)(1+h_1^{-2} b_1^2)}$ Rectangular cross section: $n_1=1-\frac{6\nu_{kch}}{1+\nu_p}$; $n_2=1-\frac{6\nu_{kch}}{1+\nu_p}b^2h^{-2}$; *** $\nu_{kch}=\frac{H_k}{Gh^3b}$.
Compression of cantilever elastic helicoid	See Fig. 2-2c	$-Q_x=-\nu_{tr}\frac{\pi^2}{4L^2\nu_\eta}$; ** $\nu_\xi<\nu_\eta;\ \vartheta_0 L>2\pi;\ \nu_{tr}\approx\frac{2\nu_\eta}{\nu_\eta+\nu_\xi}$

In the Table 2-4 formulae * After V.B. Petrov, 1967,** After B.F. Shorr, 1968. *** H_k is the section torsional rigidity with load Q.

Specific rigidities. The relations between displacement s_γ, rotation θ_γ, stretching longitudinal force F_Q, and torque M at the pretwisted beam extension and twisting (in accordance with the solution of differential equation 4, Table 1-10) may be expressed through the three rigidity features:

$$F_Q=j_{kk}\,s_\gamma-j_{\tau k}\theta_\gamma; \tag{2-35}$$

$$M=j_{k\tau}s_\gamma-j_{\tau\tau}\ \theta_\gamma, \tag{2-36}$$

where j_{kk} is the extensional rigidity at extension deformation; $j_{\tau k}=j_{k\tau}$ is the torsional rigidity at the extension deformation or vice versa; $j_{\tau\tau}$ is the torsional rigidity at the twist deformation.

The stress components in the pretwisted beam correspond to the stress tensor

$$\sigma^{\xi\eta}=\sigma_Q{}^{\xi\eta}S_\gamma+\sigma_M{}^{\xi\eta}\theta_\gamma, \tag{2-37}$$

where $\sigma_Q{}^{\xi\eta}$ are the stress tensor contravariant components at the longitudinal deformation and the absence of the external twisting moment ($M=0$); $\sigma_M{}^{\xi\eta}$ are the stress tensor components at the torsion deformation and the absence of the external longitudinal stretching force ($Q=0$). The stress tensor in matrix form for coordinates 1, 2, 3 (*see* Fig. 2-3a) is equal to:

$$\hat{\sigma}_{1T} = \begin{bmatrix} \sigma^{11} & \sigma^{12} & \sigma^{13} \\ \sigma^{21} & \sigma^{22} & \sigma^{23} \\ \sigma^{31} & \sigma^{32} & \sigma^{33} \end{bmatrix}. \qquad (2\text{-}38)$$

The indices show the direction of the corresponding component; the first number denotes the normal to the plane where stress component has action, the second one denotes the direction to which the corresponding stress component is parallel.

Transmission nonlinearity.

All of the above discussed concepts belong to linear pretwisted beam behavior. But for elastic helicoidal mechanisms it is very important to study possible nonlinear relations between beam longitudinal stretch and its angle of rotation. This will allow us to estimate the possible quasi-linear working range of the helicoidal mechanism and the systematic uncertainty in that mechanism from nonlinearity. We study these problems with the variational method on the basis of the potential energy equation by comparing its linear and nonlinear tensor components. The first step is to get a system of equations to find a linear kinematics relation between longitudinal strain $s_\gamma = s/L$ and the untwist rotation $\theta_\gamma = \theta/L$ of the elementary helicoid with unit length L=1. Let us recall that several systems of coordinates (Fig. 2-2a,b; Fig. 2-3a) are used to simplify the solution: Cartesian coordinates X, Y, Z, non-Cartesian helical coordinates $\xi(1)$, $\eta(2)$, ζ (3), and ξ_* and η_* as "following" coordinates that are rigidly connected to the initial contour of the beam's cross section. Figure 2-2b represents a pretwisted beam as an envelope at the cross-section contour's uniform rotation and uniform sliding with respect to the straight line ζ(3). The "following" Cartesian coordinates have a fixed axis X that coincides with straight line ζ and two axes $Y_a = \xi_*$ and $Z_a = \eta_*$ rotating around this line.

The total potential energy with respect to the length unit for a pretwisted beam with linear elastic material following the Hooke's law may be expressed through the tensor of deformation (strain) components

$$U_v = G \iint_A [\varepsilon^2_{11} + \varepsilon^2_{22} + \varepsilon^2_{33} + 2(\varepsilon^2_{12} + \varepsilon^2_{13} + \varepsilon^2_{23})$$

$$+ \nu_p \Theta^2 / (1\text{-}2\nu_p)]\, dA - F_Q\, s_\gamma - M\theta_\gamma, \qquad (2\text{-}39)$$

where G is the shear modulus, Θ is invariant of the tensor of deformation, A is the area of the beam's cross section, F_Q and M are a stretching force and a twisting moment(torque), respectively; s_γ and θ_γ are the stretch strain and the angle of twist per unit of the beam's length, respectively. The constituents of deformation in this case correspond to differential equation $N^{\underline{o}}$ 4 in Table 1-10.

The system of five simultaneous equations are obtained with the differentiation of potential energy U_v . Their structure is a complicated one and may be shown by blocks as follows:

$$\partial U_v / \partial \Gamma_a = f_1(S_\gamma;\ \theta_\gamma;\ \kappa_{p0};\ J_p;\ T_0);\ \iint_A (\partial\psi^0 / \partial\vartheta)^2\, dA;$$

$$\iint_A [(\partial\psi^0 / \partial\xi)^2 + (\partial\psi^0 / \partial\eta)^2]\, dA;\quad \iint_A (\partial\psi^0 / \partial\vartheta)(\xi^2 - \eta^2)\, dA;$$

$$\iint_A \xi \rho\eta (\partial\psi^0 / \partial\rho)\, dA; \ (\Gamma_a ; \Gamma_b ; \Gamma_c) = 0; \qquad (2\text{-}40)$$

$$\partial U_v / \partial \Gamma_b = f_2(S_\gamma; A; \kappa_{p0}; J_p; T_0 ; I_{20}; I_{02}; \Gamma_a ; \Gamma_b ; \Gamma_c) = 0; \qquad (2\text{-}41)$$

$$\partial U_v / \partial \Gamma_c = f_3(S_\gamma; \kappa_{p0};I_{20}; I_{02} ; I_{22};I_{40}; I_{04});$$

$$\iint_A (\partial\psi^0 / \partial\vartheta)(\xi^2 -\eta^2)\, dA; \ \iint_A \xi \rho\eta (\partial\psi^0 / \partial\rho)\, dA; (\Gamma_a ; \Gamma_b ; \Gamma_c) = 0; \qquad (2\text{-}42)$$

$$\partial U_v / \partial \theta_\gamma = f_4 (\theta_\gamma; J_p; T_0; \Gamma_a) = -M / f[G(J_p - T_0)]; \qquad (2\text{-}42a)$$

$$\partial U_v / \partial s_\gamma = f_5(S_\gamma;\kappa_{p0};A;J_p;T_0; \Gamma_a; \Gamma_b) = Q / f_6 (4GA). \qquad (2\text{-}43)$$

Here ψ^0 is the warping function in torsion of a plane straight beam with the same cross section as for the pretwisted one (initial contour); ϑ and ρ are the contour polar coordinates (ρ is a radius-vector) ; $I_{qm} = \int_A \xi^q \eta^m\, dA$ are the two-multiple (double) integrals; $J_p = I_{20} + I_{02}$ is the polar moment of the cross-section inertia; $T_0 = \int_A (\partial\psi^0 / \partial\vartheta)\, dA + J_p$ is the geometric rigidity at the twist of the plane straight beam with the same cross section; Γ_a , Γ_b, and Γ_c are the constants that should be defined from the solution of the system of simultaneous equations (40) – (43); M is the twisting moment (torque); Q is the longitudinal extension acting force. S_γ and θ_γ values for the quasi-linear system are determined from the solution of (40) - (43) by substituting the warping function ψ^0 and expressions of the double integrals $I_{qm} = \int_A \xi^q \eta^m\, d\xi d\eta$ for the corresponding cross sections. The indices in the integrals are respectively equal to $q = 0, 1, 2, 4, 6$; $m= 0, 1, 2, 4, 6$.

The nonlinear system should have the "following" Cartesian coordinate Z_a, Y_a rotation with variable speed because its helix parameter $\kappa_p = \kappa_{p0} + \theta_\gamma$ will change at the beam section rotation unlike of the constant helix parameter κ_{p0} in the linear system. It should also be remembered that at the stretching of the pretwisted beam the angle of untwist rotation is negative, $\theta_\gamma < 0$. As a result, we will find the relationship between components ε_{ik} of strain tensor for a nonlinear body in the "following" system of coordinates and the known strain tensor components ε^0_{ik} for a linear body (noted by zero indices) of the same pretwisted beam: $\varepsilon_{11} = \varepsilon^0_{11}$; $\varepsilon_{22} = \varepsilon^0_{22}$; $\varepsilon_{12} = \varepsilon^0_{12} = 0$; $\varepsilon_{33} = \varepsilon^0_{33} - \theta_\gamma \Gamma_a\, \partial\psi^0 / \partial\vartheta$;

$\varepsilon_{13} = \varepsilon^0_{13} - 2\theta_\gamma \Gamma_c\, \xi^2 \eta$; $\varepsilon_{23} = \varepsilon^0_{23} - 2\theta_\gamma \Gamma_c\, \xi\eta^2$, $\Theta = \Theta^0 - \theta_\gamma \Gamma_a \partial\psi^0/\partial\vartheta$,

where 1, 2, 3 are the coordinate axes numbers in Fig. 2-2b, 2-3a.

Potential energy at the stretching of the elementary (L=1) nonlinear pretwisted beam without an external twist moment ($M_{xL} = 0$) can be expressed as

$$R_v = G \iint_A \{ [\varepsilon^{0\,2}_{11} + \varepsilon^{0\,2}_{22} + (\varepsilon^0_{33} - \theta_\gamma \Gamma_a\, \partial\psi^0 / \partial\vartheta)^2] + 2[(\varepsilon^0_{13} - 2\theta_\gamma \Gamma_c\, \xi^2 \eta)^2$$

$$+ (\varepsilon^0_{23} - 2\theta_\gamma \Gamma_c\, \xi\eta^2)^2] + (\Theta^0 - \theta_\gamma \Gamma_a\, \partial\psi^0 / \partial\vartheta)^2 \}\, dA$$

$$- QS_\gamma = R_v^0 + R'_v, \qquad (2\text{-}44)$$

where $R_v^0 = U_v$ is the known expression (39) of the potential energy at a small deformation of the pretwisted beam and $M_{xL} = 0$, R'_v is an additional member that shows the influence of the large degree of beam untwist. By subtracting (39) from (44) with the substitution, $M=M_{xL} = 0$, we finally obtain

$$R'_\nu = G \iint_A 2\{ (- \theta_\gamma \Gamma_a \partial\psi^0 / \partial\vartheta)[2\, S_\gamma - 2\kappa_{p0}\Gamma_a (\partial\psi^0 / \partial\vartheta) \ + 4\, \Gamma_b - 4\, \Gamma_c(\xi^2 - \eta^2)]+$$

$$(\theta_\gamma \Gamma_a \partial\psi^0 / \partial\vartheta)^2 - 4\, \Gamma_c\, \theta_\gamma \xi\, \eta\, [(\Gamma_a/ 2)\left(\xi\frac{\partial\psi^0}{\partial\xi}+\eta\frac{\partial\psi^0}{\partial\eta}\right)$$

$$-2\Gamma_c\, \kappa_{p0}\, \xi\, \eta\, (\xi^2 + \eta^2)] + 4\, \Gamma^2{}_c\, \theta_\gamma^2\, \xi^2\eta^2(\xi^2 + \eta^2)\}\, dA. \tag{2-45}$$

The constants Γ_a ; Γ_b; Γ_c can be defined with the principle of stationary potential energy when deriving the solution by the method of variations

$$\partial R_\nu / \partial\Gamma_b = (\partial R_\nu{}^0 / \partial\Gamma_b); \tag{2-46}$$

$$\partial R_\nu / \partial\Gamma_c = (\partial R_\nu{}^0 / \partial\Gamma_c); \tag{2-47}$$

$$\partial R_\nu / \partial\, 0_\gamma = (\partial R_\nu{}^0 / \partial\, \theta_\gamma) + (\partial R'_\nu / \partial\, \theta_\gamma); \tag{2-48}$$

$$\partial R_\nu / \partial\Gamma_a = (\partial R_\nu^0 / \partial\Gamma_a); \tag{2-49}$$

$$\partial R_\nu / \partial S_\gamma = (\partial R_\nu{}^0 / \partial S_\gamma). \tag{2-50}$$

Then the following five equations at M=0 are found:

$$S_\gamma + 6\, \Gamma_b + 2b_1{}^2\Gamma_c + (1/3)\, (\, \kappa_{p0} + \theta_\gamma)\, \Gamma_a b_1{}^2\, (1\text{-}3\lambda) = 0; \tag{2-51}$$

$$-S_\gamma - 6\Gamma_b + 3.9b_1{}^2\Gamma_c - (3/5)\, b_1{}^2\, (1\text{-}2\lambda)(\, \kappa_{p0} + \theta_\gamma)\Gamma_a = 0; \tag{2-52}$$

$$2\, (\, \kappa_{p0} + \theta_\gamma)\, (S_\gamma \ + 2\, \Gamma_b) - (\theta_\gamma /2) + \ \Gamma_a [\, \tfrac{1}{2} + (6/5)\, b_1{}^2\, (1\text{-}\lambda)(\, \kappa_{p0} + \theta_\gamma)^2]$$
$$- (12/5\,)\, \Gamma_c (1+ \lambda)(\, \kappa_{p0} + \theta_\gamma)\, b_1{}^2 = 0; \tag{2-53}$$

$$\Gamma_a - \theta_\gamma(1{+}4\, \lambda) - 45\, S_\gamma \Gamma_a - (12/5)\, b_1{}^2\, \kappa_{p0}\, \Gamma_a^2 - 8\, \Gamma_a \Gamma_b - (24/5)\, b_1{}^2\, \Gamma_a \Gamma_c$$
$$- (12/5)\, \theta\gamma\, b_1{}^2\, \Gamma_a^2 = 0; \tag{2-54}$$

$$S_\gamma + 2\, \Gamma_b + (1/3)(\, \kappa_{p0} + \theta_\gamma)\Gamma_a b_1{}^2(1\text{-}3\lambda) - (2/3)\, b_1{}^2\Gamma_c = (2/3)\, k^0{}_Q\, . \tag{2-55}$$

The solution of these simultaneous equations relative to $k^0{}_Q$; Γ_c ; Γ_a ; Γ_b; S_γ yields the nonlinear equation (Y.M. Tseytlin *et al.*, 1968) for the average transformation ratio $i_{\theta av}$, in °/μm , for the flexible helicoids as follows $i_{\theta av}=|\theta_\gamma/2S_\gamma|$

$$= 0.0286\, (\kappa_{p0}{+}\theta_\gamma) \frac{E[1 + \theta_\gamma(\kappa_{p0}{+}\theta_\gamma)^{-1} - 3(\kappa_{p0}{+}\theta_\gamma) b_1^2 \theta_\gamma]}{4G\lambda[1 + \nu_{if}\, (\kappa_{p0}{+}\theta_\gamma)^2 b_1^2 \lambda^{-1} + 3 0_\gamma b_1^2 \lambda^{-1}(\kappa_{p0}{+}\theta_\gamma)/8]}, \tag{2-56}$$

where ν_{if} is the transformation coefficient of cross-section form that depends on the I_{qm} integrals over the cross-section area. Note that this transformation ratio is effective for the typical double elastic helicoids with length $2L$ which consists of two pretwisted beams 1 and 2 each of length L with the left and right helix, respectively, positioned on one axis and connected through the rigid rotatable body 4 (*see* Table 1-3, Fig. 2-3b, and Chaps. 4, 8).

We can find on the basis of (56) that the relationship between transformation ratios for the different forms of strip's cross section with different transformation coefficients ν_{if1} and ν_{if2} may be expressed as follows

$$\frac{i_{\theta av}(\nu_{if1})}{i_{\theta av}(\nu_{if2})} = \frac{\lambda b_1^{-2} + \nu_{if2}\, (k_{p0}{+}\theta_\gamma)^2 + 3\theta_\gamma (k_{p0}{+}\theta_\gamma)/8}{\lambda b_1^{-2} + \nu_{if1}\, (k_{p0}{+}\theta_\gamma)^2 + 3\theta_\gamma (k_{p0}{+}\theta_\gamma)/8}\ . \tag{2-56a}$$

Substitution of $\theta_\gamma - 0$ and $\chi_{kp} = \kappa_{p0}{}^2\, b_1{}^2 \lambda^1$ into (56) yields the formula of transformation ratio (Y.M. Tseytlin, 1964c, 1965) for the linear pretwisted strip system with the different cross sections

$$i_\theta = 0.0286\, \kappa_{p0}\, E / [4G\lambda\, (1 + \nu_{if}\, \chi_{kp})]. \tag{2-56b}$$

Let us show that (56) can be expressed as a cubic equation relative to the angle θ_γ, (in rad / mm) with subsequent notations:

$$A_3\theta_\gamma^3 + A_2\theta_\gamma^2 + A_1\theta_\gamma + A_0 = 0, \tag{2-57}$$

where $A_0 = -\,0.00715(ES_\gamma / G\lambda)\kappa_{p0}$;
$A_1 = 2[14.325\, \nu_{if}\chi_{kp} - 0.00715\,(ES_\gamma / G\lambda) + 14.325] + 0.02145\, E\, \chi_{kp}\, ES_\gamma / G$;
$A_2 = 28.65\, \chi_{kp}[(2\nu_{if} + 0.375) + 0.0015\, ES_\gamma\, / G]\, \kappa_{p0}^{-1}$;
$A_3 = ½\, b_1^2[57.3\,(\nu_{if} + 0.375\,) + 0.0429\, ES_\gamma\, / G]\, \lambda^{-1}$.
The instantaneous transformation ratio can be expressed as a partial derivative of untwist function θ_γ with respect to stretch of the strip S_γ per its unit length

$$\partial\theta_\gamma^\circ / 2\partial S_\gamma = i_\theta = -1.229\,(E/G)(b_1^2/4\lambda)\frac{2\theta_\gamma^3 + 4\kappa_{po}\theta_\gamma^2 + [2\kappa_{po}^2 - (4/3)b_1^{-2}]\theta_\gamma - (2/3)b_1^{-2}\kappa_{po}}{3\theta_\gamma^2 A_3 + 2\theta_\gamma A_2 + A_1},$$

where S_γ is in μm/mm. (2-58)

It is also possible to obtain a nonlinear equation for untwist angle of the pretwisted beam if in (30) one substitutes the coefficients of compliance with $\vartheta_0 = \kappa_{p0}$ by $\vartheta_0 = (\kappa_{p0} + \tau)$. In this case the corresponding cubic equation will be (*see* B.F.Shorr, 1968) as follows

$$\vartheta^3 + 3\vartheta^2\beta_\rho + 2(\,1 + \mu_Q + \beta_\rho^{\,2})\vartheta = 2\,(\vartheta_m - \mu_Q\beta_\rho), \tag{2-59}$$

where $\mu_Q = (1/\,GT_o)\,[(J_p\,/A)Q_\zeta - (J_{p\eta}\,/\,J_\eta)M_\eta + (J_{p\xi}\,/\,J_\xi)M_\xi\,]$;
$\beta_\rho^{\,2} = 2(1 + \nu_p)\,\kappa_{p0}^{\,2}\,J_r /\,T_0;\vartheta_m = (\beta_\rho/\kappa_{p0})\,M_\zeta/G\,T_0\,;\ \vartheta = (\beta_\rho/\kappa_{p0})\tau;$ and $\tau = \theta'$.
Internal moments M_η, M_ξ, M_ζ ,and force Q_ζ correspond to the matrix (25).
Equation (59) is a cubic one, as is (57). However, (57) directly corresponds to the cubical kinematical relationship between the angle of rotation and elongation of the stretching beam, unlike (59) where the coefficients contain the variable force expressions. Analysis shows that (59) can have one root at $\mu_Q << 1$ when its discriminant $\Delta = \beta_\rho^{\,2} + (2 - \beta_\rho^{\,2})^3 / 27 > 0$; it can also have three roots at $\beta_\rho^{\,2} > 7.2$ which leads to $\Delta < 0$. The latter condition still corresponds to the moderate pretwisted beam field because $\beta_o^{\,2} <<1$. Note that (57) has only one value for a certain angle of rotation and correspondingly yields an unambiguous kinematic solution. The members with ϑ^3 and ϑ^2 in (59) can be omitted if β_o, β_ρ , and ϑ are small. In this case $\vartheta = (\vartheta_m - \mu_Q\beta_\rho)\,/\,(1 + \mu_Q + \beta_\rho^{\,2})$,and the angle of rotation with the average stress $\sigma_{av} = Q_\zeta / A$ at a pure extension is equal to:

$$\theta = G^{-1}T_0^{-1}[1 + (Q_\zeta A^{-1}J_p/GT_0) + \beta_\rho^2]^{-1} LQ_\zeta J_p k_{p0}/A\ , \tag{2-60}$$

where L is the length of the pretwisted beam and A is the area of its cross section. This equation still shows the nonlinear relationship between the angle of rotation and the stretching force. However, this nonlinearity is small at a small initial extension.

The torsional rigidity at the pretwisted beam extension $j_\theta = Q_\zeta/\theta^\circ$ in cN/(angle degree) for a linear system may also be expressed (Y.M. Tseytlin, 1965) through the coefficient of form rigidity ν_{jf} analogous to the transformation ratio i_θ presentation in (56b) as follows

$$j_\theta = 69.6\; GA\lambda\,(1 + \nu_{jf}\,\chi_{kp})\,/\,L\kappa_{p0}\,. \tag{2-61}$$

We estimate the values of form coefficients ν_{if} and ν_{jf} for the cross sections of pretwisted beam with different shapes in the following section.

2-3. Motion Transformation in a Thin Helicoidal Strip

Structure and shape. The greatest motion transformation in the helicoidal strip occurs between the rotation of its "output"section and the strip's longitudinal extension, or at the change of distance between the strip's ends. The thin helicoidal strip practically does not have transverse rigidity. Therefore it cannot be used in the cantilever version. It usually consists of two sequentially connected pretwisted strip segments 1, 2 (Fig. 2-3b) with opposite direction of helix lines and length L each (double flexible helicoid), flat crosspiece 4 between them, and flat segments 3, 5 on both ends. End 3 is connected to the input lever 6 in point A. One of the pretwisted segments has left-handed helix lines and another one has right-handed helix lines. Crosspiece element 4 is the rotational output of the strip's mechanism to which a pointer, mirror, or the secondary transducer can be connected. The ratio between the width $b = 2b_1$ and thickness $h = 2h_1$ of the strip for precision mechanisms and transducers is usually within $\Lambda_\lambda = 1/(\sqrt{\lambda}) = 15$ to 25, and the thickness h is in the range 0.002 to 0.02 mm. However, the functional relations between these geometric parameters, material characteristics, pretwisted helix parameter $\kappa_p = 2\pi/ S_o$ with pitch S_o, and the kinematic and rigidity features of a flexible thin helicoid are complicated and will be studied in this section on the basis of above discussed theories (*see* Section 2-2). The double helicoidal strips are usually formatted from longitudinally rolled flat strips by the plastic twist through its crosspiece middle section rotation. This process will also be described at the end of this section.

The cross-section form of such strips depends on the method of their manufacturing. These forms include: ellipse (Fig. 2-3c) at the rolling of round wires by the cylindrical rolls, oval with a waist (Fig. 2-3d) at the rolling of a thin wire by the balls, and a rectangular one (Fig. 2-3e) at the cutting of strips from rolled wide thin sheets. The punched cross section (punched, perforated strip) is also available (Fig. 2-3f). The small accurate holes in the narrow and thin strip can be provided with the laser ray. The geometric features of elliptical and rectangular contour are widely known. The oval with waist and punched strip geometric features are more complicated and may be expressed with the quartic equation of the fourth degree similar to oval of Cassini:

$(y^2 / b_1^2) + (z^2 / h_c^2) + Ç\,(h_c^{-2} - b_1^{-2})\,(y^2 - z^2) - Ç\,(y^4 - 6y^2z^2 + z^4)/(b_1^2 h_c^2) = 1 - Ç$.

Here $Ç$ is a constant defined under the condition

$h_1^2 /(b_1^2 + h_c^2) > Ç > -\tfrac{1}{2}(\sqrt{2} - 1)$. For a punched section (punched or perforated strip), there should be an additional condition, $h_c^2 < 0$. Let us recall that Cassini oval is an anallagmatic curve, which is invariant under inversion. Such oval contour for the simplification of calculations may also be approximated with the arcs of a circle and segments of straight lines. The equation of the first arc with radius r_1 (Fig. 2-3d) can be expressed as:

$\xi_{(1)} = [-\eta^2 + 2\eta q_1 + (r_1^2 - q_1^2)]^{0.5}$. The equation of the second arc r_2 is as follows

$\eta_{(2)} = [\xi^2 + 2\xi U_q + (r_2^2 - U_q^2)]^{0.5}$. The I_{qm} integrals over oval with waist cross sections approximated by arcs and segments of straight lines converge with Abel's integrals.

Materials. The materials used for the strips under consideration are shown in Table 1-14.

Form coefficients.

Let us evaluate the coefficients of form for rigidity ν_{jf} and for transformation ν_{if} of different cross-section profiles. The rigidity of the pretwisted beam under extension related to the untwist angle, on the basis of (60), is equal to:

$j_\theta = Q_\zeta / \theta = GT_0 A\ [1 + (\sigma_{av}\, J_p / GT_0) + \beta_\rho^2] / (L\, J_p\, \kappa_{p0}).$ (2-62)

Rigidity. If the strip's cross section is an ellipse with a ratio of thickness $h=2h_1$ to width $b=2b_1$ that equals $\sqrt{\lambda}\ = h/b << 1$, then

$J_p / T_0 = (1+\lambda)^2 / 4\lambda;$

$J_r / T_0 = (b_1^2/16)(1\text{-}2/3\lambda + \lambda^2)\ (1+\lambda)/\ \lambda \approx b_1^2/16\lambda;$ and

$\beta_\rho^2 = 2(1+\nu_p)\ \kappa_{p0}^2\ J_r / T_o \approx (1+\nu_p)\ \kappa_{p0}^2\ b_1^2/8\lambda.$

The function J_r for the elliptical cross section is equal to:

$J_r = \pi b^6(\sqrt{\lambda})\ (1\text{-}2/3\lambda + \lambda^2)/1024 \approx \pi b^6(\sqrt{\lambda})/1024.$ Substitution of these expressions into (62) yields

$j_\theta \approx 4G\lambda A\ \{[1 + \sigma_{av}(1+\lambda)^2 / 4G\lambda] + [(1+\nu_p)\ \kappa_{p0}^2\ b_1^2/8\lambda\ ]\} / [L\kappa_{p0}(1+\lambda)^2].$

Hence, at $\sigma_{av} \to 0$ at the beginning of the stretch of the pretwisted strip with an elliptical cross section and the expression of j_θ in cN/o, we finally have an equation similar to (61):

$j_{\theta 0} \approx 69.6\ G\ A\ \lambda\ (1 + \nu_{jf}\,\chi_{kp}) / L\kappa_{p0},,$ (2-63)

where $\nu_{jf} = \nu_{jfel} = (1+\nu_p)/\ 8$; area $A = \pi b_1 h_1$; the λ ratio usually corresponds to $\lambda^{0.5} \leq 1/15\ldots1/25$, and if $\nu_p = 0.333$, as for many metals, $\nu_{jfel} \approx 0.17$.

If the strip's cross section is a rectangular one with similar geometric characteristics, then $T_0 / J_p \approx 4\lambda/(1+\lambda)$; $J_r = 0.00555 b^6(\sqrt{\lambda})/[1+1.067\ (0.5)^2]$;

$J_r / T_o = 0.0525 b_1^2/\lambda$; $\beta_\rho^2 = 0.1052(1+\nu_p)\kappa_{p0}^2\ b_1^2/\lambda.$; and area $A = 4b_1h_1$.

In this case, at $\sigma_{av} \to 0$, the rigidity form coefficient for the rectangular cross section of the helicoid strip extension is $\nu_{jf} = \nu_{jfrc} = 0.1052(1+\nu_p)$, and if $\nu_p = 0.333$, then $\nu_{jfrc} = 0.14$; at $\nu_p = 0.279$ for the zinc bronze (*see* Tables 1-14) $\nu_{jfrc} \approx 0.135$.

Transformation. The pretwisted strip with a thin rectangular cross section and constant thickness h has compliance which for the "following" coordinates with directions ξ, η, ζ, and ϑ are shown (B.F. Shorr, 1968) in Table 2-5.

Table 2-5. The compliance ν_{ij} of the pretwisted strips with rectangular cross section

	ζ	ξ	η	ϑ
ζ	$\frac{1}{EA}\left[1+\frac{0.0417\,(1+\nu_p)\,b^4\vartheta_0^2}{h^2+0.0263\ (1+\nu_p)\,b^4\vartheta_0^2}\right]$	$\frac{1}{GT_0}\frac{0.0111\,b^4\theta_0^2 h}{1.267\,h^2+0.0333\,(1+\nu_p)\,b^4\vartheta_0^2}$	0	$-\frac{0.0833\,b^2\vartheta_0\,h^2}{GT_0[h^2+0.0263\ (1+\nu_p)\,b^4\vartheta_0^2]}$
ξ		$\frac{1}{EJ_\xi}\left[1+\frac{0.0089\,(1+\nu_p)\,b^4\vartheta_0^2}{1.267\,h^2+0.0333\,b^4\vartheta_0^2}\right]$	0	$\frac{1}{GT_0}\frac{0.133\,b^2\vartheta_0 h}{1.267\,h^2+0.0333\,(1+\nu_p)\,b^4\vartheta_0^2}$
η	0	0	$\frac{1}{EJ_\eta}$	0
ϑ		$\frac{1}{EJ_\xi}\left[1+\frac{0.0089\,(1+\nu_p)\,b^4\vartheta_0^2}{1.267\,h^2+0.0333\,b^4\vartheta_0^2}\right]$	0	$\frac{1}{GT_0}\frac{h^2}{[h^2+0.0263\ (1+\nu_p)\,b^4\vartheta_0^2]}$

Note: $i=\zeta,, \eta,\xi, \vartheta$; $j=\zeta,, \eta,\xi, \vartheta$; Cell $\zeta\zeta$ contains expression for ν_ζ; cell $\xi\zeta$ contains $\nu_{\xi\zeta}$ etc. $\nu_{ij} = \nu_{ji}$.

The corresponding compliances ν_x, $\nu_{x\theta}$, and ν_θ in Cartesian coordinates have direct correlation with compliances in the "following" coordinates:

$\nu_x = \nu_\varsigma =$

$(EA)^{-1}$ $\{1 + [(0.0417(1 + \nu_p)b^4 \vartheta_o^2) / (h^2 + 0.0263 (1 + \nu_p) b^4 \vartheta_o^2)] \}$;

$\nu_{\theta x} = \nu_{\varsigma\vartheta} = - (GT_o)^{-1}$ $[0.0833\, b^2\, \vartheta_o\, h^2 / (h^2 + 0.0263 (1 + \nu_p) b^4 \vartheta_o^2)]$;

$\nu_\theta = \nu_\vartheta = (GT_o)^{-1}$ $\{ h^2 / [h^2 + 0.0263 (1 + \nu_p) b^4 \vartheta_o^2]\}$.

Matrix equation (26) for the pretwisted beam under the pure extension in this case yields expressions for the derivatives of displacements $S'_x = S_\gamma$ for extension and $\theta'_x = \theta_\gamma$ for untwist , as follows

$$S_\gamma = \nu_x Q = E^{-1}A^{-1}\left(1 + \frac{0.0417(1+\nu_p)b^4\kappa_{po}^2}{h^2 + 0.0263(1+\nu_p)b^4\kappa_{po}^2}\right)Q; \tag{2-64}$$

$$\theta_\gamma = \nu_{\theta x} Q = -\frac{0.08333\, b^2 h^2\, \kappa_{po}}{GT_0[h^2 + 0.0263(1+\nu_p)\, b^4\, \kappa_{po}^2]}Q, \tag{2-65}$$

where $T_0 = bh^3/3$ for very thin rectangle ($b/h > 10$).

Hence, the transformation ratio in (angle degree) / μm for the pretwisted strip in liner deformation state (remember that the thin strip is a double one with left-hand and right-hand helix segments) with a rectangular cross section in accordance with (64) and (65) after elementary cancellations is equal to:

$$i_\theta = 57.295\cdot 10^{-3}\theta_\gamma / 2S_\gamma =$$

$$-\frac{57.295\cdot 3(0.0833)E\kappa_{po}}{2G\lambda[1 + 0.272(1+\nu_p)b_1^2\kappa_{po}^2/\lambda]1000} = -\frac{0.0286\, E\,\kappa_{po}}{4G\lambda(1+\nu_{ifrc}\chi_{kp})}, \tag{2-66}$$

where the kinematic transmission form coefficient $\nu_{ifrc} = 0.272\ (1 + \nu_p)$; and, for example, at $\nu_p = 0.279$ for the zinc bronze (see Table 1-14), $\nu_{ifrc} \approx 0.345 \approx 29/84$. (66) is the same expression which was presented in Y.M. Tseytlin (1964c and 1965) publications. We can also find the transformation form coefficient ν_{ifel} for an elliptical cross section, where

$$\nu_\varepsilon = \frac{4}{E\pi b^2\sqrt{\lambda}}\{1 + \frac{(1+\nu_p)\beta_m^2(1-\lambda)^4}{8\lambda(1+\lambda)}[1 + \frac{(1+\nu_p)\beta_m^2(1-\lambda)^2}{8\lambda(1+\lambda)}(1 + 2/3\,\lambda + \lambda^2)]^{-1}\}$$

$$\nu_{\varepsilon\theta} = -\frac{2\beta_m(1-\lambda)^2}{G\pi\, b^3\lambda^{3/2}}[1 + \frac{(1+\nu_p)\beta_m^2(1-\lambda)^2}{8\lambda(1+\lambda)}(1 + 2/3\,\lambda + \lambda^2)\,]^{-1}; \quad i_\theta = \nu_{\varepsilon\theta}/\,2\,\nu_\varepsilon,$$

and i_θ, in angular degree/μm, similar to (66) after certain substitutions and cancellations is equal to: $i_\theta = -\dfrac{0.0286\, E\kappa_{po}}{4G\lambda\,(1+\nu_{ifel}\,\chi_{kp})}$,where the form transformation coefficient for the elliptical cross section is equal to $\nu_{ifel} \approx (1 + \nu_p)/\,4$; and at $\nu_p = 0.250$--0.4, as for many materials, $\nu_{ifep} = 0.32 - 0.35$.

More on rigidity. The expression for the extension rigidity of a pretwisted strip with a rectangular cross section related to the untwist angle (in cN/ angular degree) can be found by using (65) as well. In this case,

$j_\theta = (1000/57.295)Q/(\theta_\gamma L) = \dfrac{17.453\,GT_o\,[h^2+0.0263(1+\nu_p)b^4\kappa_{po}^2]}{0.0833b^2h^2\kappa_{po}}$, and after elementary cancellations and substitutions, we finally have the same expression as in (63):

$$j_\theta = 69.6GA\lambda\,(1 + \nu_{jfrc}\,\chi_{kp})/\,L\kappa_{p0}. \tag{2-67}$$

For the cross section of the oval with waist, the rigidity form coefficient ν_{jfoc} is theoretically and practically close to the one for the elliptical cross section, but its area is close to the rectangular cross section. This is because the oval with a waist cross section of thin strips practically has a very shallow slope (less than 0.0025) on the long side and rounding with the radius equal to half of the thickness on the short side. The depth of a "crater" on the long side of such a cross section is not more than 3-5 interference fringes, or less than 1.5 μm (60 μin). We recall that the rigidity coefficient for the pretwisted strip with elliptical cross section is $\nu_{jfel} \approx 0.17$. Therefore, we accept that ν_{jfoc}=5/28≈ 0.178 which value corresponds to experimental data of the author.

The longitudinal rigidity of the pretwisted strip slightly increases after its comparatively large untwist $\theta°$ when the actual helix pitch
$S_0 = 360L\,/\,[360(L/\,S_0) - \theta°]$ increases and pretwist helix parameter value $k_p = 2\pi/\,S_0$ correspondingly decreases in comparison with its initial value k_{p0}.
Dangerous points. The stress state in the most critical of the elliptical cross-section points A_o and B_o (Fig. 2-3a) may be approximately presented by the following expressions (*see* Y.M. Tseytlin, 1972) at the absence of an external twisting moment:

$$\sigma_{QAo}^{\;11} \approx \sigma_{QAo}^{\;22} \approx -\,(9/112)\,\chi_{kp}\,C_\sigma^{\;-1}\,\sigma^{33}{}_0;\;\; \sigma_{QAo}^{\;12} = 0; \tag{2-68}$$

$$\sigma_{QAo}^{\;13} = (h_1\kappa_{p0}\,/\,2\lambda)\,C_\sigma^{\;-1}\,\sigma^{33}{}_0;\;\; \sigma_{QAo}^{\;23} = 0; \tag{2-69}$$

$$\sigma_{QAo}^{\;33} = [1 + (69\,/224)\,\chi_{kp}\,]\,C_\sigma^{\;-1}\,\sigma^{33}{}_0\,, \tag{2-70}$$

where $C_\sigma = 1 + 5/28\,\chi_{kp}$; $\sigma^{33}{}_0 = F_Q/\,A$, and $\chi_{kp} = b_1^{\;2}\,\kappa_{p0}^{\;2}\,/\,\lambda$.
This system may be represented by the tensor

$$\hat{\sigma}_{2T} = \begin{bmatrix} \sigma_{QA0}^{11} & 0 & \sigma_{QA0}^{13} \\ 0 & \sigma_{QA0}^{22} & 0 \\ \sigma_{QA0}^{31} & 0 & \sigma_{QA0}^{33} \end{bmatrix}. \tag{2-71}$$

The component σ_{QA0}^{22} of tensor $\hat{\sigma}_{2T}$ (as follows from the tensor's view) is one of the three principal components of the stress tensor: σ_{qAo}; σ_{iAo}; and σ_{jAo}. If $\sigma_{QA0}^{22} = \sigma_{qAo}$, the two other principal components are defined with the known formula

$$\sigma_{i,j} = ½\,(\sigma^{11} + \sigma^{33}\,) \pm [(1/4)\,(\sigma^{11} - \sigma^{33}\,)^2 + (\sigma^{13}\,)^2\;]^{0.5}. \tag{2-72}$$

Hence, after substituting into (72) corresponding components and eliminating small in comparison with one members, we will have

$$\sigma_{iAo} = -\,(16/224)\,\chi_{kp}\,C_\sigma^{\;-1}\,\sigma^{33}{}_0; \tag{2-73}$$

$$\sigma_{jAo} = [1 + (67/224)\,\chi_{kp}]\,C_\sigma^{\;-1}\,\sigma^{33}{}_0. \tag{2-74}$$

We can conclude that on the basis of the comparison values σ_i; σ_j and σ_q, the following sequence $\sigma_j = \sigma_1 > \sigma_i = \sigma_2 > \sigma_q = \sigma_3$ appears, and therefore the stress status in point A_o is characterized by the expression

$$\sigma_{1Ao} - \sigma_{3Ao} = [1 + (85/224)\,\chi_{kp}\,]\, C_\sigma^{-1}\, \sigma^{33}{}_0. \tag{2-75}$$

In point B_0 $\eta = 0$; $\xi = b_1$. Then the stress tensor in this point will have the following components:

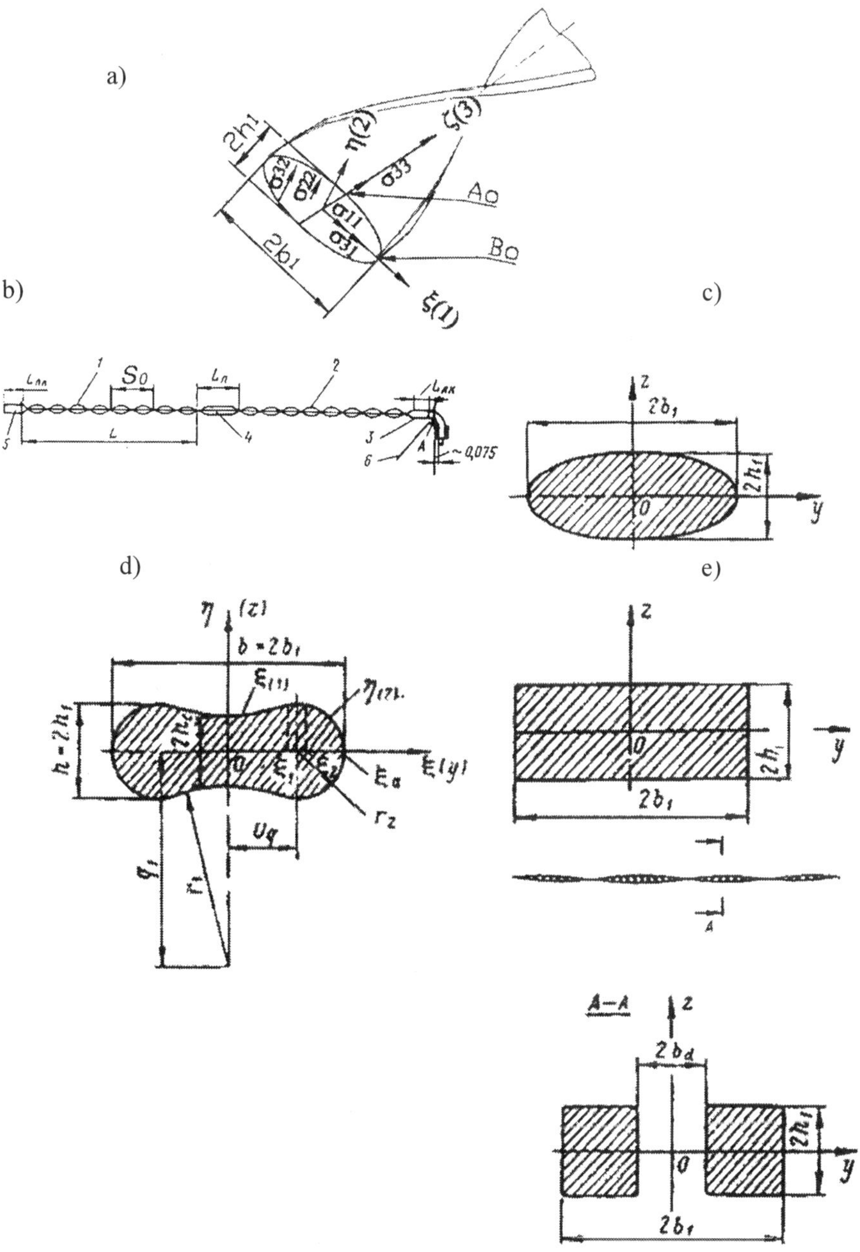

Fig. 2-3. Pretwisted thin strip, its typical stress components, structure, and cross sections.

$\sigma_{qBo} = \sigma_{QBo}{}^{11} = (19/112)\, \chi_{kp}\, C_\sigma{}^{-1}\, \sigma^{33}{}_0;$ (2-76)

$\sigma_{iBo} = (6/56)\, \chi_{kp}\, C_\sigma{}^{-1}\, \sigma^{33}{}_0;$ (2-77)

$\sigma_{jBo} = [1 + (69/224)\, \chi_{kp}]\, C_\sigma{}^{-1}\, \sigma^{33}{}_0.$ (2-78)

We can conclude that on the basis of comparison of values σ_i; σ_j and σ_q, the following sequence $\sigma_j = \sigma_1 > \sigma_i = \sigma_2 > \sigma_q = \sigma_3$ appears, and the stress status in point B_o is characterized by expression

$\sigma_{1Bo} - \sigma_{3Bo} = [1 + (63/224)\, \chi_{kp}\,]\, C_\sigma{}^{-1}\, \sigma^{33}{}_0.$ (2-79)

Hence, the stress state in point A_o is more dangerous because

$(\sigma_{1Ao} - \sigma_{3Ao}) > (\sigma_{1Bo} - \sigma_{3Bo})$.

Effective cross section. Derived for the elliptical cross section conclusion is effective for a rectangular cross section as well. However, the most critical point (with the largest tangential stress) in the cross section of the oval with waist type (Fig. 2-3d) under torsion is not in the middle of a long side. This point in accordance with B. Saint -Venant theory (*see* Saint-Venant, 1858) may be split to two and positioned at the distance of ~ ½ b_1 from the center of cross-section profile that distances it from the center fiber with the largest normal stess. The reader can find more detailed investigation of this problem in L.N.G. Filon (1900) publication.

Pretwist limitation. The pretwisted strips for double flexible helicoids are produced, as mentioned above, from the flat blank strips by the elastic-plastic twist. This twisting is usually provided in a cold state at room temperatures with sequential low annealing (temperature approximately 170°C), which is simpler than the twisting of hot strips. In the latter case, special measures are required to prevent the strip's oxidation. However, the process of cold twisting requires knowledge of the relationship between the necessary number $n_{v1}$of turns of the middle crosspiece link during the manufacturing procedure, and the number of turns n_{vo} that remains on the pretwisted sections of the flexible double helicoid after its release from the fixture. This process occurs in the following manner.

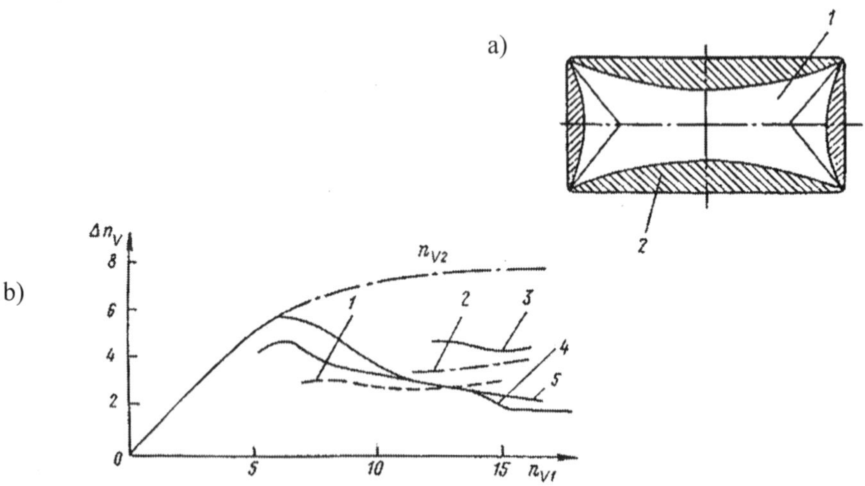

Fig. 2-4. Thin strip cold pretwist's features.

At the first several turns, elastic deformation 1 (Fig. 2-4a) takes place in the strip's body. Then, the material on the surface begins to convert to a state of plasticity. After that, plastic deformation 2 gradually penetrates deeper inside the cross section.

And finally, the whole strip's material converts to the plastic state, and further cold twisting becomes impossible. If the strip after elastic-plastic deformation is released, then the elastic part of deformation will disappear, i.e., the strip will leave out part of the turns n_{v2} corresponding to the ratio between the excluded torque, supplied by the fixture, and the torsion rigidity of the strip. At excluding simultaneously the longitudinal force Q_0 that was supplied by the fixture to both ends of the strip during the twisting procedure, the pretwisted strip will shrink in length and the middle crosspiece link will rotate on n_{v3} turns in the direction that increases the strip's twist. Therefore, the remaining twist deformation of the strip can be defined by the superposition of the above described partial deformations

$$n_{v0} = n_{v1} - n_{v2} + n_{v3}. \quad (2\text{-}80)$$

The proper adjustment of the twisting procedure usually requires finding the necessary number of turns n_{v1} supplied by the fixture to obtain the specified remaining turns n_{vo} on the strip. Partial deformations n_{v2} and n_{v3} for that purpose should be expressed through the strip's parameters, where

$$n_{v3} = Q_0 / j_\theta = Q_0\, n_{vo} / \{4G\lambda A[1 + 5/7(b_1^{\,2}\pi^2\, n^2_{v0}) / (\lambda L^2)]\}\ ; \quad (2\text{-}81)$$

and the elastic untwist n_{v2} in accordance with the definition equals

$$n_{v2} = LM_{tq} / (2\pi T_{0H}\, G), \quad (2\text{-}82)$$

where $T_{0H}\, G \approx (16/3)\, b_1 h_1^{\,3} G$ is the torsion rigidity of thin strip with the rectangular cross section. The torque at the elastic-plastic deformation of strip (*see* Y.M. Tseytlin, 1964e) in this case is equal to:

$$M_{tq} = \tau_T\, \{(2h_1)^2 - (1/3)\,[\tau_T\, L / (2\pi G\, n_{v1})]^2\}\, b_1, \quad (2\text{-}83)$$

where $\tau_T = \{\sigma_T\}/\sqrt{3}$ is the tangent and $\{\sigma_T\} = \sigma_e$ is the normal stress at the yield stress limit of material. The substitution of the expressions for M_{tq} and $T_{0H}\, G$ into (82) yields: $n_{v2} = (3/8)\,[\tau_T\, L / (\pi G h_1)] - [(\tau_T L / \pi G h_1)^3 / (128\, n_{v1}^{\,2})]$. (2-84)
Then, substitution of expressions (81) and (84) into (80) finally yields the sought relationship between numbers n_{v1} and n_{v0}:

$$n_{v1}^3 - n_{v1}^2\left\{n_{v0} + \frac{3}{8}\frac{\tau_T L}{\pi G h_1} - \frac{n_{v0} Q_0}{4G\lambda A[1 + (5/7) b_1^2 \pi^2\, n_{v0}^2 / (\lambda L^2)]}\right\} + \frac{1}{128}\left(\frac{\tau_T L}{\pi G h_1}\right)^3 = 0. \quad (2\text{-}85)$$

This expression has been yielded by the author on the basis of general relations and can be used not only for strips, but also for the pretwisted thin beams with rectangular, elliptic, and oval cross sections. Expression (85) is the cubic equation in relation to the number of treatment turns, n_{v1}. It allows us to get the explicit solution relative to n_{v1}, but n_{v1} can also be calculated by the iteration method. The convenient iteration formula may be represented as follows:

$$n_{v1;\,k+1} = \sqrt[3]{B_n n_{v1;k}^2 - C_n}\ , \quad (2\text{-}86)$$

where

$$B_n = n_{v0} + (3/8)\frac{\tau_T L}{\pi G h_1} - \frac{n_{v0} Q_0}{4G\lambda A\,[1 + (5/7)\, b_1^2\, \pi^2\, n_{v0}^2 / (\lambda L^2)]}\ ; \qquad C_n = \frac{1}{128}\left(\frac{\tau_T L}{\pi G h_1}\right)^3\ ;$$

$n_{v1;k}$ is the k-th approximation of sought value n_{v1}, and $n_{v1;\,k+1}$ is the $(k + 1)$ approximation of value n_{v1}. The iteration process converges to the sought value n_{v1} with necessary precision on the third or fourth step of approximation at the successful zero-order approximation. It is also possible to find from (85) the necessary value of extension force Q_0 if the relationship $(n_{v1} - n_{v0}) = \Delta n_v$ is specified. The obtained value should be verified for the condition of strip's strength. It was assumed in the theoretical estimation that the elastic-plastic twist procedure has been provided under constant force Q_0. This indeed takes place in the special twisting fixture, which has the mechanism for the moving of strip, mechanism for the rotation of the middle crosspiece, and clamping device, which prevents the rotation of the strip's ends. Value Q_0 is supplied by special weights, which are connected to clamping devices. These devices are movable in the longitudinal direction of the strip and allow translational motion of the strip's ends at strip's twisting. Despite the simplifications used and the elementary superposition of deformations from twist and tension, (85) adequately reflects the links between n_{v1} and n_{v0}. This was proven by the experimental results. The crosspiece of each microstrip sample received n_{v1} turns in the fixture. After the sample's release, the number of remaining turns n_{vo} on the pretwisted segment was determined. The charts in Fig. 2-4b show calculated 1, 2 and experimental 4, 3 difference $\Delta n_v = (n_{v1} - n_{v0})$ in relation to n_{v1} for zinc bronze strips 0.008 x 0.12 and 0.006 x 0.1, respectively. The first number in the strip's features description shows the thickness in mm and the second is the width in mm of the strip's cross section. All of the calculations were produced with the iteration formula. The zero-order approximation in this case was accepted as $n_{v1,0} = n_{v0} + 3$. It is seen from the charts in Fig. 2-4b that the difference between the corresponding calculated and experimental values is not more than 15%, and at some points it is much smaller. Without superposition of the elastic self-twisting n_{v3} turns after the strip's release, typical for pretwisted strips, the calculated values $\Delta n_{v} = n_{v2}$ will be different from the experimental ones by more than 50% with the significant functional difference of the calculated and experimental relations. This equation allows us to make the following notes: with the increase of technological deformation n_{v1} under twist, value Δn_v after approaching the maximum (see curve 5) begins to decline; and for the strips with a smaller thickness, value Δn_v is larger. The ratio between the thickness and the width of the strip also has some influence on the relation between n_{v1} and n_{vo}. The greatest possible number n_{vo} on the strip with cold plastic twisting can be calculated using the formulae of ultimate torque for the "elastic-plastic joint" and the formulae for ultimate material deformations with the condition of the material's continuity. The ultimate torque can be determined from the following expression $M_{utq} + 3Q_o / \mu_a\psi_{utq} = [12\, b_1 h_1 \sigma_T /(\mu_a \psi_{utq})]\sqrt{1+\mu_a \psi_{utq}^2 h^2/9}$, where $\mu_a = 3$, and ψ_{utq} is defined from the transcendent equation $(1/6)\sqrt{\mu_a}\;\psi_{utq} h_1 = \sinh[\sqrt{\mu_a}\;\psi_{utq} Q_o / (12 b_1 \sigma_T)]$, σ_T is the normal stress after the yield strength limit of the material. Hence, the greatest untwist n_{v2} is equal to: $n_{v2lim} = M_{utq} L / (2\pi T_{0Q} G)$, where the geometric torsion rigidity for the rectangular cross section of the strip under tension with force Q_0 is equal to: $T_{0Q} = (16/3)\, b_1 h_1^3 + (1/3G)\, Q_0 b_1^2$. (2-87) Now it is possible to obtain the corresponding ultimate value for n_{v1} from (83) and, after substituting the latter into (85), to find the ultimate value $n_{vo,\,ult}$ of turns for the "elastic–plastic joint" action. However, the condition of continuity of the material requires the limitation of strain by specified value $\{\delta_{ed}\}$. The twisting of the strip with the free moving in axial direction ends follows by shortening the central fiber with compression stress σ_x and extending the outer fibers of the strip. The difference in length of these fibers in relation to their initial length should not exceed 2 $\{\delta_{ed}\}$:

$$2\,\{\delta_{ed}\} \geq (L_{out} - L_c)/\,L_c = \sqrt{b_1^2\kappa_p^2 + 1} - 1, \tag{2-88a}$$

and the corresponding limit of value $n_{v0,ult}$ is equal to:

$$n_{v0,ult} = L\,\sqrt{(2\,\{\delta_{ed}\} + 1)^2 - 1}\,/\,(2b_1\pi). \tag{2-88b}$$

Hence, the greatest achievable number of turns on the plane strip at its cold twist may be defined as the smaller one derived from the condition of material continuity and the ultimate torque of the "elastic-plastic joint". The results of the experimental estimations are shown in Table 2-6, where $n_{v0,ult}$ are equal to the average data for these parameters, and the data for pitch $S_{o.ult}$ are equal to the minimal experimental values for different dimensions of the strips' cross section from tin-zinc bronze (4Tin, 3Zink). The length of all the samples was approximately equal to $2L + L_n$, where the length of each twisted segment $L = 18$ mm and the length of crosspiece link $L_n = 4$ mm. The applied extension force was equal to Q_0, whose values are also shown in the table.

Table 2-6. The ultimate number of turns and minimal pitch of the strip's cold twist

Cross-section of microstrip dimensions h x b, mm	Ultimate number of turns, $n_{v0,ult}$	Minimal pitch of cold twist, $S_{o,ult}$, mm	Extension force Q_0 at twisting, cN
0.008 × 0.12	16	1.1	10
0.006 × 0.10	18	1.0	5
0.007 × 020	14	1.3	10
0.004 × 0.08	18	1.0	2
0.006 × 0.122	12.5	1.4	5

Length of the strip is within 36-40 mm.

The twisting of a microstrip in the absence of constant external extension force (Q_0=0) and immovable fixed ends is non-rational. The type of relation between n_{v1} and Δn_v in this case is changed because at the immovable ends the central fiber of the strip cannot become shorter; as a result, the condition of material continuity becomes more limited:

$$\{\delta_{ed}\} \geq = \sqrt{b_1^2\kappa_p^2 + 1} - 1. \tag{2-88c}$$

Hence, the ultimate number of turns with cold twist of the strip will be smaller:

$$n^1_{v0,ult} = \mathrm{L}\,\sqrt{(\{\delta_{ed}\} + 1)^2 - 1}\;/\,(2\mathrm{b}_1\pi), \text{ and } n^1_{vo,ult} < n_{vo,ult}.$$

The contraction $\Delta\varepsilon^{33}_m$ of the longitudinal axial filament of the microstrip at the elastic twist of angle θ can be estimated by the following approximate expression

$$\Delta\varepsilon^{33}_m \approx \frac{2h}{A}\int_0^{b_1}\left(\sqrt{b_{1y}^2(\theta/L)+1} - 1\right)db_{1y} - 0.33b_1^3(h/A)(\theta/L)^2 .$$

Parameters deflection. If the non-uniformity is a small and random one (with normal distribution as the result of many factors' influence), then the motion transformation in the group of pretwisted strips with the same nominal parameters may be presented in the following form:

$$\theta = \theta_\gamma L \pm \tfrac{1}{2}\sqrt{\sum_k (\Delta\theta_k L_k)^2}\,, \tag{2-89}$$

where $\Delta\theta_k$ is the range of possible deflection of the untwist angle on each k-th segment of the pretwisted strip; L_k is the length of the k-th strip's segment; $\Delta\kappa_p$; ΔE_k; ΔG_k; Δb_{1k}; Δh_{1k} are the corresponding deflections of the parameters on each segment L_k of the strip from the nominal value. The calculation and the manufacturing practice show that the random deflections of all parameters, which are in the range of 5 to 15% from nominal, change the transformation ratio by up to 5 to 10%. This influence can be seen in the data of Table 2-7. The strongest influence on the transformation ratio has the deflection of thickness because thickness has the smallest absolute value.

Table 2-7. The deflections of pretwisted strip parameters

Nominal value of parameter	Manufacturing deflection of parameters	Deflection of transmission ratio* Δi_θ, o/ μm from nominal value	
		Experimental	Calculated
$2h_1 = 0.008$ mm $2b_1 = 0.12$ mm	$\Delta h = \pm 0.009$ mm $\Delta b = \pm 0.01$ mm	+ 0.2	+ 0.28
$\kappa_p = \pi$ rad mm^{-1} $G = 44$ GPa	$\Delta\kappa_p = \pm 0.5$ rad mm^{-1} $\Delta G = \pm 3$ MPa	- 0.3	- 0.28

* The nominal transformation ratio $i_\theta = 3.4$ o/μm

Pendulum application. There were attempts to use the torsional pendulum method for identification of the flat blank strips, similar to that used successfully for the torsion suspension. However, this simple method is not representative for the pretwisted strips mechanisms.

Table 2-8. Period T_τ of the torsion pendulum oscillations with the strip suspension

Blank No.	Dimensions of microstrip in mm		m_w, gr	R_{cyl}, mm	T_τ, s	i_θ, °/μm
	$2b_1$	$2h_1$				
1	0.08	0.004	14	11	140	7.25
2	0.1	0.006	18	11	105	4.76
3	0.12	0.008	54	15	130	3.45
4	0.122	0.006	18	11	90	3.82
5	0.122	0.006	54	15	169	3.82

Table 2-8 data show that pretwisted strips (blank No 2 and 4) made from flat strips with very close periods have different transformation ratios at the pretwist parameter $k_p=\pi$ mm^{-1}. At the same time, the pretwisted strips with close transformation ratios were made from flat blanks No.3 and 4 with the different period values. Blank No 5 and No 4 with the same cross section dimensions have different periods in the pendulum with different mass m_w. Let us look at this problem in more detail. The oscillation period T_τ of the flat straight strip depends on its torsion rigidity $T_\tau = 2\pi\sqrt{I_w L_{pm}/(GJ_{tr})}$. The torsion rigidity J_{tr} of the strip with a rectangular cross section is equal to:

$$J_{tr} = [(16/3)\, b_1 h_1^3 + (1/3G) Q_a b_1^2]\, G = (16/3)\,)\, b_1 h_1^3 G + (1/3)\, Q_a b_1^2\,. \tag{2-90}$$

$$I_w = (1/2)\, m_w\, R_{cyl}^2,$$

where $Q_a = m_w g$ is the force of longitudinal tension that is produced by mass m_w of a cylindrical torsion pendulum with radius R_{cyl}, and L_{pm} is the length of the pendulum suspension. The comparison of (90) with any expression for the transmission ratio or the longitudinal rigidity of the pretwisted strip shows that the condition of T_τ = constant does not necessarily mean that i_θ = constant or j_θ = constant; and conversely, the condition of the constant working features does not require one certain value of torsion rigidity. This is evident because the pretwisted strip is characterized by the three different rigidity features (j_{kk}, $j_{\tau k}$, and $j_{\tau\tau}$). Moreover, there is no representation of the pretwist factor in (90). Hence, the identification of the blank strips to obtain the specified transformation ratio and longitudinal rigidity should be based on the measurement of the strip's dimensions by objective methods. Part of these parameters indeed depends on the specifics of the manufacturing procedure: constant thickness and variable width at the cutting of strips from the sheet with a certain thickness h, and constant relation λ at the rolling of wire by cylindrical or spherical rolls. It is impossible to draw conclusions about the quality of the blanks on the basis of the above described torsion pendulum. However, the torsion pendulum method is helpful in some cases. The following equation reflects the elliptical cross section parameters relations: $(4\lambda)^{-1} = (J_p / T_0) - 1/2$, where J_p is the polar moment of cross-section inertia; T_0 is the geometric torsion rigidity of a flat strip with the same cross section as for the pretwisted one. The transformation ratio in this case may be expressed as follows:

$$| i_\theta | = \frac{\kappa_p E}{G[(J_p/T_0)-1]+(29/21)(b_1^2\,\kappa_p^2 G)} \quad . \tag{2-91}$$

There are effective relationships for the relatively thick strips and beams ($\chi_{kp} < 1$) with any shape of cross section whose parameters may be defined with the torsional pendulums

$$| i_\theta | = \kappa_p\,(E/G)\,[1-(J_p/T_0)]; \qquad F_Q = EAS_\gamma + E(J_p - T_0)\,\kappa_p\,\theta_\gamma\,.$$

It is known from the theory of torsion pendulum oscillations that $M/\theta = I_w\,(2\pi / T_{\tau 1})^2$, where M is the torque, θ is the amplitude angle of rotation at oscillations, $I_w = ½\, m_w\, R_{cyl}^2$ is the dynamic moment of inertia of a cylindrical pendulum with radius R_{cyl}, mass m_w, and $T_{\tau 1}$ is the period of pendulum oscillations. As a result, $J_{tr} = M\,L_{pm} /(\theta G) = I_w\,(2\pi / T_{\tau 1})^2\,L_{pm} /G$, where L_{pm} is the length of the pendulum suspension. The rigidity at pure torsion GT_0 of the strip can be defined from the known J_{tr} in accordance with equation (90) after subtraction of a part, $(1/3)Q_a b_1^2$, caused by the pendulum mass action (longitudinal tension $Q_a = gm_w$). The best way to subtract is to use a pendulum with a small mass whose influence may be neglected. The value of G in suspension wire material is determined from the system's period $T_{\tau 2}$ of torsional oscillation as $G = 16\,(2\pi / T_{\tau 2})^2\,L_f m_w\,R^2_{cyl}/(\pi\,d^4)$. (2-92).

The estimation of polar moment J_p for any symmetric section can also be done with the torsion pendulum if necessary (*see* Y.M. Tseytlin, 1965).

Important second member. Equation (90) has two members, one of which is dependent on the strip shear modulus and another one is an independent of strip material elastic parameters, but dependent on the longitudinal force applied to the pendulum. It was shown that the latter member should be eliminated for the pretwisted strip's blank correct identification. This equation was published by the author of this book (Y. M. Tseytlin, 1965) with the purpose to explain the necessity of elimination the influence of the second member of the equation. This problem may even have negative influence in the new micromechanical system device with hanged mass (*see* P.J. Cumpson *et al.*, 2004) for calibration of AFM cantilever spring constants because the elastic element of the beam rotation is not round, unlike the carbon fiber with diameter 7 μm in their preceding model.

However, many years later the second member of equation similar to (90) was successfully used in novel torsion balance for the measurement of Newtonian gravitational constant (T.J. Quinn *et al.*, 1997). This example shows how many useful sides the nature of precision elasticity has when the used model is correct.

3.

Elastic Functions in Beams, Plates, and Shells by Force and Pressure

This chapter is dedicated to the analysis of the kinematic behavior and stress of prismatic beams, plates, diaphragms, bellows, shells and their possible nonlinearity. All of these elastic elements are usually relatively thin for larger possible deflections from the initial unloaded position and smaller stresses. Corresponding methods of effective structural synthesis are developed.

3-1. Geometric Nonlinearity of Thin Beam Motion

Thin flexible beam's features.

Thin beam. The flexible beam is called a thin one if the thickness h of its cross section is significantly smaller than its length L and radius ρ of curvature ($h / L < 0.1$; $h / \rho < 0.05$). It is possible at planar bending of this beams to neglect all the components of stress tensor σ_{ik} except the component $\sigma_x = E\varepsilon_x$ along the beam's axis. These beams can be straight or curved in their initial unloaded state. The thin rings also belong to this group. The flexible strips represent a version of thin beams with an elongated cross-section whose thickness h is significantly smaller than width b. The initially straight flexible strips in their unloaded state are often called leaf springs. Their useful applications for the guides and mechanisms are shown in Chap.4.

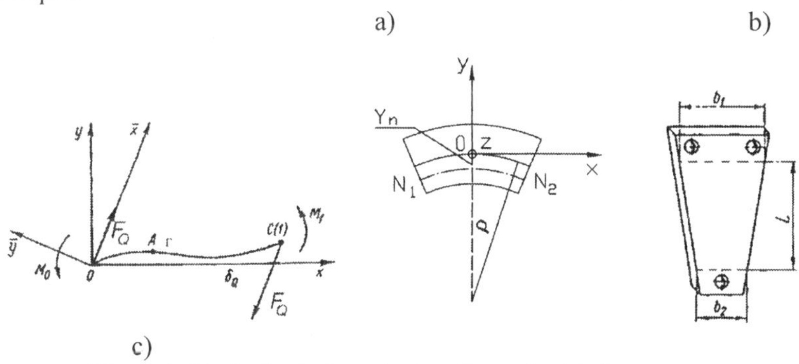

Fig. 3-1. Typical relations for bending of a thin beam

Strain. The relative elongation ε_x or strain of the bending beam's layer positioned at distance Y_n from the neutral layer is defined as a difference between the length of its element after deformation (Fig. 3-1a)

$\overline{N_1' N_2'} = dl\,(1 - Y_n / \rho)$ and before deformation $\overline{N_1 N_2} = dl\,(1 - Y_n / \rho_o)$, related to the length of the layer before deformation. This yields

$$\varepsilon_x = (\overline{N_1' N_2'} - \overline{N_1 N_2}) / \overline{N_1 N_2} = \frac{Y_n}{1 - Y_n/\rho_0}\left(\frac{1}{\rho} - \frac{1}{\rho_0}\right), \tag{3-1}$$

where *dl* is the length of an element in the neutral layer arc, ρ_o is the radius of initial curvature of the neutral layer before deformation, and ρ is the radius of that layer after deformation. Let us remember that the length of the neutral layer in the bending process is usually considered constant. If the fraction Y_n / ρ_o is significantly less than one, i.e., $(h/2\rho_o) \ll 1$, then at $(h/2\rho_o) \leq 1/20$ the value $(1-Y_n/\rho_0)^{-1} \approx 1.00$ with a high enough accuracy. This limitation at the longitudinal force $F_{Qx}=0$ yields a condition under which the neutral layer of the beam coincides with the center of gravity in its cross section, and the double integral $\iint_A \frac{y^2}{1-y/\rho_0} dA$ is equal to the section's moment of inertia J. As a result, the bending moment M is related to the change in the beam's curvature by the known expression

$$M = \left(\frac{1}{\rho} - \frac{1}{\rho_0}\right) EJ = \left(\frac{1}{\rho} - \frac{1}{\rho_0}\right) H, \tag{3-2}$$

that is an approximate and effective at the above mentioned limitations.
If at pure bending of initially straight beam $\rho = L/\theta$, then we have $M/\theta = EJ/L$.
Elemental rigidity. The product $EJ = H$ is called the elemental flexural rigidity of the cross section. However, H in some cases has a different value. In the case of the wide b >h strip, $H = JE/(1-\nu_p^2)$ as a result of two-dimensional strain, and for preliminary curved strips with a constant cross section is accepted that

$$H = \frac{EJ}{1-\nu_p^2}\left[1 - \frac{2\nu_p^2(\cosh A_\nu - \cos A_\nu)}{A_\nu(\sinh A_\nu - \sin A_\nu)}\right], \text{ where } A_\nu = 1.3\,(b/h)\sqrt{h/\rho_0}\,.$$

The section flexural rigidity for the plane strip with a linearly changing width from b_1 to b_2 (Fig. 3-1b) at transverse bending by force is considered conditionally equal to $H = E\,(18^{-1}b_1 + 36^{-1}b_2)\,h^3$.

Nonlinear bending kinematics.

Differential equation. Equation (3-2) after substitution of the approximate expression for the curvature of a flat curve in Cartesian coordinates (x, y) resembles the regular nonlinear differential equation (Table 1-10, No2) for the planar bending of a thin beam

$$\frac{M}{H} = \frac{y_x''}{[1+(y_x')^2]^{3/2}}\,. \tag{3-3}$$

The approximate solution for (2) and (3) was already derived by L. Euler in 1744. The known solutions for equation (3) at certain expressions of the ratio M/H are shown in Table 3-1. Special functions are required to obtain an accurate solution of these equations in some cases.

Exact solution. The accurate analytic and graph-analytic general solution of (2) was obtained with the elliptic integrals by A.N. Krylov (1931) and E.P. Popov (1948). The elliptic integrals of the first and second kind are expressed respectively as

Table 3-1. Solutions of nonlinear equation (3) for bending

No	The function M / H	Solution
1	a_{QM} = const	$(y - C_1)^2 + (x - C_2)^2 = a_{QM}^{-2}$
2	$2\,a_{QM}\,x$	$y - C_1 + \int \frac{a_{QM}x^2 + C_2}{\sqrt{1-(a_{QM}x^2+C_2)^2}}dx$
3	$a_{QM}\,y$	$y'_x = \frac{1}{a_{QM}y^2 - C}\sqrt{4-(a_{QM}y^2-C)^2}$
4	$2\,a_{QM}\,(y + b_p x + c)$	$u(x) = y + b_p x + c$; x is expressed through u with elliptic integrals

Note: C_1;, C_2, and C are constants of integration.

$$F(\psi, k) = \int_0^{\psi} (1 - k^2 \sin^2\varphi)^{-1/2}\, d\varphi \text{ and } E(\psi, k) = \int_0^{\psi} (1 - k^2 \sin^2\varphi)^{1/2}\, d\varphi.$$

Here ψ is amplitude and k is a module ($k < 1$). Popov's method is applicable to the planar bending of flexible thin beams and strips with a small rigidity under any large elastic displacement. However, some problems exist in its practical use for synthesis of precision elastic kinematical systems. This method is worked out in detail for the so-called problems of the fundamental class, to which are related the cases of bending a beam with the constant initial curvature $K_0 = 1 / \rho_o$ = constant. Hence, the initial axis of the beam is a straight line or an arc of a circle. In addition, the form and size of the cross section should be constant across the whole length of the beam (flexural rigidity $H(x)$ = constant), and the bending should occur under the action of concentrated force F_Q at angle δ_Q and moments M_o and M_1 applied to ends 0 and C(1) of the beam (Fig. 3-1c). The elastic line of the deformed beam can have points of inflection A_I or can be described by the equations, which allow the presence of the points of inflection (the so-called form of the inflectional type). In the opposite case, if the form of the elastic line cannot have points of inflection, it is called the form of the inflectionless (non-inflection) type. The fundamental class includes all cases of bending where the elastic curve may be divided into a finite number of segments, each of which belongs to the fundamental class conditions. Hence, the fundamental class includes problems with the step change of initial curvature, internal forces, moments, and rigidity. The problems, which do not belong to this class, include cases of straight or curved beam bending with a distributed load (including the uniform one), and an arbitrary changing initial curvature along the beam's length or its cross section rigidity. This classification shows that many problems with elastic strips in ELKS' calculation belong to the fundamental class problems whose solution we will examine in some detail.

Dimensionless elastic parameters. The initial differential equation is derived by substituting in (2) an accurate expression for the curvature of planar curve $1/\rho = d\vartheta / dl$, which yields $(d\vartheta / dl) - (d\vartheta_o / dl) = M / H,$ (3-4)

where ϑ and ϑ_o are the angles between the tangent to the elastic line and the initial curve, respectively; l is the length of the arc of the elastic line from the origin. The differential equation of the second order derived by differentiation of (4) with respect to l is thus equal to:

$$(d^2\vartheta / dl^2) - (d\vartheta_o^2 / dl^2) = dM / H\,dl = - (F_Q / H) \sin(\vartheta + \delta_Q), \qquad (3\text{-}5)$$

where δ_Q is an angle between coordinate axis x and the line of action of whole internal force F_Q in the section under consideration. The integration of (5) yields the system of simultaneous equations for the elastic line in the parametric representation:

$$x - x_o = (\bar{x} - \bar{x}_0)\cos\delta_Q + (\bar{y} - \bar{y}_0)\sin\delta_Q; \qquad (3\text{-}6a)$$

$$y - y_o = (\bar{y} - \bar{y}_0)\cos\delta_Q - (\bar{x} - \bar{x}_0)\sin\delta_Q, \qquad (3\text{-}6b)$$

where x; y and x_o; y_o are the coordinates of the current and initial points on the elastic curve in fixed Cartesian coordinates, respectively; $\bar{x}, \bar{y}$, and $\bar{x}_0, \bar{y}_0$ are the same coordinates but oriented in the direction of force whose origin coincides with the origin of the fixed coordinates; and axis $\bar{x}$ is parallel to the line of action of the acting force in the origin of the elastic line (point 0).

The forms of the inflectional type have in this case the following relations:

$$(\bar{x} - \bar{x}_0)/L = (2/\beta)\,[E(\psi; k) - E(\psi_o; k)] - (l/L); \qquad (3\text{-}7a)$$

$$(\bar{y} - \bar{y}_0)/L = (2/\beta)\,k\,(\cos\psi_o - \cos\psi), \qquad (3\text{-}7b)$$

and for the forms of the inflectionless type we can use the following expressions

$$(\bar{x} - \bar{x}_0)/L = \frac{2}{k\beta}[E(\psi; k) - E(\psi_o; k)] - [(2/k^2) - 1](l/L), \qquad (3\text{-}8a)$$

$$(\bar{y} - \bar{y}_0)/L = \frac{2}{k\beta}\,[\mathrm{dn}\,(\psi_o;k) - \mathrm{dn}\,(\psi;k)\,], \qquad (3\text{-}8b)$$

where L is a length of the thin beam under bending; $\beta = \sqrt{F_Q L^2/H}$ is the coefficient of force similarity, $E(\psi; k)$ and $E(\psi_o; k)$ are the elliptic Legendre's integrals of the second kind for the current and initial points on the elastic line, respectively; k and ψ are the module and elliptic amplitude (the arguments of elliptic integrals) for the current point, ψ_o is the elliptic amplitude for the initial point on the elastic line; and $\mathrm{dn}\,(\psi;k); = \sqrt{1-k^2\sin^2\psi}$; $\mathrm{dn}\,(\psi_o;k) = \sqrt{1-k^2\sin^2\psi_0}$ are the delta-amplitudes for the current and initial points on the elastic line, respectively. In addition to the coefficient of force similarity, the coefficient of moment similarity $\omega_M = M/\sqrt{F_Q H}$ and the angular coefficient of similarity $\zeta_\delta' = \vartheta + \delta_Q$ are involved.

The table value of argument ψ is noted as φ: At $2n90^\circ \le \psi \le (2n+1)\,90^\circ$ value $\psi = 180^\circ n + \varphi$; $0^\circ \le \varphi \le 90^\circ$; at $90^\circ(2n-1) \le \psi \le 2n90^\circ$ value $\psi = 180^\circ n - \varphi$; $90^\circ \ge \varphi \ge 0^\circ$.

Parametric equations. The equation of the same elastic line in the dimensionless coordinates for the solution of problems with Popov's elastic parameters has the following view

$$(x - x_o)/L = \beta^{-1}(\xi - \xi_o)\cos\delta_Q + \beta^{-1}(\eta - \eta_o)\sin\delta_Q, \qquad (3\text{-}9a)$$

$$(y - y_o)/L = \beta^{-1}(\eta - \eta_o)\cos\delta_Q - \beta^{-1}(\xi - \xi_o)\sin\delta_Q, \qquad (3\text{-}9b)$$

where ξ; ξ_o and η; η_o are the dimensionless coordinates of current and initial points on the elastic line, respectively, which are correspondent to dimensionless elastic parameters ξ'; η' or ξ''; η'' shown on the diagrams of elastic parameters (Fig. 3-2a).

They may also show the values of dimensionless parameters for force λ'; $\lambda''$and moment ω_M'; ω_M''.

a) b)

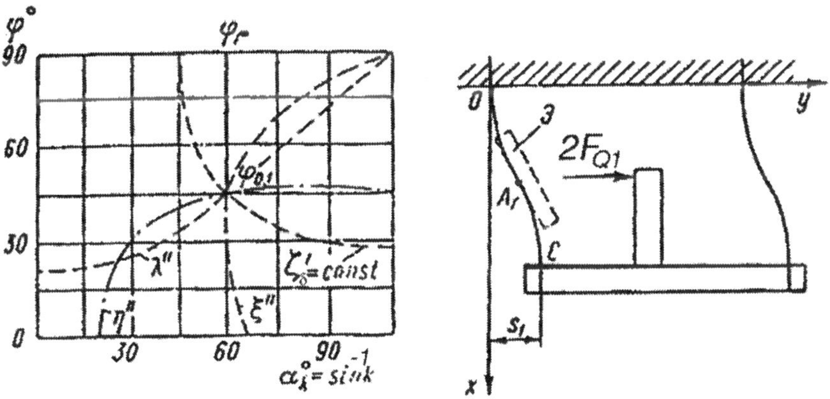

Fig. 3-2. The diagram of elastic parameters and the spring parallelogram model

Diagram of functional relationships. The equilibrium state of the elastic line on the diagram of elastic parameters (Fig 3-2a) is shown as a straight line with the argument k = constant. Ordinate φ_o corresponds to initial point O; ordinate φ_1 corresponds to the final point 1 of the elastic line; $\eta'' = \eta'_{90} - \eta'$; $\xi'' = \xi'_{90} - \xi'$; $\beta = \lambda'_1 - \lambda'_o$; $\lambda'' = \lambda'_{90} - \lambda'$. Index 90 corresponds to the parameters with ordinate $\varphi = 90^\circ$, coordinate φ_Γ corresponds to the inflection point.

The comparison of equations (6) through (9), which describe the same elastic line, shows that between dimensionless elastic parameters, elliptic integrals, and their arguments there are definite functional relationships. For the forms of inflectional type

$\lambda' = F(k; \varphi)$, $\eta' = 2k\,(1-\cos\varphi)$, $\zeta_\delta' = 2\sin^{-1}(k\sin\varphi)$; $\xi' = 2E\,(k;\varphi) - \lambda'$, and $\omega_M' = 2k\cos\varphi$; For the forms of inflectionless type

$\lambda' = kF(k; \varphi)$; $\eta' = 2(1 - \sqrt{1-k^2\sin^2\varphi})/\,k$;

$\xi' = [2E\,(k;\varphi)\,/\,k] - [\,(2/k^2) - 1]\,\lambda'$; $\zeta_\delta' = 2\,\varphi$; $\omega_M' = (2/k)\,\sqrt{1-k^2\sin^2\varphi}$.

The designer of precision ELKS is practically interested in the parameters and coordinates, which describe the mutual position of initial 0 and final 1 sections of the beam or the strip where the real or conditional rigid elements are to be clamped. Solution to any such problem of the fundamental class using the accurate method requires one to specify any three parameters of similarity or three parameters for initial 0 and final 1 points of the elastic line (two for one point and one for another). Other parameters and coordinates can be defined in the way of solution. The trajectory of motion for a point on the elastic line can be built on the basis of a sequential set of its discrete coordinate values at the different applied loads. Equations (6) – (9) are convenient for calculations at known values of the elliptic integrals. However, it is difficult to use them for the functional analysis of ELKS accuracy and synthesis of ELKS with the separation of the linear constituents. These obstacles and isolation of the theory lead to the limited application of an accurate Popov's method for the synthesis of ELKS. Let us try to install a "bridge" between the "accurate method" and the approximate methods that are more convenient for the analysis and synthesis of ELKS. The benefit of this "bridge" will be in the better classification and more universal approach in solving the corresponding problems. *Parallelogram model.* For example, we will choose an elastic spring parallelogram (Fig. 3-2b) with driving force $2F_{Q1}$as a model for the forms of the inflection type. The elastic state of strips in this device at the symmetric load and ideal accuracy of its elements reflects on the diagram of elastic parameters (Fig. 3-2a) by a segment of straight line whose ordinates φ_o and φ_1 coincide, and the inflection point A_Γ has ordinate $\varphi_\Gamma = 90^o$ (ω_M=0; $M = 0$). Coordinates x_{cp} and y_{cp}=s_1 of the moving end C of the strip in an ideal parallelogram ($\delta_o = \zeta_o = \zeta_c$) can be defined by formulae (9a,b) after substitution of the following values which correspond to the conditions of restraints and the form of the elastic line

$$x_0 = 0;\ y_0 = 0;\ \eta'_c - \eta'_0 = 2\,\eta''_0;\ \xi'_c - \xi'_0 = 2\xi''_0;\ \beta = 2\lambda''_0;\ \delta_0 = \zeta_0;$$

$$x_{cp} / L = \beta^{-1} (2\xi''_0 \cos \delta_0 + 2\eta''_0 \sin \delta_0); \qquad (3\text{-}10a)$$

$$y_{cp} / L = \beta^{-1} (2\eta''_0 \cos \delta_0 - 2\xi''_0 \sin \delta_0). \qquad (3\text{-}10b)$$

Finite rotational motion. Let us examine the trajectory of motion for the conditional rigid body Э (in Fig. 3-2b shown by dashed lines), which has zero mass and is connected to inflection point A_Γ without any influence on the strip's function. Body Э moves in a plane–parallel motion relative to a fixed plane, i.e., it translates together with point A_Γ and rotates around this point. The trajectory of motion of the inflection point in accordance with the feature of symmetry corresponds to the following expressions

$$x_\Gamma / L = \tfrac{1}{2}\, x_{cp} / L = \beta^{-1}(\xi''_o \cos \delta_o + \eta''_o \sin \delta_o), \qquad (3\text{-}11a)$$

$$y_\Gamma / L = \tfrac{1}{2}\, y_{cp} / L = \beta^{-1} (\eta''_o \cos \delta_o - \xi''_o \sin \delta_o). \qquad (3\text{-}11b)$$

If we define by these equations the coordinates of two subsequent positions of point A_Γ $x_\Gamma^{(1)}; y_\Gamma^{(1)}$ and $x_\Gamma^{(2)}; y_\Gamma^{(2)}$, then it is easy to define the angle ϑ_{12} of rotation for body Э. Hence, we know the two finite positions of the movable body but this is enough only for unambiguous linear interpolation. However, it is possible in each finite position to define coordinates x_{P_1} and y_{P_1} of *ICR* in accordance with (1-13, 1-14) as

$$x_{P_1} = x_\Gamma - d\,y_\Gamma / d\,\vartheta_\Gamma;\quad y_{P_1} = y_\Gamma + d\,x_\Gamma / d\,\vartheta_\Gamma,$$

where ϑ_Γ is a slope angle of the strip section at point A_Γ of inflection. To use these equations we need first to differentiate (11a,b) that yields (Y.M.Tseytlin,1972)

$$\frac{1}{L}\frac{dx_\Gamma}{d\vartheta_\Gamma} = \frac{1}{L}\frac{\partial x_\Gamma}{\partial\varphi_0}\frac{d\varphi_0}{d\vartheta_\Gamma};\quad \frac{1}{L}\frac{dy_\Gamma}{d\vartheta_\Gamma} = \frac{1}{L}\frac{\partial y_\Gamma}{\partial\varphi_0}\frac{d\varphi_0}{d\vartheta_\Gamma} \qquad (3\text{-}12a,b)$$

All members of (12a,b) are defined with the expression of dimensionless elastic parameters through elliptic integrals and their arguments k and φ_o. Then after differentiation with respect to φ_o, we obtain

$$\partial\eta''_o / \partial\varphi_o = -2k\sin\varphi_o = -2\,\sin\zeta_o/2 \text{ (at } k = \text{constant)},$$

$$\partial\xi''_o / \partial\varphi_o = -2\,\text{dn}\,(\varphi_o , k) + 1/[\text{dn}\,(\varphi_o, k)],$$

$\partial\beta/\partial\varphi_o = -\chi_\varphi / [\mathrm{dn}(\varphi_o, k)]$; $\partial\delta_o/\partial\varphi_o = \eta''_o / [\cos(\zeta_o/2)]$ (at k = constant);
$\partial\delta_o/\partial\varphi_o = \partial\zeta_o/\partial\varphi_o = 0$ (at $\zeta_o = \zeta_o'$ = constant);
$\partial\eta''_o/\partial\varphi_o = -2\sin(\zeta_o/2)\csc^2\varphi_o$ (at ζ_o = constant).

Number χ_φ depends on the form of the elastic line and in this case is equal to 2; number $\chi_\varphi = 1$ at the calculation of the cantilever beam under axial-transverse bending. Angle ϑ_Γ of body Э rotation may be expressed through the angular coefficients of similarity

$$\vartheta_\Gamma = \zeta_\vartheta - \zeta_o = 2\sin^{-1}k - 2\sin^{-1}k\sin\varphi_o. \tag{3-13}$$

Then $d\vartheta_\Gamma/d\varphi_o = 2k\cos\varphi_0/\sqrt{1-k^2\sin^2\varphi_0} = \partial\zeta_o/\partial\varphi_o$ (at k = constant),
or finally, at k = constant $d\varphi_o/d\vartheta_\Gamma = \mathrm{dn}(\varphi_o, k)/\eta''_o$, and at ζ_o = constant

$$d\varphi_o/d\vartheta_\Gamma = -\frac{\sqrt{1-\sin^2(\varsigma_0/2)\csc^2\varphi_0}}{2\sin(\varsigma_0/2)\cot\varphi_0\csc\varphi_0}.$$

It is easy to show with the help of the derived expressions that for large angles ϑ of rotation the value of vector-radius r_Γ of the inflection point is equal to:

$$r_\Gamma = \sqrt{(dy_\Gamma/d\vartheta_\Gamma)^2 + (dx_\Gamma/d\vartheta_\Gamma)^2},$$

and its value is not constant (*see* Table 3-2). However, for a small angle of rotation in some cases, it is possible to consider $r_\Gamma \approx$ constant. The calculation of r_Γ yields in this case the constant value $r_\Gamma = 0.667\,L$. It is possible to assume that at a small angle of rotation $\vartheta \ll 1$ for the moving body, the position of the *ICR* also does not change, and therefore

$$y_{P_1} = 0;\ y_\Gamma \approx r_\Gamma\vartheta_\Gamma,\ \text{and}\ x_\Gamma \approx r_\Gamma(1 - \vartheta_\Gamma^2/2). \tag{3-14}$$

Table 3-2. Calculated radius r_Γ at the transverse bending of a cantilever strip

ϑ_Γ	ζ_ϑ'	φ_o	ζ_o'	$\frac{1}{L}\left\lvert\frac{dx_\Gamma}{d\vartheta_\Gamma}\right\rvert$	$\frac{1}{L}\left\lvert\frac{dy_\Gamma}{d\vartheta_\Gamma}\right\rvert$	r_Γ/L
10°	100°	67°23'	90°	0.092	0.661	0.667
50°	140°	48°48'	90°	0.439	0.516	0.678
80°	170°	45°13'	90°	0.668	0.284	0.726

Approximate solutions.

The approximate solutions of nonlinear differential equation (3) can be obtained by different methods and particularly: the series expansion and empirical approximation. Both methods are effective in the certain problems.

Expansion method. Series expansion of the nonlinear differential equation of bending

$$1/\rho = y_x''/[1+(y_x')^2]^{3/2} = (M_1/H) + (F_Q/H)[(x_c - x)\sin\delta_Q - (y_c - y)\cos\delta_Q] \tag{3-15}$$

yields (V. N. Zheludkov, 1972; P. Zenow, 1970) the approximate relations with the acting forces in the third degree for the more general conditions of elastic beam loading under bending $x_c/L = x_1\beta^2 + x_2\beta^4 + x_3\beta^6 + \ldots$;
$y_c/L = y_1\beta^2 + y_2\beta^4 + y_3\beta^6 + \ldots$; $\vartheta_c/L = \vartheta_1\beta^2 + \vartheta_2\beta^4 + \vartheta_3\beta^6 + \ldots$

The comparison of this approximate solution with accurate Popov's solution at force factors $M_cL/H < 1$; $F_QL^2/H < 1.2$, and $[(M_cL/H) + (F_QL^2/H)] < 0.8$ shows that the errors of coordinate calculation for displacements (y_c; x_c; s_{cx}) and stress σ_x are less than 5%. Approximate solutions of (15) with expansions are the unwieldy ones that make it significantly difficult to use them in practice. It is clear that the series expansion is convenient to use only at a relatively small nonlinearity. Therefore other empirical methods of the solution were developed (A.I. Gots *et al.*, 1969; L.M. Tsukernik, 1976) for the differential equation of bending $1/\rho = y_x'' / [\,1 + (y_x')^2\,]^{3/2} = (M_c/H) - [F_Q(x - x_c)/H] + 1/\rho_o = f(x)$.

Empirical method. It was shown that because

$$y'_x = \tan\vartheta = \sin\vartheta / \sqrt{1-\sin^2\vartheta}\,,\ \sin\vartheta = \int f(x)dx + C_1 = \Lambda(x)\,,\ \text{then}$$

$$y = \int \Lambda(x)\,dx / \sqrt{1-\Lambda^2(X)} + C_2, \tag{3-16}$$

where C_1 and C_2 are the constants of integration. If $f(x)$ = constant (the polynomial of zero order), then $\Lambda(x)$ represents the polynomial of the first order. This means that integral (16) has a solution with elementary functions. In all other cases as it was above shown, the problem is solved with the elliptic integrals. Integral (16), however, can be solved approximately if the integrand $\Xi(x) = \Lambda(x) / \sqrt{1-\Lambda^2(x)}$ is substituted with the empirical expression

$$\Xi_{app}(x) = -0.89823 - 1.305134 / [\Lambda(x) - 1.385159]. \tag{3-17}$$

The difference between approximate $\Xi_{app}(x)$ and accurate $\Xi(x)$ values at $\Lambda(x) =$ 0.00 to 0.95 is not more than 3%. Integral (16) with the substitution of (17) may be presented as

$$y = -0.89823x - 1.305134 \int \frac{dx}{\Lambda(x) - 1.385159} + C_2,\ \ C_2 = y_c. \tag{3-18}$$

The integral in the right-hand side of (18) can be expressed through the elementary functions. The approximate evaluation of displacements in the axis x direction at invariable length of the beam's elastic line with equation

$$L = \int_{x_a}^{x_c} dx / \sqrt{1-\Lambda^2(x)} \tag{3-19}$$

can analogously be done by using the following empirical expression

$$\sqrt{1-\Lambda^2(x)} = 1.08 - \frac{0.0684 + 0.2016\Lambda(x)}{1.2817 - \Lambda(x)}. \tag{3-20}$$

The difference between an accurate and approximate value of the square root at Λ= 0.00 to 0.95 is not more than 5%. The approximate calculations with (17) and (20) expressions may be provided in the range from the linear theory of bending (the lower boundary) to the value $\Xi(x) \le 0.9$ (the upper boundary).

Integral of probability. Some practical interest may be in the use of the integral of probability for the approximate solution of bending nonlinear differential equation (3). It is not difficult to show that this equation with an uncertainty of order $(y')^4$ corresponds (Y.M. Tseytlin, 1972) to the expression

$$\sqrt{\pi/6}\,\Phi\,(y'\sqrt{3/2}\,) = \int (M/H)\,dx, \tag{3-21}$$

where $\Phi\,(y'\sqrt{3/2}\,)$ is the integral of probability.

An example of using the method of best approximation for the calculation of nonlinear bending the cantilever beam is shown in section 8-4.

Small elastic deformation.

Linearization. The linearization of differential equation (3) is possible if the elastic deformation is small and the squares of the beam's axis slopes are negligible, i.e., $(y')^2 \ll 1$. Then the differential equation simplifies and becomes a linear one

$$M(x) / H = y_x''. \tag{3-22}$$

The bending moment $M(x)$ (*see* Fig. 3-1c) at $M_c = M_1$ is equal to:

$$M(x) = M_1 + F_{Qy}(L - s_x - x) - F_{Qx}\,[y_c - y(x)], \tag{3-23}$$

and is usually expressed as follows

$$M(x) = M_o - F_{Qy}\,x - F_{Qx}\,y. \tag{3-24}$$

Here F_{Qx} and F_{Qy} are the coordinate projections of force F_Q.

Uncertainty. Analysis of the linear bending theory shows the volume of its force and kinematical information. The following are the sources of uncertainty with calculations on the basis of the simplified linear theory, which neglects:

a) members of differential equation with $(y')^2$, b) transverse displacements of the elastic line's points in the expression of the bending moment, and c) longitudinal displacement of the beam's ends ($s_{cx} = 0$).

By using the method of sequential approximation and series expansion of the denominator in equation (3), it is possible to show that neglecting of members with $(y')^2$ yields the uncertainty which is proportional to the force coefficient $\beta < 1$ in the power $\nu_\beta \geq 6$. The uncertainty from neglecting the transverse displacement of points on the elastic line is proportional to β in the power $\nu_\beta \geq 4$. Neglecting the longitudinal displacement s_{cx} for the point of the force application yields an uncertainty that is proportional to β in the power $\nu_\beta \geq 12$. The general solution of (22), at the bending moment in accordance with (24), can be obtained by the following steps (J.E. Plainevaux, 1953-1956a,b). Involving the dimensionless variables, we have expressions

$$\tilde{x} = \frac{x}{L};\; 4\beta_y^2 = F_{Qy}\,L^2/\,H;\; \beta_x = F_{Qx}L^2 / H;\; \beta_y = F_{Qy}L^2 / H,\; \chi_M = M_oL / H.$$

Substitution them into (22) and (24) yields the differential equation

$y_x'' + 4\beta_y^2\, y = L(\chi_M - \beta_x \tilde{x})$, whose general solution for displacement at presence of compression force F_{Qx} is equal to:

$$y = C_1 \sin 2\beta_y \tilde{x} + C_2 \cos 2\beta_y \tilde{x} + L\,(\chi_M - \beta_x \tilde{x}) / (4\beta_y^2)\,,$$

and the slope of the elastic line equals $y_x' = (1/L)\, d\tilde{x} \,/\, dy = (2\beta_y\, C_1/ L)\cos 2\beta_y \tilde{x}$

$$- (2\beta_y\, C_2/ L) \sin 2\beta_y \tilde{x} - \beta_x / (4\beta_y^2). \tag{3-25}$$

Constants of integration C_1 and C_2 are defined through specified initial conditions:

$y = y_x' = 0$ at $\tilde{x} = 0$. Then, using the boundary-value conditions at $x = L$ with $y = s_y$, and $y_x' = \tan\vartheta$, one can get the following calculation relationships

$$s_y / L = (\beta_x/ 8\beta_y^3)\,(\sin 2\beta_y - 2\beta_y) + [\chi_M /(4\beta_y^2)](1 - \cos 2\beta_y),$$

$$\tan\vartheta = (\beta_x/ 4\beta_y^2)(\cos 2\beta_y - 1) + (\chi_M / 2\beta_y) \sin 2\beta_y.$$

This yields the definitions for $\beta_x (F_{Qx}, L, H)$ and $\chi_M (M_o, L, H)$.

Displacement s_x of the beam's end in the direction of axis x is equal to the integral

$$s_x = \tfrac{1}{2}\int_0^L (dy/dx)^2\, dx. \qquad (3\text{-}26)$$

After substitution of the expression for derivative (25) into integral (26) and some necessary cancellations, one gets the relationship (*see* Y.M. Tseytlin, 1972) from which we can obtain the simplified calculation relationships by a series expansion with neglecting the higher order members. These relationships are effective only at a very small displacements s_x/L. Moment M_c is also defined from the linear differential equation

$$M_c / H = y_x'' |_L = (1/L^2)(d^2y/ d\tilde{x}^2 |_1. \qquad (3\text{-}27)$$

The expression for s_x/L in this case is as follows:

$$\frac{s_x}{L} = \frac{tg^2\vartheta}{15}\left[1 \pm \frac{11\beta_x}{420} + ...\right] - \frac{1}{10}\frac{s_y}{L} tg\vartheta\left[1 \pm \frac{\beta_x}{70} + ...\right] + \frac{3}{5}\left(\frac{s_y}{L}\right)^2\left[1 \pm \frac{\beta_x}{420} + ...\right]. \qquad (3\text{-}28)$$

It is evident from (28) that the linear bending theory allows one to define the poles of the first (*ICR*) and second order only. If the longitudinal forces are applied to the beam in addition to the transverse forces, the beam deflections increase at a compression longitudinal force(upper signs) and decrease at a tension one (lower signs).

Combined load. Fig. 3-3 shows: a) displacements $v_{n\text{-}n}$ under transverse force P and longitudinal extension and compression forces F_s ; b) the *ICR* coordinate ξ_{ICR} of strip's end rotation; c) flexural center C_z and pole P_M of the movable plane Э motion on the elastic strips, d) EKS with the flexural center C_z , translational $v_{x,y}$ and rotational (torsional)Φ_c compliance ; e) Lotze's symbolic presentation of the point-like spring, f) diagram for the estimation of a mutual position of the flexural center and pole of motion, and g) the flexural center's coordinates.

The deflection $y = v_{n\text{-}n}$ (Fig. 3-3a) for a combined transverse-longitudinal load can be determined by using Timoshenko's approximate formula

$$y = v_{TR} / (1 + F_s / F_{eul}) = v_{TR} / (1+v_s)\,, \qquad (3\text{-}29)$$

where $v_{TR} = PG(x,\xi)$is the deflection caused by the transverse force P alone at known Green function $G(x,\xi)$and F_s is the longitudinal force (taken as positive for tension and negative for compression). This formula is an accurate enough even at comparatively large ratio $v_s = F_s / F_{eul}$. Its error equals 0.7% at v_s=1and 1.7% at v_s =10. However, it is clear that for application of this formula the following limitation is evident: $v_s > -1$.

One can determine the Euler's critical force F_{eul} from the formula $F_{eul} = \pi^2 EJ/ (v_{eu} L)^2$, where v_{eu} is a coefficient which depends on the method of mounting the beam's (strip's) extremities. The deflection v_{TR} for distributed $f(\xi)$ load is equal to $v_{TR} = \int_0^L G(x;\xi) f(\xi) d\xi$.

If v_{eu}=2 (case 2, Table 3-3), $F_s = F_{Qx}$, and moment M on the beam's moving end causes its rotation ϑ_0, then

$$M/\vartheta_0 \approx A_{si}\, EJ(1 \pm 0.405 F_s L^2/EJ)/L \approx A_{si}\, EJ(1 \pm 0.405\alpha_s^2 L^2) / L\,, \qquad (3\text{-}29a)$$

which may be transformed with the precision of few percents at $L\alpha_s = L\sqrt{F_s / EJ}$ to
$M/\vartheta_0 = EJ\alpha_s / \tanh(L\alpha_s)$ (3-29b)
for F_s in tension(sign +) with $A_{s1} = (1\ 0.405\,\alpha_s^2 L^2/6)$ at $L\alpha_s \leq 1.7$and $A_{si} \approx 1$ at $L\alpha_s \leq 0.8$; $A_{s2} \approx (5/6 + A_{s1})/2$ at $2.6 \geq L\alpha_s > 1.7$, and $M/\vartheta_0 = EJ\alpha_s / \tan(L\alpha_s)$ (3-29c)
for F_s in compression (sign -) at $L\alpha_s \leq 1.56$ with $A_{si} = (1 + 0.405\,\alpha_s^2 L^2/6)$,

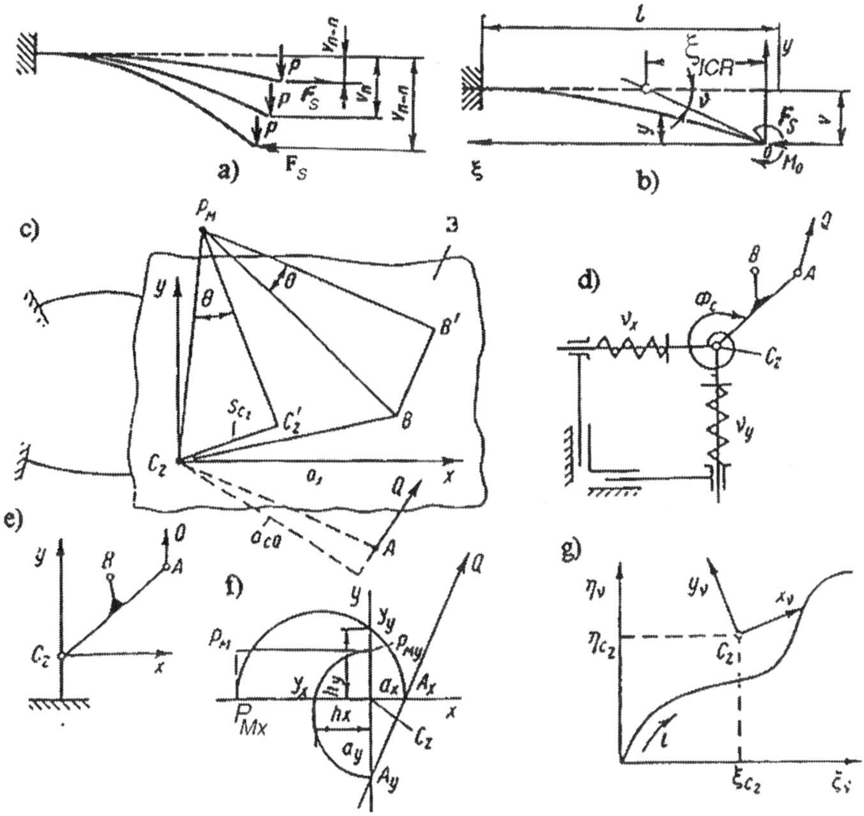

Fig. 3-3. Kinematics of elastic strips bending and their parameters (a, b after L.E. Andreeva, 1962; c-g W. Lotze, 1963)

$A_{si} \approx 1$ at $L\alpha_s \leq 0.8$; and $A_{si} \approx 1\frac{1}{6}$ at $L\alpha_s \geq 1.5$.

Expressions (29b,c) are similar to the known solutions of linear differential equations for bending a thin cantilever strip by moment and longitudinal force (*see* below and in F.S. Eastman, 1937). However, (29a) does not require special functions and yields solution in one step (compare with W.D. Weinstein, 1965, p. 155-157). The principle of tension and bending moment dependence was used by F.S. Eastman (1935) in flexure pivots with self-compensating restoring moment (force compensator).

ICR and other poles. The *ICR* for a flat elastic spring loaded and fastened as in Figure 3-3b corresponds to the first kinematic pole (*see* section 1-2) whose abscissa coordinate can be defined from the trigonometric relationships as

$$\xi_{ICR}= dy / d\vartheta|_v \approx v/ \tan\vartheta ,$$

where v and ϑ are the deflection and angle of rotation of the initially flat strip 's end.

Table 3-3. Values of the critical buckling load

<table>
<tr><th>#</th><th>Mounting method</th><th>Load type</th><th>Coefficient</th><th>Critical load</th></tr>
<tr><td>1</td><td>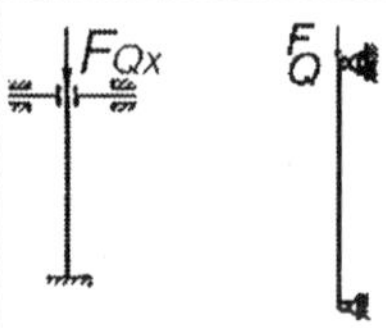</td><td rowspan="4">Longitudinal force</td><td>$\nu_{eu}=1$</td><td>$F_{eul}=\pi^2 EJ/L^2$</td></tr>
<tr><td>2</td><td>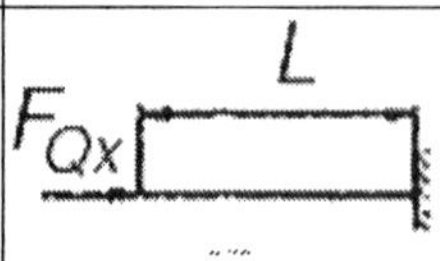</td><td>$\nu_{eu}=2$</td><td>$F_{eul}=\pi^2 EJ/(2L)^2$</td></tr>
<tr><td>3</td><td></td><td>$\nu_{eu}=0.7$</td><td>$F_{eul}=\pi^2 EJ/(0.7L)^2$</td></tr>
<tr><td>4</td><td></td><td>$\nu_{eu}=0.5$</td><td>$F_{eul}=\pi^2 EJ/(0.5L)^2$</td></tr>
<tr><td>5</td><td>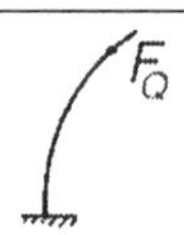</td><td>Following force</td><td>* $\nu=0.22$</td><td>$F_{cr}= 20.19\, H/ L^2$</td></tr>
<tr><td>6</td><td>Lateral strip's bending
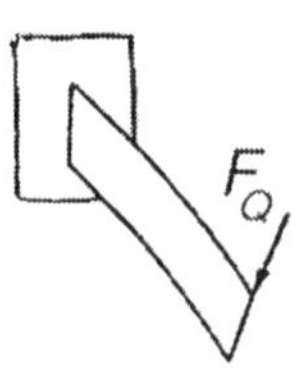
$J_\omega=\int\omega^2_A dA$ - sectorial moment of inertia

$J_\omega\to 0$</td><td>Force (following, dynamic stability)

(static stability)</td><td>...</td><td>$F_Q=\frac{1-\Psi_\omega}{2\sqrt{\Psi_\omega}}F_{Q^{**}}$
$F_{Q^{**}}=\frac{45.61}{L^2}\times$
$\sqrt{(GT_o+\frac{\pi^2}{L^2}EJ_\omega)EJ_y}$
$\Psi_\omega=\frac{\omega^2_{nx}}{\omega^2_{n\theta}}$; $\omega_{nx}=\frac{n^2\pi^2}{L^2}\sqrt{\frac{EJ_x}{m_\gamma}}$
$\omega_{n\theta}=\frac{n\pi}{\rho_j L}\sqrt{\frac{GT_o}{m_\gamma}}$; $n=2$
$F_{Qcr}=\frac{4.01\sqrt{EJ_y GT_o}}{L^2}$</td></tr>
</table>

Table 3-3. Cont'd.

#	Mounting method	Load type	Coefficient	Critical load
7	Pure bending 	Moment	...	$M = n\pi \frac{\sqrt{EJ_y GT_o}}{L}$

Note: Pos. 2, 5, 6 are cantilever beams. It is more accurate to suppose that L is equal to the beam's length in a moment of loss of stability so that critical buckling force is equal to $F_{eulb} = \pi^2 EJ/ [\nu_{eu} (L_o - F_{eul}L_o/ EA]^2$ where L_o is its initial length; a free supported beam with uniformly distributed vertical load q has critical buckling value as follows: $q_{cr} = 18.6EJ/ L^3$ (Y. G. Panovko et al., 1979). In pos.6, ρ_j is a polar radius of gyration, m_y is the mass of the strip per unit length. Pos. 3,4 After L.E. Andreeva, 1962; pos 5-7 and *after V.V. Bolotin, 1956.

It is clear from the state of the strip's elastic line at a pure bending that $\vartheta = L / \rho$, where $\rho = L/\vartheta$ is the radius of the elastica's curvature and L is the length of the neutral line of the flexed strip. The deflection v of the free end of the strip (that is determined from the geometric relationship) is $v = \rho(1 - \cos\vartheta)$. If angle ϑ is very small and less than $\vartheta \le 0.1$ rad , then $\cos\vartheta \approx 1 - 0.5\vartheta^2$ and $\tan\vartheta \approx 1.002\vartheta$ (*see* Table 1-9). In this case, $\xi_{ICR} = 0.5\,\vartheta^2\rho / 1.002\vartheta = 0.499\,\vartheta\rho \approx 0.5\,L$. The same result can be obtained by formula (1-13) for the coordinate of pole P_1, where $x_{p1} = x_0 - dv/d\vartheta = \rho\sin\vartheta - d(0.5\rho\vartheta^2)/ d\vartheta = 0.999L - 0.5\,L = 0.499L \approx 0.5\,L$. If the longitudinal forces F_s are applied, we consider the position of the beam after the deformation and draw axes $yO\xi$, making origin O coincide with the free end of the beam. The bending moment at any section of the beam is $M = -(M_o + F_s y)$. (a)

The sign of the moment is determined from the sign of the beam elastic line's curvature. In the given case, the beam curvature is negative. Substitution (a) into (22) gives $Hy'' = -(M_o + F_s y)$. (b)

Introducing the designation

$F_s / H = \alpha_s^2$, (c)

we obtain from (b) the following relation:

$y'' + \alpha_s^2 y = - M_o / H$. (d)

The general solution of this equation is

$y = a_\xi \sin\alpha_s\xi + b_\xi \cos\alpha_s\xi - (M_o / F_s)$. (e)

The integration constants a_ξ and b_ξ are determined from the boundary-value conditions. In the given case at $\xi = 0$, $y = 0$, and at $\xi = L$, $\vartheta = y' = 0$. This gives the following values for the constants $a_\xi = (M_o / F_s) \tan\alpha_s L$ and $b_\xi = M_o / F_s$. Substitution of these values into (d) finally yields

$y = (M_o / F_s)(\tan\alpha_s L \cdot \sin\alpha_s\xi + \cos\alpha_s\xi - 1)$. Hence, in the given case (*see* L.E. Andreeva, 1962; F.S. Eastman, 1935)

$y_{\xi=L} = (M_o / F_s)(\tan\alpha_s L \cdot \sin\alpha_s L + \cos\alpha_s L - 1) = (M_o / F_s)(1 - \cos\alpha_s L) / \cos\alpha_s L$;

$y'_{\xi=0} = \tan\vartheta_o = (M_o\,\alpha_s / F_s)(\tan\alpha_s L \cdot \cos\alpha_s\xi - \sin\alpha_s L)_{\xi=0}$

$= (M_o\,\alpha_s / F_s)\tan\alpha_s L$ and at $\vartheta_o < 1$ $M_0/\vartheta_0 = EJ\alpha_s/\tan\alpha_s L$.

The latter expression is similar to (3-29c) which was derived with the help of (3-29) approximation. The initial position of the *ICR* is, therefore,

$\xi_{ICR} = (1 - \cos\alpha_s L) / (\alpha_s \sin\alpha_s L)$. If the second member in (d) has minus sign that corresponds to extension axial force F_s, its general solution is equal to $y = C_1\cosh\alpha_s\xi + C_2\sinh\alpha_s\xi - (M_0/F_s)$ and the corresponding stiffness ratio is

$M/\vartheta_0 = EJ\alpha_s/\tanh(L\alpha_s)$ as in (3-29b). The calculation formulae of the *ICR* position for the flat strip hinge at other schematics of loading are shown in Table 3-4,where $H = EJ$.

Table 3-4. Position of the initial center of rotation for a flat strip hinge

No	Loading schematic	Initial position ξ_{ICR} of the *ICR*	Note
1	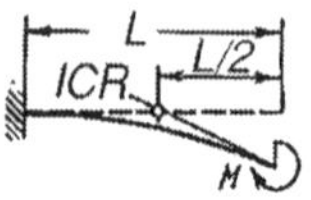	$0{,}499L \approx 0.5\ L$	$\vartheta \le 0.1$ rad, $P=0$, $F_s=0$, $M \ne 0$
2	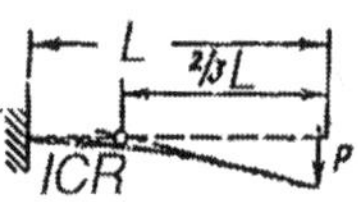	$0.667\ L$	*See* Table 3-2, $\vartheta° < 10°$ $P \ne 0$, $F_s = 0$, $M=0$
*3		$(L/3)[2 + 3\ M/(PL)]/[(1 + 2\ M/(PL)]$	$P \ne 0$, $F_s = 0$, $M_0 \ne 0$
*4	Fig 3-3b	$(1-\cos\alpha_s L)/(\alpha_s \sin\alpha_s L)$	$P=0$, $F_s < 0$, $M_0<0$
*5		$(\cosh\alpha_s L - 1)/(\alpha_s \sinh\alpha_s L)$	$P=0$, $F_s > 0$, $M_0<0$

Note: $\alpha_s^2 = |F_s|/H$; * After L. E. Andreeva, 1962.

If one limits the problem to the external forces with mutually related

components $F_{Qy} = F_Q$; $F_{Qx} = F_Q\bar{\delta}_Q$; and $M_z = F_Q L(\mu_o + \mu_1\vartheta)$,

then the equation for the bending moment is expressed (W. Lotze) as follows

$$M(x) = F_Q[L(\mu_o + \mu_1\vartheta + 1) - \bar{\delta}_Q s_y - x + \bar{\delta}_Q s_y(x)]. \tag{3-30a}$$

This corresponds to the following differential equation

$$s''_y(x) - s_y(x)\,\bar{\delta}_Q F_Q/H = (F_Q L/H)\,[(\mu_o + \mu_1\vartheta + 1) - \bar{\delta}_Q s_y/L] - F_Q x/H = (F_Q L/H)\pi_y - F_Q x/H. \tag{3-30b}$$

The solution of (30b) is represented in the form of a power series.
Then, after excluding F_Q value, one will get (W. Lotze) with $\mu^* = (2\mu_o + 1)$ that

$$s_y = L\{(½ + \mu^{*-1}/6)\vartheta - [\bar{\delta}_Q\,(\frac{1}{12}\mu^{*-1} - \frac{7}{90}\mu^{*-2} - \frac{1}{36}\mu^{*-3}) + (\mu_1/3)\,\mu^{*-2}]\,\vartheta^2\}; \tag{3-31a}$$

$$s_x = -L\,[(1/6) + (1/12)\,\mu^{*-1} + (1/60)\,\mu^{*-2}]\,\vartheta^2 + \ldots \tag{3-31b}$$

The substitution of these equations into (1-13) and (1-14) at $\vartheta = 0$ allows one to define the initial position of *ICR* (P_1)

$$\xi_{P_1} = -L\,[\,½ + (1/6)\,\mu^{*-1}], \tag{3-32a}$$

$$\eta_{P_1} = 0. \tag{3-32b}$$

Coordinates ξ_{P_2} and η_{P_2} for the position of the second order pole P_2 are defined through (1-15) and (1-16) as equal to:

$$\xi_{P_2} = -L\,[(1/3) + (1/30)\,\mu^{*-2} + (1/6)\,\mu^{*-1}]; \tag{3-32c}$$

$$\eta_{P_2} = -L\,\{\bar{\delta}_Q\,[(1/6)\,\mu^{*-1} - (7/45)\,\mu^{*-2} - (1/18)\,\mu^{*-3}]$$

$+ (2/3)\mu_1\mu*^{-2}\};$ (3-32d)

coordinates η_{P_3} for the position of the third order pole P_3 are defined through equation (1-12) as follows

$$\eta_{P_3} = L\{-\mu_1 [\mu*^{-2} + (2/5)\mu*^{-3}]$$

$$+ \bar{\delta}_Q [-(4/15)\mu*^{-1} + (2/15)\mu*^{-2} + (37/210)\mu*^{-3} + (1/30)\mu*^{-4}]\}. \quad (3\text{-}32e)$$

It is easy then to define the components of the diameter of the rotational circle, which are equal to:

$$d_\xi = L[(1/6) - (1/30)\mu*^{-2}]; \quad (3\text{-}32f)$$

$$d_\eta = -L\{\bar{\delta}_Q [(1/6)\mu*^{-1} - (7/45)\mu*^{-2} - (1/18)\mu*^{-3}] + (2/3)\mu_1\mu*^{-2}\}. \quad (3\text{-}32g)$$

And further, d_ξ is expressed by W. Lotze through ξ_{P_1} as follows

$$d_\xi = L\{(1/6) - (6/5)[(\xi_{p1}/L) + ½]^2\}, \quad (3\text{-}33)$$

where coordinates ξ and η are oriented along the initial axis of the straight strip and the direction of its normal, respectively (in more general cases they are oriented in the direction of the pole's tangent and its normal).

Artificial point spring method.

The calculation of strips and beams' kinematics in the planar ELKS may be significantly simplified with the method of the so-called artificial point springs. This method of point elastic quasi-elements was developed by W. Lotze (1963) for the calculation of a small elastic displacement at the bending of thin strips and beams. The method is based on the assumption that translational motion and rotation of a movable plane connected to the flexures are the linearly independent functions of forces and moments applied to the definite point that is called the flexural center C_z (Fig. 3-3c). This condition is expressed by the following formulae

$$s_{C_z x} = \nu_x Q_x; \quad (3\text{-}34a)$$

$$s_{C_z y} = \nu_y Q_y; \quad (3\text{-}34b)$$

$$\theta = \Phi_c M. \quad (3\text{-}34c)$$

Let us remind that, as shown in section 1-2, the finite displacement of plane Э may be achieved by the pure rotation around some pole P_M,, which in general is different from the *ICR*. At the same time it may be performed by the sum of the translational displacement together with some flexural center C_z and angular displacement θ relative to this point C_z. If force Q acts in point A of plane Э, the torque around point C_z is equal to:

$$M = Q_y x_A - Q_x y_A = Q a_{cQ}. \quad (3\text{-}35)$$

The small magnitude of the considered elastic deformations and displacements allows one to accept x_A and y_A values as equal to the similar values in the initial non-deformed system. If in the motion, the point of the movable plane that coincides in initial position with flexural center C_z has displacement $\overline{C_z C_z'} = s_{C_z}$, then the projections of motion of any other point B on the coordinate axes can be expressed through the following relationships at a small angle of rotation θ as

$s_{Bx} = s_{CzX} - y_B\theta; \quad s_{By} = s_{CzY} + x_B\theta.$ (3-36a,b)

The EKS may be represented in accordance with these relationships as a lever system (Fig. 3-3d), which has the hinge in flexural center C_z. It is supposed that the three springs in the hinge with flexibility compliance coefficients ν_x, ν_y, and torsional compliance Φ_c are independent. Lotze's symbolic image of the point spring is shown in Figure 3-3e where axes y and x correspond to the direction of flexibility ν_x and ν_y, respectively. It may be shown that the coordinates of the fixed pole P_M (Fig.3-3c) for rotation of the movable plane are easy to define on the basis of conditions (36a,b) at $s_{Bx} = s_{By} = 0$, where

$0 = s_{CzX} - y_{PM}\,\theta = \nu_x\, Q_x - y_{PM}\,\Phi_c a_{cQ}; \quad 0 = s_{CzY} + x_{PM}\,\theta = \nu_y\, Q_y - x_{PM}\,\Phi_c\, a_{cQ}.$

Then we can find the coordinates of the pole P_M, which always resides behind the flexural center C_z if we look from the point of the force application. Therefore,

$$y_{P_M} = (\nu_x / \Phi_c)(1/a_y); \qquad (3\text{-}37a)$$

$$x_{P_M} = (\nu_y / \Phi_c)(1/a_x), \qquad (3\text{-}37b)$$

where a_y and a_x (Fig. 3-3f) are the segments that are cut by the force Q line of action on the coordinate axes (x- and y-intercepts). If we put the segments with length $h_y = \sqrt{\nu_y/\Phi_c}$ and $h_x = \sqrt{\nu_x/\Phi_c}$ on the coordinate axes and through their ends Y_y ; Y_x and points A_x; A_y provide semicircles , then points P_{Mx} and P_{My} correspond to the values of the coordinates for the pole of rotation P_M. For the practical calculations it is necessary to know the flexibility ν_x , ν_y , Φ_c , and also the position of the flexural center C_z. Parameter Φ_c for the flexible strip of any shape (Fig. 3-3g) in coordinates η_ν and ξ_ν is defined from the linear differential equation of bending (22) through the evident expression

$$\Phi_c = y_x'/M = \int_0^L dl\,/H,$$

where in general the flexural section area stiffness H is not constant along the elastic line and is a function of its current length $H(l)$. The coordinates of the flexural center C_z are defined from the action of the force components Q_ξ and Q_η

$Q_\xi\xi_{Cz}\Phi_c = \int_0^L (Q_\xi\xi_\nu/H)dl\,; \quad Q_\eta\eta_{Cz}\Phi_c = \int_0^L (Q_\eta\eta_\nu/H)dl$, from which one obtains

$$\xi_{Cz} = (1/\Phi_c)\int_0^L (\xi_\nu/H)dl\,; \quad \eta_{Cz} = (1/\Phi_c)\int_0^L (\eta_\nu/H)dl. \qquad (3\text{-}38a, b)$$

Then the directions of the main axes of flexibility x_ν and y_ν are established from the condition of displacement projections' independence

$$\int_0^L (x_\nu y_\nu/H)\,dl = 0. \qquad (3\text{-}39)$$

The displacement of points on the elastic line of the strip relative to the flexural center is described (as the first approximation) by the following differential equations

$ds_{C_z x} = y d\theta = (Q_x y^2/H)dl; \quad ds_{C_z y} = x d\theta = (Q_y x^2/H)dl.$

From these equations we can define v_x and v_y after the integration along the length of the elastic line and division of its result by Q_x and Q_y, respectively. Hence,

$$v_x = s_{CzX} / Q_x = \int_0^L (y^2/H)dl; \quad v_y = s_{CzY} / Q_y = \int_0^L (x^2/H)dl . \qquad (3\text{-}40a, b)$$

The problem is a simpler one at elemental flexural rigidity H = constant. In this case, flexural center C_z coincides with the centroid (center of gravity) of the elastic line of the flexible beam (strip). The advantage of the *point-spring method* is in the integration that depends on the form of the strip (beam) and does not depend on the load conditions. Therefore, the results of integration for some widely used elementary forms of the beams (strips) can be tabulated (Table 3-5).

The calculation of ELKS with sequentially and parallel connected elastic elements can be provided by the superposition of the partial displacements for elements with known point flexural parameters or by definition of point flexural parameters for the whole composite system. The calculation formulae for the set of composite ELKS are shown in pos. 9-12 of Table 3-5, where Φ_c, v_x , and v_y are the flexural parameters of the whole composite system, and Φ_{ci}, v_{xi} , and v_{yi} are the partial flexural parameters for each elastic element.

Table 3-5. Elastic parameters of the point-spring elements and their joints

Form of element or joint	Schematic	Flexural parameters
1. Rectangular, planar	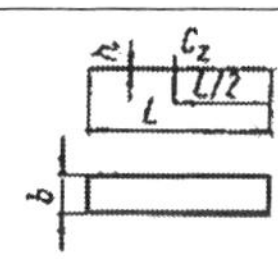	$\Phi_c= L/(EJ)$; $v_x=0$; $v_y= L^3/(12EJ)$; at extension: $v_x = L/EA$ at tension; $A=bh$
2. Rectangle, wrinkled	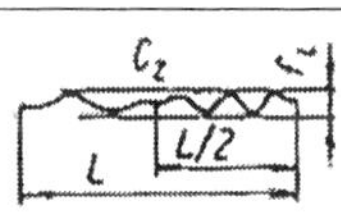	$\Phi_c= L/H$; $v_y= L^3/(12H)$ $v_x=(1/12\ldots1/6)(f_l/L)^2L^3/H$
3. Preliminary curved	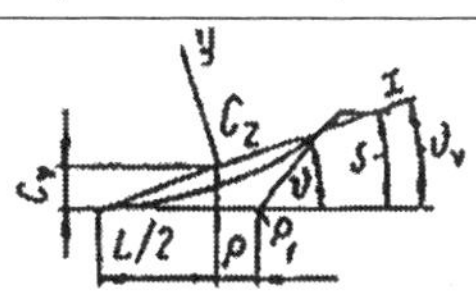	$c_v=[(L/6)-\rho/2]\vartheta$; $\vartheta_v=(1-12\rho/5L)\vartheta/2$; $v_x=(L^3/H)[(1/720)+\rho^2/700L^2]\vartheta^2$; $v_y= L^3/(12H)$
4. Trapezoid	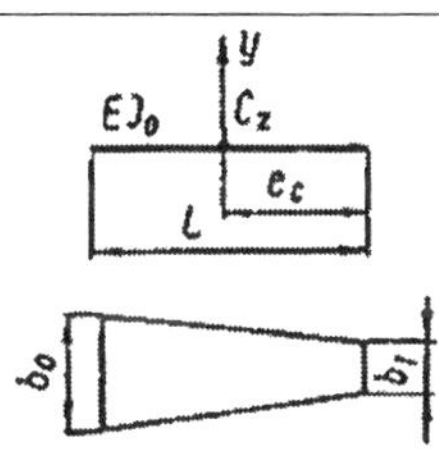	$e_c=L\{1-\beta_v^{-1}-[\ln(1-\beta_v)]^{-1}\}$; $\Phi_c=(L/EJ_o)\beta_v^{-1}\ln(1-\beta_v)^{-1}=Lk_\Phi/EJ_0$ $v_y=\frac{L^3}{\beta_v EJ_0}\left[\frac{2-\beta_v}{2\beta_v}+\frac{1}{\ln(1-\beta_v)}\right]$ $=\frac{L^3}{12EJ_0}k_v$; $v_x\approx0$; $\beta_v=(b_0-b_1)/b_0$
5. Segment of circle $2\alpha < 90^o$	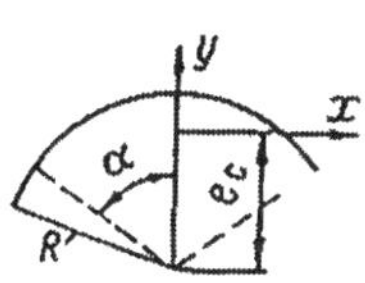	$e_c= R\sin\alpha/\alpha$; $\Phi_c=2\alpha R/H$; $v_x=(R^3/H)[\alpha+ \frac{1}{2}\sin2\alpha-2\sin^2\alpha/\alpha]$; $v_y=(R^3/H)(\alpha - \frac{1}{2}\sin2\alpha)$
6. A quarter of circle $2\alpha=90^o$		$e_c= 0.900R$; $\Phi_c=1.571R/H$; $v_x=0.01216R^3/H$; $v_y= 0.2854R^3/H$

Table 3-5. Cont'd.

Form of element or joint	Schematic	Flexural parameters
7. Semicircle	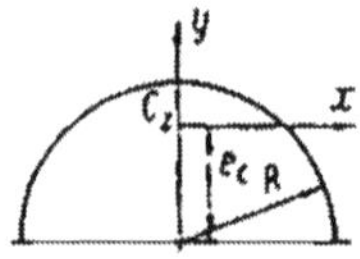	$e_c = 0.6366R$; $\Phi_c = 3.141R/H$; $\nu_x = 0.2976R^3/H$; $\nu_y = 1.571R^3/H$
8. Helix spiral $2\alpha = 360°n$	-	$e_c = 0$; $\Phi_c = 2\pi nR/H$; $\nu_x = \nu_y = \pi nR^3/H$
9. Two sequentially connected initially flat elements (1&2) with initial common axis	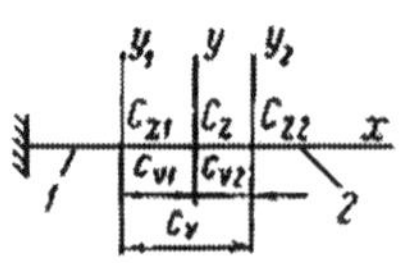	$\Phi_c = \Phi_{c1} + \Phi_{c2}$; $\nu_x = \nu_{x1} + \nu_{x2}$; $c_{v1} = c_v - c_{v2} = (\Phi_{c2}/\Phi_c)\, c_v$; $\nu_y = \nu_{y1} + \nu_{y2} + c_{v1}^2\Phi_{c1} + c_{v2}^2\Phi_{c2}$
10. Two sequentially connected equal symmetrically positioned elastic elements (1, 2)	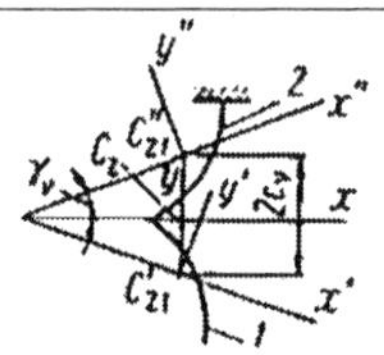	$c_{v1} = c_{v2} = c_v$; $\Phi_c = 2\Phi_{c1}$; $\nu_x = 2[c_v^2\Phi_{c1} + \nu_{x1}\cos^2\gamma_v + \nu_{y1}\sin^2\gamma_v]$; $\nu_y = 2[\nu_{x1}\sin^2\gamma_v + \nu_{y1}\cos^2\gamma_v]$
11. Two connected in parallel elements (1,2) with common axis x	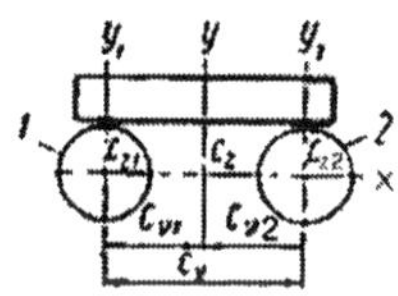	$c_{v1} = c_v/(1 + \nu_{y2}/\nu_{y1})$; $\nu_x^{-1} = \nu_{x1}^{-1} + \nu_{x2}^{-1}$; $\nu_y^{-1} = \nu_{y1}^{-1} + \nu_{y2}^{-1}$; $\Phi_c^{-1} = \Phi_{c1}^{-1} + \Phi_{c2}^{-1} + c_v^2/(\nu_{y1} + \nu_{y2})$
12. Two symmetrically positioned equal (1,2) elements	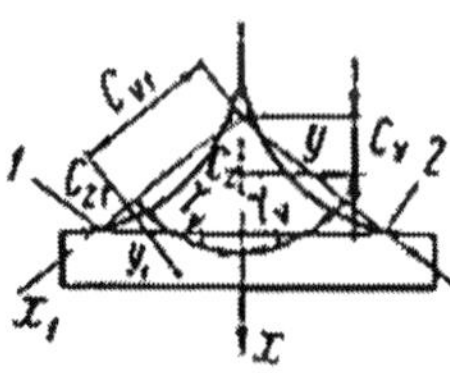	$c_{v1} = c_v\,(\nu_{x1}/\nu_{y1}\cos\gamma_v) \times [\sin^2\gamma_v + (\nu_{x1}/\nu_{y1})\cos^2\gamma_v]^{-1}$ $\Phi_c^{-1} = 2\Phi_{c1}^{-1}\{1 + (\Phi_{c1}/\nu_{y1}) \times c_v^2\,[1 + (\nu_{x1}/\nu_{y1})\cot^2\gamma_v]^{-1}\}$; $\nu_x = (\nu_{x1}/2)[\cos^2\gamma_v + (\nu_{x1}/\nu_{y1})\sin^2\gamma_v]^{-1}$ $\nu_y = (\nu_{y1}/2)[\sin^2\gamma_v + (\nu_{x1}/\nu_{y1})\cos^2\gamma_v]^{-1}$

After W. Lotze.

3-2. Concave Elastic Hinges and Rings Effective Theory

The concave elastic monolithic hinges are known as the notch hinges. Notch hinges are produced by drilling or milling two holes with a small separation t between them to form a circular (with radius R, Fig. 3-4a) or elliptical (with semi-axes a_x and a_y in Fig3-4b) notch or web which serves as a flexible element. Filleted notch hinge (Fig. 3-4c) in structure is similar to a leaf spring with thickness t, however, it is more convenient in assembly and helps to eliminate a stick-slip friction in the junction to the rigid elements of base and lever.

Leaf type hinge typically consists of a slender member connected at each end by two rigid bodies to provide a compliant coupling. Because of relatively high off-axis stiffness, the notch hinge is correspondingly more stable to applied parasitic forces. The leaf type hinge distributes the deflection over the length of the hinge, thus lowering stress and allowing greater deflection for a given hinge length. But in contrast to that of the notch hinge, the rotation axis of the leaf hinge is not localized and moves along the hinge as it deflects. The notch hinges with rectangular profile and usually filleted with radius R_f corners (Fig. 3-4c), which are similar to the leaf spring, are not as accurate as the circular notch hinges. Comparison of precision and motion options of the corner-filleted and elliptical notch hinges was discussed by W. Xu and T. King (1996) on the basis of the finite element analysis application. It was shown that corner-filleted notch hinge is more suitable for displacement oriented applications whereas the elliptical and circular notch hinges are more effective for precision oriented structures with smaller deflection of their central point. While each type of hinge is frequently used, there are many applications (*see* Chaps. 4 and 8) where notch hinges are preferable in comparison with the leaf hinge; the advantages are in their more stable and precise kinematics, simpler assembly, and overall smaller dimensions. But their stress conditions are more complex and require special analysis.

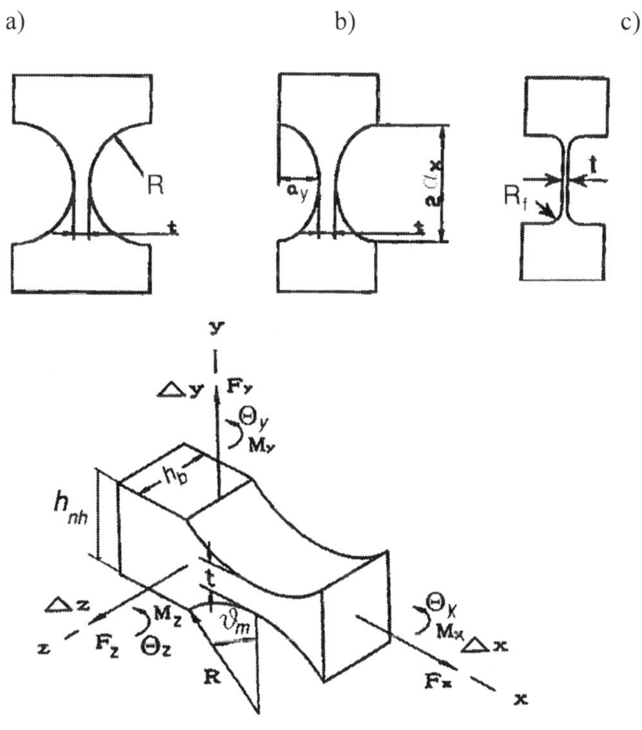

d)

Fig. 3-4. Notch hinges with different geometric parameters.

Circular notch hinge, theoretical relationships revisited.

There have been many attempts to derive satisfactory theoretical equations for the compliance, stiffness, and stress predictions of notch hinges, including traditional methods with integration of linear differential equations over the actual hinge contour.

"Exact solution". The full theoretical equation (J.M. Paros and L. Weisbord, 1965) for the rotational compliance around Z-axis (Fig. 3-4d) of notch hinges with a circular profile is:

$$Clrpw\gamma = \theta_z/M_z = 3(2Eh_bR^2)^{-1}(2\beta+\beta^2)^{-1}\{[(1+\beta)/\gamma^2)+ \frac{3+2\beta+\beta^2}{\gamma(2\beta+\beta^2)}][\sqrt{1-(1+\beta-\gamma)^2}]+[6(1+\beta)(2\beta+\beta^2)^{-3/2}]\times [\tan^{-1}(\sqrt{\frac{2+\beta}{\beta}}\frac{(\gamma-\beta)}{\sqrt{1-(1+\beta-\gamma)^2}})]\}, \quad (3\text{-}41)$$

where t is the thickness and $\beta = t/2R$ is the relative thickness of the hinge crosspiece; h_{nh} (Fig. 3-4d)is the width and $\gamma = h_{nh}/2R$ is the relative width of the hinge; h_b is the depth of the hinge, and R is the radius of the hinge contour. This "exact" equation is "unwieldy", as noted by its authors J.M. Paros and L. Weisbord in 1965. For full semicircular notch (right circle) the dimensionless parameter γ is equal to $\gamma = 1+\beta$ and the expression (41) for hinge compliance slightly reduces (S.T. Smith, 2000) but still remains an "unwieldy" one

$$Clrpw=\theta_z/M_z=3(2Eh_bR^2)^{-1}(2\beta+\beta^2)^{-1}\{[(3+4\beta+2\beta^2)(1+\beta)^{-1}(2\beta+\beta^2)^{-1}] +[6(1+\beta)(2\beta+\beta^2)^{-3/2}]\tan^{-1}(\sqrt{(2+\beta)/\beta})\}=3(2Eh_bR^2)^{-1}\,f(\beta). \quad (3\text{-}41a)$$

Derived later, Slocum's integral formula (A.H. Slocum, 1992) for semicircular notches is also complicated one as it is clear from the following slightly reduced presentation $Clrsl = \theta_z/M_z$

$$= 3R(2Eh_b)^{-1}\rho_{n2}^{-1}\{\rho_{n1}^{-1}+[\frac{2R^2+\rho_{n1}^2}{\rho_{n1}}+\frac{3R\rho_{n1}\left(\frac{\pi}{2}-\tan^{-1}\left(-R/\sqrt{\rho_{n2}}\right)\right)}{\sqrt{\rho_{n2}}}]\,\rho_{n2}^{-1}\}, \quad (3\text{-}42)$$

where $\rho_{n1}=0.5t+R$; $\rho_{n2}=\rho_{n1}^{\;2}-R^2$.

Paros and Weisbord derived a more concise approximate equation for circular thin notch hinge compliance

$$Clr_{SMP} = \theta_z/M_z = 9\pi R^{1/2}/(2Eh_bt^{5/2}). \quad (3\text{-}43)$$

This equation corresponds to the full theoretical formula (41) within 1% for circular hinges with relative thickness $\beta = t/2R$ in the range from 0.01 to 0.05, and within 5-12% for thicker circular hinges with β in the range 0.1 to 0.3. However, comparison of the calculations obtained from these theoretical equations with experimental data and finite element analysis results shows a much larger difference of up to 25% or more (*see* S.T. Smith, 2000).

A different version of equations for design calculations of circular notch flexure hinges were presented in 2002 by Yingfei Wu and Zhaoying Zhou (WZ) with parameter $s = R/t$. The general equation for angular compliance $Clrwz = \theta_z/M_z$ of a circular notch hinge in this publication is even more unwieldy (see Y. Wu and Z. Zhou , 2002) than (41) and (42). These authors showed that calculation results with their equations are identical to Paros and Weisbord full (or what they called "exact") formula for the same notch hinges. However, it was found by Smith that circular hinges represent the worst-case error between Paros and Weisbord's theoretical and finite element models. A direct finite element analysis method (S.T. Smith, 2000) is more reliable and better corresponds to experimental data, but it is not convenient for a synthesis of new hinge systems. For that purpose, S.T. Smith has derived the following empirical equation (*see* S.T. Smith and D.G. Chetwynd, 1992) from the finite element analysis results

$$Clr_{SEM} = \theta_z / M_z = 2K_{tR}R/EI_{zz};\ K_{tR} = 0.166 + 0.565t/R,\ I_{zz} = (1/12)h_b t^3 \quad (3\text{-}44)$$

This equation is concise and adequate, but only for thick circular hinges with relative thickness β in the range 0.1 to 0.5. Notch hinges in real structures have β in the range from 0.01 to 0.3 and may have different actual profiles. So it is clear that some different kind of theoretical equations, specifically based on the continuum mechanics for the whole range of the relative thickness and profiles, are necessary. The initial differential equations are not shown in Wu and Zhou's (2002) article. In Paros and Weisbord's paper as well as in Slocum's book the initial differential equation is presented as $y''_x = \theta' = M_z(x)dx/EJ(x)$ that is typical for small bending of straight elastic beams. Moreover the compliance calculation results by Wu and Zhou's formula, Slocum's formula (42), and Paros and Weisbord's formula (41a) are identical. If it is the case, we suppose that all three sources are based on the same theoretical foundation of a straight beam bending that they claim to be exact. But it can be shown that the behavior of the circular notch hinge, especially the thick one, is rather similar to the superposition of two bending imaginary parallel and opposite directed curved beams for which the initial differential equation is different than for straight beam. Note that section flexural rigidity constant for a curved beam is equal to:

$J_{zz} = \int_{-h/2}^{h/2} y^2 h_b dy/(1+y/R)$. When h/R is less than 0.1 or $\beta<0.05$, the value of J_{zz} is very close to the section moment of inertia, which is $J_z = \int_{-h/2}^{h/2} y^2 h_b dy$.

However, when h/R approaches value 1, the difference between J_{zz} and J_z becomes significant. Therefore, it is possible that the exact solution of the previously used differential equation is not the exact solution of the problem. With this in mind, we use here a different theoretical approach to the problem (*see* Y.M. Tseytlin, 2002).

Inversion of approximating contour.

Contour maps. The contour maps of circular notch hinges consist of two separated by $2\Delta b = t$ non-contiguous circular arcs with radius R that correspond to concave hinge surfaces. It is evident that each concave contour may be approximated by two contiguous circles with radius $R_c = l_x^2/2\Delta b$ shifted a distance, l_x (Fig. 3-5a). This approximation is necessary for possible application of inverse conformal mapping with the transformation of the circular contours to rectilinear ones and excluding an infinite point from the calculation field.

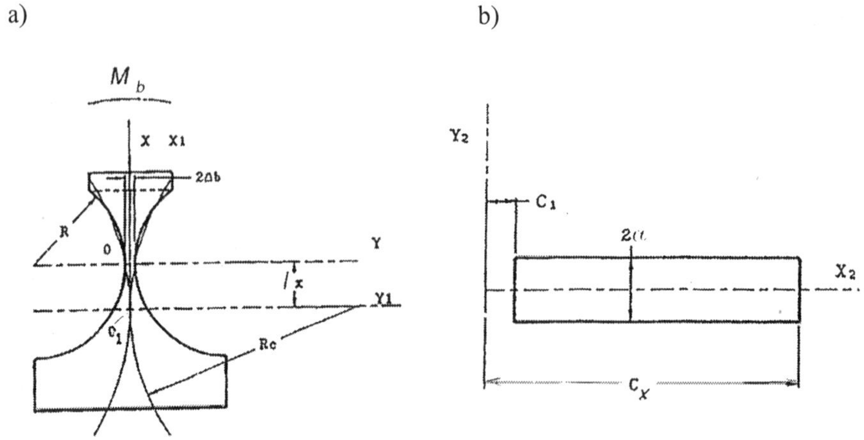

Fig. 3-5. Effective concave approximating contour and its conformal inversion

Without such approximation, inverse conformal mapping (J.H. Michell, 1901) would not be effective, unlike that developed for sharp arc sectors by A. Timpe (1905), as a result of the shift of real profiles relative to the center of inversion. In this case, circles would be transformed again into circles and that is not helpful. If the origin O_1 of new coordinates is positioned at the point of contiguity, the resulting coordinates will be equal to: $X_1 = X + l_x$ and $Y_1 = Y$. The approximating contour of the hinge will be converted with inversion to the rectangular band defined by X_2 and Y_2 (Fig. 3-5b) coordinates with dimensions

$$a = 1/2R_c,\ C_x = 1/\, l_x,\ C_1 = 1/\,(l_x + X_{\max}). \tag{3-45}$$

The relative thickness of the hinge crosspiece is expressed here as $\beta = t/2R = \Delta b/R$, where thickness $t = 2\Delta b$. Now the elastic deformation of the circular hinge can be effectively analyzed by using an inversion of the approximating contour. The two-dimensional mode displacements in plane elasticity U_x and V_y along axes X_1 and Y_1 are equal to:

$$U_X = (1/2G)\,[(1-\nu_p)\, S(X_1,Y_1) - \partial F_{sm}/\partial X_1] + U_o - \vartheta_o Y_1, \tag{3-46a}$$

$$V_y = (1/2G)\,[(1-\nu_p)\, T(X_1,Y_1) - \partial F_{sm}/\partial Y_1] + V_o + \vartheta_o X_1, \tag{3-46b}$$

where G is the shear modulus and ν_p is Poisson's ratio.
Under loading moment M Airy's stress-functions (*see* A.I. Lurie, 1970; A.Timpe) are

$$F_s = -\, M\, Y_2^{\,3}/\,(4\, a^3), \tag{3-47}$$

$$F_{sm} = -M\, Y_1^{\,3}\,/\,[(4a^3)(X_1^{\,2}+Y_1^{\,2})^2] \tag{3-48}$$

in inversion and the hinge fields, respectively (*see* Table 3-6).

Table 3-6. Airy's stress-function in the hinge and inverse fields

Type of Load	Inverse stress-function, F_s	*Hinges'stress-function, F_{sm}
Extension force P; $\Omega_\kappa = P/2a$	$(\Omega_\kappa/2)Y_2^2$	$(\Omega_\kappa/2)Y_1^2/r_1^2$
Moment, M; $\Omega_\kappa = 3M/2a^3$	$(\Omega_\kappa/6)Y_2^3$	$(\Omega_\kappa/6)Y_1^3/r_1^4$
Transverse force Q; $\Omega_\kappa = 3Q/4a^3$	$\Omega_\kappa\,(a^2Y_2 - Y_2^3/3)(C_x - C_1 - X_2)$	-

*in polar coordinates r_1, θ; ; $r_1 = (X_1^2+Y_1^2)^{1/2}$; $2a$ is the inverse strip height (*see* Fig. 3-5a, b).

In addition,

$$S(X_1,Y_1) + iT(X_1,Y_1) = \int_z f(Z)\, dZ\,, \tag{3-49}$$

and $f(Z) = s_\nabla + it_\nabla$. (3-50)
After yielding the Laplacian, $s_\nabla = \nabla^2 F_{sm}$, and conjugate harmonic function, t_∇, we have found by integration the expressions for the functions U_x, V_y, $\partial F_{sm}/\partial X_1$, and $\partial F_{sm}/\partial Y_1$ at $U_0 \approx V_0 \approx \vartheta_0 \approx 0$ in accordance with the boundary-value conditions. Let us recall that concentrated forces at inversion are expressed as a single-valued unambiguous function:

$M_b / h_b \Delta b = M / a$ and $M = M_b a / (h_b \Delta b)$, (3-51)

where M_b is the bending moment in the hinge field. Stress σ_x in the hinge body under bending moment is equal to the second partial derivative of the stress function F_{sm}. Therefore, at the side surface center point with $Y_1 = \Delta b$, $X_1 = l_x$, we have

$\sigma_x = \partial F_{sm}{}^2 / \partial Y_1{}^2 |_{\Delta b,\, lx} \approx 1.5\, M_b / (\Delta b^2 h_b)$. (3-52)

It is known that the boundary in the initial system, which is free of distributed tension before the inversion, is transformed into a boundary, which is subject to normal distributed tension in the inverse system. This tension rate has the same value at all points of the transformed boundary, and its effect on the hinge's kinematics is negligible. The stress concentration factor will be discussed later in this section. The tilt angle of hinge center line may be derived from (46) such that

$\theta = \partial V_y / \partial X_1 |_{Y1=0} = 3M_b(1-\nu_p) / (8Ga^2 h_b \Delta b^3 X_1)$. (3- 53)

The rigid body rotation angle θ_h on the hinge equals to two times the center line tilt at the center hinge point ($X_1 = l_x$, $Y_1 = 0$) and therefore

$\theta_h = 3M_b(1-\nu_p) l_x / [4G(\Delta b)^3\, h_b]$. (3-54)

Hence, the rotational compliance of the hinge is consequently equal to:

$Clr = \theta_h / M_b$. (3-55)

Circular notch hinge.

Contour approximation. The circles' effective shift l_x and effective radius R_c of the contour map may be found by using the least maximum absolute error approximation with Chebyshev's polynomial of the second order $R_2(Z)$ that has the smallest deflection $\tilde{\gamma}(Z)$ from zero in the range of $Z \subseteq [0,1]$. If the ratio δ_b of the effective circle deflection from the real profile to the crosspiece thickness of the hinge is assumed to be uniformly small, then the following equation with this condition can be used

$\delta_b = (1/\Delta b)\{[X^2/(RA_\xi \xi^2)] + \Delta b - (X + l_x)^2 / (R_c A_\xi \xi^2)\} \subseteq \{\gamma(X)\}$, (3-56)

where $\{\gamma(X)\}$ is the allowed relative error of the contour approximation. Values of the constants A_ξ and ξ depend on the range of the hinge's active contour polar angle (Fig. 3-4d), $\vartheta_m = \sin^{-1}(X_{max} / R)$, (3-57)
on the hinge body at the profile under consideration, as well as its presentation in accordance with the corresponding trigonometric function expansion by a series with one variable member (minimal polynomial). Note that the overall width of the hinge body is equal to: $W = h_{nh} = 2R(1-\cos\vartheta_m) + t$.
Here we have three options:
(i) thin notch hinges where the most significant part of deformation occurs within the boundaries of the hinge body corresponding to the polar angle $\vartheta_m \subseteq \pm 0.8$ rad,

(ii) thick notch hinges where the boundaries of approximation pertain to the whole profile at $X/R = 1.0$ and $\vartheta_m \subseteq \pm (\pi/2)$ rad; the width of the hinge body in this case is $W = 2R + t$ and $W/2R = 1+ \beta$,
(iii) intermediate notch hinges where the polar angle of the hinge body is considered within the limits $\vartheta_m \subseteq \pm 1.2$ rad.
Therefore, we have in accordance with Table 1-9 the following solutions:
(i) For the thin notch hinge with small ratio, $2\beta = t/R$ in the range $2\beta \leq 0.07$, the best approximation of the functions $\sin\vartheta$ and $(1-\cos\vartheta)$ by a series with one variable member corresponds to: $A_\xi = (1/0.478) = 2.092$ and $\xi = 0.922$,
(ii) for the thick notch hinge with ratio, $2\beta = t/R$ in the range $0.20 < 2\beta \leq 1.0$, the best possible expansion of the functions $\sin\vartheta$ and $(1 - \cos\vartheta)$ by series with one variable member corresponds to $A_\xi = (1/ 0.421) = 2.375$ and $\xi = 0.72$, and
(iii) for the intermediate notch hinges with ratio, $2\beta = t/R$ within $0.07 < 2\beta \leq 0.20$, the effective expansion of the functions $\sin\vartheta$ and $(1-\cos\vartheta)$ by series with one variable member corresponds to $A_\xi = 2.375$ and $\xi = 0.82$.
These approximations give us the option to derive concise formulae with comparatively moderate uncertainty. Maximum uncertainty of approximation for angle $\vartheta \subseteq \pm 0.8$ in the sin function is within ± 0.02 and in the $(1 - \cos\vartheta)$ function is within ± 0.003. We will see their effectiveness in comparison with the "exact" but "unwieldy" formulae of Slocum, Paros and Weisbord, their simplified representation for thin circular notch hinges, and the empirical Smith's formula derived from the finite element analysis data.

Substitution $X = X^{o}X_{max}$ into (56) yields an equation that may be approximated by Chebyshev's polynomial $R_2(X^{o})$ of second order from which a system of following equations is derived

$$R_c = l_x^2 / Q_b, \tag{3-58}$$

$$l_x = \{R + [R^2 + (0.8284)^2 R X_{max}^2 / Q_b]^{1/2}\} / (0.8284\, X_{max} / Q_b), \tag{3-59}$$

where $Q_b = A_\xi \xi^2 \Delta b$.
General equation. Substitution of l_x expression from (59) into (54) yields the general equation for angle θ_h of the circular notch hinge rotation under bending moment M_b as

$$\theta_h = \frac{3M_b(1-\nu_p^2)\{1+[1+0.6862(X_{max}/R)^2(R/\Delta b)/(A_\xi \xi^2)]^{1/2}\}}{2(0.8284)Eh_b(\Delta b)^2(X_{max}/R)/(A_\xi \xi^2)}. \tag{3-60}$$

Let us recall from Chap. 1 that Young's modulus $E = 2G(1+\nu_p)$. Substituting Poisson's ratio by $\nu_p = 1/3$ (typical for many metals) we finally obtain

$$\theta_h = \frac{4M_b\{1+[1+0.6862(X_{max}/R)^2(R/\Delta b)/(A_\xi \xi^2)]^{1/2}\}}{3(0.8284)Eh_b(\Delta b)^2(X_{max}/R)(A_\xi \xi^2)}. \tag{3-61}$$

Therefore, for thin circular notch hinges after the substitution of A_ξ and ξ into (61),

the rotational compliance in the thin hinge case (i) is equal to:

$$Clri = 4\{1+[1+0.1986\,(R/\Delta b)]^{1/2}\}/[Eh_b(\Delta b)^2]. \tag{3-62}$$

Coefficient 0.1986 may be changed to 0.215 if we assume that angle

$\vartheta_m \subseteq \pm 0.9$.
For thick notch hinges (ii), the rotational compliance is

$$Clrii = 4\{1+[1+0.5573(R/\Delta b)]^{1/2}\}/[2Eh_b(\Delta b)^2]. \quad (3\text{-}63)$$

If Poisson's ratio $\nu_p \neq 0.333$, multiply *Clrii* obtained from (63) by factor $(1-\nu_p^2) / 0.889$.
For intermediate notch hinges (iii), we have

$$Clriii = 4\{1+[1+0.373(R/\Delta b)]^{1/2}\}/[1.45Eh_b(\Delta b)^2]. \quad (3\text{-}64)$$

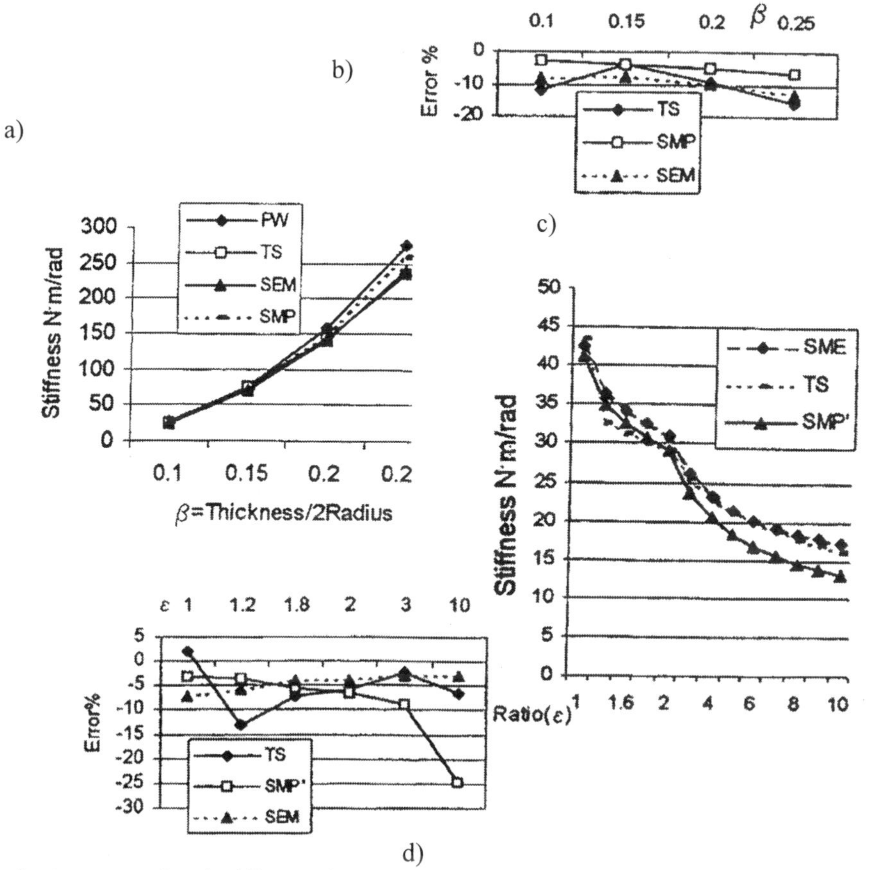

Fig. 3- 6. Rotational stiffness of notch hinges and its percentage error with different models

Comparison of the results for circular hinges. Several examples of the hinge stiffness $Kr = 1/Clr$ calculations with different formulae obtained for circular hinges by different authors are shown on the charts in Fig. 3-6a, b. These calculations have been made for the circular notch hinge with radius R=10 mm, depth h_b= 1mm, and Young's modulus E= 207 GPa. It is evident that stiffness calculations with Tseytlin's formulae *TS* (i, ii, iii) have good precision over certain ranges in comparison with results obtained through analytical and empirical formulae by other authors: Paros and Weisbord's (1965) full accurate formula (*PW*) at the hinge's relative width $W/2R = 1+\beta$; their simplified formula (*SMP*), Smith's (1987) empirical formula (*SEM*) from finite element analysis for thick circular hinges. Paros and Weisbord's accurate (PW) and Slocum's (1992) integral (*SL*) "unwieldy" formulae give identical calculation results: $PW = SL$. But *TS* charts are much closer to *SEM* than to others that shows its actual effectiveness. Let us recall that *SEM* formula, $\theta_h = 2KMR/EI_{zz}$ with $I_{zz} = h_b t^3/12$, $K_R = 0.166 + 0.565t/R$, looks concise but has limited application only to circular notch hinges with $0.1 < \beta \le 0.5$. All known formulae as stated have different forms, but it is possible to show that *SMP*, *SEM*, and *TS* (ii) formulae may be converted to a similar form through the following substitutions. For the *SMP* compliance: $\theta_{hSMP} / M_b = 28.274\, R^{1/2} / (2Eh_b t^{5/2})$. (3-65)

For the *SEM* compliance:

$\theta_{hSEM} / M_b = 33.94\,(1.13\,\beta^{1/2} + 0.166 / \beta^{1/2}) R^{1/2} / (2Eh_b t^{5/2})$. (3-66)

For the *TS*(ii) compliance:

$\theta_{hTSii} / M_b = 40.048\,(1.13\,\beta^{1/2} + 0.125 / \beta^{1/2}) R^{1/2} / (2.02Eh_b t^{5/2})$. (3-67)

It is seen from these equations that the structure of Tseytlin's formula (ii) is close to Smith's empirical formula for thick circular hinges, unlike the structure of Paros and Weisbord's simplified formula.

Elliptical notch hinge.

Profile approximation. In the Cartesian coordinates the elliptical profile may be specified by two semi-axes, a_x and a_y (*see* Fig. 3-4b) along *X* and *Y* coordinates at elliptical ratio, $\varepsilon = a_x / a_y$. It is easy to show that for elliptical notch hinge's compliance (stiffness) the radius *R* in (60) – (64) should be replaced by the ratio $a_x^2 / a_y = \varepsilon\, a_x = R_e$, because for an ellipse $Y_e = a_y \cos\vartheta$, $X_e = a_x \sin\vartheta$, $\vartheta = X / (a_x\xi)$, and $Y = a_y - a_y \cos\vartheta = a_y(1 - \cos\vartheta) = a_y X^2 / a_x^2 A_\xi \xi^2$. Hence, the relative deflection for the elliptical profile is

$$\delta_{be} = (1/\Delta b)\{[a_y X^2 / (a_x^2 A_\xi \xi^2)] + \Delta b - (X + l_x)^2 / (R_e A_\xi \xi^2)\} \subseteq \{\gamma(X)\}, \qquad (3\text{-}68)$$

and the general and thickness-dependent (i) equations for elliptical hinge compliance remain the same as those for the circular hinges, if we accept that $a_x^2 / a_y = R_e$ and radius *R* is substituted by a conditional radius, $R_e = \varepsilon\, a_x$. In this case, we have

$$\theta_{he} = \frac{3M_b(1-\nu_p^2)\{1+[1+0.6862(X_{max}/R_e)^2(R_e/\Delta b)/(A_\xi \xi^2)]^{1/2}\}}{2(0.8284)Eh_b(\Delta b)^2(X_{max}/R_e)/(A_\xi \xi^2)}. \qquad (3\text{-}69)$$

At $\nu_p = 1/3$, compliance for thin elliptical hinges will be predicted by equation

(i)$_e$ $Clrie = 4\{1+[1+0.1986(R_e/\Delta b)]^{1/2}\}/[Eh_b(\Delta b)^2]$. (3-70)

For thick elliptical hinges, compliance will be predicted by equation (ii)$_e$
$Clriie = 4\{1+[1+0.5573(R_e/\Delta b)]^{1/2}\}/[2Eh_b(\Delta b)^2]$. If Poisson's ratio $\nu_p \neq 0.333$, *Clriie* should be multiplied by the factor $(1-\nu_p^2)/0.889$. (3-71)

For intermediate elliptical hinges, compliance will be predicted by equation (iii)$_e$ $Clriiie = 4\{1+[1+0.373(R_e/\Delta b)]^{1/2}\}/[1.45Eh_b(\Delta b)^2]$. Relative thickness ($\beta = t/2R$) for elliptical hinges should be replaced by $\beta_e = t / 2R_e$.

The thin elliptical hinge equation (i)$_e$ is recommended at $\beta_e < 0.007$ and elliptical ratio $\varepsilon > 5$; the thick elliptical hinge equation (ii)$_e$ is for hinges with $0.1 < \beta_e < 0.3$; the intermediate elliptical hinge equation (iii)$_e$ is effective for hinges with

$0.007 < \beta_e \leq 0.1$. If $\varepsilon = 1$ and a_x = R, any of these equations [(69), $(i)_e$, $(ii)_e$, $(iii)_e$] will converge to similar equations for circular hinges (60-64).

Comparison of results for elliptical hinges. Comparison of calculations with different known equations for elliptical hinges with $\beta_x = t/2a_x = 0.12$, a_x = 10 mm, h_b =1 mm, ε=1to 10, and E=206.4 GPa is shown in (Fig. 3-6c,d), where: SME corresponds to Smith's model for elliptical notch hinges, TS to $(i, ii, iii)_e$ Tseytlin's approximation, SMP' to the simplified Paros and Weisbord's formula with Tseytlin's modification to expand it for elliptical profiles. Smith's theoretical formula for elliptical hinges has been derived (S.T. Smith, 2000) from Paros and Weisbord's "accurate" formula substituting β with $\beta_y = t / 2a_y$. However, as a result of such transformations, this new Smith's formula may be considered "unwieldy". With this in mind, we tried to modify Paros and Weisbord's simplified formula for circular notches with the substitution $R = R_e = a_x^2 / a_y$ which yields a similar transformation to the simplified formula for elliptical hinges $\theta_{he} = 9\pi\varepsilon M_b / 2Eh_b a_y^2(\beta_y)^{5/2}$. (3-72)
But this concise *SMP'* formula for elliptical hinges is not accurate enough as it is seen in Figures 3-6c,d. The difference between calculations with this modified *SMP'* formula, Tseytlin's approximate (*TS*) and Smith's (*SME*) accurate formulae for the same hinge is up to 25% at the elliptical ratio ε=10. It is clear from Figures 3-6c,d that all other results except *SMP'* are close enough, within 12% or less. Therefore, the modified equation *SMP'* may be used only for preliminary estimations if it is helpful. *SEM* corresponds to Smith's data from his finite element analysis of elliptical hinges.

Revisited conclusion. Computerized calculations are provided with a specially built program prepared by the author of this book for all known equations of circular and elliptical hinge rotational stiffness. From these calculations it is clear that inverse conformal mapping of the approximating contour provides a more effective theoretical and practical solution with respect to finite element analysis and experimental data of rotational stiffness especially for circular hinges. The general equation is common for all types of hinges with a possible approximation of their profiles by two shifted contiguous circles and includes material parameters with Young's modulus and Poisson's ratio. The latter is totally omitted in known theoretical solutions by other authors. New concise equations are derived from the original general equation on the basis of trigonometric functions' simplified series expansion in certain ranges of hinge crosspiece relative thickness and should be used accordingly: (i) at $\beta <= 0.035$; (ii) at $0.1 < \beta \leq 0.3$; (iii) at $0.035 < \beta \leq 0.1$ for the thin, thick, and intermediate circular hinges, respectively,$(i)_e$ at $\beta_e = t / 2R_e = (ta_y / 2a_x^2) \leq 0.007$, $\varepsilon > 5$; $(ii)_e$ at $0.1 < \beta_e \leq 0.3$; $(iii)_e$ at $0.007 < \beta_e \leq 0.1$ for the thin, thick, and intermediate elliptical hinges, respectively, with elliptical ratio ε up to10.
In these cases, the difference between theoretical rotational stiffness, properly built finite element analysis, and experimental models should not be more than 10%.

Virtual center of rotation. An elastic hinge body doesn't have actual guiding surfaces, and its center of rotation is a virtual instantaneous one. The position of this center in the hinge body can be estimated by an experimental study of its motion (*see* Y.M. Tseytlin, 2002). The hinge rotation virtual center also can be estimated by calculations because its position corresponds to the first kinematic pole of the rigid body connected to the hinge. The position of the instantaneous circular hinge virtual rotation center theoretically derived from *TS* formulae corresponds to the central point of the hinge. Its possible displacement should be less for thinner hinges with a smaller radius. Additionally, the elliptical hinge with the same length as the circular one, $2a_x = 2R$, has a lager conditional radius, $R_e = (a_x^2 / a_y)$. Therefore, this profile may have a smaller concentration stress and a larger limit of possible elastic rotation, but probably does not provide the same level of *ICR* stability in the center of the hinge body. The elliptical hinge represents some kind of intermediate solution between the circular hinge and leaf spring. And the leaf spring hinge does not provide a stable position for the center of rotation, which depends on the applied load and spring deflection from its initial position. N. Lobontiu *et al.* (2001) showed that as a consequence of being more compliant than the right circular flexure hinges, the corner-filleted flexures are also less precise in keeping the center of rotation with minimum offset. The shift of the rotation center caused by bending effect only can be 10 to 15 times larger in a corner-filleted flexure than the shift experienced by a similar right circular flexure hinge for the same amount of loading. For small/micro-scale applications where two links connected by a flexure must replicate pure rotation as close as possible, this aspect might pose serious limitations. It is clear that the problem with the virtual center of notch hinges is important and should be investigated further in depth.

Compliance around different axes.

Sometimes it is necessary to calculate displacement and rotation caused by auxiliary or parasitic forces (in relation to *X*, *Y*, and *Z* axes, *see* Fig. 3-4d). The compliance in these cases (Table 3-7, 3-7a) can be estimated on the basis of the elastic beam's small deformation theory. For example, we can use here the known differential equation, $EJ_y(x)Y'' = M_y$, to obtain by integration a formula for the compliance of notch hinges around axis *Y* at the application of moment M_y, because angle $\theta_y = 2Y'$ $= 2\int_0^{X\max} M_y \, dx/EJ_y(x)$ and the moment of section flexural inertia in relation to axis *Y* is equal to $I_y(x) = y(x)h_b^3/12$. The polar version of this equation may be obtained by the following substitutions: 1) for circular profile $dx = R\cos\vartheta d\vartheta$; $y = t + 2R(1-\cos\vartheta)$, where *R* and ϑ are the polar arguments for *y* contour of the notch, and $\vartheta_{\max} = \pi/2$ for semi-circle contour (so-called "right" circular hinge); 2) for elliptical profile $dX = a_x \cos\vartheta \, d\vartheta$, $y = t + 2a_y(1-\cos\vartheta)$, and $\vartheta_{\max} = \pi/2$ where a_x and a_y are the semi-axes with elliptical ratio $\varepsilon = a_x / a_y$. For the torsion compliance estimation, one can use the partial differential equation $\frac{\partial s_u^{(2)}}{\partial x} - \frac{\partial s_u^{(1)}}{\partial y} = 2\theta_\gamma z$ (see Table 1-10) in connection with the inverse conformal mapping of the approximating contour as it was above shown for the notch hinge bending. The compliance in relation to *X*, *Y*, and *Z* axes for the circular notch hinges are studied by J.M. Paros and L. Weisbord, and by S.T. Smith for notch hinges with elliptical contour. Paros and Weisbord (1965) presented the compliance equations for two-axis flexure with the "exact" expression for $(\theta_y/M_y)_{ta} = (\theta_z/M_z)_{ta}$ and their simplified formulae as well.

Table 3-7. Auxiliary compliance of circular notch hinges

Limit	$Cl_{\theta zFy}=\theta_z/F_y$	$Cl_{\delta yMz}=\delta Y/M_z$	$Cl_{\delta yFy}=\delta Y/F_y$	$Cly=\theta_y/M_y$	$Cl_{\theta yFy}=\theta_y/F_z$	$Cl_{\delta zMy}=\delta Z/M_y$
General	$(Clr)\ R\sin\theta_m$	$(Clr)\ R\sin\theta_m$	*	*	$(Cly)\ R\sin\theta_m$	$(Cly)\ R\sin\theta_m$
^Simplified	$(Clri)\ R\sin\theta_m$	$(Clri)\cdot R\times\sin\theta_m$	$(Clri)\ R^2\times(2\gamma-\gamma^2)$	$*(12/Eh_b^3)\{-2\tan^{-1}[\gamma/(2-\gamma)]^{1/2}+(2/\beta)^{1/2}\tan^{-1}[2\gamma/\beta\times(2-\gamma)]^{1/2}\}.$	$(Cly)\cdot R\sin\theta_m$	$(Cly)\cdot R\sin\theta_m$
Semicircle	$\mathbf{(Clrii)\cdot R}$	$\mathbf{(Clrii)\cdot R}$	$(Clrii)\ R^2$	$*(12/Eh_b^3)[\pi(1/2\beta)^{1/2}-2.57]$	$(Cly)\cdot R$	$(Cly)\cdot R$

Table 3-7a. Auxiliary compliance of circular notch hinges

Limitation	$Cl_{\delta zFz}=\delta Z/F_z$	$Cl_{\delta xFx}=\delta X/F_x$	$Cl_{\delta zFzs}=(\delta Z/F_z)_s$	$Cl_{\delta yFys}=(\delta Y/F_y)_s$	$Clx^{-1}=M_{xk}/\theta_x$
General	*	$(h_b^2/12)\ (Cly)$	*	$Cl_{\delta zFzs}$	-
Simplified	*	*	*	$Cl_{\delta zFzs}$	-
Semicircle (right)	$(12\pi R^2/Eh_b^3)[(1/2\beta)^{1/2}-0.25]$	$[\pi(1/2\beta)^{1/2}-2.57]/Eh_b$	$*[-2.57+\pi(1/2\beta)^{1/2}]/Gh_b$	$*[-2.57+\pi(1/2\beta)^{1/2}]/Gh_b$	$Gt^2\beta(\xi)h_b f_1(\beta)$ **

Note: $\theta_m = 2\tan^{-1}[(\gamma-\beta)/\sqrt{1-(1+\beta-\gamma)^2}]$; $Clr=\theta_z/M_z$ is the appropriate rotational compliance from equations (41);(41a);(42); (60); or (61); *Clri* is the rotational compliance from the thin notch hinge equation (62) that may be substituted by simplified equation (43); *Clrii* is the rotational compliance from thick notch hinges' equation (63) that may be substituted by the empirical equation (44); $Cly=\theta_y/M_y$ is the compliance around axis Y; postscript s corresponds to the shear deflection; $\gamma=h_{nh}/2R$; $\beta=t/2R$; $\sin\theta_m\approx 0.95\,\theta_m$ at $\theta_m<0.8$ (*see* options in Table 1-9) for simplified ^versions. Compliances which are marked with asterisk * correspond to J.M. Paros and L. Weisbord, 1965, presentation; the stiffness expression which is marked with double asterisk ** corresponds to C.-S. Han *et al.*, 2002, publication where $\beta(t/W_T)<\beta(\xi)<(t/h_b)$ at $W_T=(2R+t)$ is designated as a "geometric related" constant and $f_1(\beta)$ is the special function of β.

It is possible to use some of the equations from Y. Wu and Z. Zhou's (2002) article for the similar compliance calculation instead of the Paros and Weisbord's equations for Cly, Cl_{yFy}, Cl_{zFz}, Cl_{xFx}, Cl_{zFzs} Cl_{yFys}. But they are not very convenient for tractable synthesis of thin hinges because parameter involved in Wu and Zhou's (WZ) equations $s = R/t \geq 1$, while parameter $\beta = t/2R \leq 0.5$ in Paros and Weisbord's formulae. The latter parameter is more digestible and tractable in the theoretical analysis and synthesis. When t/R approaches one, the proportional difference between simplified Paros and Weisbord's (SMP) and (WZ) equations for θ_z/F_y is about 15 %; the proportional differences for $\delta Y/F_y$ and $\delta Z/F_z$ are about 25%. The similar differences for other compliances between the results from (*SMP*) and (*WZ*) equations are very large. But at $\beta \leq 0.3$ they approach only 30%, which is still permissible (*see* Chap. 6) for estimations by *SMP* because the accuracy of these auxiliary compliances is not as important as for the rotational compliance of the hinge. (*WZ*) equations may be used in numerical calculation procedures for right circular hinges as slightly more concise models in comparison with the full (*PW*) equations. All these relations may and should be verified on the basis of the correctness of the used theoretical foundation and experimental tests.

The stress concentration factor. Equation (52) defines the normal stress on the edges of a circular hinge crosspiece at its bending. However, the real stress in the beam's notch at its deformation increases by the so-called stress concentration factor K_t and is equal to: $\sigma_x = 6\ M_b\ K_t /\ t^2 h_b$. (3-73)

The problem with the stress concentration factor is indeed a complex one because this factor depends on the shape and parameters of notches in relation to the type of applied loads (*see* Y.M. Tseytlin, 1972). The simplest expression for the circular notch hinge stress concentration factor is presented by S.T. Smith (2000), where $K_t = (1+\beta)^{9/20}$. If we use this expression for K_t and derive the expression for moment M_b from equation (43), then (73) after K_t and M_b substitution and simple cancellations can be presented as follows

$$\sigma_x = 4\,(\sqrt{2})\,E\,\theta(1+\beta)^{9/20}\beta^{1/2} / 3\pi. \tag{3-74}$$

It is clear from (74) that stress (at a specified angle of rotation θ) is smaller for relatively thin notch hinges with smaller values of β. Furthermore, the stress concentration factor is not to exceed 1.12 for a typical range of notch hinge's relative thickness $\beta = 0.01$-0.3. If we use $K_t = \sqrt{(1+\beta)}$ instead of $K_t = (1+\beta)^{9/20}$, the error will be less than 2% in the specified range of β. In this case, it is easy to derive from (43), (73), and (74) the expression for the maximum allowable thickness t_{max} of the circular hinge crosspiece that is equal to

$$t_{max} = -R + \sqrt{R^2 + (\sigma_e / \theta)^2 9\pi^2 R/(16E^2)}\ , \tag{3-75}$$

where maximum angle θ of the hinge's rotation , yield stress σ_e, Young's modulus E of the material, and radius R of the hinge's profile should be specified. But for typical alternating stress in the notch hinge at work conditions, one should use endurance stress limit σ_{en} instead of σ_e value in the calculation of the maximum allowable thickness of the crosspiece. Therefore,

$$t_{max} = -R + \sqrt{R^2 + (\sigma_{en}/\theta)^2 9\pi^2 R/(16E^2)}\ . \tag{3-76}$$

Elastic ring kinematics.

The circular elastic ring is one of the most common elements in machinery and instrumentation. Rings in instrumentation are usually thin and can be used as an elastic link. The theory for the calculation of thin elastic rings is well developed and will not be discussed in this book. Some typical loading schematics and calculation formulae are shown in Table 3-8 where forces P, bending moment M_1 in the ring plane, moment M_2, force P_2, and torque M_k in the perpendicular plane sections are applied. N, Q_1, and Q_2 are the normal and shear internal forces in the ring's plane and off plane, respectively.

Table 3-8. Load and displacement of ring elements

Loading schematic**	Calculation formula**
 1	δ_x=-0.137 PR^3/EJ; δ_y= + 0.149 PR^3/EJ M_1=-PR(0.3183-½ sinφ); N=½ Psinφ; Q_1=-½ Pcosφ
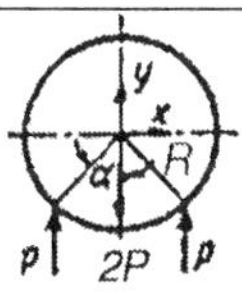 2	$\delta_x = (PR^3/EJ)\,[-\tfrac{1}{2}\sin^2\alpha + 0.6366(\alpha\sin\alpha + \cos\alpha - 1)]$; $\alpha \le \varphi \le \pi$; $\delta_y = (PR^3/EJ)\,[\tfrac{1}{2}(\alpha + \sin\alpha\cdot\cos\alpha) + 0.6366(\alpha\sin\alpha + \cos\alpha - 1) - \sin\alpha]$; at $0 \le \varphi \le \alpha \Rightarrow$ $M_1 = PR[0.3183(\alpha\sin\alpha + \cos\alpha + \sin^2\alpha\cos\varphi - 1) - \sin\alpha + \sin\varphi]$; at $\alpha \le \varphi \le \pi \Rightarrow$ $Q_1 = -0.3183\, P\sin^2\alpha\sin\varphi$; $N = 0.3183 P\sin^2\alpha\cos\varphi$
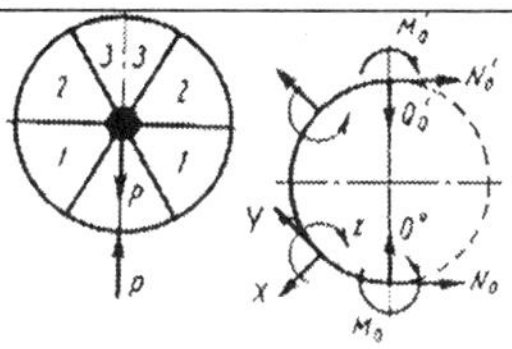 3	$X_{Q1} = \tfrac{1}{2}(\nu_1 + \nu_2)$; $Y_{Q1} = \tfrac{1}{2}(\phi_1 + \phi_2)$; $Z_{M1} = \tfrac{1}{2}(\wp_1 + \wp_2)$; $N_0 = \tfrac{1}{2}(S_{Q1} + S_{Q2})$ $X_{Q3} = \tfrac{1}{2}(\nu_1 - \nu_2)$; $Y_{Q3} = \tfrac{1}{2}(\phi_1 - \phi_2)$; $Z_{M3} = \tfrac{1}{2}(\wp_1 - \wp_2)$; $M_0 = \tfrac{1}{2}(\iota_1 + \iota_2)$ $X_{Q2} = S_{Q1} - \tfrac{1}{2}\nu_1 - 0.8660\phi_1$; $Y_{Q2} = 0.8660\nu_2 - \tfrac{1}{2}\phi_2 - P/2$; $Z_{M2} = \iota_2 + S_{Q2}R - \wp_2 - + \phi_2 R - Y_{Q2}R$; $M'_0 = \tfrac{1}{2}(\iota_1 - \iota_2)$; $N'_0 = \tfrac{1}{2}(S_{Q1} - S_{Q2})$; $\chi_3 = L^3 E_r J_r/(R^3 E_s J_s)$; *
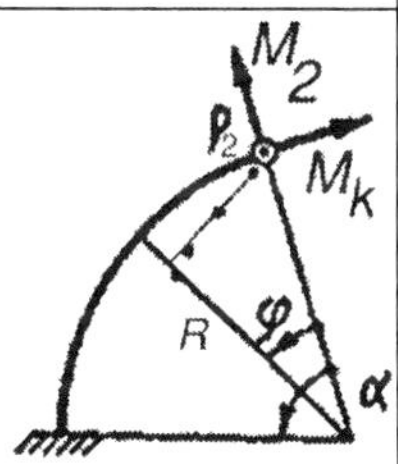 4	$u = (R^2/EJ)\{P_2R(\tfrac{1}{2}(3\lambda_k + 1)\alpha + \tfrac{1}{4}(\lambda_k - 1)\sin 2\alpha - 2\lambda_k\sin\alpha) +$ $M_2[\tfrac{1}{2}(1 - \lambda_k)\sin^2\alpha - \lambda_k(1 - \cos\alpha)] +$ $M_k[\lambda_k\sin\alpha - \tfrac{1}{4}(\lambda_k - 1)\sin 2\alpha - \tfrac{1}{2}(\lambda_k + 1)\alpha]\}$; $\lambda_k = EJ_2/GJ_k$; $\vartheta_2 = (R/EJ)\{P_2R[\tfrac{1}{2}(\lambda_k - 1)\sin^2\alpha - \lambda_k(1 - \cos\alpha)] + M_2[\tfrac{1}{2}(\lambda_k + 1)\alpha$ $- \tfrac{1}{4}(\lambda_k - 1)\sin 2\alpha] + \tfrac{1}{2}M_k(\lambda_k - 1)\sin^2\alpha\}$; $\vartheta_k = (R/EJ)\{P_2R[\tfrac{1}{2}(\lambda_k - 1)\sin 2\alpha + \tfrac{1}{2}(\lambda_k + 1)\alpha - \lambda_k\sin\alpha]$ $+ \tfrac{1}{2}M_2(\lambda_k - 1)\sin^2\alpha + M_k[\tfrac{1}{2}(\lambda_k + 1)\alpha + \tfrac{1}{4}(\lambda_k - 1)\sin 2\alpha]\}$
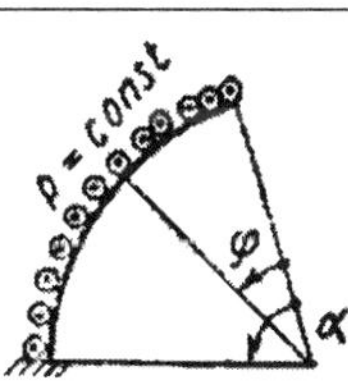 5	$M_2 = -pR^2(1 - \cos\varphi)$; $M_k = -pR^2(\varphi - \sin\varphi)$. $u = \dfrac{pR^4}{EJ}[(1 - \cos\alpha)^2 + \lambda_k(\alpha - \sin\alpha)^2]$; $\vartheta_2 = \dfrac{pR^3}{EJ}[(\lambda_k + 1)\times$ $(\sin\alpha - \alpha/2) + \dfrac{\lambda_k - 1}{4}\sin 2\alpha - \lambda_k\alpha\cos\alpha]$;

Table 3-8 Cont'd.

Loading schematic**	Calculation formula**
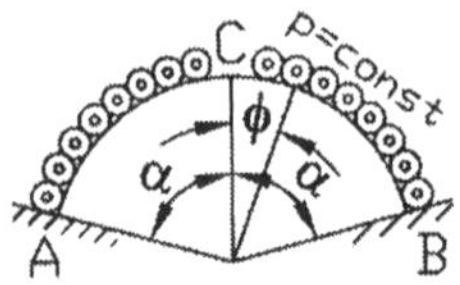 6	$Q_{2C}=0$; $M_{kC}=0$; $M_{2C}=\dfrac{-pR^2\{\sin\alpha-(1/4)\sin 2\alpha-(1/2)\alpha+\lambda_k(-\alpha\cos\alpha+\sin\alpha+(1/4)\sin 2\alpha-\alpha/2)\}}{(1/4)\sin 2\alpha+\lambda_k[-(1/4)\sin 2\alpha+\alpha/2]+\alpha/2}$

** After I.A. Birger, N.A. Fedorov, 1968, where formulae * for parameters ν_1, ν_2, ϕ_1, ϕ_2, ρ_1,ρ_2,

S_{Q1}, S_{Q2}, l_1, and l_2 are presented with the assumption that L/R=0.9.

Certain changes in diameters δ_x, δ_y, linear displacements w, v, u in radial, tangential, and binormal directions, angular displacement ϑ_2, and twist ϑ_k may be estimated on this basis. The plus sign in formulae corresponds to an increase in diameter or to displacement in the direction of applied force. Vectors shown by the circle with a dot or cross are perpendicular to the plain of the ring; EJ, GJ_k are the flexural and twisting stiffness of the ring's (circular beam's) cross section, and $\lambda_k= EJ/GJ_k$. If the ring has initial bending v_{in}, w_{in}, and κ_{vwin} deflection in its plain, the displacement of its points from circular form V, W and alteration of curvature κ_{vw} will be as follows $V = V_{in} + V_{el}$; $W = W_{in} + W_{el}$; $\kappa_{vw} = \kappa_{vwin} + \kappa_{vwel}$, (3-77)
where V_{el}, W_{el}, and κ_{vwel} are the tangential, radial elastic displacements, and elastic change of curvature under applied loads. A ring with six rigidly connected to it straight spokes is shown in schematic 3 (Table 3-8), where X_i, Y_i, Z_i are the normal and shear forces and the bending moment in the ith spoke, L is the length of a spoke, R is the radius of the ring, subscripts s and r designate values of Young's modulus E and section moment of inertia J related to spokes and to the ring, respectively. Note that the "hub" for the spokes connection in the center point is assumed to be absolutely rigid and the active length of the spoke is equal to 0.9R. These formulae show that even in the comparatively simple problem, the direct synthesis procedure can be unwieldy. This approach is acceptable to analysis where now the computerized finite element model may be preferred. Therefore we try to find digestible and tractable ways, which may be more effective in the structural synthesis. One of them is the conformal mapping, which we applied to the notch hinge's problem and will discuss for another problem in Section 3-3.
Schematic 4 in Table 3-8 represents loading of the cantilever ring segment by concentrated force P_2 in the perpendicular to its plane direction, bending M_2 and twisting M_k moment on the free end of the cantilever. This loading causes displacement u in the perpendicular to the segment's initial plane direction, bending of angle ϑ_2 and twisting ϑ_k of its free end. Schematic 5 in Table 3-8 represents loading of the cantilever ring segment in the perpendicular to its plane direction by distributed load with the intensity p. In addition to displacement u and bending rotation ϑ_2 of the free end, the latter will have twist of angle

$\vartheta_k=\frac{pR^3}{EJ}[(\lambda_k+1)(1-\cos\alpha)-\frac{\lambda_k-1}{4}(1-\cos2\alpha)-\lambda_k\alpha\sin\alpha]$. Schematic 6 in Table 3-8 represents loading of the ring segment which is clamped on both ends. Distributed load p is applied in the perpendicular to the segment's initial plane direction and causes bending moment M_{2C} in the central point C. If the segment's angle α=90°, then (*see* I.A. Birger, N.A. Fedorov, 1968) the segment does not have in perpendicular to its initial plane direction the shear stress and twisting moment in the C section, $Q_{2C}=0$, $M_{kC}=0$, but the bending moment in this case is equal to:

$$M_{2C}=pR^2(1-4/\pi). \tag{3-78}$$

These values in the other sections of the ring segment are as follows:

$$M_{2A}=M_{2B}=pR^2; \tag{3-79}$$

$$M_{kA}=-M_{kB}\approx 0.3pR^2; \tag{3-80}$$

$$Q=pR\phi; \tag{3-81}$$

$$M_2=M_{2C}\cos\phi+pR^2(1-\cos\phi); \tag{3-82}$$

and $$M_k=M_{2C}\sin\phi+pR^2(\phi-\sin\phi). \tag{3-83}$$

3-3. Diaphragms, Plates, Bellows, Shells, and Manometric Tubes Relations

Diaphragms. In many manometric instruments a diaphragm is a round flexible plate to which a considerable elastic deflection w is imparted by pressure p or concentrated force. Diaphragms can have either linear or nonlinear pressure characteristics depending on their geometry. Diaphragms of the simplest shape are the flat ones whose characteristics are regressive (decreasing or negative monotonic) for large deflections. Corrugation of the diaphragm increases its possible deflections. The latter diaphragm characteristic can be varied by changing the shape and dimensions of the corrugations. They can be made with linear, regressive, or progressive (increasing or positive monotonic) characteristics. Another option to increase the deflection of the flat diaphragm is the perforation of the plate. However the perforated elastic plate looses the sealing ability and can be used only in kinematics function, most often as a guidance device. Convex (spherical or conical) diaphragms find application in various relays and signaling devices. When such a diaphragm is loaded from the convex side a loss of stability occurs and the diaphragm inverts (deflects by jump). This phenomenon is applied, e.g., to the opening and closing of electrical contacts. Nonmetallic diaphragms have a very low spring rate and accordingly operate as a rule in conjunction with helical measuring springs. The diaphragm transforms the pressure into a force and the measuring spring than takes the force up.

Flat diaphragms. The diaphragm is usually stretched on its perimeter when fastened to the instrument body, particularly by conical flange. Initial tension is sometimes made use of in order to influence the characteristic; the latter can be varied within wide limits by varying the value of the initial tension of the flat diaphragm.

Fig. 3-7shows: a) deformation under pressure p, b) radial σ_r and tangential (peripheral) σ_t stress in a diaphragm segment under action of pressure p with the developed internal forces Q_r, radial M_r and tangential (peripheral) M_t moments; and c) internal forces Q_r and radial moments M_r in the central segment. Consider a flat diaphragm fastened along its perimeter and loaded by pressure (Fig. 3-7a). The pattern of the diaphragm deformation (Fig. 3-7b) depends on the magnitude of its deflection under a load. Small diaphragm deflections are mainly due to bending deformations (stresses $\sigma_{r.b}$, $\sigma_{t.b}$). The median (equidistant from the diaphragm surfaces) plane suffers no elongation, and it can be assumed that the diaphragm has a linear characteristic.

a) c)

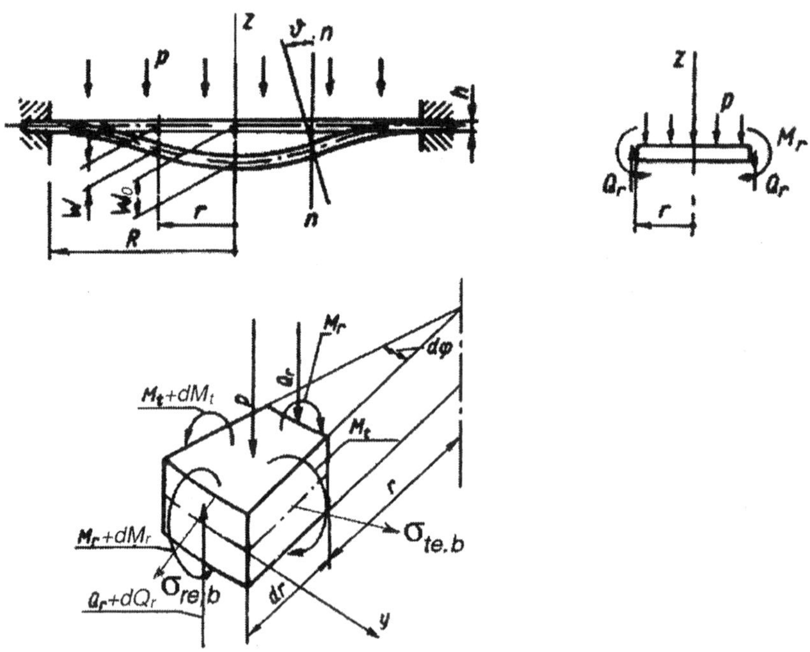

b)

Fig. 3-7. Strain and stress in a flat diaphragm

The linear theory of bending of round plates can be used in the calculation of diaphragms in the range of small deflections. With the increased load, the deflections of the diaphragm reach a value roughly equal to its thickness, the median plane is subjected to elongation and tensile stresses $\sigma_{r.e}$, $\sigma_{t.e}$ of a magnitude commensurate with the bending stresses σ_b are set in the median plane. As a result, the diaphragm deflections increase more slowly than the load, and the characteristic becomes decreasing. Calculation of the diaphragm in the region of large deflections must be based on the nonlinear theory, which takes into account both the bending and the stretching of the diaphragm in the median plane. Further increases in deflection result mainly from stretching. Here the calculation can be conducted using the theory of the absolutely flexible diaphragms (membranes) and neglecting the flexural rigidity. The shape of the elastic surface of the diaphragm changes as the deflection increases. The inflection point on a diametric diaphragm section moves toward the perimeter with the increase in deflection, and the elastic surface approaches a spherical shape. This variation in shape of the elastic surface is the main reason for the difficulties in obtaining an accurate solution embracing both the ranges of small and large deflections. The active diameter of the diaphragm is determined by the available space in the structure of application, the required effective area, and the specified deflection of its center. The diaphragm thickness is selected in accordance with the required sensitivity, strength, and other technical parameters including safety factor.

Table 3-9. Calculation of a flat diaphragm in the range of small deflection

Schematic, Load	Formulae*
1.Fig.3-7a,b, c; pressure p, edge clamped	$\vartheta = -(pr/16D)\rho_{rd2}$; $w = p\rho_{rd2}^2/64D$; $w = w_o(1-\rho^2)^2$; $w_o = w_{r=0} = pR^4/64D$; $\sigma_{rb} = \pm(3p\rho_{Rh2}/8)[(3+\nu_p)\rho^2 - (1+\nu_p)]$; $\sigma_{tb} = \pm(3p\rho_{Rh2}/8)[(3\nu_p + 1)\rho^2 - (1+\nu_p)]$; $\sigma_{eq\rho=1} = (3p\rho_{Rh2}/4)(1-\nu_p+\nu_p^2)^{1/2}$; $\sigma_{eqo} = 3p(1+\nu_p)\rho_{Rh2}/8 \approx 0.488p\rho_{Rh2}$ $p_{ald} = 1.5[\sigma_{ald}]\rho_{Rh2}$; $w_{ald} = 0.256[\sigma_{ald}]\rho_{Rh2}/E$; $D = Eh^3/12(1-\nu_p^2)$; $w_{\rho=1} = 0$; $\rho_{rd2} = R^2 - r^2$; $\rho_{Rh2} = R^2/h^2$; $V = (1/3)\pi R^2 w_0$
2. As in 1, but edge simply supported**	$w = \frac{pR^4(1-\rho^2)}{64D}[\frac{(5+\nu_p)}{1+\nu_p} - \rho^2]$; $w_O = \frac{(5+\nu_p)pR^4}{64(1+\nu_p)D}$; $M_r = p(3+\nu_p)\rho_{rd2}/16$; $M_t = pR^2[(3+\nu_p) - \rho^2(1+3\nu_p)]/16$; $\sigma_{r\max} = \sigma_{t\max} = 3(3+\nu_p)p\rho_{Rh2}/8$
3. With rigid center r_o, Fig.3-8a, pressure $p\neq0$, $Q=0$, $M=0$	$w_o = A_p pR^4/Eh^3$; $A_p = 3(1-\nu_p^2)(k^4 - 1 - 4k^2\ln k)/16k^4$; $k = R/r_o$; $\sigma_{rout} = \sigma_{rin} = \pm B_p Ehw_o/R^2$; $B_p = 4k^2(k^2-1)(k^4 - 1 - 4k^2\ln k)^{-1}/(1-\nu_p^2)$
4.With rigid center r_o, Fig. 3-8b, and concentrated force Q, $p=0$, $M=0$	$w_0 = A_Q QR^2/Eh^3$; $Q_r = Q/(2\pi r)$; $A_Q = 3(1-\nu_p^2)\{[(k^2-1)/4k^2] - \ln^2 k/(k^2-1)\}/\pi$, $\sigma_{qrout} = \pm B_{Qex} Ehw_0/R^2$; $\sigma_{qrin} = \pm B_{Qin} Ehw_0/R^2$; $B_{Qex} = 2k^2(k^2-1-2\ln k)/[k_m(1-\nu_p^2)]$; $B_{Qin} = 2k^2(2k^2\ln k - k^2 + 1)/[k_m(1-\nu_p^2)]$; $k_m = (k^2-1)^2 - 4k^2\ln^2 k$ $A_{eff} = Q/p = R^2 A_p/A_Q$
5. With rigid center r_0 Fig. 3-8c,external bending moment M, $p=0$, $Q=0$	$\varphi = A_\varphi M/Eh^3$; $A_\varphi = 3(1-\nu_p^2)[\ln k - (k^2-1)/(k^2+1)]/\pi$, $\sigma_{rout} = B_{\varphi ex}\varphi Eh/R$; $\sigma_{rin} = B_{\varphi in}\varphi Eh/R$; $B_{\varphi ex} = 3A_\varphi^{-1}\pi^{-1}(k^2-1)/(k^2+1)$; $B_{\varphi in} = B_{\varphi ex}k$

*After L.E.Andreeva, 1962, except ** which is after J.T. Oden *et al.*, 1981.

Here: h is the thickness of the diaphragm; R is its external (outer) radius; r is the current radius for calculations, $\rho = r/R$; E and ν_p are Young's modulus and Poisson's ratio of the diaphragm's material; w is the diaphragm deflection at the circle with radius r; w_o is the deflection of the diaphragm center (inner contour); D is the cylindrical flexural rigidity; σ_{rb} and σ_{tb} are the bending stresses in the radial and peripheral directions at points $\pm h/2$, respectively; σ_{eqo} is the equivalent stress in the diaphragm's center, $\sigma_{eq\rho=1}$ is the equivalent stress in the diaphragm's edge where at $\nu_p = 0.3$ $\sigma_{eq\rho=1} \approx 0.666p\rho_{Rh2} \leq \sigma_{ald}$; w_{ald}, p_{ald}, and σ_{ald} are the allowed deflection, pressure, and stress, respectively, r_o is the radius of the rigid center; σ_{rout} and σ_{rin} are the maximum stress at points of outer and inner perimeters; B_{Qex}, B_{Qin}, A_p, B_p, A_Q, $B_{\varphi ex}$, $B_{\varphi in}$, and A_φ are the coefficients whose numerical values are given in Fig. 3-15 a-d as a function of argument $k = R/r_o$; φ is the angle of rotation of the rigid center that is loaded by external moment M; M_r and M_t are the internal radial and the peripheral moments per the unit length of circle with radius r, respectively, Q is the transverse force, V is the volume between the initial plane of diaphragm and its elastic surface, A_{eff} is the effective area.

Small deflection. Calculation formulae for internal stresses and displacement in the diaphragm at small deflections are presented in Table 3-9.
The differential equation for the round plate subjected to small deflection with a uniform distributed load can be expressed as follows

$$\frac{d}{dr}\left(\frac{d^2w}{dr^2}+\frac{1}{r}\frac{dw}{dr}\right)=-Q_r/D\text{, or} \tag{3-84}$$

$$r\vartheta'' + \vartheta' - \vartheta/r = Q_r r / D, \tag{3-85}$$

where $Q_r = pr/2$, $\vartheta = dw/dr$ is an angle of section's rotation (*see* Fig. 3-7a) , w is the deflection of the diaphragm's surface ,and D is the cylindrical rigidity.

a) c)

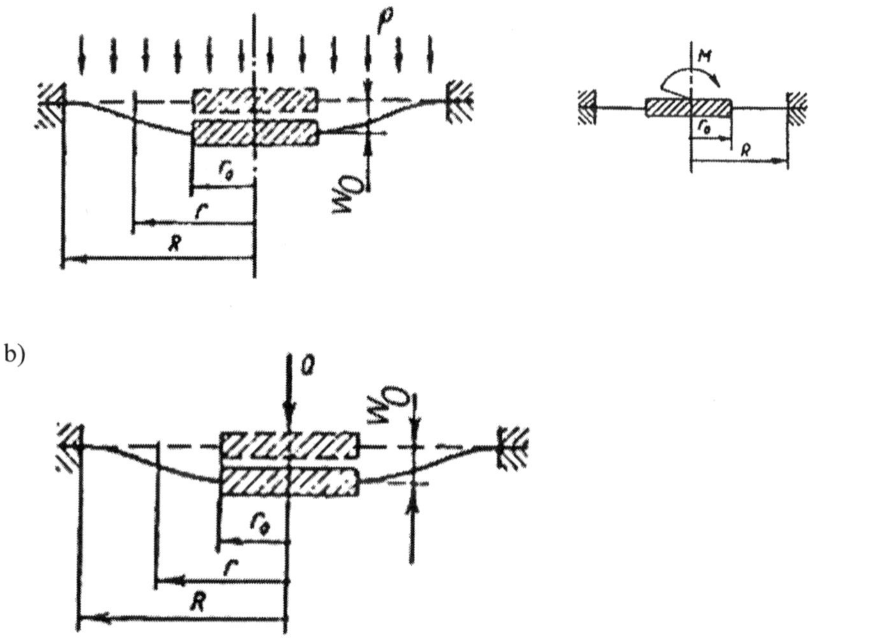

Fig. 3-8. Strain in flat diaphragm with rigid center.

It is convenient to characterize the ability of a manometric element to produce driving force $Q=2\pi r Q_r$ (Fig. 3-7c, Fig. 3-8a, b) by the ratio of this force to the corresponding pressure p. This ratio has the area's dimension and therefore is called effective area. Since flat diaphragms have a linear characteristic $w=f(p,Q_r)$ in the region of small deflections, their effective area can be expressed as follows:

$$A_{eff} = Q/p = A_p / A_Q\, R^2, \tag{3-86}$$

where A_p and A_Q are the coefficients whose numerical parameters are presented in Fig. 3-15a (*see* notes to Table 3-9). The effective area for small deflection of a diaphragm with the rigid center can be expressed with sufficient accuracy by the approximate formula $A_{eff} = \pi R_m^{\,2}$, where $R_m = ½\,(R + r_o)$ (3-87)
and r_o is the radius of the diaphragm's rigid center.
Large deflections. The determination of characteristics for the flat diaphragm in the range of large displacements is based on the "superposition" method. The basic idea

behind the superposition method is to treat the resistance of the diaphragm to an external load as the sum of the resistance to bending and tension. The resistance of the diaphragm to bending is determined using the linear theory of the bending of a round plate while the resistance to tension is determined from the calculations for an absolutely flexible diaphragm, which operates in tension only. A thin diaphragm operating with very large deflections ($w_o >> h$) can be considered as being absolutely flexible membrane. The tensile stresses in such a diaphragm will be much higher than the bending ones, and the flexural rigidity can therefore be neglected in the calculation. The corresponding calculation formulae for the flat diaphragms in the range of large displacements are shown in Table 3-10.

Table 3-10. Flat diaphragm in the range of large deflection

Schematic, Load	Formulae*
1. Absolutely flexible, Fig. 3-7a, pressure p	$\vartheta = C\rho$; $C = -2w_o / R$; $p\rho_{Rh2}^2/E = (7-\nu_p)(1-\nu_p)^{-1}w_o^3/3h^3$; at $\nu_p = 0.3$ $p\,\rho_{Rh2}^2/E = 3.58w_o^3/h^3$; $\sigma_{re} = (Ew_o^2/4R^2)[(3-\nu_p)(1-\nu_p)^{-1}-\rho^2]$; $\sigma_{te} = (Ew_o^2/4R^2)[(3-\nu_p)(1-\nu_p)^{-1}-3\rho^2]$; $\sigma_{r,t\,\max r=0} = (3-\nu_p)(1-\nu_p)^{-1}Ew_o^2/(4R^2)$
2. Absolutely flexible with rigid center, Fig. 3-8a	$p\rho_{Rh2}^2/E = B_{pa}w_0^3/\mathrm{h}^3$; $B_{pa} = [3^{-1}(7-\nu_p)(1+\rho_0^2+\rho_o^4) + (3-\nu_p)^2(1+\nu_p)^{-1}\rho_0^2]/B_m$; $B_m = (1-\nu_p)(1-\rho_0^4)(1-\rho_0^2)^2$
3. Absolutely flexible with pretension U_1 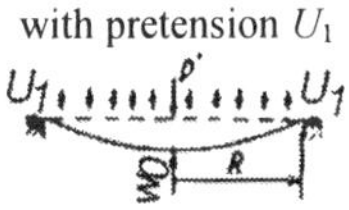	$p\rho_{Rh2}^2/E = [(7-\nu_p)(1-\nu_p)^{-1}w_o^3/3h^3] + 4(1-\nu_p)^{-1}w_o\,U_1R/h^3$
4. Regular under pressure p, by SM-superposition method	$\vartheta = C(\rho^{\check{r}} - \rho)$; $p^{\sim} = p_b^{\sim} + p_t^{\sim}$; $p^{\sim} = p\rho_{Rh2}^2/E$; $p_b^{\sim} = 16\,w_o(1-\nu_p^2)^{-1}/3h$; $p_t^{\sim} = (7-\nu_p)(1-\nu_p)^{-1}w_o^3/3h^3$; @ $\nu_p = 0.3 \rightarrow p\rho_{Rh2}^2/E = 5.86\,(w_o/h) + 3.58w_o^3/h^3$; $w = w_o\{(\check{r}-1)^{-1}[2\rho^{\check{r}+1} - (\check{r}+1)\rho^2] + 1\}$
5. Regular, under pressure p, rigid center, by SM	$p\rho_{Rh2}^2/E = (A_p^{-1}w_o/h) + B_{pa}w_0^3/h^3$

* After L.E. Andreeva. Here $\rho_o = r_o/\mathrm{R}$; $\check{r}$ is a parameter which defines the shape of the elastic surface and depends on the deflection of the diaphragm; $\check{r} = 3$ corresponds to small deflections, $\check{r} \rightarrow \infty$ corresponds to the nearly spherical surface in the case of an absolutely flexible diaphragm; C is a constant which together with parameter $\check{r}$ can be defined from the integration of corresponding differential equations (94); $\rho_{Rh2} = R^2/h^2$. The expressions of coefficient A_p is the same as in Table 3-9 for schematic 3.

The system of differential equations for a real regular flat diaphragm subjected to large deflections (V.I. Feodos'ev, 1949) has a superposition form

$$\rho\psi'' + \psi' - \psi/\rho = \tfrac{1}{2}\,\vartheta^2;\ \psi = -N_r\rho/(Eh);\ \rho = r/R, \tag{3-88}$$

$$\rho\vartheta'' + \vartheta' - \vartheta/\rho = (R^2/D)\,[-Eh\psi\vartheta + \tfrac{1}{2}\,pR\rho^2], \tag{3-89}$$

where ψ is the function of radial force N_r. The later study (L.E. Andreeva, 1981) shows that while the superposition method is accurate enough for getting the diaphragm deflection nonlinear characteristic, it is not accurate for stress calculation in the range of large displacements. The latter can be expected because the superposition does not work properly in nonlinear systems.

Nonlinearity. In many cases the nonlinearity should be minimized. It is possible to show that the nonlinearity of an absolutely flexible diaphragm can be very small at the certain radial pretension U_1. Let us find the connection between expected approximate quasi-linear characteristic $pR^4/Eh^4 = F_o\,(w_o/h)$, the radial pretension U_1, and the expected remaining relative nonlinearity $\gamma(w_o/h)$ of the diaphragm characteristic. At substitution ν_p=1/3 (typical for many metals) and the quasi-linear expression for the left hand side member in the formula of clause 3 from Table 3-10 for the absolutely flexible diaphragm with pretension, we will obtain the equation that can be approximated by Chebyshev's polynomial $P_3(Z)$ from Table 1-7. In this case,

$$(w_o/h)^3 + (w_o/h)\,3.33^{-1}(6U_1R/h^2 - F_o) = 0 \qquad \text{(3-90)}$$

and the corresponding Chebyshev's polynomial with minimal absolute deflection from zero is $P_3(Z) = Z^3 - (3/4)\,Z$. Its maximum deflection from zero in the range $Z=\pm1$ equals $\gamma_{max} = ¼$. If the deflection of the diaphragm is in the range $(w_o/h) \le 1$, the uniform minimum of its nonlinearity will be achieved when

$$3.33^{-1}(6U_lR/h^2 - F_o) = -3/4 \qquad \text{(3-91)}$$

or pretension $U_1R/h^2 \approx (1/6)\,(F_o - 2.5)$, (3-92)

and the relative remaining nonlinearity of the diaphragm's characteristic will be equal to: $\gamma_{max}\,(w_o/h) = 1/4\,(1/\,F_o)$, (3-93)

or $F_o \approx 6\,U_lR/h^2 + 2.5$ and $F_o = (1/4)\,[1/\gamma_{max}\,(w_o/h)]$. (3-94)

In this case, at allowed relative nonlinearity $\gamma_{max}\,(w_o/h) = 1.4 \cdot 10^{-4}$, expected quasi-linear function $F_o\,(w_o/h) = 1786\,(w_o/h)$ and the corresponding pretension should be equal to: $U_1R/h^2 \approx 300$.

If, for example, the pretension is $U_1R/h^2 \approx 100$, the expected quasi–linear factor $F_o \approx 602$ that corresponds well to the charts presented for this type of diaphragm's loading by L.E. Andreeva. It can also be shown that to achieve simultaneously uniform and quadratic minimization of nonlinearity in accordance with the principle of energy similarity (*see* Section 2-1) the pretension of the absolutely flexible diaphragm should be nearly doubled.

Features of corrugated diaphragms.

Advantage. Corrugated diaphragms function reliably at much larger deflections than flat diaphragms. Depending on the profile shape (generatrix of the median plane), the diaphragm characteristic as a function of pressure and displacement may be linear, increasing, and decreasing. This constitutes an advantage of corrugated diaphragms over other types of manometric elastic elements (bellows, Bourdon tubes, etc.) whose characteristics almost linear with pressure. Corrugated diaphragms permit the easy measurement of magnitudes, which are related nonlinearly with pressure (such as flow rate of fluid through a pipeline, speed of an aircraft, its altitude, etc.).

Corrugation's profile. The frequently used profile shapes of corrugated diaphragms: trapezoidal, saw-tooth, and sinusoidal (Table 3-11).

The change in the diaphragm corrugation profile has a much smaller influence on the characteristic than the change of its depth. Diaphragms with a fine saw-tooth profile are simple to manufacture and stable at small overloads. Diaphragms with deep corrugation usually have a trapezoidal or sinusoidal profiles, which have a smaller stress concentration along the crests and troughs.

Table 3-11. Profile and geometric coefficients of corrugated diaphragm

Profile*	k_1	k_2
1.Trapezoidal 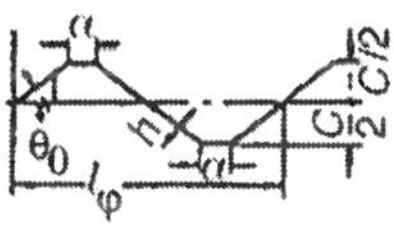	* $[(1-2a/l_\varphi)/\cos\theta_0]$ $+ 2a / l_\varphi$; <u>Optional approximation</u> at $\theta_0 \leq 30°$ $1+0.489\theta_0^2(1-2a/ l_\varphi)$	$\frac{*C^2}{h^2}\left[\frac{1-\frac{2a}{l_\varphi}}{\cos\theta_0}+\frac{6a}{l_\varphi}\right]+(1-\frac{2a}{l_\varphi})\cos\theta_0$ $+\frac{2a}{l_\varphi}$; <u>Optional approximation</u> at $\theta_0 \leq 30°$ $(1-2a/ l_\varphi)[(C/h)^2(1+0.489\ \theta_0^2)-(1-0.489\ \theta_0^2)]+6(C/h)^2[a/ l_\varphi]+2a/ l_\varphi$
2. Saw-tooth 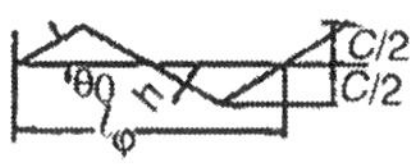	* $1/ \cos\theta_0$ <u>Optional approximation</u> at $\theta_0 \leq 30°$ $1+0.489\theta_0^2$	*$[(C^2/h^2)/ \cos\theta_0] + \cos\ \theta_0$
*3.*Shallow sinusoidal (θ<15°) 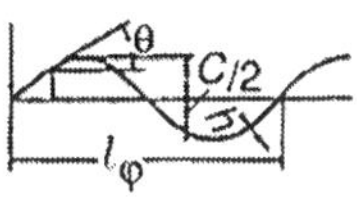	* 1	* $3/2\ (C^2/h^2) + 1$

* In accordance with L.E. Andreeva, 1962.

After the diaphragm bottoms on the plane, it is capable of withstanding considerable overloads from the convex side of the beads. The diaphragm profile is sometimes of variable depth. This decreases the diaphragm rigidity and simplifies the design of the stops used for overload protection. The corrugation bead farthest from the diaphragm center has a shape different from that of the others and is called the edge bead. It can be toroidal or cylindrical. The use of an edge bead is dictated mostly by purely design considerations. It is, for example, necessary when two diaphragms are connected to form a box. The edge bead passes into a narrow flat flange or a cylindrical rim that serves to fasten the diaphragm. Single diaphragms, fastened to the case by the flange, are used relatively rarely.

Box structure. The synthesis of the diaphragm assembly is considerably simplified if two identical diaphragms are connected along their flanges to form a diaphragm box. The deflection of the box under a given pressure is twice as large as that of a single diaphragm, which is an advantage of this type of structure. In addition, the installation of a diaphragm box in the device is much simpler than the installation of a single diaphragm. The diaphragms are joined to form a box by welding or soldering along the peripheral flange. Since both diaphragms are heated under identical conditions in the process of being welded, brazed, or soldered, no noticeable thermal stresses are set up, as they are when a single diaphragm is welded to its solid mount. Welded boxes are stronger than soldered boxes and they have a better seal, lower damping, and a higher stability of the elastic properties. In order to increase the strength of the joint, the diaphragms are sometimes soldered according to the mushroom shape (Fig. 3-9a). Welding along the cylindrical rim also ensures better constancy of the active diameter of the diaphragm seam and, therefore, smaller scatter in the value of the effective area. If a minimum volume of the inner cavity is required, collapsible diaphragm boxes (capsule) are used (Fig. 3-9b). They have another advantage in that they do not lose their properties if overloaded by external pressure. In order to ensure maximum contact between the diaphragms when overloaded, both diaphragms are formed at the same time in the same die. When absolute pressure is to be measured (with absolute-pressure manometers, barometers, altimeters, etc.) a vacuum is set up inside the box. Such boxes are called aneroid boxes (capsules), in contradiction to the manometric boxes, the inner cavity of which is connected to the pressure to be measured. Several boxes may be connected to increase the deflections (Fig. 3-9c). The rigid center 1 is fastened to one of the diaphragms, and a mounting nipple 2 joined to the other fulfills the dual function of fastening the box to the case and transmitting the pressure to the box cavity.

a) b)

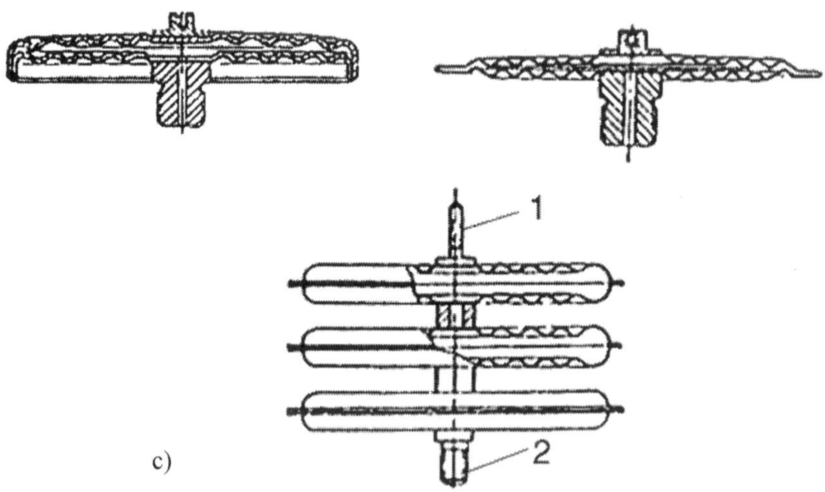

c)

Fig. 3-9. Boxes of corrugated diaphragms (After L.E. Andreeva).

Stops. To protect the diaphragm box against overloading, special stops are installed which restrict the diaphragm deformation (see L.E. Andreeva, 1981).

Deflections. Just as in the case of a flat diaphragm, the resistance of a corrugated diaphragm to external loading can be represented as the sum of the resistance to bending and tension. The resistance to bending predominates at the beginning of the loading, while the resistance to tension predominates in the range of large deflections. A diaphragm with deep corrugation has a higher resistance to bending, and therefore its characteristic will be closer to the linear than the characteristic of a diaphragm with shallow corrugation in which the stretching of the beads, and therefore the nonlinearity of the characteristic, are manifested at smaller deflections.

Initial tension. Corrugated diaphragms are very sensitive to initial tension forces and this is especially for low spring–rate diaphragms. If compression forces are set up in the initial diaphragm plane, the diaphragm rate drops, and the characteristic is liable to be of an increase type. On the other hand, initial tension increases the diaphragm rate. A diaphragm box with a fine saw-tooth profile and cylindrical rim shows smaller deflection under a pressure applied to the cavity than under the same pressure applied from outside. The configuration of the end bead determines whether the contour along which the central corrugated part of the diaphragm connects with the end bead will suffer a radial shift under the load. If this shift is toward the diaphragm center, compression forces are set up which increase the deflections, and the characteristic can become increasing. Shifting of the contour in the opposite direction causes tension in the diaphragm and an increase in its rate. The diaphragm characteristic varies very sharply if an initial convexity (spherical or conical) is imparted to the diaphragm during manufacture. If the pressure is applied from the convex side, the diaphragm rate drops. This is more pronounced the larger the initial convexity whose magnitude is characterized by height A_s (*see* case 8, Table 3-12). If value A_s is sufficiently enlarged, the diaphragm will change its deflection abruptly. A small convexity is imparted to corrugated diaphragms used in measuring instruments in those cases where it is necessary to reduce the rate on the initial portion of the characteristic, the diaphragms of an altimeter aneroid box can serve as an example. Their sensitivity must be sufficient when operating at great altitudes where pressure varies very little.

Calculation model. The corrugated diaphragm can theoretically be considered a structural anisotropic (orthotropic) plate that has different rigidity in radial and peripheral directions. Such a calculation schematic allows one to get calculation relationships in their most simple analytical form. The nonlinear system of differential equations for the flat anisotropic diaphragm that is equivalent to the specified corrugated one (L.E. Andreeva, 1962) are as follows

$$\rho\psi'' + \psi' - \alpha_k^2\psi/\rho = \tfrac{1}{2}\,\vartheta^2; \qquad (3\text{-}95a)$$

$$\rho\vartheta'' + \vartheta' - \alpha_k^2\vartheta/\rho = (R^2k_1/D)[-Ehk_1\psi\vartheta + \tfrac{1}{2}\,pR\rho^2]. \qquad (3\text{-}95b)$$

Small deflections. At small deflections the function of extension force is $\psi = 0$ and the differential equation becomes linear

$$\rho\vartheta'' + \vartheta' - \alpha_k^2\vartheta/\rho = \lambda_p\rho^2, \qquad (3\text{-}96)$$

where $\lambda_p = \tfrac{1}{2}pR^3k_1/D$; $D = \dfrac{Eh^3}{12(1-\nu_p^2/\alpha_k^2)}$ is the flexural rigidity, and $\rho = r/R$ is the dimensionless radius of the diaphragm. The solution of this equation at the boundary conditions $\vartheta_{\rho=0} = 0$ and $\vartheta_{\rho=1} = 0$ is as follows $\vartheta = [\lambda_p/(9-\alpha_k^2)](\rho^3 - \rho^{\alpha_k})$. The dimensionless parameter α_k is connected to the geometric coefficients k_1 and k_2 by the relationship $\alpha_k^2 = k_1 k_2$. (3-97)

These coefficients are defined by the equations shown in Table 3-11. If there is no corrugation on the diaphragm, then $k_1 = k_2 = 1$, $\alpha_k = 1$, and (95a, b) become equivalent to the similar differential equations (88),(89) for a flat isotropic diaphragm in the range of large deflections. The calculation formulae for the corrugated diaphragms at small and large deflections are shown in Table 3-12. In Table 3-12 F_{ef} is a relative effective area; A_{efp} and F_{efp} are the practical formulae for effective and relative effective areas of corrugated diaphragms with a rigid center, respectively; K_Q and K_p are the force and pressure spring rates; $n_{s,c}$ and $q_{s,c}$ are the values of coefficients for spherical (subscript s) and conical (subscript c) snap-action diaphragm, respectively. The plus sign in schematic 8 (Table 3-12) is used at pressure from the concave side of the snap-action diaphragm and the minus sign for the pressure action from the convex side of snap-action diaphragm.

Table 3-12. Corrugated and snap diaphragms calculation formulae

Schematic, Load	Formulae*
1.Small deflections under pressure p	$p\rho_{Rh2}^2/E = a_k(w_o/h)$; $a_k = 2(3+\alpha_k)(1+\alpha_k)/[3k_1(1-\nu_p^2/\alpha_k^2)]$; $V = 2(1+\alpha_k)\pi R^2 w_o/(9+3\alpha_k)$
2.Large deflections under pressure p	$p\rho_{Rh2}^2/E = a_k(w_o/h) + b_k(w_o/h)^3$; $b_k = 32k_1(\alpha_k^2-9)^{-1}[(1/6)-(3-\nu_p)(\alpha_k+3)^{-1}(\alpha_k-\nu_p)^{-1}]$
3. Deflection under force Q	$QR^2/(\pi Eh^4) = a_Q(w_o/h) + b_Q(w_o/h)^3$; $a_Q = (1+\alpha_k)^2/[3k_1(1-\nu_p^2/\alpha_k^2)]$; $b_Q = [k_1/(\alpha_k^2-1)]\{½ - (1-\nu_p)/[(\alpha_k-\nu_p)(\alpha_k+1)]\}$
4.With rigid center of radius r_0 under pressure p	$p\rho_{Rh2}^2/E = \eta_p a_k(w_o/h) + \zeta_p b_k(w_o/h)^3$; $\eta_p = (3-\alpha_k)(1-\alpha_k)/\phi_m$; $\phi_m = (3+\alpha_k^2)(1-\rho_o^4) + 4\alpha_k(1-\rho_o^{2\alpha_k})^{-1}[2\rho_o^{\alpha_k+1}{}^{+1}(1+\rho_o^2)$ $-(1+\rho_o^{2\alpha_k})(1+\rho_o^4)]$; $\zeta_p = (1-\rho_o^2)^{-4}(1+\rho_o^2)^{-1}[(1/6)-(3-\nu_p)(\alpha_k+3)^{-1}(\alpha_k-\nu_p)^{-1}]^{-1}\upsilon_m$; $\upsilon_m =$ $\{(1-\rho_o^6)/6-(3-\nu_p)(1-\rho_o^{2\alpha_k})^{-1}[(1-\rho_o^{\alpha_k+3})^2(\alpha_k+3)^{-1}(\alpha_k-\nu_p)^{-1}$ $+(\rho_o^{\alpha_k}-\rho_o^3)^2(3-\alpha_k)^{-1}(\alpha_k+\nu_p)^{-1}]\}$; $\rho_0 = r_0/R$
5.With rigid center of radius r_0 under force Q	$QR^2/(\pi Eh^4) = \eta_Q a_Q(w_o/h) + \zeta_Q b_Q(w_o/h)^3$; $\eta_Q = (1-\alpha_k)^2\{(1+\alpha_k^2)(1-\rho_o^2) + [2\alpha_k/(1-\rho_o^{2\alpha_k})][4\rho_o^{\alpha_k+1}$ $-(1+\rho_o^{2\alpha_k})(1+\rho_o^2)]\}^{-1}$; $\zeta_Q = (1-\rho_o)^{-4}\{½-(1-\nu_p)/[(\alpha_k-\nu_p)(\alpha_k+1)]\}^{-1}u_Q$; $u_Q = ½(1-\rho_o^2)-$ $[(1-\nu_p)/(1-\rho_o^{2\alpha_k})]\{(1-\rho_o^{\alpha_k+1})^2[(\alpha_k-\nu_p)(\alpha_k+1)]^{-1}$ $+(\rho_o^{\alpha_k}-\rho_o)^2/[(\alpha_k+\nu_p)(1-\alpha_k)]\}$; $\rho_0 = r_0/R$
6.Effective A_{ef} and relative F_{ef} area at small linear deflection	$A_{ef} = dQ/dp = K_Q/K_p$; $K_Q = Q/w_o = \eta_Q a_Q \pi Eh^3/R^2$; $K_p = p/w_o = \eta_p a_k Eh^3/R^4$; $F_{ef} = A_{ef}/\pi R^2$; Approximate expressions: $A_{efp} = \pi(R^2 + Rr_o + r_o^2)/3$; $F_{efp} = (1+\rho_o+\rho_o^2)/3$
7.Relative effective area F_{ef} at large deflection	$F_{ef} = [\eta_Q a_Q + 3\zeta_Q b_Q(w_o/h)^2]/[\eta_p a_k + 3\zeta_p b_k(w_o/h)^2]$
8. Snap-action 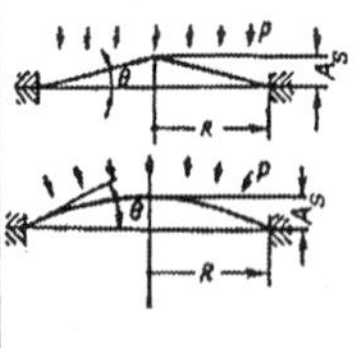	$p\rho_{Rh2}^2/E = a(w_o/h) \pm b(w_o/h)^2 + c(w_o/h)^3$; $a = 5.86 + n_{s,c}(A_s/h)^2$; $b = q_{s,c}A_s/h$; $c = 2.76$; $n_s = 4.88$; $n_c = 3.46$; $q_s = 7.72$; $q_c = 6.52$ at $\nu_p = 0.3$

*After L.E. Andreeva. Note that $\rho_{Rh2} = R^2/h^2$, where R is the outer radius and h is the thickness of diaphragm; V is the change of volume under elastic surface of the diaphragm at action of pressure p; $A_{ef(p)}$ and $F_{ef(p)}$ are approximate expressions for effective and relative effective area, respectively, (A_s/h) is the convexity characteristic, and parameter $\alpha_k = k_1k_2$ which are presented in Table 3-11. Subscript s corresponds to a spherical diaphragm and subscript c corresponds to a conical one. The charts of coefficients η_Q, ζ_Q see in Fig. 3-15e, f and coefficients η_P, ζ_P are in Fig. 3-19c, d.

Large nonlinear deflection. The approximate superposition method for calculation of corrugated diaphragms and the numerical solution of (95a, b) system has a difference up to 30%. At small deflections the difference should not exceed 8%. Experimental stress data and their calculated by numerical integration values are satisfactory except for the radial tension which is small and in the order of experimental uncertainties. Bending stress in the peripheral direction σ_{2b} is approximately three times smaller than the radial σ_{1b} one: $\sigma_{2b} = \nu_p \sigma_{1b}$. At small displacements, bending stresses σ_{1b} are distributed along the radius of the diaphragm with a frequency that is twice the frequency of corrugation. In the range of large displacements, the extension of corrugations appears which causes the additional bending stress with a frequency equal to the frequency of corrugation. If the latter is dominated, the frequency of radial stress will be equal to the frequency of corrugation, or it is twice less than at the small displacements. The position of the critical stress point for the diaphragm with quasi-linear characteristics is close to the crest of external corrugation on the outer surface of the diaphragm. For the diaphragm with a large nonlinearity, the position of the critical stress point depends on the load. At the beginning of loading, the critical stress point is located at the clamping on the outer surface of diaphragm. If the corrugation (*see* Table 3-11) is absent (C/h=0) or, on the other hand, is so deep that the characteristic of the diaphragm is close to a linear, the change of pressure direction (its sign) has a very small influence on the values of bending and stress. It changes only the sign of bending and stress. The sign of pressure has the maximum influence on the characteristic and stress of the diaphragm at the relative depth of corrugation in the range C/h = 2 to 6. The non-uniform corrugation with a smaller relative depth at the center of diaphragm can have a positive effect in the more uniform stress' distribution. The non-uniformity of the corrugations' depth can be expressed as the ratio $\alpha_C = (C_{iii} - C_i)/C_{ii}$, where $C_{i,ii,iii}$ are the depths of sequential corrugations beginning from the center *RC* (*see* Fig. 3-10a). The α_C has a plus sign if the depth of corrugation in the center is less than in the outer corrugation at the base *BC*, and a minus sign in the opposite relationship. The deformation of the central and the ending waves of corrugation are more equal at $\alpha_C > 0$ than at α_C=0 or $\alpha_C < 0$.

The influence of the shift in clamping. If wavelength is l_w and the clamping plane in the base contour's (*BC*) and rigid center (*RC*) clamping area coincides with the point 1 of corrugation (Fig.3-10a), the shift is considered to be equal to zero (δ_{sh}=0), and the diaphragm has profile with the coincidence of trough and clamping planes. If $\delta_{sh} = \frac{1}{4} l_w$ (the diaphragm contour is clamped along the circle of points 3), the diaphragm is called " neutral" because its crests and troughs are at equal distance from the clamping plane. The change of shift by $\delta_{sh} = \frac{1}{2} l_w$ may be seen as equivalent to the change of sign of pressure that can accordingly change the diaphragm characteristic and stress. The type of diaphragm contour's clamping also has an influence on its behavior. A comparison of diaphragms' characteristics shows that the largest deflection has a diaphragm with moving articulated support.

The diaphragm with the non-moving rigid support (as with welding or soldering to the rigid base) has the lesser deflection by approximately 1.6 times. The influence of the different boundary conditions is less at a larger depth of corrugation: at C/h = 18 the equivalent stress σ_{eq} changes only within 10-20%, and the deflection changes are within 20-30%.

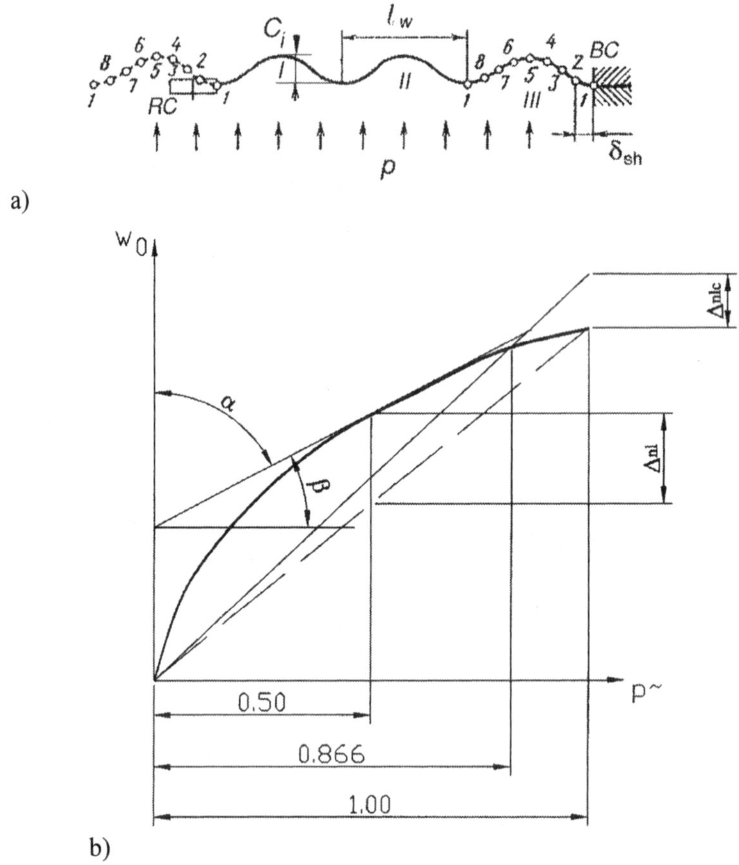

Fig. 3-10. Influence of clamping spots (after L.E. Andreeva, 1981) and nonlinearity estimation

The influence of the ending bead on the deflection and stress depends on its type. The cylindrical ending bead has a small influence, and the toroidal one has much larger influence, because it allows larger radial displacements at the clamping area. The change of an angle ϑ_0 in the toroidal bead from $\vartheta_0 = 0$ to $\vartheta_0 = -45^\circ$ can lead to twice the change of the diaphragm's deflection. The stability of the diaphragm can be verified by the snap-action formulae in Schematic 8 of Table 3-12 for the spherical and conical versions.

Nonlinearity. The linearity issue is very important at the synthesis stage of corrugated diaphragms. The nonlinearity of elastic characteristic is usually estimated by the value $\eta_{nl} = (\Delta_{nl} / w_{omax})100\%$,

where Δ_{nl} is the largest difference between the nonlinear function (Fig. 3-10b) and the straight line, which connects the origin of coordinates with the end point of the nonlinear characteristic $w_o(p^{\sim})$ corresponding to the pressure working range $p^{\sim}_{end}$=1.00. However, it is not the only one possible definition of nonlinearity in this case. It is also possible to express nonlinearity of the corrugated diaphragm on the basis of the polynomial interpolation because its elastic characteristic in the large range of deflection is represented by the polynomial of the third degree (*see* Tables 3-10 and 3-12). It is easy to show that the minimal "uniform" nonlinearity of a such characteristic will be found if the mentioned straight line connects the origin of coordinates with the point $p^{\sim}_{rt} = p_{rt}R^4/Eh^4$ that corresponds to the root of Chebyshev's polynomial $P_3(Z)$ of the same third order (*see* Table 1-7). In this case $p^{\sim}_{rt} = 0.866\, p^{\sim}_{end}$, and the largest difference Δ_{nlc} between the straight line, which connects the origin of coordinates with the root point (0.866), and the nonlinear characteristic $w_o(p^{\sim})$ is equal to ½ | Δ_{nl} | at the points $0.5\, p^{\sim}_{end}$ and $p^{\sim}_{end}=1$. Hence, the nonlinearity value will be smaller than at traditional definition because

$$\eta_{nlc} = \Delta_{nlc} / w_{\text{omax}} = \tfrac{1}{2}\, \eta_{nl}. \tag{3-98}$$

The chart in Fig. 3-10b shows Δ_{nl} (dashed line) and Δ_{nlc} (solid line) values. Chebyshev's polynomial with the minimum relative deflection $P^{o}_{3} = ZP_{n-1}(Z)$ or the principle of energy similarity (*see* Section 2-1) can also be used for the synthesis of corrugated diaphragm with the minimum nonlinearity in the elastic characteristic at corresponding conditions.

Effective area. If the diaphragm is used for the transforming pressure into force at the conditions of force compensation (as in balance-type instruments, in piezoelectric and frequency transducers, in contact-pressure resistance transducers, etc.), the deflection of the diaphragm center is very small. It is important in such cases that the diaphragm has a strictly constant effective area. The calculation problem for this case is solved on the bases of the numerical integration of differential equations (95a, b). If the diaphragm is used for transforming the pressure into a deflection, it must have sufficient effective area in order to successfully overcome the resistance of the instrument mechanism. The effective area must be higher than some minimum value required for normal functioning of the instrument over entire active range of the characteristic. An approximate assessment of the effective area and of its variation with deflection is sufficient in this case. This problem is solved with the equations from Table 3-12. All arguments of initial rigidity $K^{\sim} = K_Q R^2/Eh^3$, $K_Q = Q/w_o$, stress in the critical point $\sigma^{\sim} = \sigma_{eq}R^2/Eh^2$, equivalent stress in the same point $\sigma_{eq} = p(\sigma^{\sim}/p^{\sim})R^2/h^2$, the ratio $\sigma^{\sim}/p^{\sim} = \sigma_{eq}\, h^2/pR^2$, and the tilt angle of the tangent to initial characteristic of rigidity $p^{\sim}/w^{\sim} = (pR/Ew_o)(R/h)^3$ are dimensionless. The thickness of the corresponding diaphragm can be found through the expression

$$h = (K_Q R^2/\, EK^{\sim})^{1/3}.$$

Plates with shaped holes.

The methods of strain-stress calculation for solid plates with different contours and plates with periodical circular holes are developed in several publications. The hypothesis of the similarity for latter plates' deformed surfaces is successfully used. The cylindrical flexural rigidity of solid D^{o} and perforated D^{o}_{φ} plates are related by the coefficient of perforation Λ_{φ} as follows $D^{o}_{\varphi} = \Lambda_{\varphi}D^{o}$.

The coefficient Λ_{φ} can be defined on the bases of experimental measurement of natural frequencies for solid and regular perforated plates that are equivalent in the outer contour, overall sizes, and materials. The empirical expressions for coefficients Λ_{φ} for the same plates are also known. If the plate with holes is similar to the pulley with a rigid rim and hub, the approximate calculation is provided as for the gear (pulley) with spokes (see Table 3-8). In this case, it is enough to calculate the elastic deformation of the separate spokes. The problem of calculation for the plates with the shaped holes is very complicated and cannot be solved in all cases by hypothesis of similarity.

Spoke calculation with Mohr's integral. The deformation of separate spoke can be found as for curvilinear thin beam loaded with the forces that perpendicular to its plane (*see* Table 3-8, schematic 4).

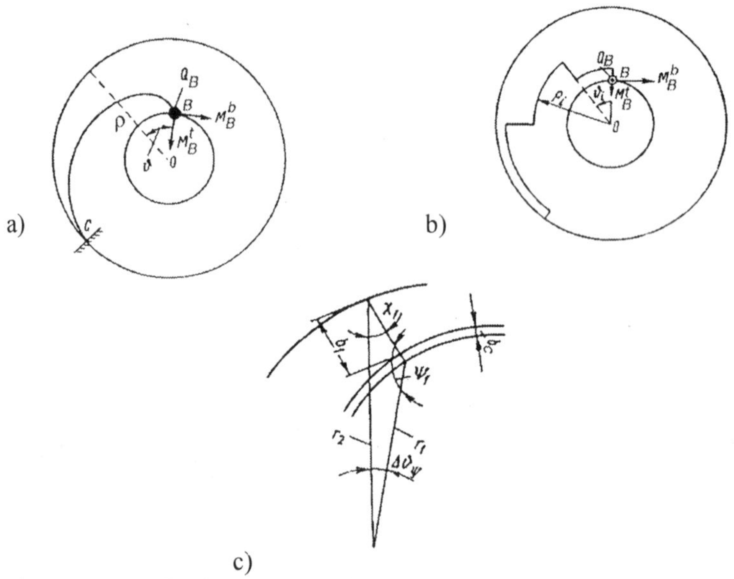

Fig. 3-11. Elastic plates with spokes of different shape.

The changes in the spoke tension are considered negligible at the range of a small bending deformation. This allows us to get calculation relations with Mohr's integral in the matrix form. The axis of the flat curvilinear beam, expressed in polar coordinates as $\rho=f(\vartheta)$, are shown in Fig. 3-11a . One end C of the beam is rigidly clamped to the frame, another movable end B is loaded by force Q_B, bending M^b_B and torsion M^t_B moments whose positive direction corresponds to the clockwise motion. Matrices of bending M^b_{ij} and torsion M^t_{ij} moments in the arbitrary sections of the beam are equal to:

$$M^b_{ij}=\begin{bmatrix} \rho\cos\mu-\rho_1\cos(\mu+\vartheta) & 0 & 0 \\ 0 & 0 & 0 \\ 0 & \sin(\mu+\vartheta) & \cos(\mu+\vartheta) \end{bmatrix}\begin{bmatrix} Q_B \\ M^t_B \\ M^b_B \end{bmatrix}=$$

$$[\,M^b_{ij}(e)\,]\,[Q_{Bi}\,], \tag{3-99a}$$

$$M^t_{ij}=\begin{bmatrix} 0 & 0 & 0 \\ -\rho\sin\mu+\rho_1\sin(\mu+\vartheta) & \cos(\mu+\vartheta) & -\sin(\mu+\vartheta) \\ 0 & 0 & 0 \end{bmatrix}[Q_{Bi}]=$$

$$[\,M^t_{ij}(e)]\,[Q_{Bi}\,]. \tag{3-99b}$$

Members M^b_{ij} (e) and M^t_{ij} (e) of square matrices of the third rank are equal to the corresponding moments in the current section from the external unit-loads.

The calculation relations for the combined (Fig. 3-11b) "beam-spoke"(that consists from the joint radial straight and arc-like parts) can be got with the similar way as in above shown cases (see Y.M. Tseytlin, 1972).

Singular points. The form of spoke should not have singular points (a double point, cusp or self-osculation point), should not touch or superpose other n periodical spokes. If the spoke lies in the segment whose angle ϑ is larger than $\vartheta_n = 2\pi/n$, it is necessary to verify the absence of a superposition of the spokes. In this case, in accordance with Fig. 3-11c, the following inequality should be satisfied

$$\{[r_1^2(\vartheta + \vartheta_n)^2 + r_2^2(\vartheta_{n+1} + \Delta\vartheta_\psi)^2]$$
$$-2r_2r_1(\vartheta_{n+1} + \Delta\vartheta_\psi)(\vartheta_{n+1} + \vartheta_n) \cos \Delta\vartheta_\psi\}^{1/2} - b_1 \geq b_c > 0, \quad (3\text{-}100)$$

where r_1 $(\vartheta + \vartheta_n)$ is the current value of a vector-radius of the spoke mean line, r_2 $(\vartheta_{n+1} + \Delta\vartheta_\psi)$ is the corresponding current value of a vector-radius of the next spoke mean line, $\Delta\vartheta_\psi = \pi - (\chi_1 + \psi_1)$ is an angle between the vector-radius and the normal to the spoke mean line, b_c is the width of a cut, and

$$\chi_1 = (\pi/2)\text{-}\tan^{-1}(r_1\, d\vartheta / dr_1), \text{ and } \vartheta_{n+1} = (\vartheta + \vartheta_n).$$

If to request that b_c = constant, then the equality in (100) will be obtained with the definite form of curves r (ϑ, ϑ_n). It can be shown that these curves are the involutes with the base radius $r_o = (b_1 + b_c)\, n_{sp} / 2\pi$, $\vartheta = \text{inv}\ \alpha_x$, and $r_{1;2}$ $(\vartheta, \vartheta_n) = r_o/\cos \alpha_x$. The similar condition can be got by the curve that is built from the straight lines and arc segments.

Typical punched structures. The plates with the punched shaped holes have larger displacements and lesser stresses than the solid ones with the same size.

The large variety of plates with shaped holes is in use: round plates with annular arc-like cuts (Fig. 3-12a), with segment-like cuts (Fig. 3-12 b, c, d), with sector (Fig. 3-12 e) and circular-like recesses (Fig.3-12f), and with spiral cuts (Fig. 3-12g). Other forms of strips with shaped holes are possible (Fig. 3-12h). Each form has the definite working features. The experimental charts of the compliance characteristics (displacement S_1 at force increment ΔQ) are shown in Fig. 3-12k for plates with outer diameter 20 mm. The center inner contour of the plate and its outer contour are usually rigidly clamped to a rigid movable slider and rigid base, respectively. Hence, the shown plates with the segment, circular, and spiral cuts can be assumed as the pulleys (gears) with the elastic spokes and a rigid rim and hub. In addition, the plates with the circular recesses are similar to the periodically punched grids. Thus, the plates with the shaped holes can be divided into two categories: spoke-like, in which each spoke is independently connected with a rim and hub, and the grid-like ones. It is evident, that the guided slide will not spontaneously rotate at the loading system directed along its axis if the spoke-like plates have the even number of spokes and the even number of congruent spokes with different direction. The rectilinear motion can be got with the periodical (double periodical) grid-like plates and with the spoke-like periodical plates. The physical realization of the grid-like plate requires that the grid does not have an ultimate dense, i.e., it does not have the conjugated cuts.

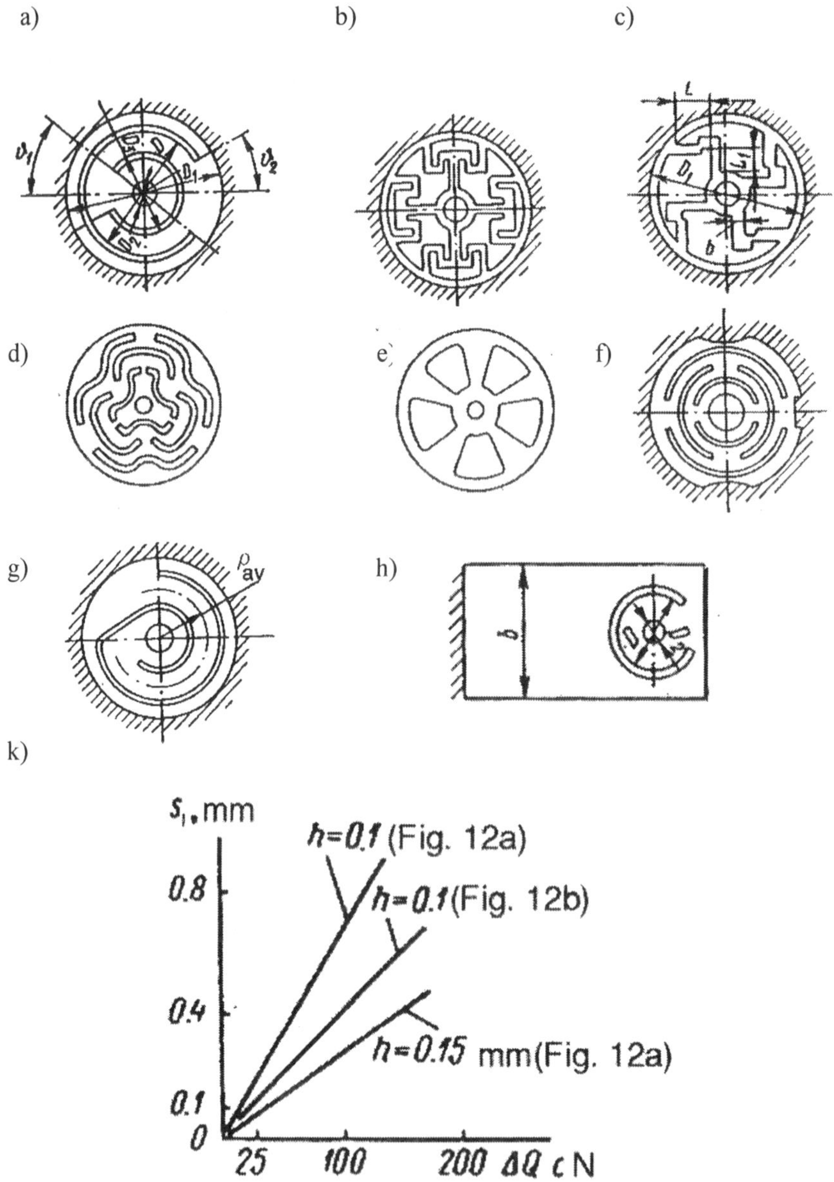

Fig. 3-12. Types of the punched plates and corresponding compliance charts.

Conformal mapping. The conformal mapping of the multiple-connected field of a perforated plate to the canonical field with the same multiple connections (e.g., to the plane or the circle with the straight or arc cuts) is one of the more general ways for solving this problem. We can show that the exponential function can be used for the conformal mapping of the periodically perforated circular plate $w_{\zeta} = e^{z} = e^{x}e^{iy}$

to the mesh of rectangles z on the strip $\pm\pi$. Thus, the plate with the shaped holes transforms in the straight beams' mesh whose problem solution is known (S.P. Timoshenko *et al.*, 1959).

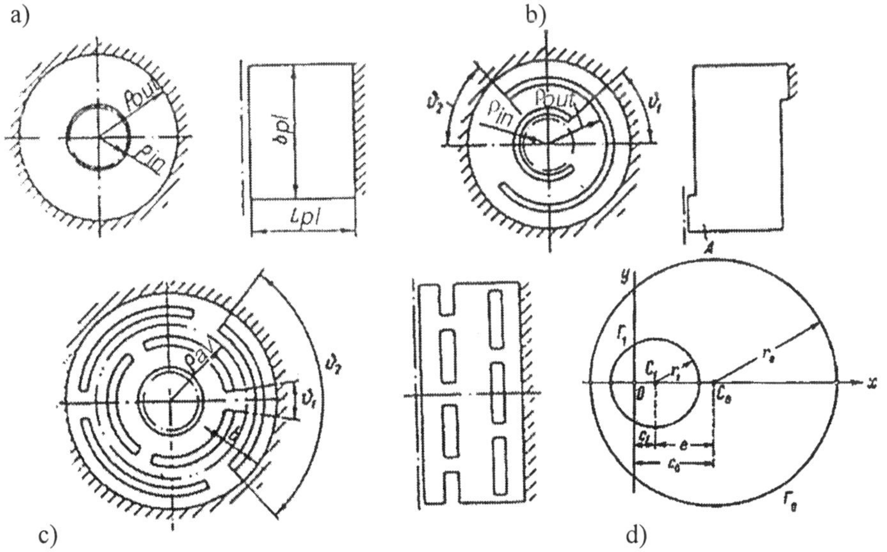

Fig. 3-13. Image and conformal mapping of punched plates and eccentric ring

In this case, e^x is equal to the ratio of outer and inner radii of the ring (ρ_{out}/ρ_{in}). The round grid-like plates are convenient for the converting to the rectangular strip with cuts (a mnemonic rule). The solid round ring converts by the exponential function to the solid flat strip (Fig. 3-13a). If the round plate is clamped at the outer contour and has translational displacement of the inner contour, the conversion has the same displacement conditions on the ends. Displacement s_{1k} of the end of the rectangular flat strip is defined by the following expression $s_{1k} = QL^3_{pl} / Eb_{pl}h^3$. (3-101)

The sought relationship after backward conversion to the round ring in the case ρ_{in} = 1 is represented by the equation

$$s_1 = Q(\rho_{out})^2 (\ln\rho_{out})^2 / (\pi e^2 E h^3) = \nu_{Qk} Q (\rho_{out})^2 / Eh^3, \quad (3\text{-}102)$$

where $\nu_{Qk} = (\ln\rho_{out})^2/(\pi e^2)$. (3-103)

In the last expression at $\rho_{in} \neq 1$, one should substitute $[\ln(\rho_{out}/\rho_{in})]^2$ instead of $(\ln\rho_{out})^2$. Table 3-13 shows ν_{Qk} data that are calculated with expression (103) and similar data borrowed from S.P.Timoshenko *et al.*, 1959.

Table 3-13. The values of ν_{Qk} for solid round plates

ρ_{out}/ρ_{in}	1.5	2	2.71	3	4	5
ν_{Qk}(103)	0.0071	0.0208	0.0435	0.052	0.086	0.113
ν_{Qk} *	0.0064	0.0237	-	0.062	0.092	0.114

* After S.Timoshenko *et al.*, 1959.

The correspondence of these values at the ratio $(\rho_{out} / \rho_{in}) > 1.5$ is satisfactory. The working part of the round plate with the circular cut can be converted by the same way to the solid rectangular plate with the clamping tab shown by hatching (Fig. 3-13b). The movable tab *A* translates. The plate with the round recesses (Fig. 3-13c) converts to the form of rectangular plate with the rectangular windows. The clamping spots of the plate and its reflection marked by hatching, and translationally moving ends are designated by the dash-dot lines. Displacement s_1 of this plate at $\vartheta_1 \leq \vartheta_2$ and the rigid enough crosspieces can be found with the approximate formula $s_1 = 2Q\rho^3_{av}/(eEbh^3)$. The round plate with the spiral cuts (*see* Table 4-3, ELKS of type TG-VIII-*sr*) also converts to rectangular plate with the cuts.
The approximate formulae for calculation of the typical plates used in elastic mechanisms' guides are shown in Chap. 4. We, however, should remember that the general approach to the problem requires inverse conformal mapping (*see* A.E.H. Love, 1944).

Conformal mapping of a ring. It is helpful to use conformal mapping for rings with special forms: eccentric *e* (*see* Fig. 3-13d), elliptic, etc. This method allows us transform of plane stress-strain problem of a complicated body to the body with more simple shape. If the two-dimensional region occupied by the body is conformally (with preserving the angle relations between corresponding contour lines) represented upon a different region by means of a functional relation between complex variables, a new state of plain strain and stress function in a body of different shape from that originally treated can be found. Since poles and logarithmic infinities are conserved in such conformal transformations, the points of application of isolated forces in the two states are corresponding ones. If we know the conformal transformation function $z=\omega(\zeta)$ of the circular concentric ring on the plane $\zeta=\rho e^{i\vartheta}$ limited by circles γ_0 and γ_1 with radii ρ_0, ρ_1 $(\rho_0 \leq |\zeta| \leq \rho_1)$, onto the two-dimensional field *z* with the body under consideration, then one can find corresponding stress and strain functions for the latter problem (A.I. Lurie, 1970). It is supposed that the outer contour Γ_0, center C_0 with radius r_0 (inner contour Γ_1, C_1 with radius r_1) in the field *z* corresponds to inner circle γ_0 (outer circle γ_1) in the field ζ, respectively.
The simplest relationships are between the stress-function ψ, satisfying the differential equation $\nabla^4\psi = 0$, and the inverted stress-function $\psi' = \psi\rho^2$, which is also satisfying to the similar differential equation $\nabla^4\psi' = 0$. Let us recall that many complicated problems in the theory of elasticity were solved on the basis of the conformal method. This method is effectively used by the author of the book for solving corresponding problems as well. It is known (A.E.H. Love), however, that free from traction in the contour (r, θ) system is transformed into a boundary subject to normal tension in the converted (r', θ') system. This tension has the same value at all points of the transformed boundary, its effect is known and can be allowed for in many cases.

The bellows specifics.

Structure and features. Bellows are the thin-walled cylindrical shells with the transverse corrugations (Fig. 3-14a). Bellows are capable of considerable deflections with a near-linear characteristic and a constant effective area.

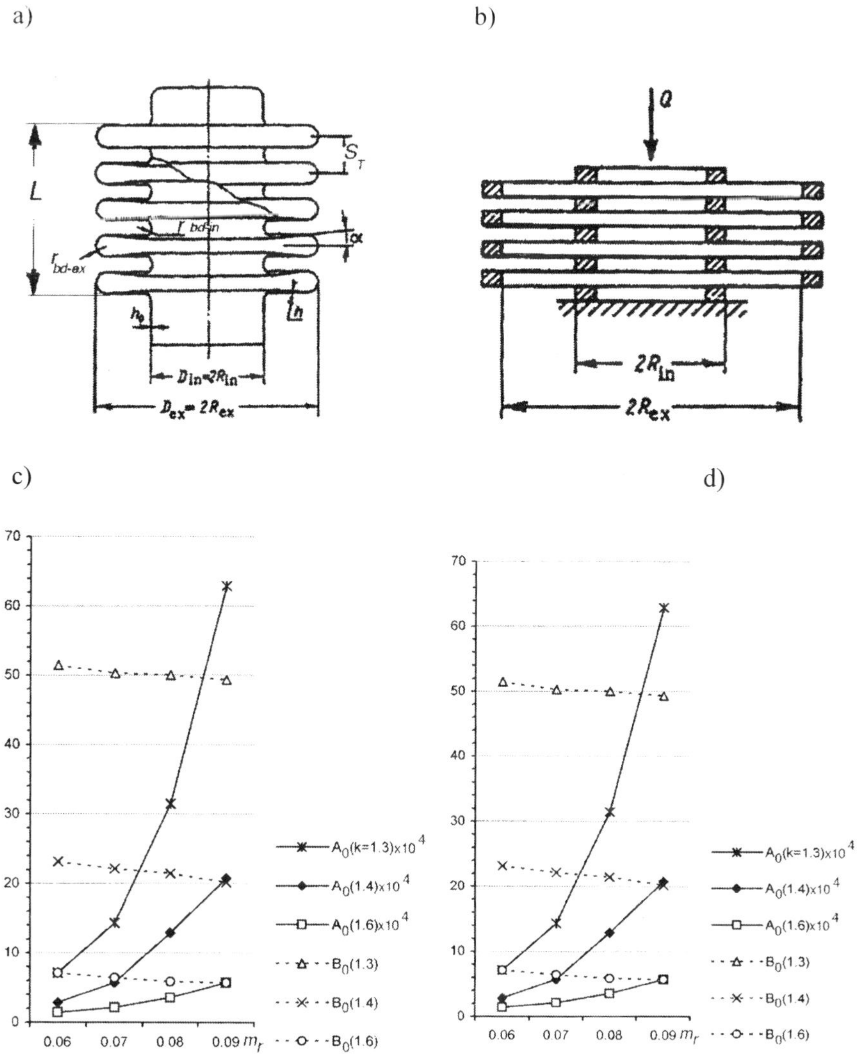

Fig. 3-14. Bellows typical dimensions, multi-plate model, and charts of coefficients for stiffness calculation (after L.E. Andreeva, 1962).

Process and material. Bellows are corrugated by hydraulic or by mechanical-hydraulic methods. Damping in bellows made from precipitation-hardening alloys does not exceed 0.5-1% within the limits of the active segment of the characteristic. At a very high pressure, when the strength of a thin-walled bellows is insufficient, laminated (multilayer) bellows made from a multilayer tube are used. The stress in the laminated bellows is lower and the sensitivity is higher, than in a single-layer bellows of the same thickness and with the same deflections. The damping, however, increases sharply in the laminated bellows because of the friction between the layers. *Calculation methods*. The calculation methods for strain and stress of the seamless bellows may be split to the three category: (i) simplified plate's model (Fig, 3-14b) with linear relations; (ii) more accurate Ritz's energy method with linear relations; (iii) numerical nonlinear method with the numerical integration of differential equations (8, Table 1-10) for large displacements and case of changeable thickness along the bellows contour. The corresponding calculation formulae are shown in Table 3-14.

Table 3-14. Formulae for calculation of seamless bellows

Load, Schematic	Method, calculation formulae*
1. Force Q 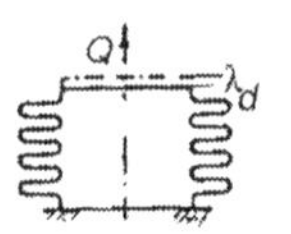	(i) $Q/\lambda_d = K_Q = Eh^3/(2nA_Q R^2_{ex})$; $\sigma_{rin} = B_{Qin} Eh_{in} \lambda_d / 2n R^2_{ex}$; $\sigma_{rex} = B_{Qex} Eh_{ex} \lambda_d / 2n R^2_{ex}$; $k = R_{ex}/R_{in}$ (ii) $K_Q = \dfrac{Eh_0}{n(1-\nu_p^2)}(A_0 - \alpha A_1 + \alpha^2 A_2 + B_0 \delta_h^2)$; $r_{bd\text{-}in} = r_{bd\text{-}ex} = r_{bd}$ $\alpha = \dfrac{4m_r - S_T/R_{in}}{2(k-1-2m_r)}$; $\delta_h = h/R_{in}$; $m_r = r_{bd}/R_{in}$ (*see* Fig. 3-14)
2. Pressure p 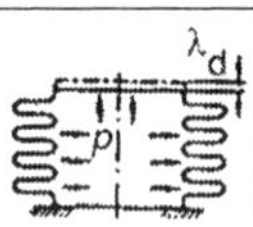	(i) $\lambda_d = 2nA_p pR^4_{ex}/Eh^3$; $\sigma_{rin} = B_p Eh_{in} \lambda_d /(2n R^2_{ex})$; $\sigma_{rex} = B_p Eh_{ex} \lambda_d /(2n R^2_{ex})$
3. Moment M 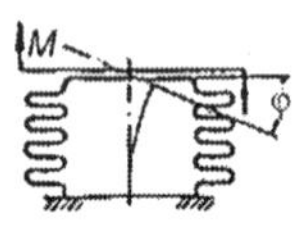	(i) $\varphi = 2nA_\varphi M/Eh^3$; $\sigma_{rin} = B_{\varphi in} Eh_{in} \varphi /(2n R_{ex})$; $\sigma_{rex} = B_{\varphi ex} Eh_{ex} \varphi /(2n R_{ex})$;
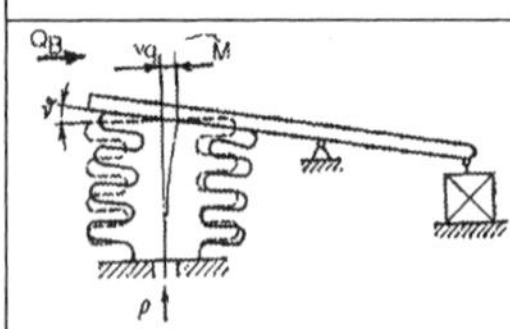 4. $v_q = v_{qM} + v_{qQ}$	(ii)' $M \neq 0$, $Q_B = 0$ $v_{qM} = ML^2/(2H_{BE}) = L(1-\cos\vartheta)/\vartheta$ at $\vartheta \leq 0.1$ $v_{qM} \approx L\vartheta/2$ $H_{BE} = Eh_o^3 L/[0.24(1-\nu_p^2)\,\bar{\vartheta}\, n]$; $\bar{\vartheta} = \vartheta E h_o^3/[0.24(1-\nu_p^2) Mn]$; (iii) $M=0$, $Q_B \neq 0$; $v_{qQ} = Q_B L^3/3H_{BE}$
5. Transverse force Q_B *see* p.4 with $M=0$; $Q_B \neq 0$	(i) $v_q = Q_B L^3/3H_B$; $H_B = Eh^3 L/(2n A_\varphi)$; $\vartheta = Q_B L^2/2H_B$
6. Effective area A_{ef}	(i) $\lambda_d = p/k_p = p A_{ef}/K_Q$; $\Delta V = A_{ef}\lambda_d$; $A_{ef} = R^2 A_p / A_Q$ $A_{ef} \approx \pi R^2_m$; $R_m = ½ (R_{ex} + R_{in})$;

Table 3-14 Cont'd.

Load, Schematic	Method, calculation formulae*
7. Critical force Q_{cr}	(i) $Q_{cr} = \pi^2 Eh^3/(\nu_{eu}^2 L2nA_\varphi)$
8. Critical pressure p_{cr}	(i) $p_{cr} = (Q_{cr} / A_{ef}) + p_o = p_o + \pi^2 Eh^3/(\nu_{eu}^2 L2nA_\varphi A_{ef})$
9. Cyclic work	(i) $\sigma^o = (\sigma^2_b/\sigma_a)(1-b_\sigma^2)(-1 + \sqrt{1+[2a_\sigma/(1-b_\sigma^2)]^2}$); $\sigma^0_{eq} = \sqrt{(\sigma^0_1)^2 + (\sigma^0_2)^2 - \sigma^0_1\sigma^0_2}$. $b_\sigma = \sigma_m / \sigma_b$; $a_\sigma = \sigma_a / \sigma_b$; $\sigma_a + \sigma_b < \sigma_{pr} \approx \sigma_e$, $\sigma^0_{eq} < \sigma_{en}$

*After L.E. Andreeva, 1962, 1981. The analytical expressions of coefficients A_Q, A_φ, A_p, B_p, B_{Qin}, B_{Qex}, $B_{\varphi in}$, $B_{\varphi ex}$ see in Table 3-9, their charts are presented in Fig 3-15a-d as a function of argument $k=R_{ex}/R_{in}$; L is the length of corrugated part and n is the number of active corrugations in the bellow, h_0 is the thickness of a blank-tube; coefficient ν_{eu} see in Table 3-3 and ν_{eu} =0.5 if angle ϑ=0 for end sections; coefficients A_0 (typical values in the range of 0.0001-0.008), A_1 (within 0.004-0.035), A_2 (in the range 0.04-0.16), B_0 (within 5-80) as a function of m_r (Fig. 3-14c, d)and dimensionless compliance $\bar{\vartheta}$ (Fig. 3-18c) as a function of the changeable along the contour relative thickness δ, $k= R_{ex}/R_{in}$ and $m_r = r_{bd\text{-}ex}/R_{in}$ are presented in detail by L.E. Andreeva,1981.

In Table 3-14, Q, M, and p are the force, the acting bending moment, and the pressure in the bellows, respectively; p_o is the pressure in the bellows at the touching of a rigid stop; Q_{cr} and p_{cr} are the critical value of the force and pressure that causes the loss of bellow's form stability; A_{ef} is the effective area of bellows; λ_d is the bellows longitudinal displacement under acting force or pressure; n is the number of corrugations (beads); R_{ex} and R_{in} are the external and internal radii of bellows; h is the thickness of bellows wall; $r_{bd\text{-}ex}$ and $r_{bd\text{-}in}$ are the radii of bead's waves rounding at external and internal surfaces , respectively; v_Q is the transverse deflection of the bellows, φ and ϑ are the angles of bellows faces rotation under bending moment; H_B and H_{BE} are the flexural rigidity of bellows defined by plates (i) and Ritz's (ii) methods; L is the initial height of the bellows, h_o is the thickness of wall of the tube-blank; ϑ' is the dimensionless flexural compliance of the semi-wave of the bellows; σ_{rin} and σ_{rex} are the stress in the internal and external points of a wave; σ^0, σ_a, σ_m, σ_b, and σ^0_{eq} (schematic 9, Table 3-14) are the stress of the symmetrical pulse cycle, stress amplitude, mean stress value, the tensile stress of material($\sigma_b < \sigma_t$), and equivalent stress of a pulse cycle with σ^0_1 and σ^0_2 principal stresses. The stress cycle may be asymmetric if $r_{cyc} \neq 0$ where $r_{cyc} = (\sigma_m - \sigma_a) / (\sigma_m + \sigma_a)$; the pulse cycle has r_{cyc} = 0. The material features of tensile stress σ_t, proportionality limit $\sigma_{pr} \approx \sigma_e$, and endurance limit σ_{en} see in Table 1-14. Charts of coefficients A_Q, B_{Qin}, B_{Qex}, A_p, B_p, A_φ, $B_{\varphi ex}$, $B_{\varphi in}$ (expressions *see* in Table 3-9) are the functions (Fig. 3-15a-d) of the ratio $k = R_{ex}/R_{in}$. Ritz method (ii) is more precise for the estimation of spring constant $K_Q = Q/\lambda_d$ in load schematic 1 (Table 3-14) because it allows to assume specifics of corrugations geometry, curvature of bellows meridian, the actual deformation of corrugations, value of the fold angle α (Fig. 3-14a).

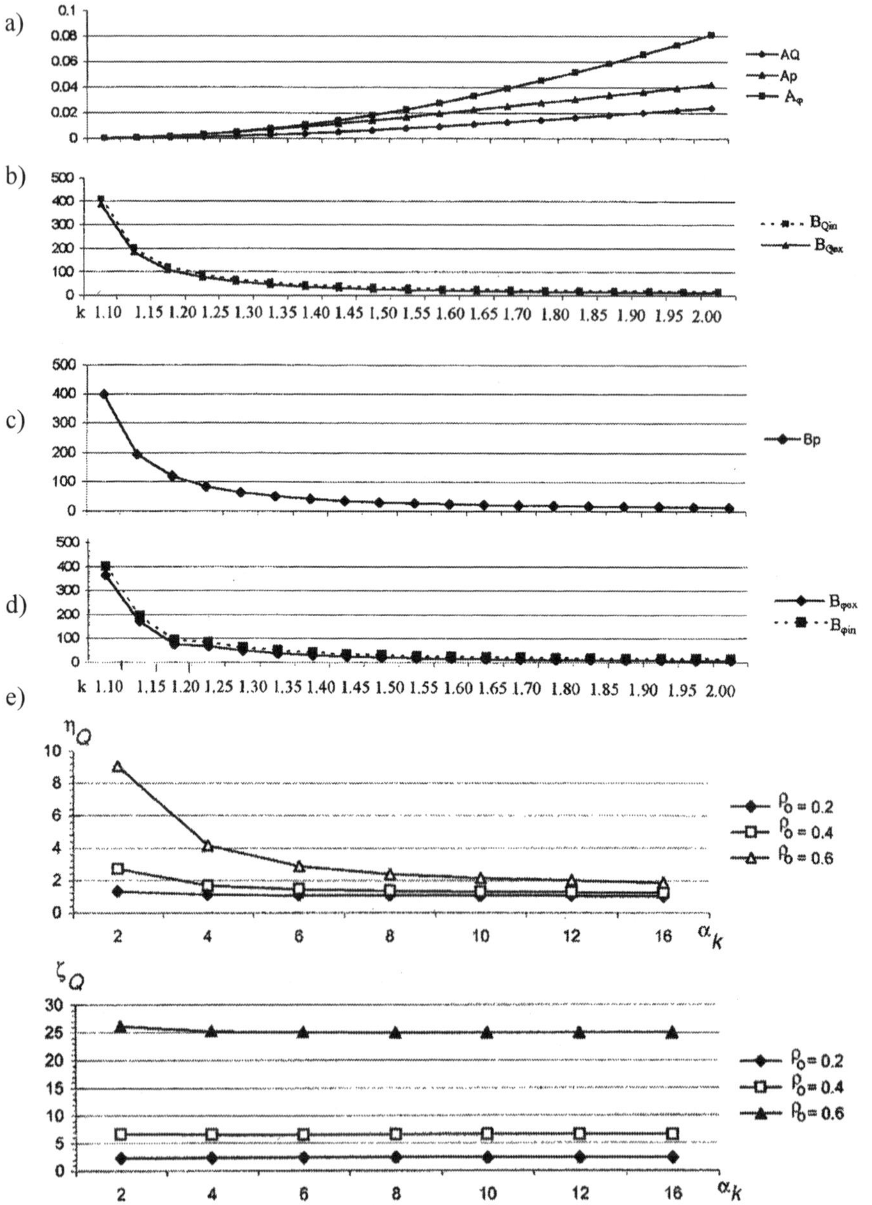

f)

Fig. 3-15. Coefficients for seamless bellows, flat and corrugated diaphragms.

This method is based on the numerical integration for coefficients A_0, A_1, A_2, B_0 (Fig.3-14c, d) whose values in more detail can be found in V.I. Feodos'ev, 1949, and

L.E. Andreeva, 1962, publications. The dimensionless compliance $\bar{\vartheta}$ (Fig. 3-18c) for the case 4 (Table 3-14) with the changeable wall thickness

$\delta=(h_0/R_{in})\sum_{j=0}^{n} a_j\cos\frac{j\pi\varsigma}{\varsigma I}$ along the contour is also obtained (iii) by numerical integration of the simultaneous differential equations (*see* *No* 8 in Table 1-10). The expression $\sum_{j=0}^{n} a_j\cos\frac{j\pi\varsigma}{\varsigma I}$, where $\zeta = s_{arc}/R_{ex}$ is a dimensionless length of the corrugation's semi-wave arc (s_{arc}) with point of symmetry ζ_I, may be found by approximation the curve of relative thickness which corresponds to the certain manufacturing process.

Basic properties of tubular manometric springs.

Types. The capacity of a hollow tube to be deformed by pressure is widely made use in manometric instruments and pneumatic actuators. The tubular manometric spring is usually a thin-walled curved tube with an oval cross section (Fig. 3-16a). When pressure is internally applied to the tube, its cross section is deformed and acquires the shape shown by the dashed line. The deformation is such that the longitudinal fiber *a-a* moves to an arc *a'-a'* of larger radius, and the fiber *b-b* to an arc *b'-b'* of smaller radius. Since the fibers tend to maintain their initial length, the tube cross section will rotate counter-clockwise. The spring will unbend, and its end will deflect by an amount λ_d.

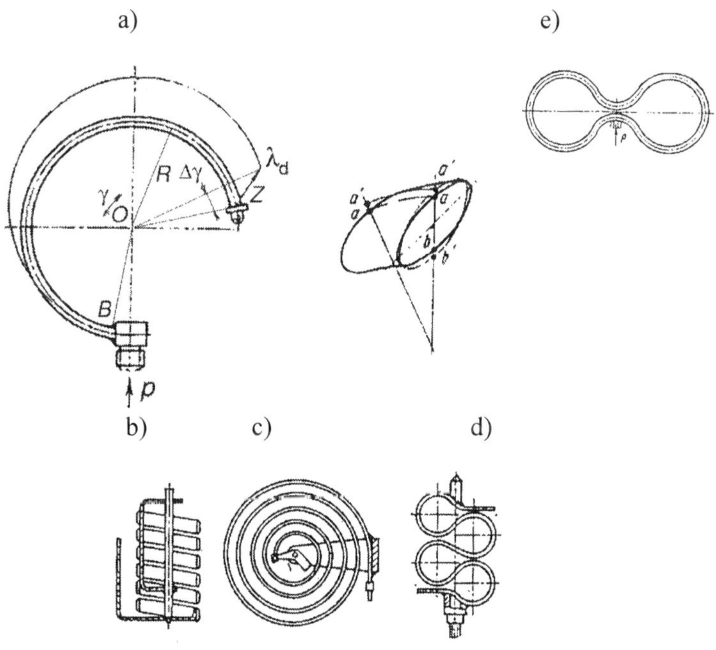

Fig. 3-16. Typical types of manometric tubes

The simplest and most widely used type of tubular manometric spring is this single-coil tubular spring, the so-called Bourdon tube. Its axis is a circular arc of a central angle between 200 and 270°. Bourdon tubes are used not only in indicating instruments, but also in recording and control manometers, in the thermometers, vacuum meters, level gages, flow-rate meters, and other instruments. Tubular springs exert a smaller tractive force than do bellows or diaphragms. This is usually a disadvantage but there are some cases (for instance, when high pressure are to be measured using a force balance scheme) when a high tractive force complicates the instrument design. Multi-coil tubular springs are used in instruments where large deflections of the elastic element are required. The helical coiled tubular spring (Fig. 3-16b) and the spiral tubular spring (Fig. 3-16c) are the typical examples. They are more difficult to manufacture than Bourdon tubes, but because of their capability of large deflections, they considerably simplify the structure of a transmission linkage in the system. The spiral springs are compact, occupy roughly the same amount of space as Bourdon tubes. Helical coiled springs are also widely used, especially when the spring diameter is restricted and its length is practically unlimited. In Bourdon tubes, the linear deflection of the tube end is utilized, while in spiral and helical coiled springs it is the angular deflection, which is used. The so-called S-shaped tubular spring (Fig. 3-16d), whose end has a translatory motion, some times provides a satisfactory structural solution. Such springs are used for the measurement of high pressure, as sensing elements of liquid manometric thermometers, and also as an element in the pressure compensators. Springs shown in Fig.3-16a-d have comparatively low natural frequency that restricts their application in pulsing pressure and vibration conditions. Manometric springs of a closed form (Fig. 3-16e) with a nipple for input pressure p are more stable to vibrations. The so-called twisted (*see* Fig.3-20a) tubular spring (the initially pretwisted tube, or the tubular quasi-helicoid) is found of use for the measurement of high pressure. The use of pretwisted tubular springs makes it possible to synthesize rugged instruments with simple transmission mechanisms. Despite that these pretwisted manometric tubes are known since 1850s, the methods of their structural synthesis were not developed with all necessary options. Therefore we discuss them below in this section in more detail.

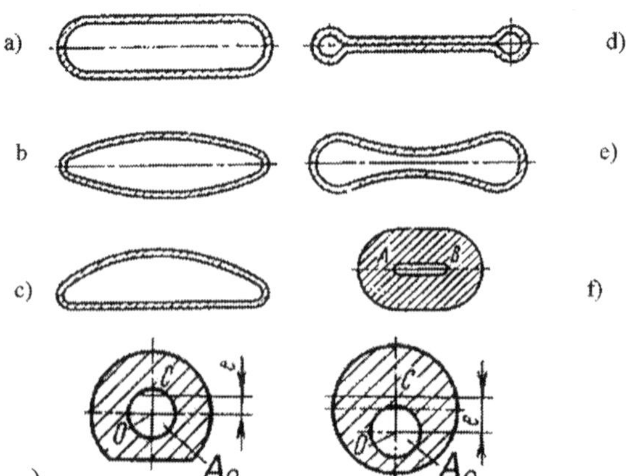

Fig. 3-17. Typical cross sections of manometric tube (After L.E. Andreeva)

Cross sections. The most frequently encountered cross-section shapes for single- and multi-coil tubular springs are flattened oval (Fig. 3-17a), elliptical (Fig. 3-17b), and *D*-shaped (Fig. 3-17c). The elliptical-section tubular spring has a higher sensitivity. The shape shown in Fig 3-17d is used where the elastic element must have a minimum initial volume. The dog-bone (other used names: oval with waist, eight-type, dumbbell, double spatula) cross section (Fig.3-17e) is stronger and is used for measuring higher pressure. In order to measure very high pressure, it is expedient to use tubular springs with the cross sections as shown in Fig. 3-17f and those with eccentric hall (Fig. 3-17g) designed by A. G. Nagatkin (*see* L.E.Andreeva, 1962) whose operation is based on the altogether different principle. While very similar to Bourdon tubes in appearance, they differ from them in that their internal bore with area A_o and center O (Fig. 3-17g) is disposed eccentrically by e relative to the external contour center C.

The effective cross sections of the twisted tubes are shown in Fig. 3-20b.

Calculation of Bourdon tubes. The calculation formulae for Bourdon tubes parameters are shown in Table 3-15. These formulae were got by Ritz's energy (i_B) and Mohr integral (ii_B) methods. Numerical integration of equations (8, Table 1-10) method (iii_B), however, gives more accurate results in comparison with the experimental data as for deflection as for stress values.

Table 3-15. Calculation formulae for Bourdon tubes

Calculation method, parameter	* Calculation formulae
1. (i_B)Rotation angle $\Delta\gamma$	$\Delta\gamma/\gamma = p[(1-\nu_p^2)R^2(1-b^2/a^2)/Ebh]\alpha_B/(\beta_B+\kappa_B^2)$; $\kappa_B = Rh/a^2$
2. (ii_B) Radial end displacement λ_r	$\lambda_r = (\Delta\gamma/\gamma)\,R\,(1-\cos\gamma)$
3. (ii_B) Tangential displacement λ_t	$\lambda_t = (\Delta\gamma/\gamma)\,R\,(\gamma-\sin\gamma)$
4. (i_B) + (i_B) End complex displacement λ_d	$\lambda_d = p[(1-\nu_p^2)R^3(1-b^2/a^2)/Ebh]\alpha_B\Gamma_p/(\beta_B+\kappa_B^2)$; $\Gamma_p=\sqrt{(1-\cos\gamma)^2+(\gamma-\sin\gamma)^2}$
5. Center X_{cr}; Y_{cr} position and radius ρ_λ of the end rotation	$X_{cr} = -R(1-\cos\gamma)/\gamma$, $Y_{cr} = R\,[1-(\sin\gamma)/\gamma]$; $\rho_\lambda = R\,\Gamma_p/\gamma$
6. (i_B)Driving moment M_D	$M_D = 24\,pRab\,(1-b^2/a^2)\zeta/(\xi+\kappa_B^2)$
7. (i_B) Radial driving force Q_r	$Q_r = pab(1-b^2/a^2)\zeta\Gamma_r/(\xi+\kappa_B^2) = M_D\Gamma_r/24R$; $\Gamma_r = 48(1-\cos\gamma)/(\gamma-\sin\gamma\cos\gamma)$
8. (i_B) Tangential driving force Q_t	$Q_t = M_D\Gamma_t/24R$; $\Gamma_t = 48\,(\gamma-\sin\gamma)/(3\gamma-4\sin\gamma+\sin\gamma\cos\gamma)$

Table 3-15 Cont'd.

Calculation method, parameter	* Calculation formulae
9. (i_B)+(ii_B) Transverse stress σ_1	$\sigma_1 = p(R^2/a^2)(1-b^2/a^2)\,3[(2\,\nu_p\Phi/\kappa_B)\pm\Omega\,]\,/\,(\beta_B+\kappa_B^{\,2})$
10. (i_B)+(ii_B) Longitudinal stress σ_2	$\sigma_2 = p(R^2/a^2)(1-b^2/a^2)\,3[(2\,\Phi/\kappa_B)\pm\,\nu_p\,\Omega\,]\,/\,(\beta_B+\kappa_B^{\,2})$
11. (i_B)+(ii_B) Change ΔV of internal volume	$\Delta V = 12p\dfrac{1-\nu_p^2}{E}\dfrac{R^3}{h}a\gamma[1-(b/a)^2]^2\dfrac{n_B}{\beta_B+k_B^2}$; $\Delta V = 12\lambda_d ab\gamma[1-(a/b)^2]n_B/\alpha_B\Gamma_p$

*After L.E. Andreeva. Parameters R, $\gamma = \angle BOZ$ and variables $\Delta\gamma$ and λ_d are shown in Fig. 3-16a. The charts of geometric functions Ω (solid lines) and Φ (dotted lines) are presented in Fig. 3-18; the charts of functions α_B, β_B, ζ, ξ, and n_B are shown in Fig. 3-19a for oval cross section and in Fig. 3-19b for elliptic cross section.

In Table 3-15, p is the applied pressure, b and a are the tube's cross section semi-axes, h is the tube's wall thickness, E and ν_p are Young's modulus and Poisson's ratio of the material, respectively; σ_1 and σ_2 are the transverse and longitudinal stress; R is the tube center line's radius, and γ and $\Delta\gamma$ are the tube's initial angle and its change under inside applied pressure.

The charts of numerically estimated functions Φ(a/b) by dashed lines and Ω (a/b) by solid lines are shown in Fig. 3-18a for manometric tubes with flat oval cross section and in Fig. 3-18b for manometric tubes with elliptic cross sections. Coefficients α_B, β_B, ζ,, n_B, and ξ values for manometric tubes with flat oval $2a \times 2b$ cross section (Fig. 3-19a) and similar coefficients for the elliptic $2a \times 2b$ cross section (Fig. 3-19b) have also been evaluated numerically. Functions Φ and Ω and the shown coefficients are presented in more detail by L. E. Andreeva (1962). In case 9 and 10 of Table 3-15, the "plus" sign is for outer surface points ($z = + h/2$) and the "minus" sign is for inner surface points ($z = -h/2$). Coefficient n_B is helpful for calculation of the change ΔV in Bourdon tube's internal volume with applied pressure p.

Helicoidal shells.

These shells are represented as (i) a coiled ribbon strips and (ii) the pretwisted tubes. A coiled spring is usually considered a helical beam. But for the coiled wide ribbon strips this model produces incorrect relations between their extension S and angular untwist θ. More precision results are received on the basis of the shell model. These strips are widely used in spatial transmission mechanisms. Therefore, they are discussed in detail in Chap 4.

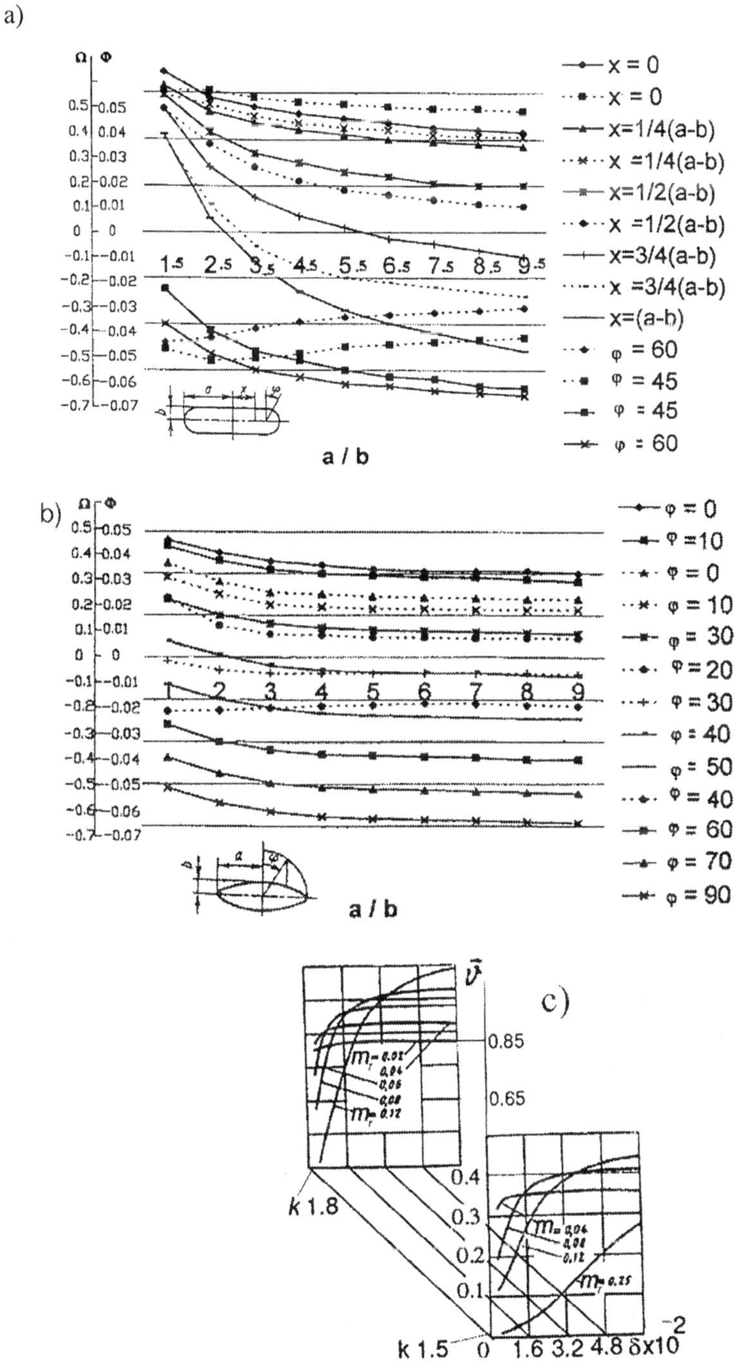

Fig. 3-18. Charts of functions Ω, Φ, and compliance $\bar{\vartheta}$ (after L.E. Andreeva).

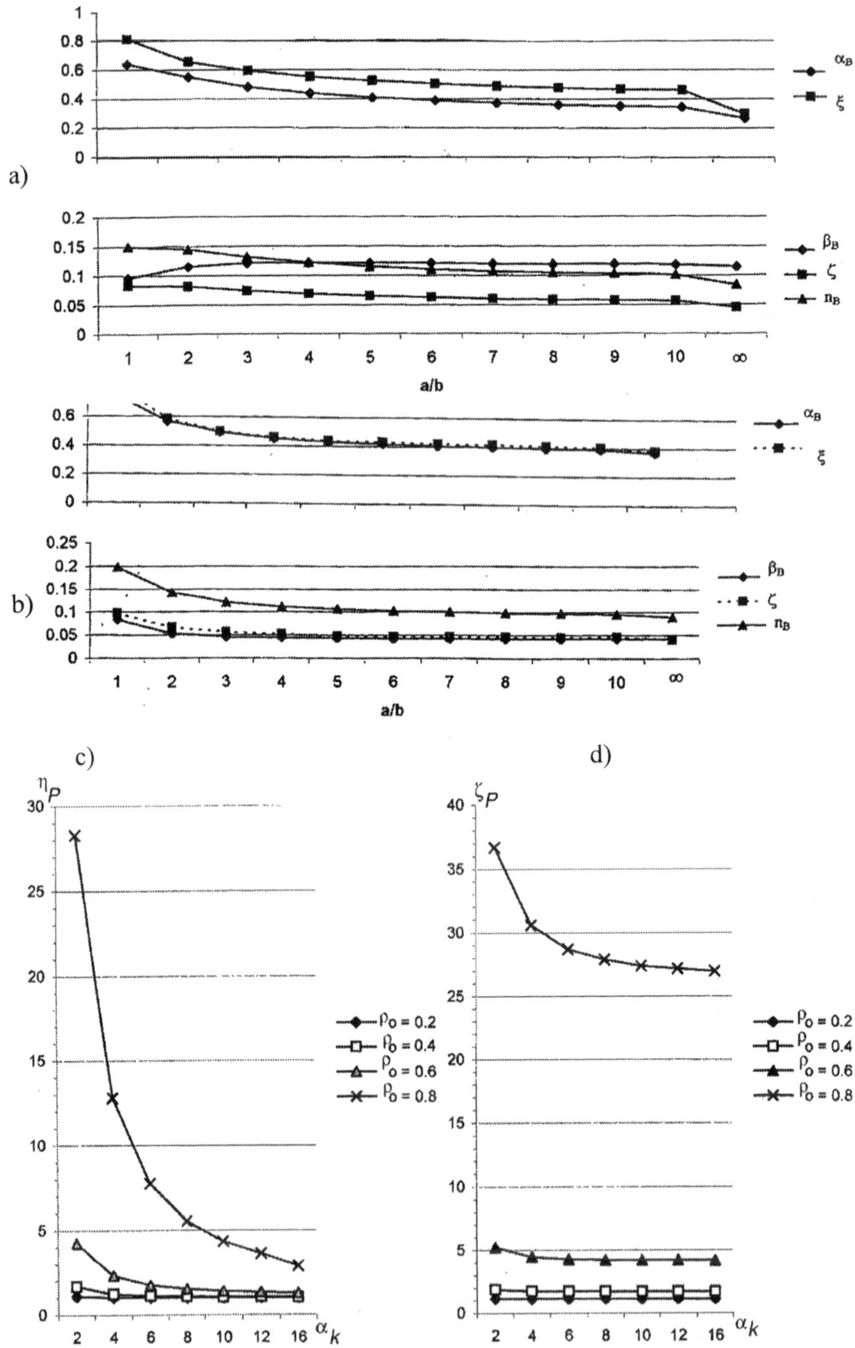

Fig. 3-19. Charts for Bourdon's tubes and corrugated diaphragms.

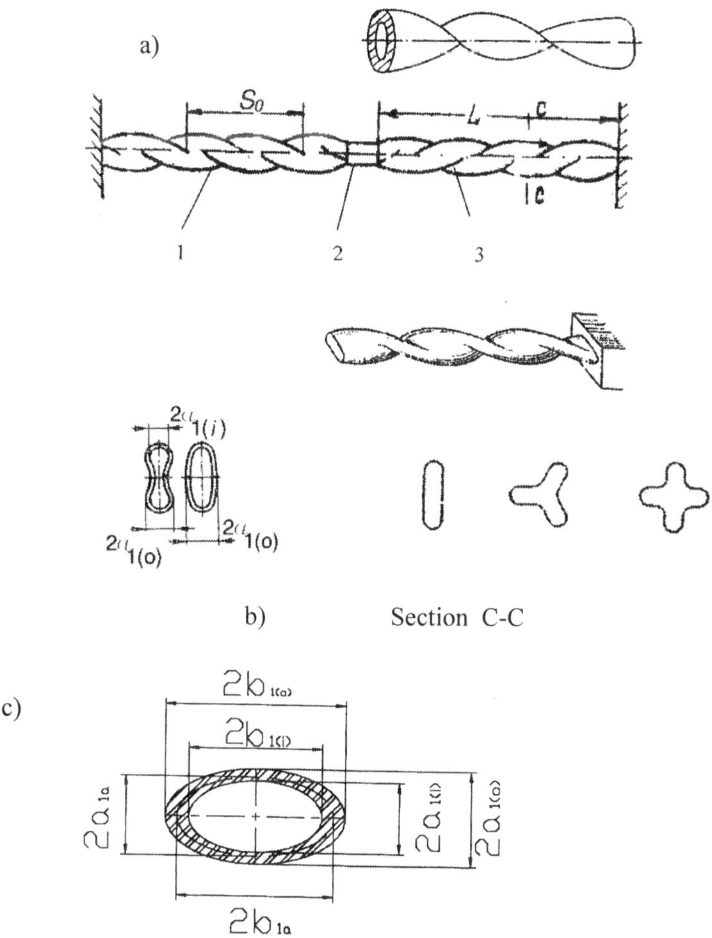

Fig. 3-20. Twisted single and double manometric tubes.

The *pretwisted helicoidal tubes* (Fig. 3-20a) find an effective application as sensitive elements in the manometers for measurement of a high pressure in the range of 250 kPa to 160 MPa (*see* Patents GB 3143883, US 2639616, SU 304459). If the pretwised tube has a plug on one end, it becomes sensitive to the air supply inside it that will untwist the tube's end.

Calculation methods. In 1950's-1970's there were published several results of pretwisted tubes experimental (L. A. Foygel *et al.*, 1959; V. P. Korsunov, 1966) and theoretical (Chen-Chu, 1952a, b; P. I. Begun, 1975) research that show the complicated nature of these elastic devices which, in addition to the untwist beam's features, have a profile change under inside pressure. Their theoretical calculations have been based on unwieldy numerical methods. The quasi-empirical equations proposed by L. A. Foygel *et al.* (1959) for untwist angle θ and conditional nonlinearity $\delta\theta$ estimation required input information about the tube's sensitivity from presented charts based on the limited experimental data. These formulae are as follows

$$\theta = \frac{\alpha_0 p}{1 - 360 L_h k_h p / \varphi_0^2}, \tag{3-104a}$$

$$\delta\theta = \pm \frac{p}{14.4 L_h} \cdot \frac{d\alpha}{d(1/S_0)}, \tag{3-104b}$$

where α and α_0 are the tube's experimental sensitivity under pressure p and without inside pressure, respectively; $k_h = (\partial\alpha/\partial S_0)h$; h, L_h, φ_0, and S_0 are the tube's specified geometrical parameters: the thickness of tube's wall, the distance between the end sections, the angle between profiles in the end sections without pressure, and the tube's helix pitch without pressure. But the development of the pretwisted helicoid calculation's methods (see Sections 2-2, 2-3, 4-3) allows us to use mentioned quasi-empirical equations on regular analytical bases.

Calculation method revisited.

Pressure action. Using (104a) and (104b) together with expressions for pretwisted helicoidal beam (see Y.M. Tseytlin *et al.*, 1989), we derived the following approximate formulae for the pretwisted tube's untwist angle θ and nonlinearity $\delta\theta$ calculation:

$$\theta = \frac{j_{\theta tuAi}^{-1}\, 100 p}{1 - q_{pi} 360 L_h k_h 100 p / \varphi_0^2}; \quad \delta\theta = \frac{p}{14.4 L_h} d(j_{\theta tuAi}^{-1}) / d(1/S_0), \tag{3-104c,d}$$

where $j_{\theta tuAi} = j_{\theta tu} / A_i = (j_{\theta(o)} - j_{\theta(i)}) / A_i$, $j_{\theta(o)}$ and $j_{\theta(i)}$ are the equivalent longitudinal stiffness for the pretwisted beams with outer and inner surface parameters for the pretwisted tube without inside pressure, respectively; A_i is the area of the inner surface cross section of the tube; p is the air pressure applied inside the tube; L_h is the distance between the edge sections of the tube which is approximately equal to the tube's length L; $\varphi_0^{\circ} = 360\, L_h / S_0$ is an angle between those sections in angular degrees; $k_h \approx (\partial j^{-1}_{\theta tuAi} / \partial S_0)h$; S_0 and h are the pitch in the pretwisted tube surface and the thickness of its wall, respectively; and normalizing factor $q_{p1} = 4$ at $p = 16000$ kPa, $q_{pi} = q_{p1}\,(p_i / 16000)$ at $25000 \le p_i = p \le 160000$ kPa. The longitudinal sensitivity of the pretwisted helicoid tube without inside pressure is approximately equal to:

$$j_{\theta tuAi}^{-1} = \frac{2Lk_p}{0.0696 G\lambda}\left[\frac{a_{1(o)} b_{1(o)}}{a_{1(i)} b_{1(i)}} (1 + \nu_{jf} b_{1(o)}^2 k_p^2 / \lambda) - (1 + \nu_{jf} b_{1(i)}^2 k_p^2 / \lambda)\right]^{-1}, \tag{3-105}$$

where $a_{1(o)}$; $b_{1(o)}$ and $a_{1(i)}$; $b_{1(i)}$ (Fig. 3-20c) are the smaller and larger semi-axes of the outer surface with subscript o and inner surface with subscript i, $\lambda = (a_1/b_1)^2$; $\kappa_p = 2\pi/S_0 = 2\pi\varphi_0^{\circ}/(360\, L_h)$, and $\nu_{jf} = 0.179$ is the approximate cross-section rigidity factor (see Chap.2) for the elliptical, rectangular, and flat-oval profiles (it can be estimated for other profiles as well). Modulus G is expressed here in kPa. Let us approximately estimate k_h factor assuming that it can be derived by the differentiation of function

$j^{-1}_{\theta uAi}$ with respect to pitch S_0. We obtain by this approximation the following expression:

$$k_h \approx -\frac{2L2\pi}{0.0696G\lambda\, S_0^2}\Big[\frac{a_{1(o)}b_{1(o)}}{a_{1(i)}b_{1(i)}}(1+\nu_{jf}b_{1(o)}^2 k_p^2/\lambda) - (1+\nu_{jf}b_{1(i)}^2 k_p^2/\lambda)\Big]^{-1}(1-\xi_q)h, \tag{3-106}$$

$$\xi_q = \frac{1}{S_0^2}[2\nu_{if}(2\pi)^2/\lambda)]\Big(\frac{a_{1(o)}b_{1(o)}}{a_{1(i)}b_{1(i)}}b_{1(o)}^2 - b_{1(i)}^2\Big)\Big[\frac{a_{1(o)}b_{1(o)}}{a_{1(i)}b_{1(i)}}(1+\nu_{jf}b_{1(o)}^2 k_p^2/\lambda) - (1+\nu_{jf}b_{1(i)}^2 k_p^2/\lambda)\Big]^{-1},$$

$$dj^{-1}_{\theta uAi}/d(1/S_0) \approx \frac{(q_{pi}/2)L2\pi 10}{0.0696G\lambda}\Big[\frac{a_{1(o)}b_{1(o)}}{a_{1(i)}b_{1(i)}}(1+\nu_{jf}\, b_{1(o)}^2 k_p^2/\lambda) - (1+\nu_{jf}b_{1(i)}^2 k_p^2/\lambda\Big]^{-1}. \tag{3-107}$$

We now estimate untwist angle θ and its nonlinearity $\delta\theta$ on the basis of (104c, d) - (107) for the pretwisted tube of stainless steel (G= 78500000 kPa) with a flat-oval cross section under high inside pressure (Table 3-16). The difference between calculated and experimental data of untwist angle θ are not more than 12 % and in many cases mush less.

It was found in an experimental study (V. P. Korsunov, 1966) that the cyclic lifetime for a pretwisted tube is proportional to the ratio of the values of pressure applied inside it in the second power.

Table 3-16. Rotation angle θ and its nonlinearity $\delta\theta$ on the end of the pretwisted tube

$L_h=L/S_0$ mm	p, kPa	h	$a_{1(o)}/b_{1(o)}$ mm	$a_{1(i)}/b_{1(i)}$	φ_0°	λ / k_p	$\theta^\circ / \delta\theta$ % Calculated (104c, d – 107)	Experiment *
40 /44	16000	0.29	1.75/3.32	1.46/3.03	327.3	0.278	3.87 / 0.8	4 / ± 0.8
60/ 44	16000	0.29	1.75/3.32	1.46/3.03	490.9	0.143	5.81/ 0.8	6 / ± 0.8
100 /44	16000	0.29	1.75/3.32	1.46/3.03	818.2		9.68 / 0.8	10 / ± 0.8
40/25	40000	0.42	1.5/2.44	1.08/2.02	576	0.377	2.23 / 1.4	2.5 / ± 1.5
60/25	40000	0.42	1.5/2.44	1.08/2.02	864	0.251	3.35 / 1.4	4 / ± 1.5
100/25	40000	0.42	1.5/2.44	1.08/2.02	1440		5.58 / 1.4	6 / ± 1.5

Table 3-16. Cont'd

$L_h = L / S_0$ mm	p, kPa	h mm	$a_{1(o)} / b_{1(o)}$ mm	$a_{1(i)} / b_{1(i)}$ mm	φ_0°	$\frac{\lambda}{k_p}$	$\theta^\circ / \delta\theta$ % Calculated (104c, d – 107)	Experiment *
40/16	100000	0.50	1.0/1.79	0.5/1.28	900	0.314	1.48 / 3.9	1.6/ ± 2
60/16	100000	0.50	1.0/1.79	0.45/1.28	1350	0.392	2.22 / 3.9	2.5 / ± 2

* After V.P. Korsunov, 1966.

4.

Elastic Guides, Transmissions, and Amplitude Transducers

This chapter presents examples of monolithic and assembled leaf spring guides, planar and spatial transmission mechanisms, along with their structure's specifics and methods of effective structural synthesis, and the amplitude transducers. Stability of bending and torsion influence, buckling and warping of elastic systems, and the dynamic functions of amplitude elastic transducers are discussed as well.

4-1. Rotary and Translational Guides

The rotary and translational elastic guides are built based on the resilient flexures with leaf spring assemblies (P.J. Geary, 1954), profiled torsion suspensions (P.J. Geary, 1960), notch hinges, and some different monolithic structures.

The typical structure.

Flexible strips. In leaf spring and strip guides (ELKS of type RG-I-RGII; TG-I- TG-VI), the strips usually have thickness h in the range of 0.1 to 0.5 mm and the width b in the range of 3 to 40 mm with ratio $b/h > 10$.

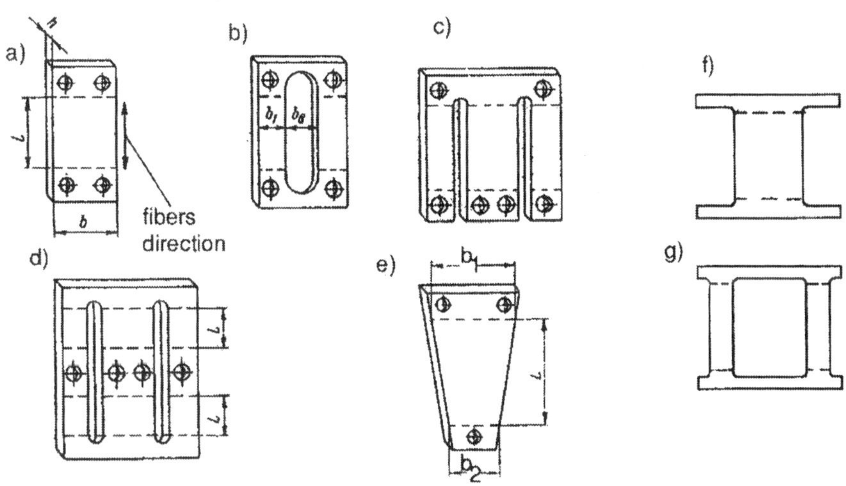

Fig. 4-1. Types of elastic strips in the precision structures

The strips that are widely used in practice are shown in Fig. 4-1 (the dashed lines separate the working portions of the strip) among which a) is a solid strip, b) has a straight recess, c) has an E- shape, d) with two symmetrical recesses, e) is a trapezoidal strip; and f) and g) have I- and D- shapes, respectively. The latter strips were used in commercially available during 1960-1990s Lucas-Troeger pivots (*see* Patent US 3,813,089, H. Troeger ,1962, A.H. Slocum,1992) where an I-shaped strip is slipped through the insertion opening in the squared-D-shaped strip.

Clamping and external friction.

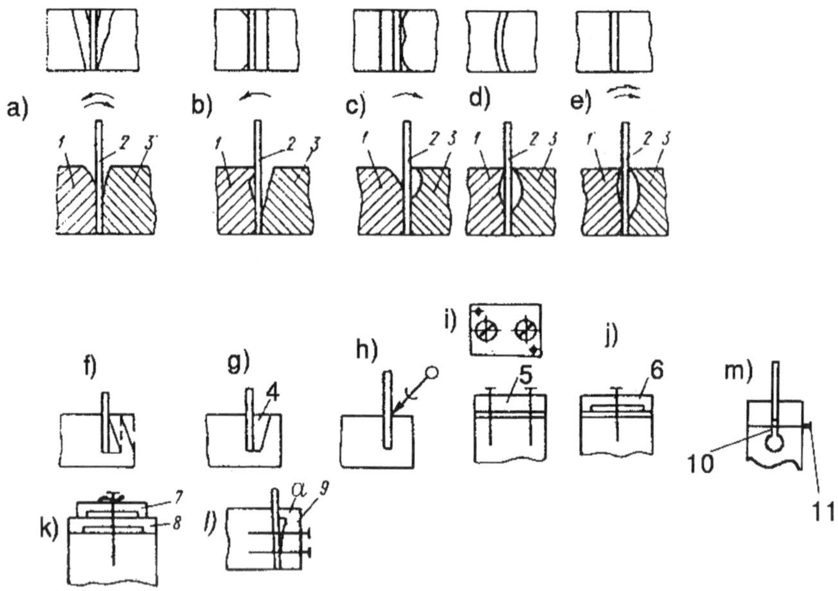

Fig. 4-2. Typical methods and conditions of clamping strips

Figure 4-2 shows several typical conditions that occur in the strip's clamping zone: a) the possible change of contact line' s position between flexing strip 2 and rigid elements 1 and 3, b) and c) provide stable position of the contact line at the strip's bending only in one direction, however, c) and d) bow, camber, and warping of the elastic strip that are caused by the curved clamping elements. The position of contact line on the flexing strip remains constant in satisfactory clamping case e) which does not cause the strip's warping. Fig. 4-2f-m show different clamping methods: by coining (Fig. 4-2 f), with wedge 4 (Fig. 4-2 g); by bonding, brazing, or welding (Fig. 4-2h); with flat clamping plates 5 (Fig. 4-2i), by clamping plate with relief 6 (Fig. 4-2 j), by two clamping plates 7 and 8 with recesses and a spring washer under a screw's head (Fig. 4-2k), by clamping plate 9 with rib *a* (Fig. 4-2*l*), and clamping in narrow slot 10 by tightening screw 11(Fig. 4-2m). The latter provides a good clamping quality at a relatively small clamping space.

A contact stress may occur in the flexing strip and in the clamping cover plate similar to that found in the contact between a cylinder and cylindrical groove. The clamping plate usually has a fillet. The clamping plate segment in the detachable joints should be flat and its specific position in the assembly can be provided by pins (Fig. 4-2i). The permanent joint (Fig. 4-2h) is a reliable one and it eliminates slippage and Coulomb friction, but it simultaneously prevents the possibility of the individual flexures' exchange in the case of their failure or their displacement for adjustment purposes. This method is very effective in the mass production of elastic pivots when the exchange of the whole elastic pivot is economically more productive. The adjustment procedure in this case is provided in the manufacturing process by an abrasive or by similar operations (*see* Patent US 4,327,527) on the pivot assembly under the parameter measurement and monitoring control of its quality in kinematics and torque.

Saint – Venant's principle. Let us recall that in accordance with this principle, the values of stresses in a solid elastic body from the statically equivalent to zero loads that are spread through the small parts ΔL_b in comparison with the size L ($\Delta L_b/L < 1$) of its boundaries diminish very rapidly. The order of diminishing at distances L_d that are larger than ΔL_b, i.e. $L_d / \Delta L_b > 1$, may be expressed as $(\Delta L_b / L_d)^{\phi}$, (4-1)
where $\phi \geq 2 - 4$ for different force conditions.

Rotary hinges. *Structural specifics.*

RG-I. In the *single leaf spring rotary hinge*, especially at non-stationary conditions, a short but relatively wide ($b > L$) strips 1 (Fig. 4-3a) are used, which provide better stability in the position of a rotational axis (*see* sec. 3-1). In detachable joint, the strip is clamped to case 4 and lever 2 by clamping plates 5 and 3 with holes for the screws and the positioning pins.

A leaf spring with the recess b_b and width $2b_1 + b_b$ (Fig.4-1b) provides better stability to a parasitic twisting moment M_{nk}. The two leaf springs with width $b_1 = ½\, b$ may be used instead. Adjustable 8 and non-adjustable 7 stops are used to avoid the breaks of the flexures in the presence of random overloads. For the same reason, conical or cylindrical limiters 6 whose axes coincide with the initial position of the axis of the hinge rotation may be used. The stops and limiters should not touch the movable element in the working state.

RG-II. In the *cross-strip hinge* there are two1 and 2 (Fig. 4-3c) or more crossing strips with a crossing angle γ in the range of 30 to 120°. The strips of length L and thickness h are clamped to stand 5 and pivoted element 3 by clamps 4. The bisymmetrical cross-strip flexural pivots (RG-IIst) with standardized dimensions are made by joining two sleeves, between which rotation is desired, with relatively thin flat *I*- and *D*-shaped flexures (*see* Fig.4-1f, g), which are positioned so that their planes are normal to each other. The flexible supports that permit self-alignment were built on the basis of a flexural pivot with two diaphragms at its sides (see Patent US 3,811,665).

The complex dual angle concept $\overset{\circ}{\Phi}$ can be used for the mathematical description of the mutual position of the strips' axes in the initially unloaded cross-strip hinge. This angle function simultaneously reflects the actual angle of crossing $\gamma = \gamma^o$ and the shortest distance $(b_b + b)$ between the axis of the consecutive strips,

i.e., $\overset{\circ}{\Phi} = \gamma^o + \varpi (b_b + b)^k$, (4-2)

where ϖ is Clifford's symbol, or dual unit (*see* Chap. 1). The rotation of the movable element on this elastic pivot occurs around axis I-I, which approximately coincides with the axis of the initial complex angle $\overset{\circ}{\Phi}$.

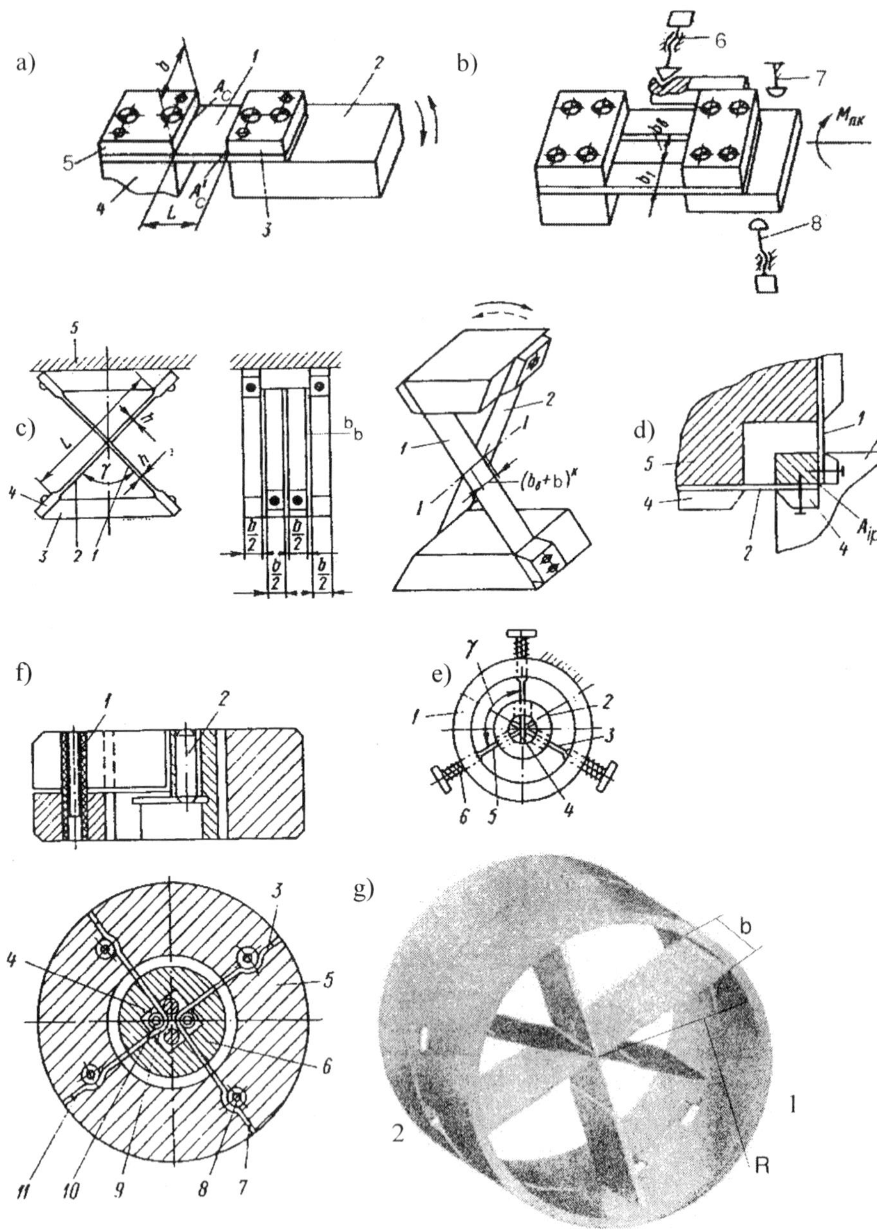

Fig. 4-3. Rotary hinges with leaf springs and "hexfoil" (latter after NASA, 2002)

RG-IIa. The *asymmetric cross-strip hinges* may have strips with equal or unequal length. The *imaginary crossing-strip hinges* (W. Aller, Patent US 2,177,398) are used in cases of small available design space. A version of such a device (RG-II*i*) is shown in Fig. 4-3d, where the two strips 1 and 2 are clamped to support 5 and lever 3 by clamping plates 4. The conditional axis for the lever rotation passes through the point A_{ip} of the strips' axially protruded projection intersection ("the imaginary point").

RG-III. Three-strip hinges. The structure of the crossing three-strip hinges (T.H. Fewster, British Patent 375354) takes different forms. Fig. 4-3e shows a version of such a structure where 1 is an annular base, 2 is a pivoted rotatable bushing, and 3, 4, 5 are the strips projections crossing in a point through which the axis of rotation passes. The strips are mutually shifted at an angle $\gamma = 120°$. One end of each strip is clamped to bushing 2, and to the other three ends tensile forces are applied through springs 6. Another structural version has both ends of the two strips fixed, and the tensile stress through the spring is applied only to the third strip.

RG-IIIm. Multi-strip pivot. Suspension of the levers can be provided with the multi-strip pivot shown in Fig 4-3f. The pivot has ring 5 in which are positioned bushing 6 with central square hole 4 and four grooves 9. Bearing strips 11 and 3 are clamped in ring 5's grooves 7. With their expansion, elastic split bushings 8 clamp the strips by the bushings' outer surfaces. The four cylindrical pins are clamped inside square hole 4; two of them are solid pins 2 and the other two pins 10 are elastic which have been built from bushings 8 and 1. Bushings 8's installation in bushings 1 is completed only after the assembly of bushing 6 with ring 5. Strips 11 and 3 should have equal tension. There are other possible versions of this multi-strip hinge (Table 1-1, ELKS of type RG-III*m*) with three and four strips positioned at angles 60 and 45°, respectively. Elastic pivots with V-shaped strips (Patent SU 298773, 1971) have extended limits of an angular displacement.

RG-VIIh. A *hexfoil rotary flexure* includes two coaxial hollow circular cylinders 1and 2 (Fig. 4-3g), which constitute a rotor and stator with radius R, respectively.

Ttranslational guides. *Structural specifics.*

Fig, 4-4 shows: a) ordinary spring parallelogram, b) spring parallelogram with cover plates; c) rectilinear spring compound movement, d) with consecutively, e) and f) with parallel positioned ordinary parallelograms, g) the version of symmetrical rectilinear movement (after S.T. Smith, 2000) with notch hinges; h) test charts of guides movement; and i) and j) are the double-spring parallelograms with auxiliary kinematical link.

TG-I. The structures of these strip guides have very different versions. In the ordinary spring parallelograms, there are solid strips (Fig. 4-1a), strips with a central recess (Fig. 4-1b), or two strips with width $b/2$ on each side. A high accuracy requirement in this case is applied to the geometric parallelepiped that is built by four lines of the strips clamping , their initial planes, and the side edges. Tolerances for the parallelism of surfaces designated with apostrophe and without it should be specified. If the lead screw with a rotating spindle drives the movable carriage, pin 3 (Fig. 4-4c) with the conical or spherical ends should be installed between the screw and the carriage to eliminate a parasitic moment from friction.

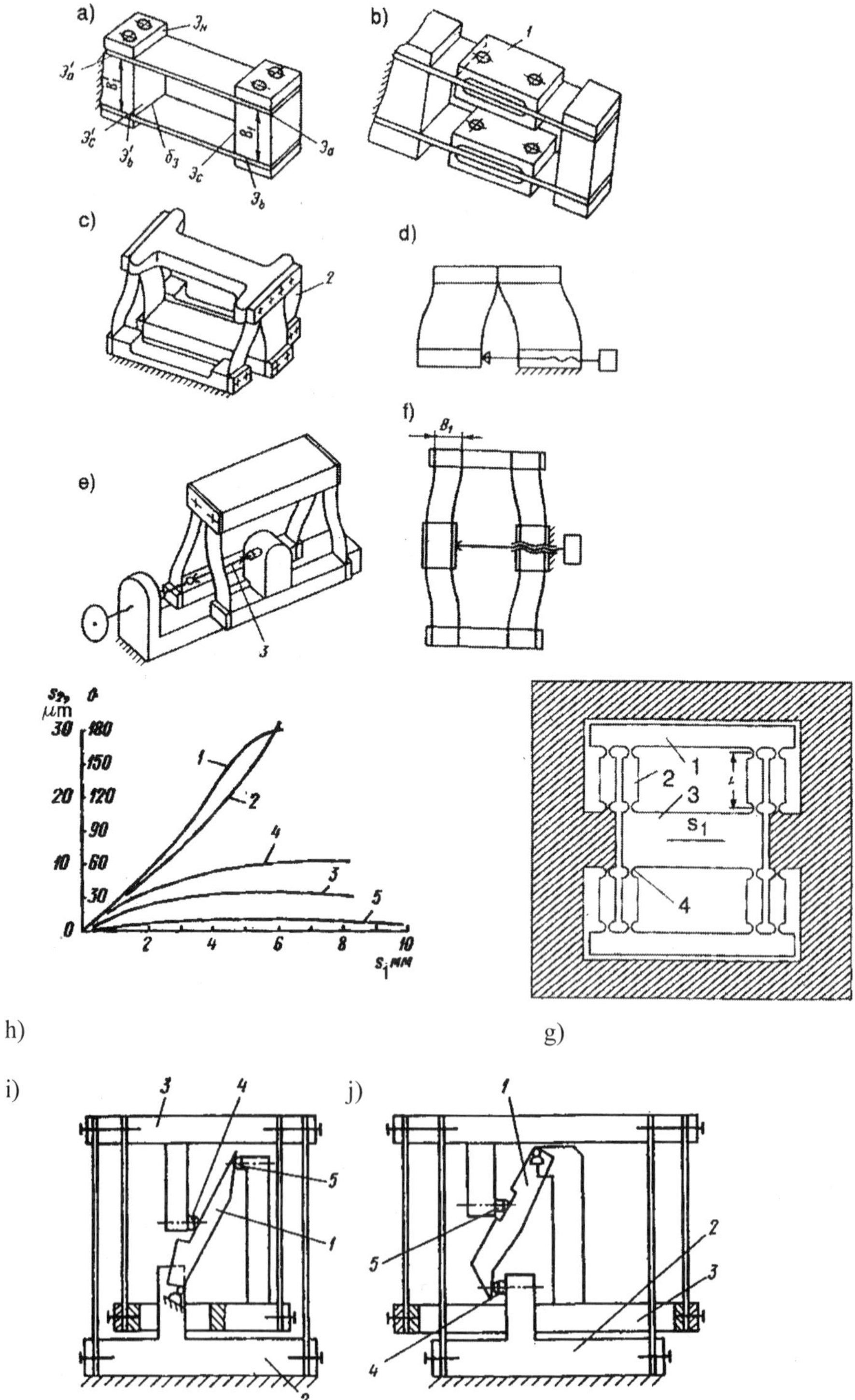

Fig, 4-4. Typical structures and features of translational guides.

TG-IV. The *rectilinear spring compound movement* has E-shaped strips 2 (Fig. 4-4c) or an equivalent system of strips. The constituent ordinary parallelograms in the rectilinear spring compound movement may be positioned consecutively (Fig. 4-4d) or in parallel (Fig. 4-4 e,f). Rectilinear version of double compound movement with two auxiliary carriages 1, eight lever L links 2, main carriage 3, and sixteen notch hinges 4 is shown in Fig. 4-4g. As a consequence of the two compound spring movements being joined together, it can be seen that a deviation from rectilinear motion s_1 of main carriage 3 is no longer theoretically possible in this structure. This is confirmed by mobility analysis (S.T. Smith, 2000) for which only one *degree of mobility*, *Mob*, remains because in this structure number of rigid links n=12 and notch joints j=16. Therefore,

$$Mob = 3(n\text{-}1)\text{-}2j\text{=}3(12\text{-}1)\text{-}2(16)\text{=}1.$$

It is clear that in this analysis the notch hinges are supposed to be non-stretchable with constant position of the center of rotation that does not correspond to their real features (*see* Chaps. 3 and 8). Indeed, every rotational and translational strip guide may have a duplicate with notch hinge version. However, the latter has a smaller range of motion as a result of larger stresses (*see* Chap. 3).

Fig. 4-4h shows the experimental charts for kinematics of certain parallelograms in the absence of longitudinal force $Q_x \approx 0$: 1 is a chart of function $S_2 = f(S_1)$ for an ordinary parallelogram (*see* Fig. 4-4a) with the length of the strips L=42 mm, width b=20 mm, thickness h =0.56 mm, and base distance B_1 = 25 mm, 2 is a chart of slope pitch angle $\vartheta = f(S_1)$ for the carriage of the same parallelogram, 3 and 4 are charts $S_2 = f(S_1)$ and $\vartheta = f(S_1)$, respectively, for the parallelogram shown in Fig. 4-4d, which is composed from the two ordinary parallelograms with the above mentioned parameters, and 5 is a chart of function $S_2 = f(S_1)$ for the symmetrical spring movement shown in Fig. 4-4f with the length of strips L = 40 mm, width b= 25 mm, thickness h= 0.35 mm, and base distance B_1 = 25 mm.

TG-V. Figures 4-4i, j show the simple structures of the *double-spring parallelogram* with the additional kinematic link (*enslaved*) in the form of lever transmission 1. The elastic pivot of the lever can be connected to case 2 (Fig. 4-4i) or to carriage 3 (Fig. 4-4j). The surface of tips 4, 5 has a cubical parabola form. Such a type of translational guide was used in the special precision measuring microscopes for the large displacement (to 25 and even 50 mm) of the measuring table.

Load-motion relationships in the strip's guides.

Fig. 4-5 shows a) spatial load schematics for leaf-spring hinge, b) the hologram of strip's surface at skewed by $\alpha_0$clamping line, c) the corresponding calculation schematic, d) chart of the calculated function $\frac{x}{b}(\alpha_0)$ on the bent strip with experimental data denoted by triangles; e) the hologram of the bent strip which clamped by triangular cover plate, f) spatial load schematic for the cross-strip hinge, and g) its planar kinematics.

Matrix equation. The elastic elements of rotational and translational guides in real conditions are loaded by a spatial system of forces. Fig. 4-5a shows the schematic of such a loaded ELKS with the single strip leaf-type hinge. If we suppose that the problem is similar to the deformation of a thin weakly pretwisted beam (*see* Chap. 2), the constituents of the leaf spring deflection in this case can be defined from the following matrix equation

$[s_x \; s_y s_z \theta_x]^T = [\nu_{mn}][Q_x M_z M_y M_x]^T$, (4-3)

where $[\nu_{mn}]$ is the matrix of compliances with members $\nu_{mn} = j^{-1}{}_{mn}$; $Q_x = Q_{xA} + Q_{Rx}$, $M_x = M_{xA} + M_{Rx}, M_y = M_{yA} + M_{Ry}$, and $M_z = M_{zA} + Q_x y + Q_R r_R$. Unlike what is shown in Chaps. 2 and 3, the redundant unknown reaction constituents Q_R, M_{Rx}, and M_{Ry} between driving body 2 and driven lever 1are added here. In addition, some parameters of these bodies' displacement and the conditional center of rotation are specified, which allows us to estimate the accuracy of rotation for the elastic pivot. As the elastic center of body 1's rotation, the point of this body should be chosen that has a small linear displacement whose value in the function of the rotation angle characterizes the errors of a rotary guide. The sought point should be close to an *ICR* whose coordinates' analytical definition is presented in Chaps. 1and 3.

Shift of clamping line. The calculation methods for flexible strips are usually obtained with the assumption that the position of their clamping line is constant and normal to their longitudinal axis. If these assumptions are not provided as a result of the clamping defects, then the displacement of the contact line will change the length of the strip and cause an action of the external friction. Let us discuss the simplest but the most common case of the skewed clamping line with tilt shifting. In accordance with Mansfield's inextensional surface theory for thin flat plates, the deformed surface of the flexed strip in the cross section, which has slope α (see Fig. 4-5c and interference fringes hologram in Fig. 4-5b) to the strip's axis, is a conical one and is developable without warping on a plane. Therefore, the curvatures K_0 in the direction of the strip's axis and K_α in the direction of the above mentioned section under angle α are related similar to the typical conical sections of an elliptical type. The list of notations for the inextensional theory of thin flat plate-strip is as follows (Fig. 4-5c): x is a distance along the reference axis; α is an angle made by a generator with the x-axis; η is a distance measured along a generator from an elemental cone apex; η_x is a value of η at the x-axis; η_1, η_2 are values of η at the strip boundaries; M_η is a moment per unit length around the generator; M_α is a total moment around the α-generator; h is the thickness of the strip; $D_\eta = Eh^3/12\,(1-\nu_p{}^2)$ is a flexural rigidity of the strip at distance η along a generator; U_n is the strain energy; σ is an outer fiber stress; $\rho_\eta = \eta_2 / \eta_1$; $\mu_p = \rho_\eta - 1$; M is an applied moment. The inextensional surface theory is yielded (E.H. Mansfield, 1955, 1989) from the condition of maximum strain energy $U_n = ½ \int F(\alpha, x, x')d\alpha$. In this case, integrand function $F(\alpha, x, x')$ for the connection between independent variable coordinate x and angle α should satisfy the nonlinear differential equation of the second order $x'' F_{x'x'} + x' F_{xx'} + F_{\alpha x'} - F_x = 0$, and $F - \alpha' F_{\alpha'} = \text{constant}$. (4-4)

An apostrophe denotes differentiation with respect to α or x. For example, $x' = dx/d\alpha$, $F_\alpha = \partial F/\partial\alpha$, $\alpha' = \partial\alpha/\partial x$, $F_x = \partial F/\partial x$, $F_{\alpha'} = \partial F/\partial\alpha'$.

The flexural compliance ν_ϑ of the strip can be defined as the value that is reciprocal to the flexural rigidity $\nu_\vartheta = D_\eta^{-1} \int K_\alpha \, dx + C_{\nu\theta}$, where $K_\alpha = M \sin\alpha / (D_\eta \ln\rho_\eta x)$; $x/b = \int dx \, (\sqrt{\rho_\eta} \ln\rho_\eta)^{-1}$, and $\alpha = \sin^{-1}[\sqrt{\rho_\eta} \ln\rho_\eta / (\rho_\eta - 1)]^{1/2}$.

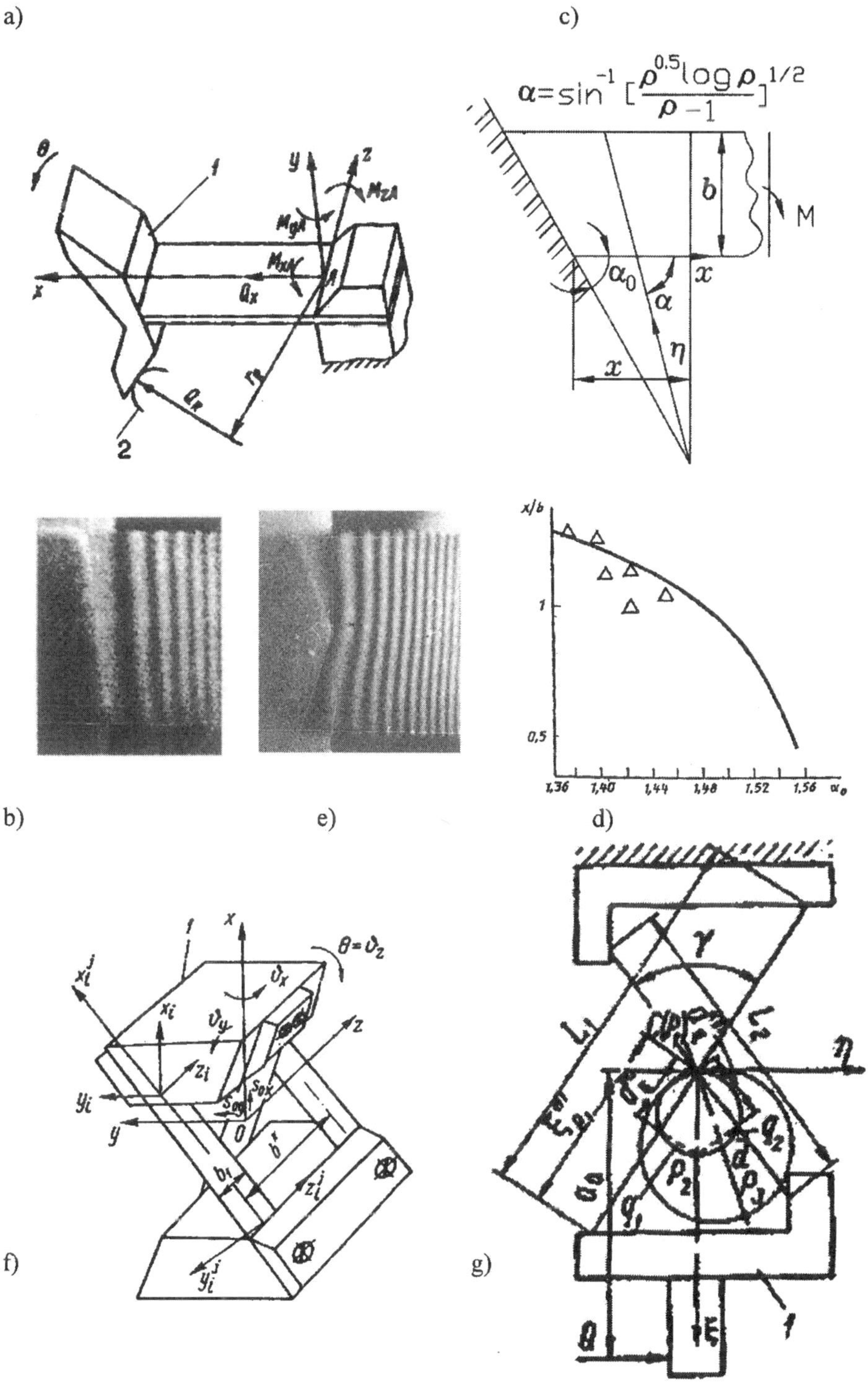

Fig. 4-5. Load-motion relations and clamping quality influence in strip's guides.

If $dx = x'\, d\alpha$, then $v_\vartheta = D_\eta^{-1} \int dx / \ln\rho_\eta + C$. The calculated values $x/b\,(\alpha_0)$ (Fig. 4-5d), representing the skewed strip zone, are obtained by numerical integration. The experimental data in Fig. 4-5d are depicted by triangles. According to the first approximation, the change of flexural compliance is

$$\delta v_\vartheta = (v_\vartheta - v_o) / v_o = \Delta\alpha_o\, b / (2L); \; x = \tfrac{1}{2} b[\ln\Delta\rho_\eta(\alpha_o) - \ln\Delta\rho_\eta(\alpha)], \quad (4\text{-}5)$$

where b is the width of the strip; $\Delta\alpha_o = (\pi/2) - \alpha_o$, and $D_\eta = D$ is assumed to be constant cylindrical stiffness of the strip. Expression (5) shows that a shift of the clamping line has a larger influence on the wide and short strips that are typical for the elastic strip pivots. This shift of the clamping line may also cause a torsion deformation in the strip in addition to its planar flexural deformation. The found features of the strip's flexure at the double-skewed clamping are clearly depicted in the hologram (Fig. 4-5e), where one can see the bending and turning of the interferometric fringes.

RG features and synthesis recommendations.

In Table 4-1 there are formulae for the rotary guides calculation, which were derived by different authors with different methods. Because they can be found in the referenced literature and their intermediate transformations are unwieldy in many cases, detailed description of these methods are not included in this book. However, the functional information constituents of ELKS general kinematic function (1-34) for each version of calculation are shown as the bold notes.

RG-II. In the cross-strip hinges, the movable ends of the strips are connected by real rigid body 1(Fig. 4-5f). Its motion under the spatial loading can be in a general case described by kinematic *motor* (screw, biquaternion) $\mathbf{S}_\kappa$ of displacement and force *motor* $\mathbf{Q}_\kappa$ linked to it. The definition of the term "*motor*" as a combination of related vectors was initially done by W.K. Clifford (1882). The displacement of the strip's ends in the accepted system of coordinates x, y, z may be found from the formula:

$$S_{\kappa i} = \mathbf{Q}_{\kappa e i}\, \mathbf{S}_\kappa . \quad (4\text{-}6)$$

Displacements $S_{\kappa i}$ of the strip's end should then be expressed in the strip's partial coordinate system x^j, y^j, z^j with transposed matrix A^T_{mn} (*see* Chap. 1and Y. Tseytlin, 1972) in which case $S^j_{\kappa i} = A^T_{mn} S_{\kappa i}$, and substituting (6) yields

$$S^j_{\kappa i} = A^T_{mn}\, \mathbf{Q}_{\kappa e i}\, \mathbf{S}_\kappa \quad (4\text{-}7)$$

Elastic displacements of the strips' ends $S^j_{\kappa i}$ are related to loads $Q^j_{\kappa i}$ applied to the strips' ends through the linear relationship by stiffness square matrix J_{mn} as follows:

$$Q^j_{\kappa i} = J_{mn}\, S^j_{\kappa i} = J_{mn}\, A^T_{mn}\, \mathbf{Q}_{\kappa e i}\, \mathbf{S}_\kappa \quad (4\text{-}8)$$

If we express (in the initial coordinate system) the obtained *motors* of partial loads applied to the end of each strip by the reciprocal matrix A^T_{nm}, the *motors* can be summed and equalized with the specified *motor* of external loads:

$$Q_\kappa = \sum_i A^T_{nm}\, Q^j_{\kappa i} = \sum_i A^T_{nm}\, J_{mn}\, A^T_{mn}\, \mathbf{Q}_{\kappa e i}\, \mathbf{S}_\kappa \quad (4\text{-}9)$$

Six equations (9) allow us to express the guided link's components of displacement in the chosen common coordinate system through the specified components of applied loads, or visa versa, to define the loads through the specified displacements. In practice we can meet the mixed conditions when part of the displacement components and the applied load components are specified.

Table 4-1. Kinematic relationships for rotary guides

Name, type, and schematic of ELKS	Calculation formulae and their application limits
Single leaf spring RG-I O_ρ is center of end rotation s_{ox}, s_{oy} are deflections from O_ρ initial position	**1. $F_k(s) + F_j(Q) + \delta F_j(Q) + \delta F_{kc}$ **** with E.P. Popov's method $\xi_{xP1}/L = L^{-1}(dy_1/d\theta) - \beta^{-1}\{-2\sin(\zeta_0/2)\csc^2\varphi\cos\delta_0 - [-2\mathrm{dn}(\varphi,k) + 1/\mathrm{dn}(\varphi,k)]\sin\delta_0 + \beta^{-1}(\chi/\mathrm{dn}(\varphi,k))\mathrm{x}$ $[\eta_0''\cos\delta_0 - \xi_0''\sin\delta_0]\}\dfrac{\sqrt{1-\sin^2(\varsigma_0/2)\csc^2\varphi}}{2\sin(\varsigma_0/2)\cot\varphi\csc\varphi}$; $\eta_{yP1}/L = L^{-1}(dx_1/d\theta) = \beta^{-1}\{[-2\,\mathrm{dn}(\varphi,k) + 1/\mathrm{dn}(\varphi,k)]\cos\delta_0 - 2\sin(\zeta_0/2)\csc^2\varphi\sin\delta_0 + \beta^{-1}(\chi/\mathrm{dn}(\varphi,k))\mathrm{x}$ $[\xi_0''\cos\delta_0 + \eta_0''\sin\delta_0]\}\dfrac{\sqrt{1-\sin^2(\varsigma_0/2)\csc^2\varphi}}{2\sin(\varsigma_0/2)\cot\varphi\csc\varphi}$; H=const; $M_1=0$; $\varsigma_0' = \varsigma_0$=const; $\delta_0 = \alpha_Q$; $\chi = 1$; $\beta = \sqrt{QL^2/H}$; $h/L \le 0.05$
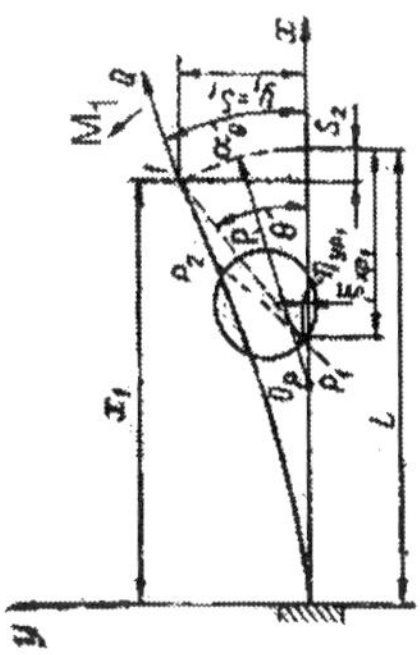	**2. $F_k(s) + F_j(Q) + \delta F_j(Q)$ **** After V.N. Zheludkov, 1971 a) $M_1 = 0$; $\alpha_Q = 90°$ $x_{ICR}/L = (1/3) - (3/140)(QL^2/EJ) = (1/3) - (3/35)\theta^2$; $y_{icr}/L = (1/15)(QL^2/EJ) - (31/3150)(QL^2/EJ)^3 = (2/15)\theta - (47/1575)\theta^3$ b) $Q = 0$; $M_1 \ne 0$ $x_{ICR}/L = ½ - (1/24)(M_1L/EJ)^2 = ½ - \theta^2/24$; $y_{ICR}/L = (1/6)(M_1L/EJ) - (1/120)(M_1L/EJ)^3 = (1/6)\theta - (1/120)\theta^3$ $H = EJ$=Const; $\left\vert\dfrac{M_1L}{H}\right\vert < 1$; $\left\vert\dfrac{QL^2}{H}\right\vert < 1$; $\left\vert\dfrac{M_1L}{H} + \dfrac{QL^2}{2H}\right\vert < 0.8$
View rotated at 90°	**3. $F_k(s) + F_j(Q)$** $\theta < 0.15$rad After A.T. Draudin, 1966 $s_{ox} = (3/16)L\theta^2 - (5/48)L[\theta(1 + k_R/5) + (2/5)k_Q\Theta_R]^2$; $s_{oy} = (1/12)L(k_R\theta - 2k_Q\Theta_R)$ $\Theta_R = Q_RL^2/2H$; $k_R = L\sin\alpha_{QR}/r_R$; See Fig. 4-5a $k_Q = (\Sigma Q_y/Q_R) + \sin\alpha_{QR}$; $\alpha_{QR} = \angle xQ_R$
	4. $F_k(s) + F_j(Q)$ After W. Lotze (1963) $\xi_{xP_1} = -L[½ + 6^{-1}(2\lambda_L+1)^{-1}]$; $\eta_{yP_1} = 0$; $d_\xi = L[(1/6) - (1/30)(2\lambda_L + 1)^{-2}]$; $d_\eta = L[(1/6)(2\lambda_L+1)^{-1} - (8/45)(2\lambda_L+1)^{-2} + (1/18)(2\lambda_L+1)^{-3}]\tan(90° - \alpha_Q)$ $\theta < 0.15$ rad; $Q_x = Q_y\tan\phi_Q$; $M_1 = Q_yL_1\cos\alpha_Q - Q_yL_1\tan\phi_Q$; $\phi_Q = 90° - \alpha_Q$; $\lambda_L = L_1/L$; L_1 is a rigid arm (*see* Fig. 4-6a)
	5. $F_k(s) + F_j(Q)$ See Section 3-1 $\Phi_c = L/H$; $\nu_x = 0$; $\nu_y = 0.0833\,L^3/H$
RG-I 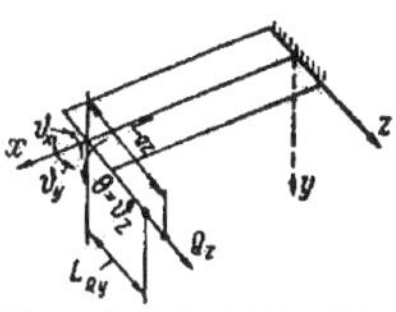	**6. $F_k(s) + F_j(Q)$** After P. Bondy (1967) $s_{Qz} = 20\times10^{-5}h^{-3}b^{-2.4}L^{1.5}(½L + L_{Qy})^{-1}\tan\theta\, L_{Qz}^{1.5}Q_z$; $\vartheta_y = 7\times10^{-9}h^{-3}b^{-1.5}L^{1.5}(L_{Qz} + ½L)\theta Q_z$; $\vartheta_x = 6\times10^{-7}h^{-3}b^{-2}L^{1.5}\theta^{0.5}L_{Qz}Q_z$ Uncertainty of this calculations within 25%

Table 4-1. Cont'd

Name, type, and schematic ELKS	Calculation formulae and their application limits
Cross-strip RG-II (Fig. 4-5g) 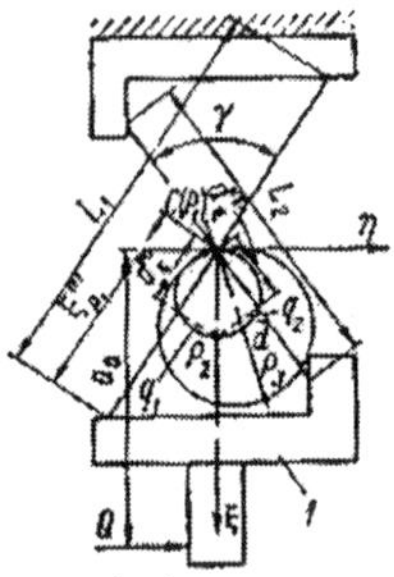	1. **F**$\mathbf{F_k(s)}$ $\xi_{P2}^{(1)(2)}$; $\eta_{P2}^{(1)(2)}$ (3-32 c, d) After W. Lotze (1963) $s_c = d\,\theta^2 + i\,\bar{d}\theta^3/(3\,!)$; $Q=0$; $\theta<0.1$ rad
	2. $\mathbf{F_k(s)}$ After A.T. Draudin, 1966 $s_{cx} = 0.707\,B_x\theta^2\,L\sec(\gamma/2)$; $B_x = 0.212\,(1.75-2\,\lambda_P)\,(2\lambda_P-0.25)$; $s_{cy} = 0.707\,B_y\theta^3\,L\sec(\gamma/2)$; $B_y = 0.6B_x(2\lambda_P-0.17)$ $L_1=L_2=L$; $0<\lambda_P<0.5$; $\theta<0.15$ rad
Symmetrical Cross-strip RG-IIs 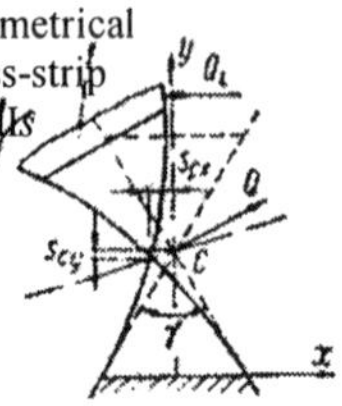	3. $\mathbf{F_k(s)+F_j(Q)+\delta\,F_j(Q)}$; $\lambda_P=0.5$; $L_1=L_2=L$; $\gamma=30$- $120°$; $\theta<20°$ $s_{cx} = \frac{L}{192\cos\gamma}\left(\frac{ML}{H}\right)^3 - \frac{L\cos\alpha_Q}{5760\sin^2\gamma}\left(\frac{ML}{H}\right)^2\frac{QL^2}{H}$; $s_{cy} = -\frac{L}{48\cos\gamma}\left(\frac{ML}{H}\right)^2 + \frac{7L\sin\alpha_Q}{1920\cos^2\gamma}\left(\frac{ML}{H}\right)^2\frac{QL^2}{H}$; $\tan\theta = \frac{ML}{2H} + \frac{ML^3Q}{24H^2}\frac{\sin\alpha_Q}{\cos\gamma} + \frac{2-\tan^2\gamma}{48}\left(\frac{ML}{H}\right)^3 - \left(\frac{ML}{H}\right)^2\frac{QL^2}{H}\frac{\cos\alpha_Q}{192\cos\gamma} +$ $\frac{11\cos^2(\gamma+\alpha_Q)+11\cos^2(\alpha_Q-\gamma)}{2880\sin^2\alpha_Q} - 20\cos(\gamma+\alpha_Q)\cos(\alpha_Q-\gamma)\frac{ML}{H}\left(\frac{QL^2}{H}\right)^2$. $\left\|\frac{ML}{H}\right\| < 1.5$; $\left\|\frac{QL^2}{H}\right\| < 1$;.2 $\left\|\frac{ML}{H}+\frac{QL^2}{2H}\right\| < 0.8$ After V.N Zheludkov**
Standardized dimensions, symmetrical cross-strip, RG-II*st* 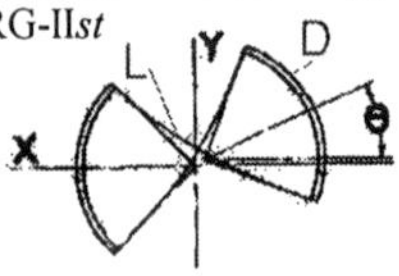	4. $\mathbf{F_k(s)+F_j(Q)}$ $L=0.6D$; $w=1.36D$ After H. Troeger, 1962 $h = \sqrt[3]{57.3\times12\,j_{\theta M}\,L/Ew} = \sqrt[3]{0.303\,j_{\theta M}/E}$; $\sigma_\theta = Eh\theta°/68.76D$; $j_{\theta M} = 33.43\left(10D/\theta°_{max}\right)^3$; $P_{t1} = 82.73D^2/\theta°_{max}$; $P_{t2} = 165.46\,D^2/\theta°$; $P_c = 12.4\times10^3D^2/(\theta°_{max})^3$; $Y_{ICR}/D = 0.00247\,\theta°$, $j_{\theta r} = 7.2\times10^4\,D^2/(\theta°_{max})^3$; $X_{ICR}/D = -1.07\times10^{-5}\theta^2$; $j_{\theta M} = \theta/M$
Two curved-strip RG-II*cu* 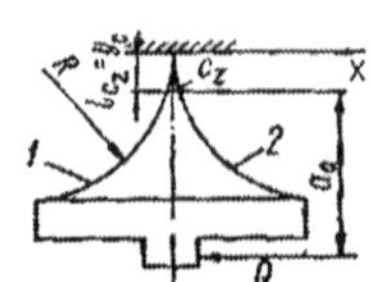	$\mathbf{F_k(s)+F_j(Q)}$ After W. Lotze, 1967-70 $l_{cz} = y_c = 0.303R$; $\Phi_c = 0.564\,RH^{-1}$; $\nu_x = \nu_y = 0.0116\,R^3H^{-1}$; $h_y = 0.144R$ $H = EJ$ $\theta < 0.1$ rad

Table 4-1 Cont'd

Name, type, and schematic of ELKS	Calculation formulae and their application limits
Three cross- strip RG-III 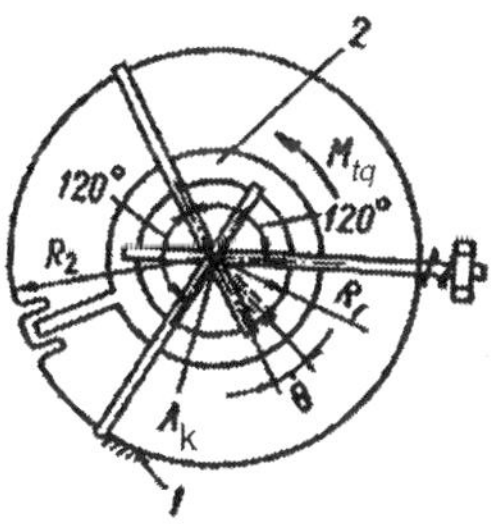	**$F_k(s) + F_j(Q)$**** After I.M. Sivokonenko *et al.*, 1964 a) $Q_s > 0$ $M_{tq}\approx 0$ @ $R_1 = \sqrt{\frac{H}{Q_s}} = \lambda_{Hs}$, $L/R_1 \geq 3$; $L = R_1 + R_2$; $\theta < 0.1$ rad $M_{tq} = \frac{3(R_1+\lambda_{Hs}\tanh\psi_M)\cosh\psi_M}{\cosh(L\lambda_{Hs}^{-1}+\psi_M)-\cosh\psi_M} Q_s\theta$; $\tanh\psi_M = \frac{R_1\lambda_{Hs}^{-1} - \sinh L\lambda_{Hs}^{-1} + R_2\lambda_{Hs}^{-1}\cosh L\lambda_{Hs}^{-1}}{\cosh L\lambda_{Hs}^{-1} - R_2\lambda_{Hs}^{-1}\sinh L\lambda_{Hs}^{-1} - 1}$, b) $Q_s < 0$; $M_{tq}\approx 0$ @ $\tan L\lambda_{Hs}^{-1} = \frac{L\lambda_{Hs}}{R_1R_2-\lambda_{Hs}^2} \approx \sqrt{\frac{Eh^2}{16.3\sigma_t}}$ *if* $R_1=R_2$ $M_{tq} = 3\frac{(R_1-\lambda_{Hs}\tan\psi_M)\cos\psi_M}{\cos(L\lambda_{Hs}^{-1}+\psi_M)-\cos\psi_M}Q_s\theta$; $\tan\psi_M = \frac{R_1\lambda_{Hs}^{-1}+\sin L\lambda_{Hs}^{-1}+R_2\lambda_{Hs}^{-1}\cos L\lambda_{Hs}^{-1}}{1-\cos L\lambda_{Hs}^{-1}+R_2\lambda_{Hs}^{-1}\sin L\lambda_{Hs}^{-1}}$
Three-strip "fork-like" RG-III*af* 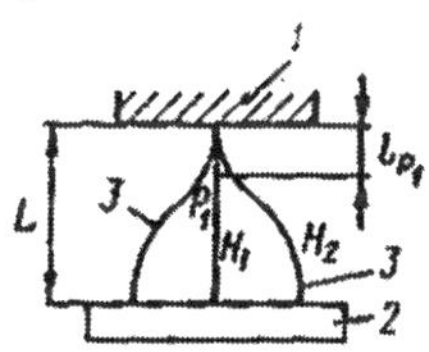	**$F_k(s)$** $H_1 = 10H_2$; $l_{P_1}\approx 0.13L$ After W. Lotze, 1963
Three-strip "lance-like" RG-III*as* 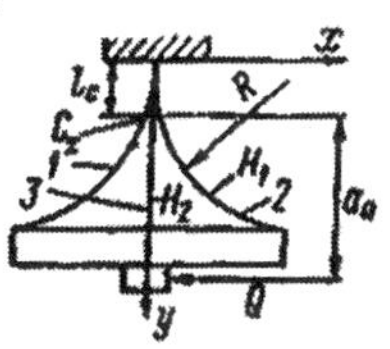	**$F_k(s) + F_j(Q)$** After W. Lotze, 1963 $l_c = 0.327R$; $\nu_y \approx 0$; $\nu_x = 0.01202R^3/H$; $\Phi_c = 0.408R/H$; $h_y = 0.18R$; $H_1 = H_2$ $\theta < 0.1$ rad
Notch hinge RG-IV	*See* Section 3-2 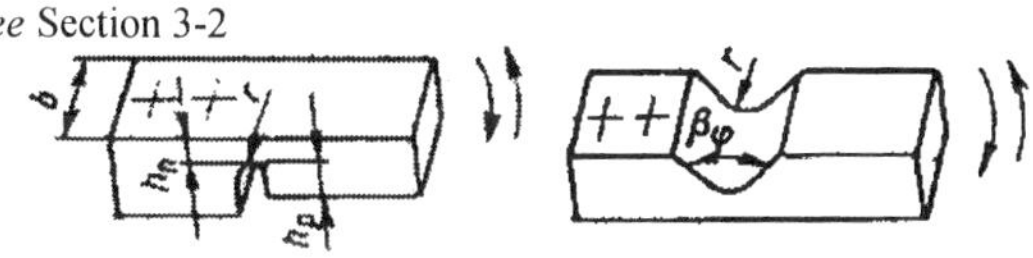
Single profiled-angle RG-VI*s* 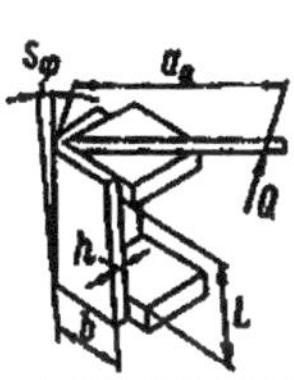	**$F_k(s) + F_j(Q) + \delta F_j(Q)$** After R. V. Jones $s_\phi = 2\varepsilon_u(1+\nu_p)^{-1}(h/b)^2 L\theta a_Q^{-1} + b\theta^2/(L\sqrt{2})$; $\varepsilon_u \approx 1$ $\theta < 0.1$ rad; $*\Gamma^{\upsilon 2} = \frac{h^3}{36}(b_1^3+b_2^3)$; @ $b_1=b_2=b \Rightarrow \Gamma^{\upsilon 2} = \frac{A^3}{144}$

Table 4-1 Cont'd.

Name, type, and schematic of ELKS	Calculation formulae and their application limits
Double profiled-angle RG-V*id* 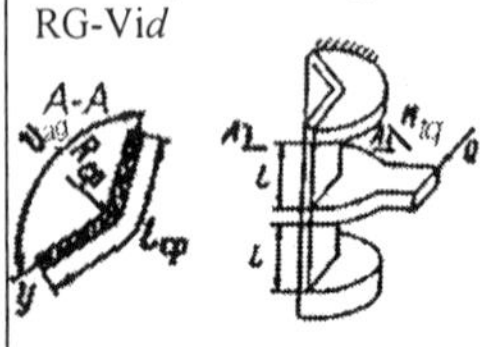	**$F_k(s) + F_j(Q) + \delta F_j(Q)$** After V.N. Zheludkov, 1963 $M_{tq} = (G h^3 L_{cp}\theta / 2L) + (hE/5L^3)\theta^3 [(\mu_m^2 - 1.4\, b\, \mu_m + b^2)^{2.5}$ $-(k_m - R_{cp})^5]$; $k_m = h\, R_{cp}\, (0.47\, b^3 - 0.02\, R_{cp}^3 - 0.15\, b^2 R_{cp}) / J_y$; $\mu_m = 1.41\, R_{cp} - k_m$; $\upsilon_{ag} = 90°$, $\theta < 0.2$ rad
Profiled-cruciform ELKS RG-VI*c* 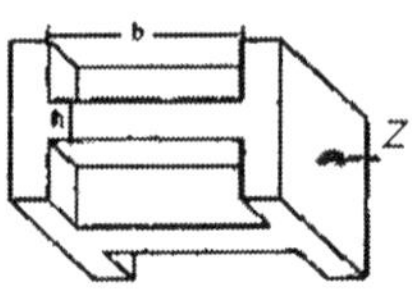	**$F_k(s) + F_j(Q)$** After S. T. Smith, 2000 $j_{\theta_z M_z} = (bh^{-1} - 0.373)\, 2Gh^4 / 3L$; $r_\beta = \sqrt{\frac{\Gamma^{\upsilon}}{I_p} + 0.0390 \frac{k^2 b^2 K}{I_p}}$; $j_{\theta_x M_x} = Gh^4 (1 + \nu_p)[(b/h) + (b/h)^3 - 1] / 6L$; $\frac{j_{\theta_x M_x}}{j_{\theta_z M_z}} = \frac{(1+\nu_p)(\chi + \chi^3 - 1)}{4(\chi - 0.373)}$; $\chi = b/h$; $\chi \leq 8$
Cartwheel hinge RG-VII*c* 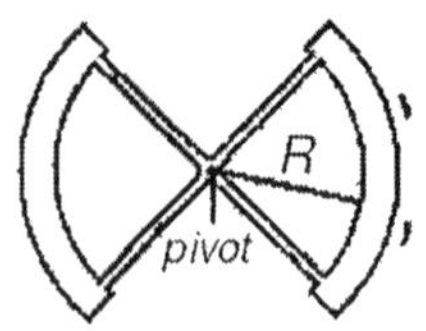	**$F_k(s) + F_j(Q)$** After S. T. Smith, 2000 $j_{\theta M} = 4EJ / R$; $\delta_p / R = \theta^2 \sqrt{2} / 30$; $\sigma_{max} = Eh\theta / R$
Tension-torsion suspension RG-IX 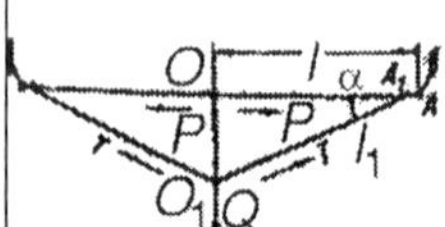	**$F_k(s) + F_j(Q)$** After Z.A. Timofeeva, 1961, P. B. Usatin, 1961 $M_{w90} = 15.7\, (A^2 / l)[(k_1 G / n) + (n/12) P \times 10^{-3} / A]$, N mm/ 90° $M_w = M_{w90}\theta^o / 90$ is a restoring torque @ angle θ^o rotation $\tau_{max} = 1.75 \times 10^{-2}\, k_2 A^{0.5}\, G\, \theta / l$, kPa $\delta_s \approx 0.5\, Ql / P$ is a sag; A is the area of cross section; Cross section: round, n=0.785; k_1=0.125; k_2 = 0.575; square, n=1.0; k_1=0.141; k_2 =0.675; rectangular, $n = b/h$; k_1=(1/3)- 0.21/n; $k_2 = 1/\sqrt{n}$; $\delta_s / l \leq 0.005$

Notes: $\theta°$max is an angle of the maximum deflection in degrees and θ is an angle in radians; dn(φ, k)= $\sqrt{1 - k^2 \sin^2 \varphi}$ is delta amplitudinis (*see* G.A. Korn and T. M. Korn, 1968) of the arguments (φ, k); J_y is a moment of cross-section inertia relative to axis of its symmetry; $j_{\theta_z M_z}$; $j_{\theta x M x}$ are the torsion and bending stiffness of the cruciform hinge; r_β is the radius of gyration with warping options in account; δ_p is the center shift of the cartwheel pivot. * Warping options $\Gamma^{\upsilon 2}$ (after J.T. Oden *et al.*, 1981) are indicated by dash-dot marquee. ** These solutions are recommended for special research. h is the strip thickness, P_{t1}, P_{t2} is the radial load rating at tension for cantilever and double end version of *RG*-II*st*, respectively, P_c is the same at compression, $j_{\theta r}$ is the radial spring rate.

Let us discuss the method for the construction of matrix $\boldsymbol{Q}_{\kappa ei}$. For this we apply to the strip's end unit-forces Q_{ei} and unit-moments M_{ei}, which are directed in parallel with coordinate axes x, y, z. This yields the components of the corresponding *motors*:

$Q_{exi}=1 \Rightarrow(0 - b_k/2\ A_k/2\ 1\ 0\ 0)$; $Q_{eyi}=1 \Rightarrow(- b_k/2\ 0\ B_k/2\ 0\ 1\ 0)$; $Q_{ezi}=1 \Rightarrow(A_k/2\ B_k/2\ 0\ 0\ 0\ 1)$;
$M_{exi}=1 \Rightarrow(1\ 0\ 0\ 0\ 0\ 0)$; $M_{eyi}=1 \Rightarrow(0\ 1\ 0\ 0\ 0\ 0)$; $M_{ezi}=1 \Rightarrow(0\ 0\ 1\ 0\ 0\ 0)$.

These values of the six unit-motors yield matrix $Q_{\kappa ei}$, where $b_k/2 = b^k/2$; A_k and B_k are coordinates of the strip's end in x, y, z system.

The loading of a movable element in most cases yields its plane-parallel motion, which is close to the rotation in planar systems. This allows us use geometrical relations between position x_{A1}; y_{A1}; x_{A2}; y_{A2} and rotation θ of the strip's ends.

RG-IV. One of the important problems in the stress analysis of notch hinges is their stress concentration factor, which depends on the type of notch profile and its loading conditions. In Section 3-2 the simplest Smith's expression for double-symmetrical circular hinge stress concentration factor at a pure bending is presented. Here we discuss more complicated relationships (*see* Y.G. Panovko, 1960, 1968; Y.M. Tseytlin, 1972). The concentration factor for the lever with a symmetric straight notch (*see* Table 4-1) at $1.05 \le h_p / h_n = \lambda_n \le 5.0$; $0.05 \le h_p/ r \le 7$ and pure bending by moment M is equal to:

$$\alpha_\sigma = \sigma_{max} / \sigma_n = 1 + [\ (\lambda_n - 1)h_n/\ 8r\ (1.07\lambda_n - 1)]^{0.85}, \qquad (4\text{-}10a)$$

where $\sigma_n = 6M / b\ h^2_n$. The concentration factor for the step rod with the circular fillet of radius r under pure bending moment M is equal to:

$$\alpha_\sigma = 1 + [\ (\lambda_n - 1)h_n/\ 9.6r\ (1.12\lambda_n - 1)]^{0.85}. \qquad (4\text{-}10b)$$

It is clear that the concentration factor increases with the decrease of the fillet's radius, and this factor decreases with the decrease of thickness h_p and increase of ratio λ_n. A rod with the circular fillet r and slope β_φ at $r \le h_p$ on the notch's side edges under pure bending has a stress concentration factor equal to:

$$\alpha_\sigma(\beta_\varphi) = 1 + (\alpha_\sigma - 1)[\ 1 - (\beta^o_\varphi / 180)^{1+2.4\sqrt{r/h_p}}\ ] \qquad (4\text{-}10c)$$

Kinematics

Two types of the kinematical interpretation are useful in the approximate evaluation of parameters for the single leaf and cross-strip hinges synthesis purposes:

1) an *EKS* of articulated mechanism with rigid elements whose length is equal to the radius of the curvature of the strip ends' trajectory;
2) an *EKS* of rolling disk with a diameter that is twice that of the rotational circle. This disk rolls along the straight guide that coincides with the guided element pole's tangent (*see* Chap. 1).

Table 4-2. Kinematic features for ELKS of type RG-I

$\lambda_L = L_1/ L$ Fig. 4-6a	$+\infty$	+1	0	-1/3	-1/2	-2/3	-1	$-\infty$
ξ_{P1}/ L	-0.500	-0.556	-0.667	-1.000	$\pm\infty$	0	-0.333	-0.500
d_ξ/ L	+0.167	+0.163	+0.133	-0.133	$-\infty$	-0.133	+0.133	+0.167

(After W. Lotze)

RG-I. If the longitudinal component force Q in the hinge is negligible $Q_x = \tan\delta_Q\ Q \approx 0$, rotational circle d in the beginning of the motion is symmetrical relative to axis x which coincides with the axis of the strip in its initial unloaded state because $d_\eta \approx 0$. The numerical data for the coordinates of the first order pole P_1 and projection d_ξ of the rotational circle on the strip's initial axis in this case are shown in Table 4-2 for different ratios $\lambda_L=L_1/L$ (Fig. 4-6a). They have been calculated on the basis of nonlinear and approximate linear relationships.

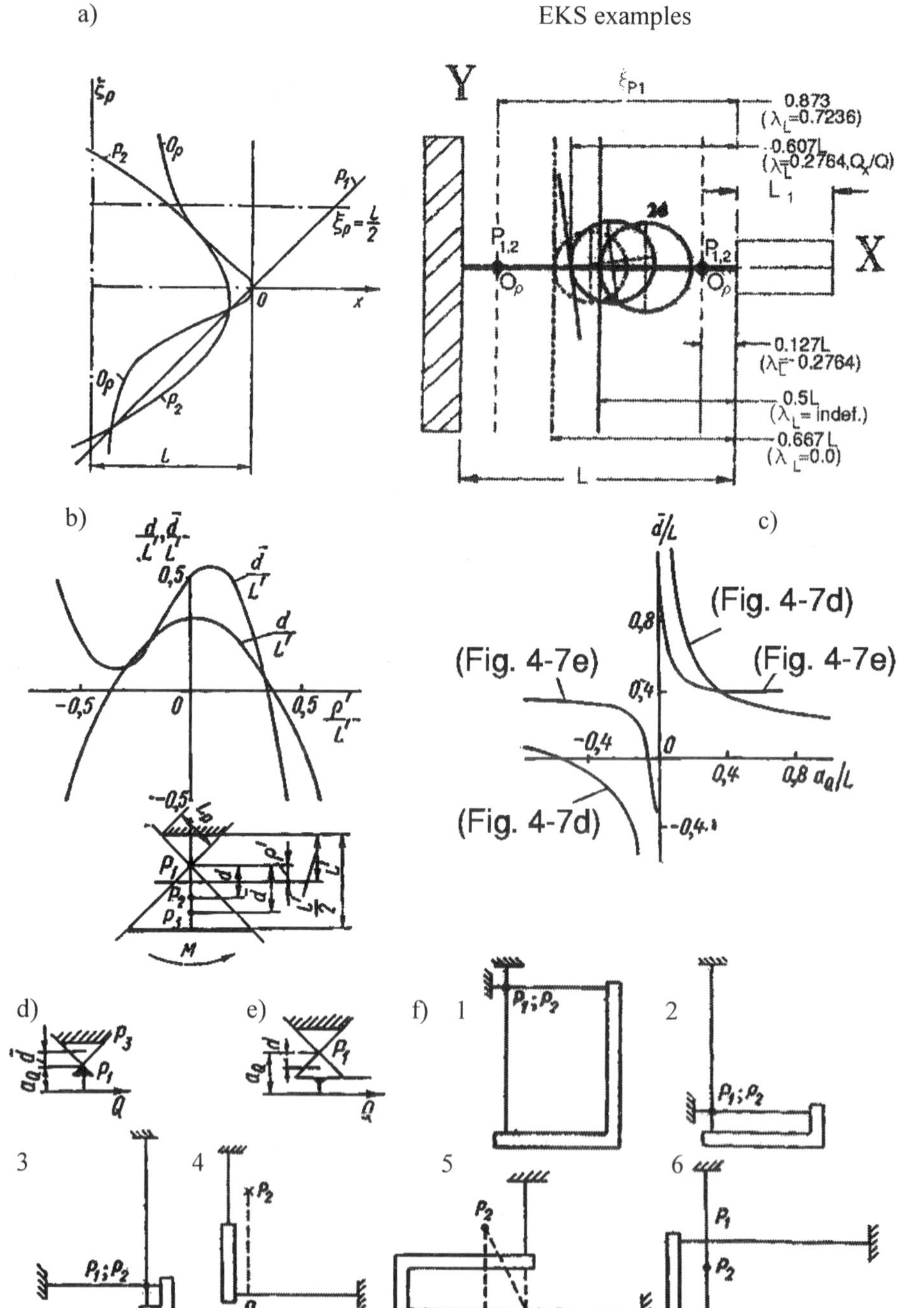

Fig. 4-6. Positions of kinematic poles in strip rotary guides (after W. Lotze)

Figure 4-6a shows charts and EKS examples of the mutual position (at tan δ_Q =0; d_η = 0) of poles P_1, P_2, and the trajectory curvature's center O_ρ in the motion of the movable strip's end. One can see that the coincidence of these three points occurs only in two special positions. . The analytical expression for the coordinates of pole P_1's position in these cases one can define from (3-33) if we assume that $d_\xi = 0$. Then $\xi_{P1}/ L= -(3 \pm \sqrt{5}) / 6$, which corresponds to the pure rotation at small angles of turn. The analogous optimal expression is also obtained on the basis of the approximate nonlinear calculation method for planar bending of a beam. The action of force longitudinal component $Q_x = Q \tan \delta_Q \neq 0$ leads to the shift of the turn's pole P_2 in the direction of a normal to the initial strip's plane. Values d_η / L at $\lambda_L = 0.2764$ and $\lambda_L = -0.7236$ are respectively equal to:

$d_\eta / L = 0.107 \tan \delta_Q$ and $d_\eta / L = -1.88 \tan \delta_Q$.

It is clear from these relations that to achieve ideal rotation in the real leaf spring hinge is very difficult even at small angles of rotation especially under longitudinal and transverse loading. The displacement of a point on the movable element in ELKS of type RG-I that coincides at initial position with the middle point on the strip (conditional center O) can achieve a significant value under unfavorable conditions of the leaf spring's loading. The displacement of point O on the movable element is equal to: $S_{ox} \approx (1/12) L\, \theta(\theta + 0.04)$ and $S_{oy} \approx (1/12) L\,(0.1\theta \mp 0.08\, k_Q)$ for some real conditions and geometric parameters: L=1 – 10 mm; $\theta \in \pm\ 0.15$ rad ($\pm$ 8.5°); $\Sigma\, Q_y \leq 10 - 30$ N; $Q_R \leq 10 - 30$ N; $r_R = 10 - 100$ mm; k_R = 1/10 to 1/15; $k_Q \in \pm\, 2$; and $\theta_R \approx 0.02$.

All of these notations are shown in Table 4-1 and Fig. 4-5a.

RG-II. The cross strip with a trajectory close to pure rotation should have the diameter of the rotational circle that is equal to zero ($d = d_\xi = d_\eta = 0$). This requirement is satisfied in the hinges where, in accordance with the substitution d = 0 into (3-33), distance L_P from the point of the strips' crossing to the end of each strip is equal to $L_p \approx 0.873L$ or $0.127\,L$. In these cases, poles P_1 and P_2 coincide.

Lotze's charts of the diameter values for the circles of second d / L and third $\bar{d} / L$ orders are shown in Fig. 4-6b in relation to the position of the crossing point for the orthogonal (γ = 90°) cross strip at $Q_x = Q_y = 0$. One can see from these charts that at the points where the second order diameter is equal to zero ($d = 0$), the diameter $\bar{d}$ of the third order rotational circle is also small. If the deflecting force Q_x (Q_y = 0) is applied to the connecting rod through the lever's arm a_Q (Fig. 4-6d for asymmetrical, and Fig. 4-6e for symmetrical case, respectively), the value of the third order circle's diameter $\bar{d}$ changes as it is shown in Fig. 4-6c in two above mentioned cases. The rational adjustment of arm a_Q's length leads to the small value of diameter $\bar{d}$ which provides the rotation of the movable element relative to pole P_1 taking in account pole P_3. The shown diagrams are correct for the statically unloaded hinge. If the movable guiding element is additionally loaded with force Q_y which causes extension force Q_{ep} in both strips, then the above mentioned optimal position of the complex angle's axis in ELKS of type RG-II yielding $d = 0$ moves to the strips' ends with the increase of Q_{ep}. The optimal position of the complex angle's axis under the compression load moves to the center of strip, coincides with this center when the load approaches a quarter of the critical value, and after that it moves to the imaginary field (W. Lotze, 1967, *Feingerätetechnik*, 113-115, 511-514; ibid. item., 1968, 440-448; 1970, 207). The position of poles P_1 and P_2 may be easily found not only in symmetrical but also in asymmetrical cross-strip hinges 1-6 (Fig. 4-6f). Hinges with the structures 5 and 6 shown in Fig. 4-6f are rarely used because they are not stable.

Stability. The relationship between the angle of rotation θ and acting in cross-strip reactive moment $M_p = M'(0)\theta + (1/6)\, M'''(0)\, \theta^3$ may be a linear one only at a certain value of the ratio $\lambda_p = L_p / L$ and the strips' crossing angle γ (W.H. Wittrick, 1948; J.A. Haringx, 1949; H. Hasselmeier, 1954; W.D. Weinstein, 1965). The strips of ELKS of type RG-II are in a pure bending, reactive moment M_p is the smallest one and proportional to angle θ of rotation with uncertainty of the third order when $\gamma = 45°$ and $\lambda_P = 0.5$. Comparison of the optimal hinges with respect to accuracy at $\lambda_P = 0.8727$ and $\lambda_P = 0.1273$ shows that the latter has smaller reactive moment. This favors hinges with $\lambda_P = 0.1273$ in the case of requirements of high kinematic accuracy. ELKS of type RG-II has angle $\theta°_{ult}$ of rotation which may be considered the limit; this is caused by: 1) the limit of strength $\theta°_{ult} \leq 2L[\sigma_y] / E\, h$; and 2) the possible reverse of reactive moment M_p's direction for cross-strip hinges with $\lambda_p > 0.5$. The latter defines some critical moment M_{cr} the exceeding of which leads to the loss in ELKS of type RG-II's stability. It is practically very difficult to keep the center of the hinge's rotation immovable (radius ρ_{P1} of its trajectory's curvature should be equal to zero) because even small deflections from the optimal values of $\lambda_p = 0.8727$ or 0.1273 yield very significant changes in ρ_{P1}. The cross-strip hinge is especially sensitive to changes in λ_p at $\gamma = 120°$. ELKS of type RG-II is an optimal one with respect to this discussion when $\gamma = 60°$ and $\lambda_p = 0.1273 \pm 0.03$. Its sensitivity to deflections from the optimal value of λ_p for this case is minimal. In addition, there is no danger of unstable work for this hinge because $\lambda_p = 0.1273 < 0.5$. In this case, however, it is possible to have nonlinearity of the third order in the relations between an angle of rotation and the corresponding torque. In a general case, the initial center of rotation (the point of a movable element which coincides in its initial position with pole P_1) in ELKS of type RG-II with strips of equal length moves at a small angle of rotation along the bisector of the strips' crossing angle in proportion to the square of hinge's angle of rotation and in the perpendicular direction in proportion to the cubic power of this angle. In particular, $s_{ox} = 0.003\, L$ at $\lambda_p / 2 = 0.25$, $\gamma = 90°, \theta < 0.15$ rad or approximately 3 μm per 1 mm of the strip's free length (A.T. Draudin, 1966); $s_{oy} = 0.0002\, L$, i.e., 0.2 μm per 1 mm of the strip's free length. The displacements of the initial center of rotation in the hinges with $2\lambda_p \geq 2$ are significantly larger than in the symmetrical cross strips: 5-25 times along axis x, 15 – 150 times along axis y. It is known that the greatest displacement of the initial axis of rotation in cross strips with $\lambda_p < 1$ occurs at $\lambda_p = ½$..

RG-IIIas. This hinge is more stable under static loading than ELKS of type RG-II*cu* as it follows from the relationships in Table 4-1. Therefore, the first structure is a preferable one. However, ELKS of type RG-II*cu* has the greater transformation of the displacement in direction s_x into rotation angle θ under force Q applied to the guided element at distance a_Q from flexible center C_z. It can be shown (*see* W.

Lotze) by a simple comparison after the elimination of force Q from the typical equations for these hinges that:

$$s_x/\theta = a_Q + \frac{v_y}{\Phi_c a_Q},\ \left(\frac{\theta}{s_x}\right)_{max} = \frac{1}{2h_y},\ \text{and}\ h_y(\text{RG-II}cu) < h_y(\text{RG-III}as). \qquad (4\text{-}11)$$

As one can see from the discussed calculation relationships for the ELKS RG-I to RG-IV (Table 4-1) that the main type of their elements deformation is flexure. The main type of elastic links deformation in ELKS RG-VI to RG-IX is torsion, which requires another methods of calculation.

RG-VI. Warping. The profiled hinges (Table 4-1, ELKS of type RG-VI) are exposed to restrained twisting, similar to thin-walled open sections. The thorough investigation of primary and secondary warping of a thin-walled twisted beam was presented by J. T. Oden and E. A. Ripperger (1981).

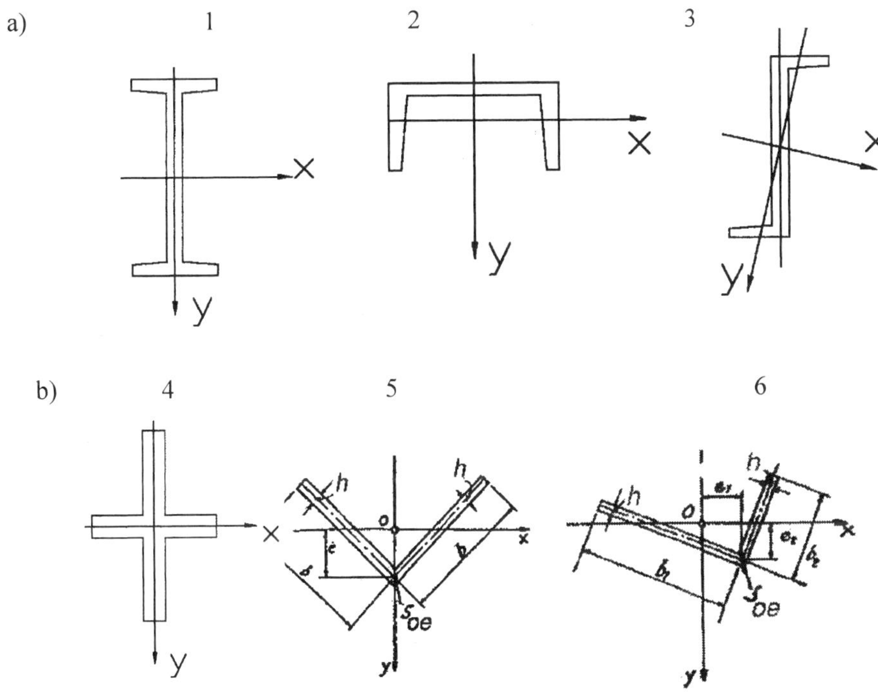

Fig. 4-7. Profiles with primary and secondary warping in a twisted beam

Fig. 4-7 shows profiles with primary a) and secondary b) warping of a thin-walled twisted beam where1is I-beam, 2-channel, 3- Z-profile, 4-cruciform, 5-angle with equal legs, 6-angle with unequal legs, distances e, e_1, e_2 are between the centroid and the shear center. Let us recall that the twist and shear centers in beams always coincide.

If torque M_T is constant in the longitudinal direction z of the twisted beam (typical for rotary hinges), the differential equation for torsion ϕ calculation is of the third order $d^3\phi/dz^3 + c_\Gamma^2\, d\phi/dz = M_T/E\Gamma^\upsilon$, where $c_\Gamma = \sqrt{GJ_T/E\Gamma^\upsilon}$; Γ^υ is a warping constant: $\upsilon = \upsilon_1$ for the primary warping effect and $\upsilon = \upsilon_2$ for the secondary warping effect; and J_T is the torsional constant of the section. Let us recall that $\Gamma^{\upsilon 1} = \int_A (2\omega_A) dA$, where ω_A is a sectorial area. It was assumed that the pole of the sectorial area is the shear center of the section and that a free edge of the section is used for the origin of the contour coordinate. We remind that the shear center is the point in the cross-sectional plane through which the plane of resultant loads must pass to prevent the development of twisting moment on the section. The secondary warping constant $\Gamma^{\upsilon 2} = (1/12)\int^{S_A} h^3 r_n ds$, where S_A is the length of the contour of the section, h is the thickness of the thin wall, and r_n is an arm from the shear center to a normal to the center line of the wall. Let us recall that in accordance with the Saint-Venant's theory of beam's torsion, the warping displacement occurs in the sections of the solid beams except the beams with circular cross sections. But this warping influence on the solid beams behavior at torsion is very small as a result of a fast diminishing of stress disturbances on the small parts of the beam that are close to their constrained sections. "An application of Saint-Venant's principle to the thin-walled beams may lead to serious errors"(*see* J.T. Oden and E.A. Ripperger, 1981, p. 207). In thin-walled open sections, stress produced by restrained warping with the self-equilibrating normal stress resultants diminish very slowly from their points of application and may constitute the primary stress system developed in the structure. As a result, the twisting moment developed on each section is no longer proportional to the rate of twist, and final shearing stress cannot be obtained by superimposing those produced by unrestrained torsion and bending. The torque in thin-walled beams is larger with warping influence and their twisting may cause bending and buckling at compression forces well below the Euler's critical force values.

According to the Vlasov's theory (*see* V.Z.Vlasov, 1961), the sections have shown in Fig. 4-7b twist without primary warping. The shear center of these sections lies at the intersection of the centerlines of the thin rectangular elements. Because of this, the resultants of shear flows produced by any type of loading must pass through these points; hence, no warping torque M_T should be developed and no longitudinal stresses are produced by torsional loads. Therefore the warping constant $\Gamma^{\upsilon 1}$ for these sections can be taken equal to zero. However, when the wall thickness h is not extremely small compared with the other dimensions, a secondary stress system with constant $\Gamma^{\upsilon 2}$ can be developed perpendicular to the contour line of the section (J.T. Oden and E.A. Ripperger, 1981). Normal stresses then vary linearly over h, the shearing stress can no longer be neglected, and a secondary warping torque M_T* is developed. In such cases, the sections undergo "secondary" warping and, consequently, possess a secondary warping rigidity $E\Gamma^{\upsilon 2}$. The secondary warping effect is significant for the sections shown in Figure 4-7b, which are similar to those used in profiled angle and cruciform elastic pivots. This effect is insignificant for the sections shown in Figure 4-7a where the main influence has the primary warping effect. But these latter sections are not in use with the structures of elastic pivots. However, the two Z -like squared structures are used in the so-called three-dimensional frictional force microscope (*FFM*) "tribolever"(*see* M. Dienwiebl *et al.*(2005), *Rev. Sci. Instr.*, **76**, 043704) with plane bending mode and practically without buckling and warping, especially if applied through the *FFM* tip torque is a small one.

The approximate calculation relationships for the deflection of an axis of rotation for a single-angle hinge (RG-VI*s*), its secondary warping effect $\Gamma^{\upsilon 2}$, and the torque for a double-angle hinge (RG-VI*d*) are shown in Table 4-1. In addition, we show the warping options for cruciform shape which may occur at the critical stress $\sigma_{cr} = \frac{\pi^2 E r_\beta}{(kb/r_\beta)}$ as a function of an equivalent slenderness ratio kb / r_β. (*see* F. Bleich, 1952). Here *kb* is the effective length of the cruciform pivot and r_β is its equivalent radius of gyration.

Translational guides' features and synthesis recommendations.

Kinematics.

Table 4-3. Kinematic relationships for translational guides

Name, type, schem	Calculation formulae and their application limits
Spring parallelogram ELKS TG-I *See* Fig. 4-8a 1 and 2 -flexible strips, 3-carriage, 4-base Q_{fr} –friction force	1. $\mathbf{F_k(s) + F_j(Q) + \delta F_j(Q) + \delta F_{kc}}$ with E. P. Popov's method $\frac{s_1}{L} = \beta^{-1}(2\eta_0'' \cos\delta_0 - 2\xi_0'' \sin\delta_0)$; $\frac{s_2}{L} = 1 - \beta^{-1}(2\xi_0'' \cos\delta_0 + 2\eta_0'' \sin\delta_0)$ $H_1 = H_2 =$ constant; $(h/L) < 0.05$; $L_1 = L_2 = L$; $\upsilon_Q = \frac{1}{2} B_1$; $\vartheta = 0$; $a_Q = x_1/2$; $h_Q = a_Q$; $Q_{fr} = 0$; $Q_v = Q_N = 0$; $Q_s = -P$; $F_Q = -F = -Q_y/2$; $Q = (P^2 + F^2)^{1/2}$
	2. $\mathbf{F_k(s) + F_j(Q)}$ * *see* Y.M. Tseytlin, 1963 $\vartheta_z = \frac{1}{350}\left(\frac{L}{B_1}\right)^2 [(\pm\upsilon_Q \mp B_1/2)\Im L H^{-1} \cos\psi_k +$ $(\pm\upsilon_{fr} \mp B_1/2) Q_{fr} L H^{-1}$ $+ a_Q L H^{-1}(Q_v + Q_N \pm \Im \sin\psi_k) + l_Q Q_N L H^{-1}$ $\mp h_Q \Im L H^{-1} \sin\psi_k - c_Q Q_v L H^{-1}]\left(\frac{s_1}{L}\right)^2$ $\mp \frac{1}{350}\left(\frac{L}{B_1}\right)^2 [\Im \cos\psi_k + Q_{fr}) L^2 H^{-1}(0.1 + 1.2 a_Q L^{-1})$ $\mp 12(1 + 2a_Q L^{-1})]\left(\frac{s_1}{L}\right)^3 + \ldots$; $s_2 = \frac{3}{5}\frac{s_1^2}{L}[1 \pm \frac{1}{420}\frac{PL^2}{EJ}]$ @ $\vartheta_z = 0$ $Q_y = Q_\upsilon + Q_N \pm Q_{fr} \sin\vartheta_z + 24\frac{s_1}{L}[\frac{H}{L^2} \mp \frac{1}{20}(\Im \cos\psi_k + Q_{fr})]$ $-12 \tan\vartheta_z [\frac{H}{L^2} \mp \frac{1}{120}(\Im \cos\psi_k + Q_{fr})] \pm \Im \sin\psi_k$ $H_1 = H_2 = constant$; $L_1 = L_2$; $\Im = m_\Im g$; $\frac{h}{L} < 0.05$; $(PL^2/EJ) \ll 1$ $S_1/L < 0.1$; **

Table 4-3. Cont'd.

Name, type, schem	Calculation formulae and their application limits
Spring parallelogram ELKS TG-I *See* Fig. 4-8a 1 and 2 -flexible strips, 3-carriage, 4-base Q_{fr} –friction force	3. **F_k(s) + F$_j$(Q) + δF$_j$(Q)** * After P. Bondy, 1961 $s_1=\frac{L^3}{24H_0}Q_y+(\frac{L^7q_B^2}{201600H_0H_1^2}-\frac{L^7}{13440H_0^3}+\frac{L^8q_B}{403200H_0H_1H_2B_1})Q_y^3$ $-\frac{L^5}{480H_0H_1}Q_yQ_s+[\frac{L^7q_B}{H_0^2}(-\frac{1}{28800H_1}+\frac{23}{1507200H_2})$ $+\frac{L^8}{H_0^2B_1}(-\frac{1}{30600H_1}+\frac{43}{2419400H_2})]Q_y^2Q_s+\frac{43L^7}{403200H_0^2H_2}Q_yQ_s^2+\ldots$ $\vartheta_z=\frac{L^7q_B}{201600H_0H_1H_2B_1}Q_y^3+\frac{L^7}{H_0^2B_1}(-\frac{1}{9600H_1}+\frac{43}{403200H_2})Q_y^2Q_s$ $s_2^{(1)}=\frac{L^5}{960H_0^2}Q_y^2+\frac{L^7q_B}{403200H_0^2H_1}Q_y^3-\frac{L^7}{9600H_0^2H_1}Q_y^2Q_s+\ldots$ $s_2^{(2)}=\frac{L^5}{960H_0^2}Q_y^2-\frac{L^7q_b}{403200H_0^2H_1}Q_y^3-\frac{43L^7}{403200H_0^2H_2}Q_y^2Q_s+\ldots$ $H_0=(H_1+H_2)/2;\ q_B=(L+2a_Q)/B_1,\ \ Q_s/Q_y=-2\ q_B$
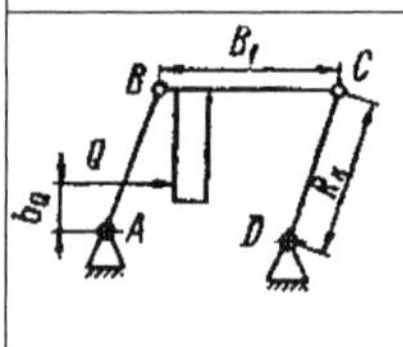	4. **F_k(s)** After R.V. Jones $R_k=5L/6;\ \vartheta_{\delta L}=\delta Ls_1^2L^{-2}B_1^{-1}/2;\ \vartheta_{\delta B_1}=\delta B_1s_1L^{-1}B_1^{-1};$ $\vartheta_z=\frac{2(L-2b_Q)h^2}{B_1^2\ L^2}s_1;s_2=\frac{3}{5}\frac{s_1^2}{L}$
Spring parallelogram ELKS TG-I *See* Fig. 4-8b,c	5. **F_k(s) + F$_j$(Q)** Y.M. Tseytlin, 1972 $\vartheta_x=\frac{2Q_z(e_Q+B_1/2)L^3}{4H_kL^2+12H_zB_1^2}$
Spring parallelogram with cover plates ELKS TG-II 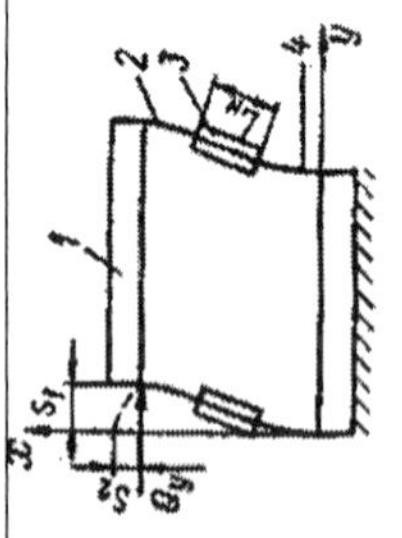 View rotated at 90°	1. **F_k(s) + F$_j$(Q)** After R. Breitinger,1969 $s_1=\frac{Q_yL^3}{24H_0}(1-m_\xi^3);\ s_2=\frac{Q_y^2L^5}{960H_0^2}(1-\frac{5}{2}m_\xi^3+\frac{3}{2}m_\xi^5);$ $y=\frac{3}{5}\frac{x^2}{L}\frac{1-\frac{5}{2}m_\xi^3+\frac{3}{2}m_\xi^5}{(1-m_\xi^3)^2}$ $m_\xi=L_H/L=\lambda_H;\ Q_x=0;\ q_B=\frac{L+2a_Q}{2B_1}$ 2. **F_k(s) + F$_j$(Q) + δF$_j$(Q)** After P. Bondy, 1961 $\vartheta_z=(16\times10^{-5}\frac{L^3q_B}{H_0B_1}Q_y+\frac{L^7q_B}{201600H_0^2B_1}Q_y^3)\psi_q;\ \psi_q\in[0;1];$ $\psi_q=1\ @\ m_\xi=0$

Table 4-3 Cont'd.

<table>
<tr><th>Name, type, schem</th><th>Calculation formulae and their application limits</th></tr>
<tr><td>Rectilinear spring compound movement ELKS TG-IV
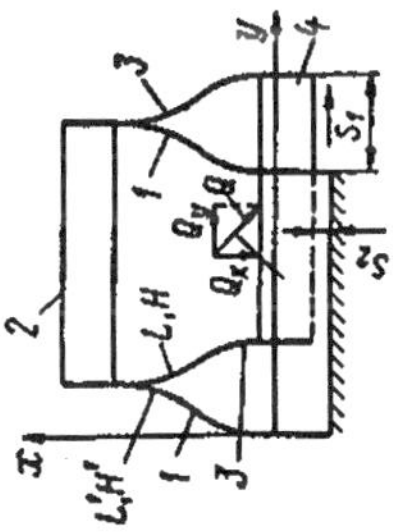
View rotated at 90°</td><td>

$F_k(s) + F_j(Q)$ * $L=L'$, $\frac{s_1}{L} < 0.25$; $\beta_\alpha = \frac{Q_x L^2}{\pi^2 H}$; $\beta'_\alpha = \frac{Q_x L'^2}{\pi^2 H'}$

$H = EJ$; $H' = E'J'$ After J.E. Plainevaux, 1953-1956

$$s_2 = \frac{3}{5}\frac{s_1^2}{L}\frac{\psi(\beta_\alpha)H'^2 - \psi_1(\beta'_\alpha)H^2}{\Phi(\beta_\alpha)HH'^2 + \Phi_1(\beta'_\alpha)H^2H'};$$

$$s_1 = \frac{Q_y L^3}{12}\left[\frac{\Phi(\beta_\alpha)}{H} + \frac{\Phi_1(\beta'_\alpha)}{H'}\right]; \lambda_{jopt} = \sqrt{\psi_1(\beta'_\alpha)/\psi(\beta_\alpha)} \ge J'/J$$

$$\Phi(\beta_\alpha) = \frac{3(\tan\frac{\pi}{2}\sqrt{\beta_\alpha} - \frac{\pi}{2}\sqrt{\beta_\alpha})}{(\frac{\pi}{2}\sqrt{\beta_\alpha})^3};$$

$$\psi(\beta_\alpha) = \frac{\frac{15}{4}\frac{\pi}{2}\sqrt{\beta_\alpha}\,(3 + \tan^2\frac{\pi}{2}\sqrt{\beta_\alpha}) - 3\tan\frac{\pi}{2}\sqrt{\beta_\alpha}}{(\frac{\pi}{2}\sqrt{\beta_\alpha})^5};$$

$$\Phi_1(\beta'_\alpha) = \frac{\frac{\pi}{2}\sqrt{\beta'_\alpha} - \tanh\frac{\pi}{2}\sqrt{\beta'_\alpha}}{(\frac{\pi}{2}\sqrt{\beta'_\alpha})^3};$$

$$\psi_1(\beta'_\alpha) = \frac{\frac{15}{4}\frac{\pi}{2}\sqrt{\beta'_\alpha}\,(3 - \tanh^2\frac{\pi}{2}\sqrt{\beta'_\alpha}) - 3\tanh\frac{\pi}{2}\sqrt{\beta'_\alpha}}{(\frac{\pi}{2}\sqrt{\beta'_\alpha})^5}.$$

At $\beta_\alpha << 1$ $\Phi(\beta_\alpha) = 1 + \frac{\pi^2}{10}\beta_\alpha + ...$; $\psi(\beta_\alpha) = 1 + \frac{17\pi^2}{84}\beta_\alpha + ...$

At $\beta'_\alpha << 1 \Rightarrow \Phi_1(\beta'_\alpha) = 1 - \frac{\pi^2}{10}\beta'_\alpha + ...$; $\psi_1(\beta'_\alpha) = 1 - \frac{17\pi^2}{84}\beta'_\alpha + ...$

</td></tr>
<tr><td>Symmetrical rectilin. spring movement ELKS TG-IVs and its sensor application
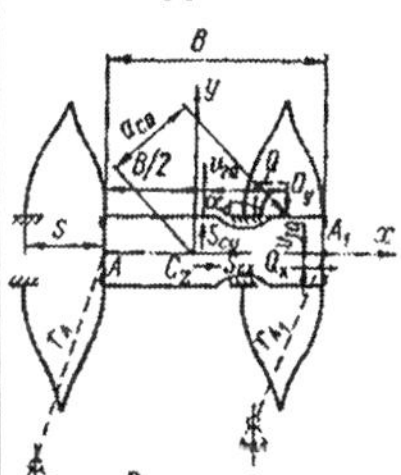
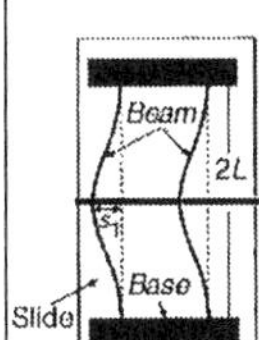
</td><td>

$F_k(s) + F_j(Q)$ $s/L < 0.25$; $\tan\alpha_Q = Q_x/Q_y$; $h/L < 0.05$ After W. Lotze, 1963

$$\Phi_C = \frac{L}{H}(\frac{L}{B})^2[\frac{13}{70}(\frac{s}{L})^2 + \frac{1}{4}(\frac{f_l}{L})^2 + 2\frac{J}{AL^2}];$$

$$\nu_x = \frac{L^3}{24H}; \nu_y = \frac{1}{4}\frac{L^3}{H}[\frac{13}{70}(\frac{s}{L})^2 + \frac{1}{4}(\frac{f_l}{L})^2 + 2\frac{J}{AL^2}];$$

$$s_{C_z x} = 8L\left\{\frac{13}{70}(\frac{s}{L})^3 + [\frac{1}{4}(\frac{f_l}{L})^2 + 2\frac{J}{AL^2}](\frac{s}{L})\right\}\tan\alpha_Q;$$

$$\vartheta = 24(\frac{L}{B})^2\left\{\frac{13}{70}\left(\frac{s}{L}\right)^2 + \left[\frac{1}{4}\left(\frac{f_l}{L}\right)^2 + 2\frac{J}{AL^2}\right]\left(\frac{s}{L}\right)\right\}\frac{a_{cQ}L^{-1} - \frac{1}{2}sL^{-1}\sin\alpha_Q}{\cos\alpha_Q}$$

$$r_A = \frac{140}{13}\frac{H}{Q_x L}\frac{1}{1 - 2\upsilon_{2Q}B^{-1}}; r_{A_1} = \frac{140}{13}\frac{H}{Q_x L}\frac{1}{1 + 2\upsilon_{2Q}B^{-1}};$$

At $a_Q = 0 \Rightarrow \vartheta = 24\frac{\upsilon_{1Q}}{L}\left(\frac{L}{B}\right)^2\left\{\frac{13}{70}\left(\frac{s}{L}\right)^3 + \left[\frac{1}{4}\left(\frac{f_l}{L}\right)^2 + 2\frac{J}{AL^2}\right]\left(\frac{s}{L}\right)\right\}$

f_l *see* in Table 3-5. $S_1 = F\nu_x/(16H)$

</td></tr>
</table>

Table 4-3 Cont'd.

Name, type, schem	Calculation formulae and their application limits
Spring parallelogram with enslaving ELKS TG-V 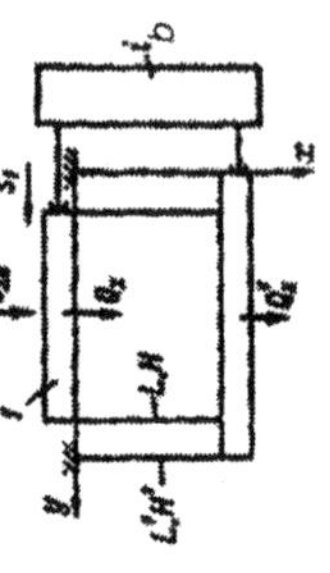	**$F_k(s) + F_j(Q)$** $s_1/L < 0.25$; After J.E. Plainevaux $s_{2b}=\frac{3}{20}\frac{s_1^2}{L}[\chi(\beta_\alpha)-\chi(\beta'_\alpha)]$; $\frac{s_2}{s_{2b}}=\frac{4[\psi(\beta_\alpha)-\psi_1(\beta'_\alpha)]}{[\Phi(\beta_\alpha)+\Phi_1(\beta'_\alpha)]^2[\chi(\beta_\alpha)-\chi(\beta'_\alpha)]}$; $\chi(\beta_\alpha)=\frac{\psi(\beta_\alpha)}{\Phi^2(\beta_\alpha)}$; $\chi(\beta'_\alpha)=\frac{\psi_1(\beta'_\alpha)}{\Phi_1^2(\beta'_\alpha)}$; $\beta_\alpha=\frac{Q_xL^2}{\pi^2H}$; $\beta'_\alpha=\frac{Q'_x(L')^2}{\pi^2H'}$ Schematic view is rotated at 90° $\lim_{\substack{\beta_\alpha\to0\\ \beta'_\alpha\to0}} s_{2b}/s_2(TGIV)\le 85$ at $i_b=2$
Spring quasi-parallelogram with plates , diaphragms TG-VIII or bellows TG-VIII*b*	1. **$F_k(s)$** Y.M. Tseytlin *et al.*, 1968 $s_1=\frac{Q(\pi-\vartheta_1-\vartheta_2)^3\rho_{out}^2}{2e^2Eh^3(\ln\frac{\rho_{out}}{\rho_{in}})^2}$; $\rho_{out}=D_{out}/2$
 Plates with annular recesses TG-VIII-*ar* 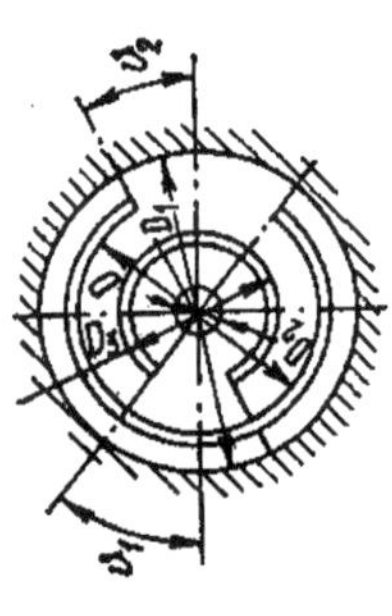	2. **$F_k(s) + F_j(Q)$;** $\frac{2s_1}{D}<0.1$ Y.M. Tseytlin *et al.*, 1968 If $\vartheta_1+\vartheta_2=\pi/2 \Rightarrow s_1=\frac{Q(R-b/2)R^2}{4EJ_n}$; $s_2=\frac{Q^2(R-b/2)^2R^3}{384(EJ_n)^2}$; $J_n=\frac{D-D_3}{24}h^3$; $R=\frac{D+D_3}{4}$; $b=\frac{D-D_3}{2}$; $W_n=\frac{bh^2}{6}$; $\sigma_{n\max}=k_n\frac{M_b}{W_n}$; $k_n=1.25-1.27$; $M_{b\max}\approx\frac{1}{10}Q(D+D_3)$; $\tau_{\max}=k'_t\frac{M_{t\max}}{W_T}+K'_b\frac{Q}{2A}$; $W_T=\nu_w bh^2$; At R/b=1.5-1.75 and $h/b\le 0.1$, ν_w=0.333-0.4; $k'_t=1.5$; $k'_b=6.5$; $M_{t\max}=Q(D+D_3)/16$
Plates with the spiral recesses 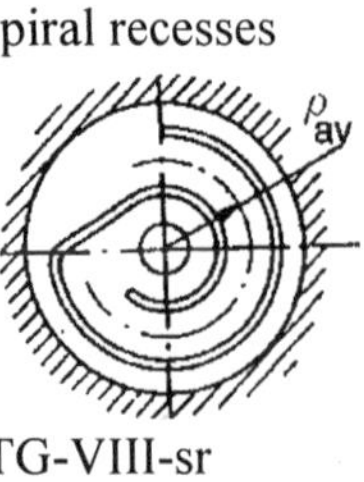 TG-VIII-sr	**$F_k(s)$** Y. M. Tseytlin *et al.*, 1968 $s_1=\frac{QL_{1c}}{2GA}\left(1+\frac{\rho_{av}^2}{\eta_{pl}h^2}\right)$; $3\eta_{pl}=1-0.63\frac{h}{b}$; $L_{1c}\approx 3\pi\rho_{av}/2$; $h<<b$

Table 4-3 Cont'd.

Name, type, schem	Calculation formulae and their application limits
Plates with segment-like recesses 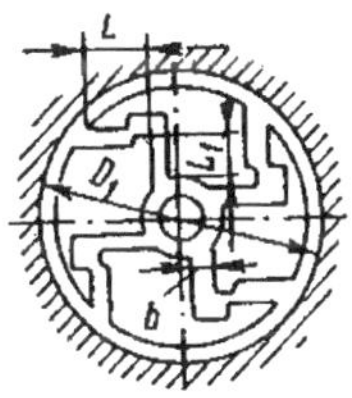 TG-VIII-ser	$F_k(s) + F_j(Q)$ Y.M. Tseytlin *et al.*, 1968 $s_1 = \frac{Q}{8}\left[\frac{L_1^3 + L^3}{3EJ} \quad \frac{L^4GJ_k}{4EJ\,(LGJ + L_1EJ)} - \frac{L_1^4GJ_k}{4EJ\,(L_1GJ_k + LEJ)}\right]$; $J = bh^3/12;\ J_k = bh^3/3;\ \sigma_{n\max} = \frac{QL}{16\,W};\ \tau_{\max} = \frac{QL_1}{16\,W_k};$ $W = bh^2/6;\ W_k \approx bh^2/3;\ \frac{s_1}{L_1+L_2} < 0.1;\ \frac{h}{b} \le 0.1$
Double-ring guide ELKS TG-IX 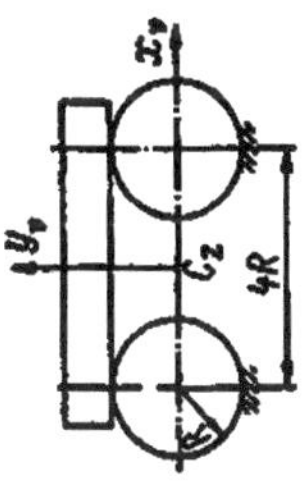	$F_k(s) + F_j(Q)$ After W Lotze, 1963 $v_x = \frac{\pi}{8}\frac{R^3}{H};\ v_y = \frac{\pi^2-8}{8\pi}\frac{R^3}{H};\ \Phi_c = \frac{\pi^2-8}{8\pi}\frac{R}{H}$

Note. *) we recommend to use for research purposes mainly; σ_n is the normal stress; $\tau_{\sigma\max}$ is the tangential (shear) stress; ϑ_z is a pitch angle around z-axis. ** The upper signs should be used for the compressive longitudinal forces and the lower signs are used for the tensile longitudinal forces in the strips.

The specific features for these structures are the larger length of strips and the possible appearance of the inflection point in the center of the elastic line. The latter is typical for the strip's flexure, which is straight in the initial position. In addition, one should remember that the *ICR* at the translational motion of a rigid body is theoretically in infinity.

The approximate nonlinear equations of variant 3 have been presented by P. Bondy (1961) without an analytical backing, but their experimental verification and comparison with the results obtained by other researchers led to positive conclusions. For example, Bondy's theoretical relationship for the slope pitch angle ϑ_z in the carriage displacement function corresponds to the analogous function of variant 2. This relationship is also expressed as the polynomial of the third degree for displacement *x* or force:

$$\vartheta_z(x) = T_{3n}\,y^3 + T_{2n}y^2. \tag{4-12a}$$

However, the empirical expression for this function is presented differently in the same publication, which corresponds to another polynomial:

$$\vartheta_{ze}(x) = T_{3n}\,y^3 + T_{1n}y. \tag{4-12b}$$

This difference may be explained as a result of the initial strips' deflection from the planar shape and a possible elongation of their elastic line at the bending under the longitudinal force. Our experimental verification gave results that are close to (12b).

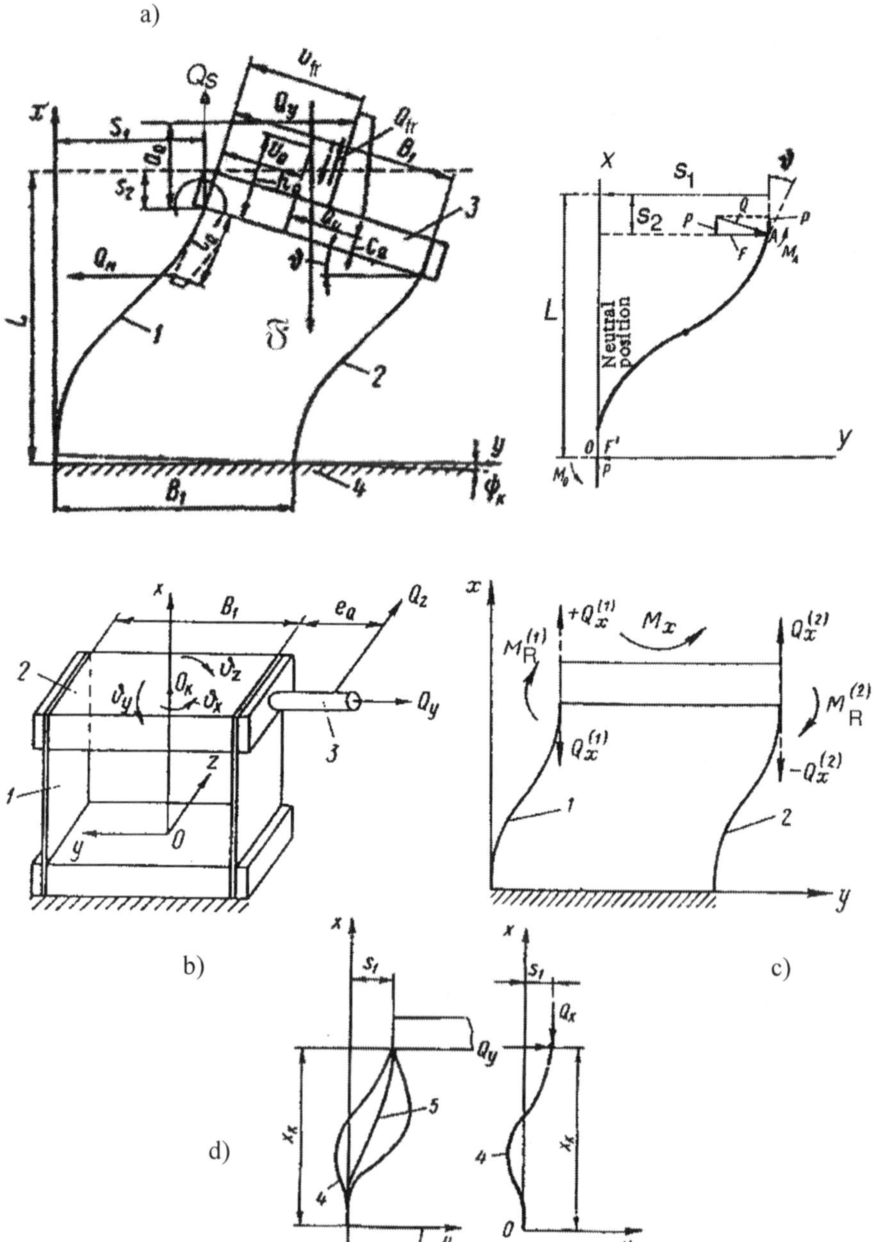

Fig. 4-8. Spring parallelogram load and kinematics relationship

Spatial loads. The action of parasitic force Q_z, which is perpendicular to the theoretical plane of the working motion of the parallelogram's carriage, is a typical example for the spatial loading. Let us express the equality of the moments relative to center O_k of the carriage for external force Q_z at $Q_y = 0$ (Fig, 4-8b), two internal

$Q_z^{(1)}$ and $Q_z^{(2)}$ reactions of the strips applied to the carriage, and two reactive moments $M_x^{(1)}$ and $M_x^{(2)}$ of the strip's twisting:

$$Q_z\,[(B_1/2) + e_Q] = [Q_z^{(1)} + Q_z^{(2)}]\,(B_1/2) + M_x^{(1)} + M_x^{(2)}. \qquad (4\text{-}13)$$

For the identical strips, therefore,

$M_x^{(1)} = M_x^{(2)} = H_k\,\vartheta_x / L$; $|Q_z^{(1)}| = 3(H_z / L^3)\,[(B_1\vartheta_x / 2) + s_{oz}]$;

$|Q_z^{(2)}| = 3(H_z / L^3)\,[(B_1\vartheta_x / 2) - s_{oz}]$; ϑ_x is an angle of carriage's turn (yaw) around x axis; s_{oz} is the displacement of carriage's center O_k in the direction of axis z; $H_k = GJ_t$ is the torsion rigidity of the strips' section; and $H_z = bh^3/12$. By substituting into (13) the expressions for reactive forces and moments, we obtain the following relationship:

$Q_z\,[(B_1/2) + e_Q] = (6H_z/L^3)\,(B_1^2/4)\,\vartheta_x + 2\,H_k\vartheta_x / L$, from which we get the expression for the carriage's slope yaw angle ϑ_x shown in Table 4-3 (variant 5 for ELKS of type TG-I). The sum of the forces applied to the carriage is equal to zero that corresponds to

$$Q_z = Q_z^{(1)} + Q_z^{(2)} = \frac{3H_z}{L^3}\frac{B_1}{2}\vartheta_x \qquad (4\text{-}14)$$

Hence, displacement s_{oz} estimated from the system of (13) and (14) is equal to:

$s_{oz} = L^3\,Q_z / 6H_z$. Calculation of the rotation yaw angle ϑ_x for the carriage under parasitic torque M_x with the formal application of the restrained twisting effects in the clamped strips' ends does not yield to any new results.

The analogous relationships are obtained earlier empirically by P. Bondy (1967) for the load with $Q_y \neq 0$ and $Q_y = 0$. Comparison of the empirical and theoretical (Table4-3, variant 5 for ELKS of type TG-I) relationships proves their qualitative correspondence. We provided special experimental investigation to verify the quantitative characteristics of the spring parallelogram with strips from beryllium bronze whose length L= 18 mm, thickness h = 0.15 mm, and width b=10 mm. The angular measurement was provided by an autocollimator with the resolution of one angular second. Angles ϑ_x and ϑ_z of the parallelogram's movable carriage 2 (Fig. 4-8b, B_1=35 mm) rotation with a mirror cemented to it around the corresponding axes was defined at the loading only by force Q_z applied to pin 3 and also at simultaneous application of forces Q_z and Q_y. The difference between the experimental and calculated data are less than 20%..

The ideal translational motion in ELKS of type TG-I may be achieved if slope pitch angle ϑ_z of the movable carriage is equal to zero. This leads to the requirement for zero value of coefficients T_{1n}, T_{2n}, and T_{3n} in polynomials (12a,b). The optimal conditions of loading in this case are as follows:

1) either $Q_x = 0$ and Q_y is applied on the level of the midpoint in the distance between the movable carriage and the stand (on the level of $L / 2$); or

2) the longitudinal internal forces in the strips are equal ($Q_{x1} = Q_{x2}$), i.e., for identical strips and the ideal quality of assembly $Q_x / Q_y = -2q_B$ where

$q_B = (L + 2a_Q) / 2B_1$. The experimental data show the significant influence of assembly quality random factors on the actual value of slope pitch angle ϑ_z. Usually angle ϑ_z is larger than that which follows from the expressions (12a,b). In addition, we cannot in all applications provide conditions 1) or 2). The wider range of acceptable parameters may be obtained on the basis of the best approximation method. To have function $\vartheta_z(y)$ with minimum deflection from zero, one should make the coefficients of polynomials (12a,b) proportional to the corresponding coefficients of the same degree Chebyshev's polynomials. Examples of these options are shown in Table 4-4.

Table 4-4. The least slope pitch angle $\vartheta_z(y)$ approximation

Polynom $\vartheta_z(y)$	The least deflection conditions	The largest deflection $\vartheta_{z\,\max}(y)$
$T_{2n}y^2 + T_{3n}y^3$	$(L/s_{1\max})(T_{2n} / T_{3n}) = -0.894$	$0.106\ (s_{1\max} / L)^3 \lvert T_{3n}\rvert \in [\ s_1/L=0;\ s_{1\max} / L]$
$T_{1n}\, y + T_{3n}y^3$	$(L/ s_1)^2 (T_{1n} / T_{3n}) = -0.75$	$0.25\ (s_{1\max} / L)^3 \lvert T_{3n}\rvert \in [-s_{1\max} /L;\ s_{1\max} / L]$

The value of ϑ_{max} decreases with the smaller value of T_{3n}, and therefore with the larger carriage's length B_1, strips' length L, and smaller strips' thickness. To provide the smallest change in the slope pitch angle, the derivative $\partial\vartheta_z / \partial s$ should have the smallest possible deflection from zero.

Choice of the strip's length can be done with the help of a diagram in accordance with specified permissible displacement s_2 at the loading of carriage by a force in direction s_1. In the first approximation, only the mean flexural stiffness of strips $H_0 = (H_1 + H_2)/2$ is important for relationships between displacements s_1 and s_2 under force Q_y. The deformation of both strips under compression force Q_x, which is transverse to working displacement s_1, can be similar to their deformation under loading only by force Q_y. If load component $Q_x = 0$, but Q_y is applied to the carriage (Fig. 4-8c) without providing an optimal condition of loading, e.g., on the level of the moving strips 1 and 2 ends, then internal longitudinal forces $Q_x^{(i)}$ appear in the strips in a direction opposite to the forces that balance reactive moments M_R acting on the carriage. Hence, the negative longitudinal force $-Q_x^{(2)}$ appears in the right strip 2 in the case of the carriage's motion to the right. This can cause probable loss of shape stability in the strip (*see* Chap. 3 and 5) under twisting spatial load, despite the fact that force $Q_x^{(2)}$ is not large in comparison with Euler's critical values. Thus we can explain the effect of the parallelogram stability's loss at the external loading with $Q_x = 0$. This negative effect occurs at large displacements of the movable carriage under application of a small twisting moment M_x. It is advisable to enlarge the base-distance of the parallelogram carriage, but without increasing its weight by choosing lighter materials and certain effective structure. This allows us to decrease the influence of defects in the clamping plates and the difference in the strips' length on the TG-I's kinematics.

TG-II. The trajectories of points on the movable carriage become more gently sloping with the increase of parameter $L_H/ L=\lambda_h$. The cover plates are clamped to the strips with some preliminary interference compression. The possible mutual shift of the cover plates and strips is prevented not only by clamping elements but also by the friction between these layers. These quasi-rigid joints can dissipate energy at the cyclic loading. One can calculate the area A_ψ of the structural damping hysteretic loop for the beam with two cover plates by the following formulae:

at a pure bending by moment M

$$A_{\psi M}=\frac{k_{FH}^{2}M^{3}(1-\mu_{cyc})^{3}}{12q_{op}h_{c}H_{0}}\,, \tag{4-15}$$

at a bending by transverse force Q,

$$A_{\psi Q}=Q^{3}k_{FH}L_{H}^{3}q_{pt}h_{c}/[6H_{0}(k_{FH}Q+q_{pT}h_{c})^{2}]\,, \tag{4-16}$$

where $k_{FH}=A_{H}h_{c}^{2}/2J$; $J=J_{0}+A_{H}h_{c}^{2}/2$, A_H and h_c are the area and thickness of the cover plates' section; J_0 is the moment of the strip section's inertia; q_{pT} is the tangent force per length unit acting between the strip's and cover plate's surface; q_{op} is the ultimate value of the tangent force per length unit which corresponds to the beginning of the sliding in the joint; and μ_{cyc} is a characteristic of the cyclic loading.

In ELKS of type TG-II a force and moment are simultaneously applied to the composite system of the beam with two cover plates. At comparatively small displacements of the parallelogram carriage, we have in accordance with the relationships shown in Table 4-3 the following:

$$|Q|=24s_{1}H_{0}/[L^{3}(1-\lambda_{H}^{3})];\ |M|=12s_{1}H_{0}L_{H}/[L^{3}(1-\lambda_{H}^{3})].$$

Substituting these expressions into (15, 16) and summing the areas of hysteretic loops yields the area of the general structural damping loop for ELKS of type TG-II: $A_{\psi}=A_{\psi M}+A_{\psi Q}$. In real structures of type TG-II $k_{FH}\approx 1$. In addition, for the preliminary estimation one can suppose that $q_{pT}=q_{op}$ and $q_{op}=\mu_{fr}p_0$ where μ_{fr} is a coefficient of friction, and p_0 is the normal pressure between the cover plate and the strip. If it is necessary to decrease A_{ψ}, we should decrease the thickness of the strips, i.e., decrease the value of H_0.

TG-III. The specifics of spring parallelogram ELKS of type TG-III depend on the behavior of the additional auxiliary strips that are compressed beyond the critical limit of buckling. Initial distance x_k between the ends of these strips 4 (Fig. 4-8d) is equal to the length of the initially straight basic strips 5. The differential equation that reflects the behavior of auxiliary buckled strip 4 may be presented (*see* also A.N. Krylov, 1931) as follows:

$$y_x^{IV}+(Q_x/H)\,y_x''=(Q_y/H)\,\delta(x), \tag{4-17}$$

where Q_x is the force of the auxiliary strip compression; - Q_y is equal to the additional working effort from the auxiliary strip that acts on the guided carriage at its motion; $\delta(x)$ is the Dirac's delta function.

The stiffness of the auxiliary strips in the direction of s_1 can be negative at the position of ELKS type TG-III that is close to the initial one within displacement $s_1=(0.1-0.2)\,L$. This option depends on the value of the preliminary longitudinal compression.

TG-IV. Experimental verification of the rigidity features in the working motion's direction for the rectilinear spring compound movement detected the deflection from the straight line at $s_1/L>0.25$ instead of the theoretically predicted ratio $s_1/L=0.4$. The theoretically predicted limit of ratio $s_1/L=0.4$ for the rectilinear motion can be achieved only in exceptional cases: at the thorough accuracy of assembly and with a carriage of a small mass (R. Breitinger, 1969).

The symmetrical connection of the two rectilinear spring movements leads to more stable rectilinear symmetrical spring movement TG-IV*s* under the changeable loading conditions.

4-2. Planar Elastic Mechanisms

Structure. The elastic elements in ELKS of type PM-I*el*, PM-II, PM-III are usually manufactured from thin metallic sheets or polycrystalline structures with good elastic properties. The increase of stiffness in some segments of the elastic body is provided by their bending in angular or channel shape, brazing, soldering, or the adhesion of ribs. The most reliable adhesion is an epoxy mixture (e.g., advanced Araldite).

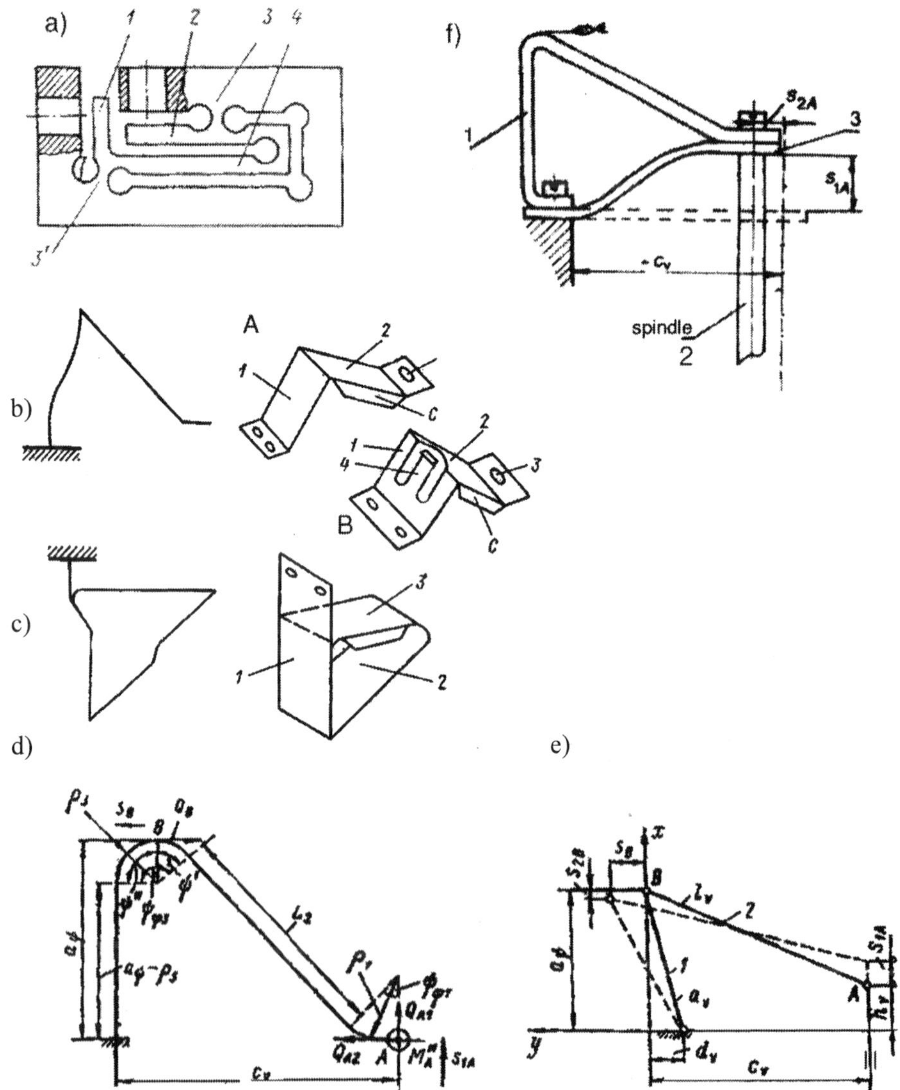

Fig. 4-9. Planar transmission structures and their *EKS* schematic

Table 4-5. Kinematical relationships for planar elastic mechanisms

Name and type of ELKS mechanism	Schematic, example	Formulae for calculation
Elastic-lever type PM-I*el*	Fig. 4-9d	**1. $F_k(s) + F_j(Q)$** *See* (18a) and (18b)
	Fig. 4-9e	**2. $F_k(s)$** After E. Ratajczyk, 1968 $s_B = \frac{-WW_1 - d_V}{W^2+1} \pm \frac{\sqrt{a_\psi^2(W^2+1) + 2WW_1 d_V + d_V^2 - W_1^2}}{W^2+1} =$ $\frac{-WW_1 - d_V}{W^2+1} \pm \frac{F_V}{W^2+1}; \quad W = \frac{c_V - d_V}{h_V + s_{1A}};$ $W_1 - \frac{s_{1A}^2}{2(h_V + s_{1A})} + \frac{2h_V(s_{1A} + a_\psi)}{2(h_V + s_{1A})};$ 3. **$F_k(s)$** $F_V = \sqrt{W^2 a_\psi + 2d_V W_1 W + a_V^2 - W_1^2}$ $s_B \approx \frac{s_{1A}(a_\psi - h_V)}{c_V - d_V + h_V d_V a_\psi^{-1}}; i_{MI} \approx \frac{a_\psi - h_V}{c_V - d_V + h_V d_V a_\psi^{-1}}$
Hinge-lever type PM-I*hl* with notch hinge	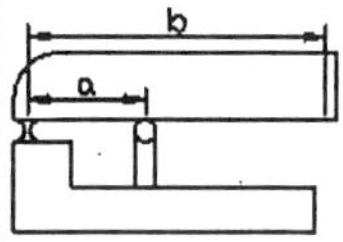 	1. **$F_k(s)$** *i=b/a* **$F_k(s) + F_j(Q)$** *see* Sect. 3-2
Two-strip reed type PM-II	Fig. 4-11a-f	**1. $F_k(s)$** $\theta < 5°$ $i_\theta \approx 0.0573\ (1/b_o)$; $j_\theta \approx 34.8\ H\ /\ (L\ b_o)$
		2. **$F_k(s) + F_j(Q) + \delta F_j(Q)$** After A.S. Valedinskii $i_{MII} = \frac{\rho_C + L/2}{b_o[1 + (h^2 / 3b_\theta^2) + (\theta\ L / 2b_\theta)\ F(\beta)]};$ $F(\beta) = (1/16)(\sinh 2\beta - \sin 2\beta) + (1/4)(\sinh \beta - \sin \beta)$ $+ (1/8)[\frac{(1+\cos\beta)^2}{\sin^2\beta}(1 - \frac{\sin^2\beta}{2\beta})$ $+ \frac{(1+\cosh\beta)^2}{\sinh^2\beta}(1 - \frac{\sinh 2\beta}{2\beta})]; b_\theta = \frac{s_1}{\theta} = i_\theta^{-1}$
Buckled-strip of type PM-III Note: *F*(*k*), *E*(*k*) are complete elliptic Integrals of the first and second kind; *M.T.A. Saif *et al.*, *RSI*,1998, **69**, 1410-1422; Table 3-3, # 4	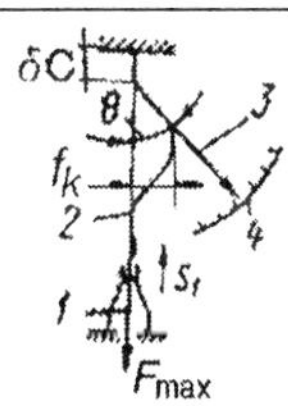 	**$F_k(s) + F_j(Q)$** Y.M. Tseytlin *et al.*, 1968 $i_{\theta av} = 2\alpha_k L^{-1}[1 - E(k) / F(k)]^{-1};$ $\delta\ cL^{-1} = (\alpha_k/4)[\frac{2E(k) - F(k)}{F(k)} - \frac{2\sin\alpha_k \cot 2\alpha_k}{F(k)}];$ $F(k) = (Q^2 L / H)^{1/2}; \quad \alpha_k = \sin^{-1} k;\ ;\ \sigma_b = 4F(k)\frac{Eh}{L}$ $\delta\ c \approx const.\ @\theta < 20^0; {}^*s_1 = \pi^2 f_k^2\ /\ (8L); F_{max} = F_{eul}$ $F/F_{eul} = (1 - f_{k0}/f_k) + (2\pi/L)^2 f_k^2 [1 - (f_{k0}/f_k)^3]/32;\ F - F_{eul} = \delta F$

Table 4-5 Cont'd.

Name and type of ELKS mechanism	Schematic, example	Formulae for calculation
Planar-cable of type PM-IV	Fig. 4-12a	1. $\mathbf{F_k(s)}$ $i_{MIV} = (1/2\vartheta_0)$; $Q_N \approx 2\, Q_s \vartheta_0$; $L_1 = L_2$; $\Delta L = 0$
		2. $\mathbf{F_k(s) + F_j(Q)}$; $H \approx 0$ $\delta\vartheta = s_2 / L$; $\vartheta_0 < 10°$ $i_{MIV} = 2^{-1}[\vartheta_0 + (\Delta L / L\delta\vartheta)]^{-1}$; $\Delta L = Q_s L / EA = Q_N L/(2\vartheta_0 EA)$; $\vartheta_{0max} \approx (0.9775 Q_N/EA)^{1/3}$ at $\vartheta_{max} < 17°$
Bridge of type PM-IV*br*	Fig. 4-12c	1. $\mathbf{F_k(s)}$ $i = -0.495/\psi_\alpha$ at $\psi_\alpha \leq 0.3$ rad
		2. $\mathbf{F_k(s) + F_j(Q)}$ *See* Chap. 8

The monolithic miniature structures are usually based on the application of notch hinges, micro- and nano-levers. An example of such a transmission structure is shown in Fig. 4-9a where recesses are made around levers 1, 2, 4, and notch hinges 3. These structures can be manufactured by electrical discharge machines and methods of lithography (*see* M. J. Madou, 2002). Fig. 4-9b shows schematics and structures for elastic-lever mechanism (ELKS of type PM-I*el*) typical in instrumentation from sheet metallic blank. Version *B* differs from version *A* in the presence of elastic cantilever tongue 4, which is used as an elastic attenuation compensator. More rigid element 2 (connecting rod) is manufactured from the same sheet material but with double-sided outward flanges *C* , which form a channel profile. Another version of elastic-lever mechanism is shown in Fig. 4-9c. Motion in the mechanisms of type PM-I*el* transmits from driving element 3 through connecting rod 2 to driven element (crank) 1, which is connected to the following element of a kinematic chain.

PM-Iel. Let us analyze the relations in typical structure shown in Fig. 4-9d. In this structure, it is necessary to find the relationships between displacements s_{1A} and s_B of driving *A* and driven *B* points when reactive force Q_B caused by the driven sequential element of the kinematic chain is applied to point B. Force Q_B can be constant or functionally dependent on displacement s_B. For example,

$$Q_B = -j_Q\, s_B\,, \tag{4-18}$$

where j_Q is the stiffness of the sequential driven system.

We can derive Mohr's integrals for the calculation of displacements

$$s_{B1} = s^{(1)}{}_{B1}(Q_{A1}) + s^{(2)}{}_{B1}(Q_{A2}) + s^{(3)}{}_{B1}(M^{b}{}_{A}), \tag{4-18a}$$

$$s_{A1} = s^{(1)}{}_{A1}(Q_{A1}) + s^{(2)}{}_{A1}(Q_{A2}) + s^{(3)}{}_{A1}(M^{b}{}_{A}) + s^{(4)}{}_{A1}(Q_B), \tag{4-18b}$$

EKS. One can provide approximate analytical research of the planar lever mechanisms by substituting them with EKS. The crank-slider mechanism is a good *EKS* for ELKS of type PM-I*el* depicted in Fig. 4-9d. The rule of a closed polygon with neglecting small deformations in the rigid enough connecting rod 2 and crank 1 (Fig. 4-9e) yields the following relationships:

$$x^2 + (y + d_v)^2 = a_v^2;\ [x - (h_v + s_{1A})]^2 + (y + c_v)^2 = l_v^2.$$

Therefore, $x^2 + y^2 + 2yd_v = a_\psi^2$; $l_v^2 = (a_\psi - h_v)^2 + (c_v - d_v)^2$; and as a result, we derive the system of equations:

$$x^2 + y^2 + 2yd_v = a_\psi^2; \tag{4-18c}$$

$$[x-(h_\nu+s_{1A})]^2+(y+c_\nu)^2=(a_\psi-h_\nu)^2+c_\nu^2. \tag{4-18d}$$

This system corresponds to the nonlinear solution for s_B shown for variant 2 in Table 4-5. We chose one value from two possible solutions that corresponds to positive x. If $x = Ws_B + W_1$ and $x > 0$, then $Ws_B + W_1 > 0$. This solution corresponds to the following ultimate conditions:

i) If $s_{1A} + h_\nu < 0$, then $W = (c_\nu - d_\nu) / (s_{1A} + h_\nu)$ and the root of the sought solution corresponds to $s_B \leq - W_1/W$; ii) If $s_{1A} + h_\nu = 0$, i.e., $s_{1A} = - h_\nu$, then the sought solution $s_B = [l_\nu^2 - c_\nu^2 + d_\nu^2 - a_\nu^2]/[2(c_\nu - d_\nu)]$.

Comparison by E. Ratajczyk of the calculation relationships presented in variant 2 (Table 4-5) and his experimental data proved their sufficient correlation. The relative difference between calculated and experimental data is less than 0.5% at $h_\nu < 4$ mm for elastic-lever mechanisms with rigid enough elements 1 and 2. The transmission ratio for ELKS of type PM-I*el* with the practically rigid crank and connecting rod is equal to the partial derivative:

$$i_{MI} = [(\partial s_B/\partial W)\, dW/ds_{1A}] + [(\partial s_B/\partial W_1)\, dW_1/ds_{1A}] =$$

$$\frac{\{[-W_1 \pm F_\nu^{-1}(a_\psi^2 W + W_1 d_\nu)](W^2+1) + 2W(W_1 W + d_\nu \pm F_\nu)\}(-W)}{(W^2+1)^2(s_{1A}+h_\nu)} + \frac{\left(1-\frac{W_1}{s_{1A}+h_\nu}\right)\left(-W \pm \frac{Wd_\nu - W_1}{F_\nu}\right)}{W^2+1}.$$

One can find the results of the experimental study and detailed recommendations on the synthesis of this mechanism in E. Ratajczyk (1968) publication. Table 4-6 shows the typical dimensions for the tested samples.

Table 4-6. Experimental samples ELKS of type RG-I*el* with typical structure in Fig. 4-9d, e

Series number	a_ψ, mm	c_ν, mm	l_ν, mm	h_ν, mm	d_ν, mm
I	12	9.5	15.3	0	-3 to +6
II	12	9.5	15.3	-4 to +8	0
III	8-16	9.5	15.3	0	0

After E. Ratajczyk (1968)

In the real mikrokator system (Table 8-1, scheme S1), distance c_ν (Fig. 4-9d, f) is restricted by the flexible strip 3, which moving end is attached to point A. As a result, transverse motion s_{2A} (in direction of force Q_{A2}) follows motion s_1 and causes additional nonlinear uncertainty in the mechanism. If the strip is initially straight in the beginning of motion s_{1A}, the change in distance c_ν is approximately equal to:

$$\frac{s_{2A}}{c_\nu} = \frac{3}{5}\left(\frac{s_{1A}}{c_\nu}\right)^2. \tag{4-19a}$$

If the strip has preliminary displacement $s_{1A0} = h_\nu$, the transverse parasitic displacement of spindle 2 (Fig. 4-9f) on strip 3 in s_{2A} direction is larger and equals

$$\frac{s_{2A}}{c_\nu} = \frac{3}{5}\left[\left(\frac{s_{1A}}{c_\nu}\right)^2 + 2\left(\frac{h_\nu}{c_\nu}\right)\left(\frac{s_{1A}}{c_\nu}\right)\right]. \tag{4-19b}$$

Fig. 4-10 a shows that mechanisms with elastic strip pivot have different charts of nonlinearity for sine 1, tangent 4 mechanisms with rigid hinges, and 2 and 3 for the same type of mechanisms with elastic hinges, respectively.

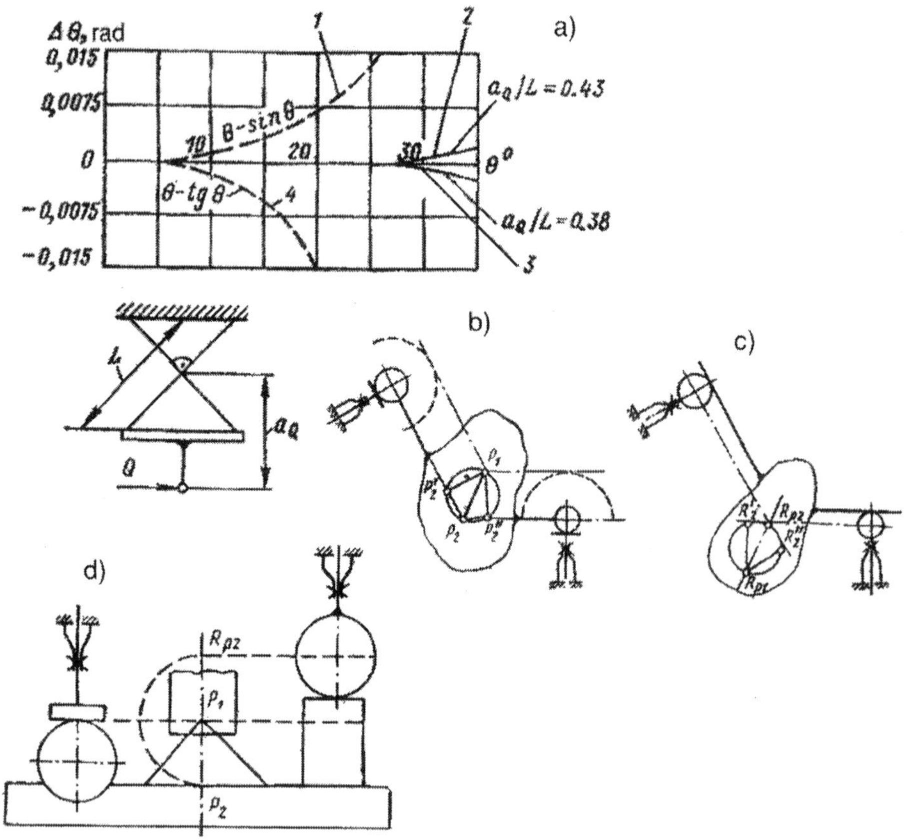

Fig. 4-10. Charts of nonlinearity and EKS for sine and tangent mechanisms (after W. Lotze)

EKS for synthesis of sine-sine (Fig. 4-10b), tangent-tangent (Fig. 4-10c), and mixed sine-tangent (Fig. 4-10d) have certain specifics. Let us recall that "sine" and "tangent" mechanisms with rigid pivots have nonlinear transmission of the third order (Fig. 4-10a, curves 1 and 4). We remember that with Taylor's series expansion $\sin\theta = \theta - \frac{1}{6}\theta^3 + \ldots$ and $\tan\theta = \theta + \frac{1}{3}\theta^3 + \ldots$ The corresponding elastic hinge-lever mechanisms with cross-strip hinges have a nonlinear $\Delta\theta(\theta)$ transmission of the 4 -5th order (curves 2 and 3 in Fig. 4-10a). In the "sine-sine" mechanism (Fig. 4-10b), the lines that follow through the centers of the levers' spheres and are parallel to the planes of the driving and driven sliders should intersect at pole P_2 of rotation. In the "tangent-tangent" mechanism, the parallels to the lever's' planes that follow through the centers of the slider's spheres should intersect in the reciprocal pole R_{p2} (Fig. 4-10c). This will eliminate an error of the second order. Fig. 4-10d shows the optimal schematic of the "sine-tangent" mechanism. The value of an acting uncertainty in ELKS of type RG-I is approximately equal to:

$\Delta_{2a} \approx (1/12)\, j_{\Delta} L(\theta + 0.02)^2$ when the base line is along the strip's axis; and

$\Delta_{2a} \leq (1/120)\theta j_{\Delta} L$ when the base line is perpendicular to the strip's axis, where $j_\Delta = \sqrt{i^2 - 2i \cos\gamma_{1a} + 1}$, i is the mechanism's transmission ratio, and γ_{1a} is an angle between the lines of action (see Fig. 1-5 with base line OA_T). The analogous uncertainty can be predicted in ELKS of type RG-II by the following formulae: $\Delta_{2a} = 0.707\, B_x j_\Delta\, \theta^2\, L/\, [\cos(\gamma/2)]$ when the base line is along with the ELKS's axis of symmetry; $\Delta_{2a} = 0.707\, B_y j_\Delta\, \theta^3\, L/[\cos(\gamma/2)]$ when the base line is perpendicular to the ELKS's axis of symmetry. The notations B_x, B_y, and γ are shown in Table 4-1 for the cross-strip rotary guide

PM-II. The two-strip reed mechanism has nonlinear relationships, which are represented in variant 2 for ELKS of type PM-II in Table 4-5.

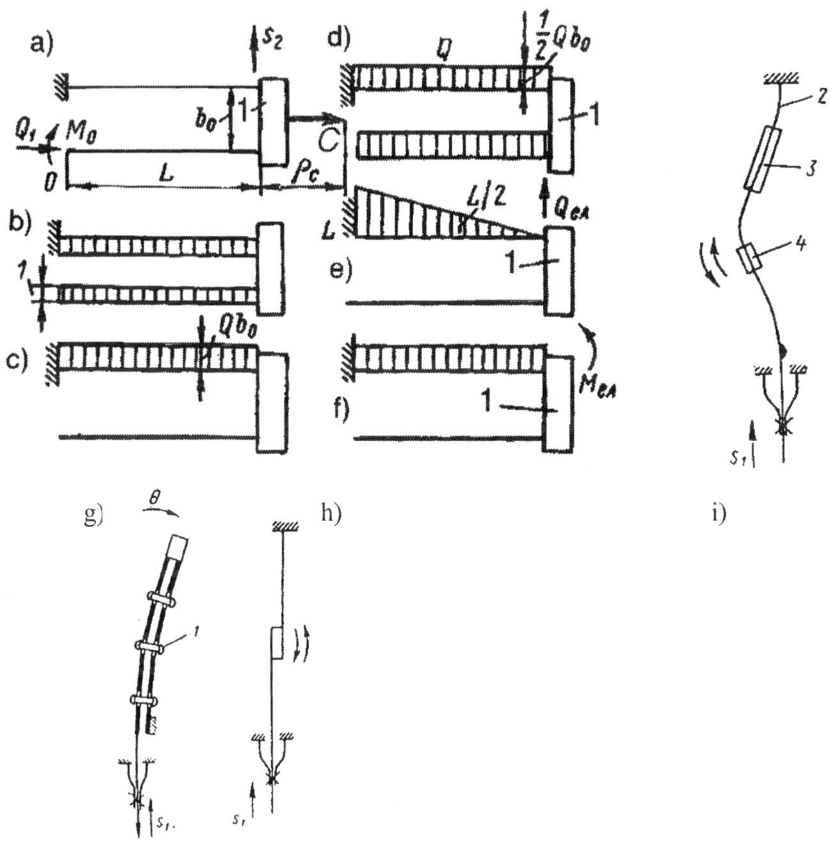

Fig. 4-11. The two-strip and buckled-strip mechanisms structures.

The simplified linear relationships in variant 1 are applicable for the small elastic strains in the strips. In this case, the mechanism may be represented as a frame shown in Fig. 4-11a for which we have the following canonical equation, $s_{11}M_o + s_{1Q} = 0$, where s_{11} is the displacement under unit moment $M_{eo} = 1$ in point O in the direction of its action; s_{1Q} is the displacement under force Q_1 in the direction of action for the moment M_{eo}'s diagram; and M_o is the bending moment at point O. From the diagrams of bending moments shown in Fig. 4-11b,c, we have the following conditions:$s_{11} = 2L$ and $s_{1Q} = -Qb_0L$. Hence, $M_o = ½\, Q\, b_0$. One can define displacement s_2 and rotation θ of carriage 1 by multiplying the summary diagram of the bending moments from force Q_1 and moment M_o (Fig. 4-11d) by the diagram of moments from unit force Q_{eA} (Fig. 4-11e) and unit moment M_{eA} (Fig. 4-11f), respectively. As a result, $s_2 = Q_1b_0L^2/(4H)$ and the angle of rotation $\theta = (Q_1/2)b_0L/H = 2s_2/L$. If we suppose that the elastic lines of the deformed strips are equidistant arcs of circles (indeed, it is true only for the pure bending of the strips), then $s_1 = b_0\theta = 2\, b_0s_2/L$.

This allows us to get the simplified relationships for the pointer's end displacement s_c and the mechanism's transmission ratio i_{MII}. Analysis of the calculated and experimental relationships of PM-II proves that this mechanism has a relatively small transmission ratio, a small angle of the driven element's rotation corresponding to $i_{MII} \approx$ constant, and a large gradient of the driving force. The transmission ratio and the driving force are less at the increased distance b_0. A decrease in the strip's thickness (usually $b/h > 30$) also significantly decreases driving force Q_1. An adjustment of the transmission ratio within a small range during the calibration is possible by changing distance b_0 near the stand side. A significant increase in the strip's length provides an increase in the transmission ratio but simultaneously requires an installation of limiters 1 (Fig. 4-11g) for the mechanism's reliability. As a result, such a version of the mechanism has an external friction, which leads to an increase in the backlash uncertainty. In some cases, the inverse type of this mechanism is used (Fig. 4 –11h).

In accordance with Fig, 4-11a for this mechanism, we have (from equation in Table 4-5) the following expressions of its uncertainty Δs_2 caused by uncertainties $\Delta\rho_c$, ΔL, and Δb_0 of elements dimensions:

$$\Delta s_2 = [(\Delta\rho_c / b_0) + (\Delta L/2\, b_0) - (2\rho_c + L)\Delta b_0/(2\, b_0^2)]\, s_1 \quad \text{or}$$

$\Delta s_2 = (A_1\,\Delta\rho_c + A_2\,\Delta L - A_3\,\Delta b_0)s_1$,where $A_1 = 1/\, b_0$; $A_2 = 1/\,(2b_0)$; $A_3 = (2\rho_c + L)/(2\, b_0^2)$.

The comparison of coefficients A_1, A_2, and A_3 shows that the predominant influence on the accuracy of PM-II has distance b_0 between the elastic strips.

PM-III. Fig. 4-11i shows a version of this mechanism's structure. Two rigid cover plates 3 and 4 are clamped to buckled strip 2. The larger cover plate 3 rotates together with pointer (see corresponding element 3 in Figure of Table 4-5). Its virtual center of rotation position δC is approximately constant when angle of rotation $\theta < 20°$. Function $f_k(s_1)$ is also useful in sensors. A maximum driving force F_{max} have very small gradient after beam-strip buckling (*see* A.N. Krylov, 1931, M. Born, 1956) that is effective in force correctors.

PM-IV. There are three cases for the calculation of planar cable mechanism: (i) a very small initial sag ($f_k < 0.05L$) when the elastic extension of the cable should be taken into account; (ii) a small initial sag ($0.2\, L > f_k > 0.05L$) when the geometric relationships can be simplified and the elastic extension of the cable may be neglected; and (iii) a large initial sag ($f_k > 0.2$L) when the used geometric relations should be accurate enough ("exact"), but the extension of the cable may be neglected. In addition, the influence of the cable's actual flexural rigidity should be verified, despite the fact that it is usually considered equal to zero.

a) b)

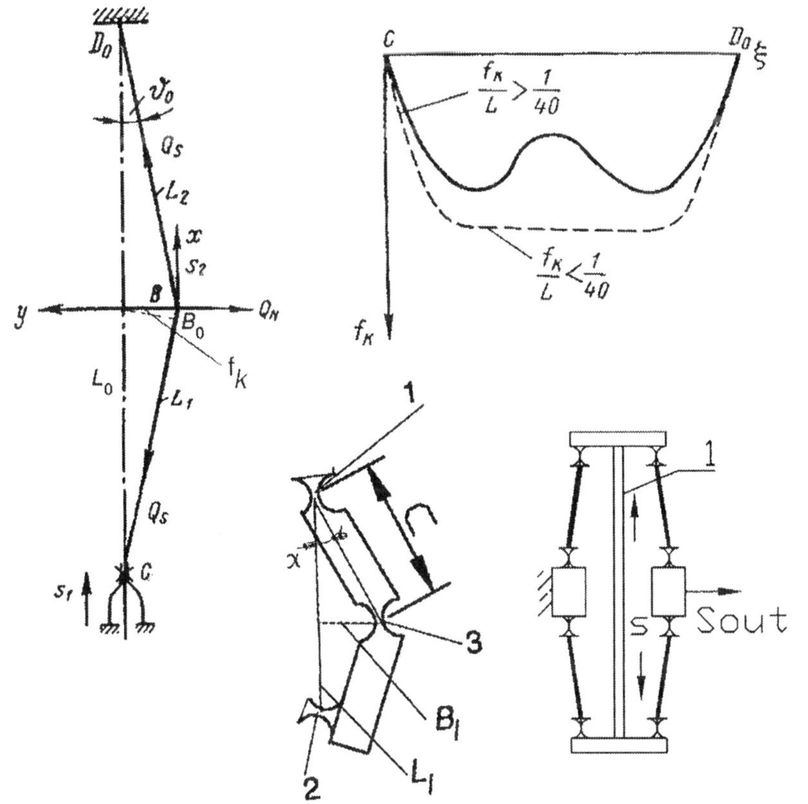

c) d)

Fig. 4-12. Cable and bridge structures and features.

It was found by test that the largest normal flexural stress in the cable does not depend on the number of wires in its cross section if their summary area is the same. A similar situation occurs in the cable mechanism with a flexible strip and flexible wire. In both cases the largest achievable transmission ratio is the same. Amplification transmission is typical for this type of mechanism. Therefore, its calculation may be based on the version with a small sag f_k (case ii). The action of the distributed load on the strip may be neglected because the mass of the strip is small. Then, transmission ratio $i = \partial s_2 / \partial s_1$ for the plane cable mechanism with a light non-stretchable strip or wire can be defined on the basis of the relationships in triangle BCD_o. In accordance with Fig. 4-12a, we have at $L_l = L_1 + L_2$ that

$$s_2 = f_{ko} - 2^{-1}(CD_o + s_1)^{-1}\sqrt{[L_l^2-(CD_0+s_1)^2][(CD_0+s_1)^2-(L_l-2L_1)^2]}\,, \qquad (4\text{-}20)$$

One can obtain the transmission ratio for the mechanism with non-stretchable cable through differentiation of (20) as: $i(s_1)=\partial s_2/\partial s_1$. We estimate the maximum of this function (Y.M. Tseytlin, 1972) by differentiating it with respect to L_l and assuming L_l= const, CD_o = const, and s_1=0. The derivative is equal to zero when $L_1 = L_l/2$, i.e., $L_1 = L_2$. This corresponds to the largest transmission ratio and a minimum of the cable's sag. Fig.4-12b shows the sag f_k values at different points of application of force Q_N to the non-stretchable cable. The form of this chart insignificantly changes with the addition of the cable extension component in cases (ii) and (iii). The elastic extension significantly changes the form of the chart (shown by a dashed line) in case (i) and leads to a decrease in the transmission ratio proven in practice. Hence, the optimal kinematic structure for the cable mechanism corresponds to the case with a small sag $(0.2L > f_k > 0.05L)$ and equal branches of the flexible element $(L_1 = L_2 = L)$. The symmetrical cable mechanism has a small parasitic displacement of driven point B in direction s_1. The approximate calculation relationships for the symmetric cable mechanism with a thin and ideally flexible non-stretchable as well as an stretchable cable may be derived on a simpler basis. In the initial position when point B is at the middle of the cable, we have the following relations:

$$L \cos\vartheta_0 = a_q;\ f_{ko} = L \sin \vartheta_0. \qquad (4\text{-}21)$$

If the initial angle ($\vartheta_0 << 1$) and the corresponding initial sag f_{ko} are small, then the above shown relationships can be further simplified as

$$a_q = L\,(1\text{-}½\ \vartheta_0^2);\ f_{ko} = L\ \vartheta_0. \qquad (4\text{-}22)$$

It is easy to show that for the symmetric stretchable cable mechanism with zero flexural rigidity $(\overline{CB} - \overline{CB}_0)/L_0 = (L_1 - L_0)/L_0 = \tan\vartheta_0 - \sin\vartheta_0 = Q_N/2EA$ (*see* fig. 4-12a and *Roark's Formulas*) that corresponds to approximate expression for $\vartheta_{0\max}$ in Table 4-5 at $\vartheta_0 < 17°$. A small change $2s_1$ in the distance between the cable's ends leads to a small change in the slope angle $\delta\vartheta$ in each cable's branch. The length of the cable may simultaneously slightly change by $2\Delta L$ as a result of the cable's extension. The angle's variation $\delta\vartheta$ in this mechanism is usually significantly smaller than the initial angle ϑ_0, i.e., $\delta\vartheta / \vartheta_0 < 1$. Thus, we derive a new expression as

$$(L + \Delta L)[1 - ½\,(\vartheta_0 - \delta\vartheta)^2] = a_q. \qquad (4\text{-}23)$$

Then, after subtracting (22) from (23) and canceling the negligible members of the second order, we get the following expression

$$s_c = s_1/2 = (L + \Delta L)[1 - ½\,(\vartheta_0 - \delta\vartheta)^2] - L\,(1\text{-}½\ \vartheta_0^2) = L\vartheta_0\delta\vartheta + \Delta L\ (1 + \vartheta_0\delta\vartheta). \qquad (4\text{-}24)$$

In addition, $f_{ko} + s_2 = f_k = (L + \Delta L)\ (\vartheta_0 - \delta\vartheta)$. (4-25)

Subtraction of (22) from (25) yields $s_2 = \Delta L\ (\vartheta_0 - \delta\vartheta) - L\delta\vartheta$. (4-26)

If extension 2 ΔL is negligible and the driving motion is only applied to one cable's end, we derive from (26) and (24) the expressions shown in Table 4-5 (*see* variant 1 for ELKS of type PM-IV).

Let us discuss the influence of the cable's flexural rigidity on the cable mechanism's function with small sag. For the calculation of deformation under the longitudinal-transverse load ($Q_x \neq 0$), we use Green's function which describes the behavior of a beam under the transverse concentrated unit force ($Q_N = 1$) applied to point $x = \xi$. This function can be found with the approximate Timoshenko's formula:

$$G(x;\xi) = \tilde{y}\,(x;\xi) / (1 + \nu_{Nx})\,, \qquad (4\text{-}27)$$

where $\tilde{y}$ (x;ξ) is the form of the elastic line at $Q_N = 1$ and $Q_x = 0$;

$$\nu_{Nx} = Q_x / Q_{xcr}, \qquad (4\text{-}28)$$

and Q_{xcr} is the critical value of the buckling force (*see* Table 3-3).

Let us recall that formula (27) is accurate enough even for large values of ν_{Nx}(see Chap. 3). We can then calculate the form of the elastic line at a specified transverse concentrated force value Q_N as follows: $y(x, \xi) = Q_N G(x;\xi)$. (4-29)
The form of an elastic line under the specified distributed load $f(\xi)$ along the beam is presented by an integral $y(x) = \int_0^{L} l\, G(x;\xi) f(\xi) d\xi.$ (4-30)

Thus, for ELKS of type PM-IV $s_2 = Q_N \tilde{y}(x; L/2)\,[-\nu_{Nx} / (1 + \nu_{Nx})]$. (4-31)

The planar chain transmission ratio.

The transmission ratio i_{ELKS} for an elastic kinematic system is equal to:

$$i_{ELKS} = ds_{ex} / ds_{in} = \frac{\partial s_{ex}}{\partial(s_{in} - s_{efb})} \frac{\partial(s_{in} - s_{efb})}{\partial s_{in}} = i_{sc}\,(1 - \frac{ds_{efb}}{ds_{in}}) = i_{sc}(1 - i_{efb}\, i_{ELKS}). \quad (4\text{-}32)$$

Hence, $i_{ELKS} = i_{sc} / (1 + i_{sc}\, i_{efb})$, where i_{efb} represents the transmission ratio for an equivalent (internal along with external) feedback. As a result, the linear ELKS's general transmission ratio with respect to (1-37) and (1-38) equals

$i_{ELKS} = s_2 / s_1 = (a_{11} + a_{12}\, j_{sr2})^{-1}$. Matrix a_{ij} is used for the calculation of planar elastic mechanisms at small relative displacements (*see* Table 4-7).

Table 4-7. Examples of matrices a_{ij} for planar ELKS

Schematic	Matrix After W. Lotze
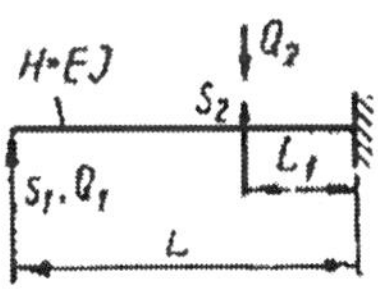	$\begin{bmatrix} \frac{1}{k\lambda}\frac{L^3}{3H} & \left(\frac{\lambda_L^3}{k\lambda} - k\lambda\right) \\ \frac{3H}{L^3}\frac{1}{k\lambda} & \frac{\lambda_L^3}{k\lambda} \end{bmatrix}$ $k\lambda = \frac{3}{2}\lambda_L^2 - \frac{1}{2}\lambda_L^3;\ \lambda_L = L_1 / L$
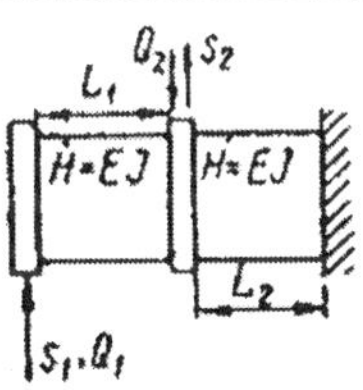	$\begin{bmatrix} 1 + \left(\frac{L_1}{L_2}\right)^3 & \frac{L_1^3}{24H} \\ \frac{24H}{L_2^3} & 1 \end{bmatrix}$

4-3. Spatial Elastic Mechanisms

The typical spatial elastic mechanisms can be divided into three structural categories: (i) pretwisted strips and tubes; (ii) coiled strips; (iii) cable structures.

Pretwisted strip mechanisms.

History. The pretwisted strip mechanism (SM-VI) which transforms the longitudinal strip's extension in the angle of its cross-section rotation was first presented and researched by J. Perry (1890) who showed its outstanding options for measuring purposes as a highly sensitive transmission system. However, despite his efforts to get adequate analytical functions for this mechanism calculation, he could not obtain them at that time even with the help of the London Physical Society. The pretwisted strip mechanism should be named Perry's mechanism, but later in the 1930s it was modified and effectively applied in the length measuring systems by H. Abramson whose name was often applied to it (*see* the explanation in Chap. 8 about the difference between Perry's and Abramson's mechanisms).

The behavior of the pretwisted strip mechanism (Fig. 2-3b) with two pretwisted parts 1 and 2 with opposite helix direction depends on a set of geometric and elastic parameters. Perry performed the first series of tests on the large steel strips, and the second series of his tests was performed on the small platinum-silver strips. The results on the glance were contradictory. The main obstacle that Perry could not overcome, in his own opinion, was the establishment of correct distribution for the normal stresses σ_x through the cross section. In the beginning, he assumed σ_x = constant, but then assumed that σ_x varies and functionally decreases with the increase of distance y from the axis of the strip. However, the analytical relationships obtained by Perry did not agree with the experimental data. Latter attempts to represent the pretwisted strip's mechanism as a bifilar model (Yu. N. Lyandon, N. N. Zyabreva, *Izmeritelnaya Tekhnika*, 1941, in Russia; N.P.L., 1956, in UK) were also not successful.

SM-VI kinematics and stiffness. The necessary analytical solutions for the pretwisted elastic double-helicoid micro-strip's mechanism (Table 4-8, ELKS of type SM-VI) were derived only in 1960s (Y. M. Tseytlin, 1964c, 1965-1972) on the basis of the developed methods of the theory of elasticity that we discussed in Sections 2-2 and 2-3.

Table 4-8. Kinematic relationships in the spatial transmission mechanisms

Name and type of ELKS	Schematic	Calculation relationships	Notes
Pretwisted strip mechanism SM-VI	Fig. 2-3	1. $\mathbf{F_k(s)}$+ $\mathbf{F_j}$ **(Q)** Y.M. Tseytlin, 1964c,d, 1965 $i_\theta = \dfrac{Ek_p}{4G\lambda}\dfrac{0.0286}{1+\nu_{if}\chi_{kp}}$; $\chi_{kp} = \dfrac{b_1^2 k_{p0}^2}{\lambda}$, *a)* $\nu_{i\,el}$=0.345; b) $\nu_{i\,rc}$=0.32-0.35, c) $\nu_{ipf} \approx 0.39$ $j_\theta = 69.6\dfrac{GA\lambda}{Lk_{p0}}\left(1+\nu_{jf}\,\chi_{kp}\right)$; $\nu_{jel,ow} = 0.179$; $\nu_{jrc} = 0.14$; $j_{\theta pf}^{-1} = 2^{-1}\dfrac{j_{\theta c}+j_{\theta rc}}{j_{\theta c}\,j_{\theta rc}}$; $j_{\tau\tau} = GT_0\left(1+\nu_{if}\,\chi_{kp}\right)$, $j_{\theta c} = \dfrac{69.6G\lambda A_c}{Lk_p}\{1+\chi_{kp}[0.5-0.3(1+\mu_b)]\}$, $A_c = (b-b_d)h$, $\mu_b = b_d/b_1 < 1$; L_c/L=0.5 *See* also (4-33), (4-34).	$\dfrac{L_{str}}{2L} \approx 0$ a)$\lambda \geq 0.4\times10^{-2}$ b)$\lambda < 0.4\times10^{-2}$ $j_{\tau k} = \dfrac{EAb_1^2 k_p}{4}$; $j_{kk} = EA$
		2. $\mathbf{F_k(s)}$+ $\mathbf{F_j}$ **(Q)** +**δF (Q)** Y.M. Tseytlin, 1964c, 1968, *see* (2-56, 2-57, 2-58)	$\nu_{if} = 0.32-0.39$

Table 4-8 Cont'd.

Name and type of ELKS	Schematic	Calculation relationships	Notes
Coiled strip helicoid ELKS SM-VII	Fig. 4-17a	1. **F**F_k **(s)** After K.N. Manuilov, 1959 $i_{\theta av} \approx 0.0573 \frac{S_{av}}{\pi D_r^2}$, $j_\theta = 6.3 \frac{bh^3 E}{D_r L}$	
		1. $F_k(s) + F_j(Q)$ After V.L. Biderman, 1963 *See* (4-61e)	
Spatial cable ELKS SM-VIII	Fig. 4-19a	F_k **(s)** Y.M. Tseytlin, 1972 $i_\theta = \frac{x_0}{2L_a L_b} \frac{\theta}{1-\cos\theta}$; $j_\theta = dQ_s / d\theta = 1.2\, EL_a d_w^4 X_o / L^3 L_b$.	θ<1

Index c in the punched strip parameters corresponds to the segments with cut, including their length L_c.

One can see the three typical segments known from experimental data of the pretwisted strip under extension s_1 (Fig. 4-13a) and rotation θ^o (Fig. 4-13a, b) transmission function: I is the initial nonlinear segment with a variable average $i_{\theta av}$ transmission ratio; II is the linear one with the constant (at deflection within 1-2%) transmission ratio; and III is the nonlinear segment with a variable transmission ratio where the strip's material can be in the elastic-plastic state (dashed line). It is natural that calculation relationships for linear segment II are based on the linear theory of elastic helicoid deformation. The transmission ratio $i_{\theta II}$ in °/ μm and driving stiffness $j_{\theta II}$ in cN/ o of the mechanism are equal to: $i_{\theta II} = 57.3\ \theta_\gamma / s_\gamma$ and $j_{\theta II} = Q / [57.3\ (\theta_\gamma L)]$.

Values of θ_γ and s_γ correspond to untwist angle and extension of the strip per unit length, respectively. Their functional magnitudes are found from the solution of the simultaneous equations (*see* Chap. 2) and presented in variant 1 for ELKS of type SM-VI (Table 4-8). However, the expressions of variant 1 in Table 4-8 do not show the influence of the flat crosspiece 4 (Fig.2-3b) and straight ends 3 and 5, which the double-helicoid strip mechanism usually has.

Straight segments. With the accounting of this straight segment extension, the expressions for $i_{\theta II}$, in °/ μm, and $j_{\theta II}$ are as follows

$$i_{\theta II} = 0.0572\left(E\kappa_p\ L/4G\lambda\right)\left[L_{str}(1+\nu_{jf}\chi_{kp}) + 2(1+\nu_{if}\chi_{kp})L\right]^{-1}, \tag{4-33}$$

$$j_{II} = EA(1+\nu_{jf}\chi_{kp})\left[(1+\nu_{jf}\chi_{kp})L_{str} + 2(1+\nu_{if}\chi_{kp})L\right]^{-1}, \tag{4-34}$$

where the longitudinal stiffness $j_{II} = j_{\theta II}\, i_{\theta II}$, $L_{str} = L_{crp} + 2L_{send}$, and $\chi_{kp} = b_1^2 \kappa_p^2 / \lambda$. The influence of the length of crosspiece L_{crp} and straight ends L_{send} in the real device's transmission and stiffness (Chap.8) may be up to 10 %. The substitution of L_{str} =0 into (4-33) converts it into the corresponding expression for i_θ shown in Table 4-8.

Comparison. We can show that for the pretwisted strips with an elliptical cross section, where ν_{if}=0.345, the expression for the transmission ratio in quasi-linear transformation part may have another more digestible form because

$i_\theta = \frac{Ek_p}{4G\lambda} \frac{0.0286}{1+\nu_{if}\chi_{kp}} \approx 0.045 \frac{E}{GS_0} \frac{1}{(h/b)^2 + 3.4(b/S_0)^2}$, where S_0 is the pitch of the helix.

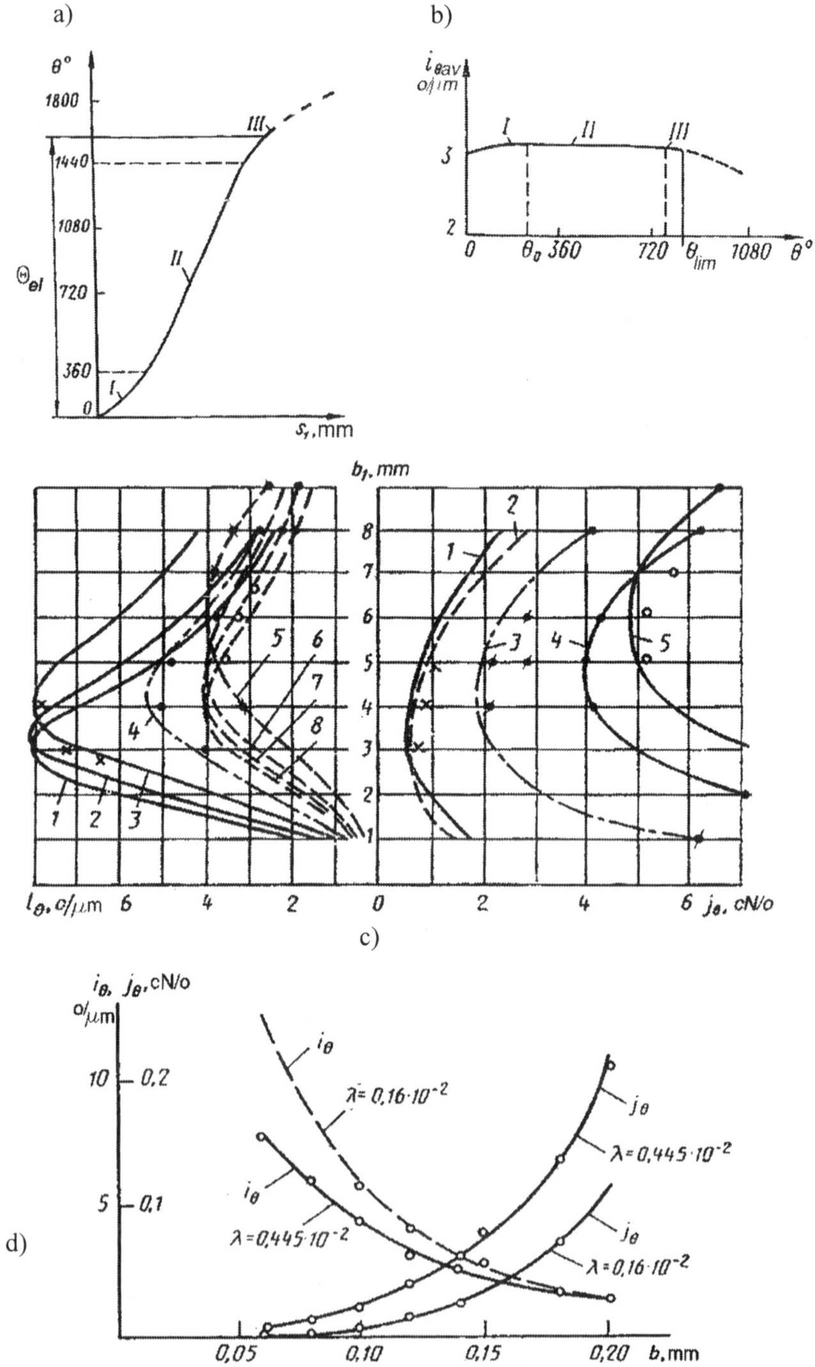

Fig. 4-13. Double pretwisted strip's typical transmission and stiffness features

Similar conversions can be done for the pretwisted strips with different cross sections. The correspondence within 3-8 % of the calculations with formulae in Table 4-8 to experimental data may be seen from the Table 4-9, where also shown the results of calculations based on the N.P.L (1956) formula. The latter results differ from the independent experimental data up to 38 %.

Table 4-9. Experimental and calculated data for elastic pretwisted microstrip transmission ratio i_θ

Strip No	Pretwisted strip dimensions, mm			Transmission ratio i_θ,°/μm; $\lvert i_\theta - i_{\theta e}\rvert 100 \lvert i_{\theta e}^{-1}\rvert$, % - a relative difference with experimental data		
	Thickness $h=2h_1$	Width $b=2b_1$	Helix pitch, S_0 (angle twist /strip length, mm)	Experiment $i_{\theta e}$	Calculated on the basis of Table 4-8	Calculated on the basis of N P L, 1956, formula**
1	0.021	0.22	2	*1.0619	1.140 ; 7.3	0.752 ; 29.2
2	0.014	0.168	1.6	*1.6690	1.613 ; 3.3	1.032 ; 38.2
3	0.010	0.13	1.33	*2.0804	2.245 ; 7.7	1.432 ; 31.2
4	0.0025	0.06	1.62-1.8 2 (4000°/ 40)	***10.0	10.4; 4.0	9.19; 8.1

* E. Ratajczyk, 1968.** amplification $i_\theta = \dfrac{d\theta}{dl} = \dfrac{-9.1L_s}{b^2 N_t}$, where N_t is the number of complete turns in the half length of the strip, $L_s \approx 2L$; $L/N_t = S_0$, minus sign designates the untwist at stretching.
*** average data by H. Abramson, 1947; we assume that the crosspiece length is approximately 4 mm as for the regular mikrokator (*see* Chap.8) and the length of twisted segment L=18 mm with the whole strip's length 40 mm.

The charts in Fig. 4-13c,d and Fig. 4-14, which are built on the basis of the relationships presented in Table 4-8, help to explain the contradictions of the conclusions obtained by J. Perry and other researchers with the limited experimental data. Fig. 4-13c presents calculated and experimental transmission ratio $i_\theta(b_1)$ and stiffness $j_\theta(b_1)$ in linear segment of transmission for pretwisted strips with parameters 1-h_1=0.002 mm, k_p =2.1 mm^{-1}; 2- h_1=0.002 mm , $k_p = \pi$ mm^{-1}, × - experimental data; 3- h_1=0.002, k_p = 2.44, ◆-experimental data for j_θ; 4- h_1=0.004, $k_p = \pi$, •-experimental data for i_θ and j_θ ; 5- h_1=0.004, k_p = 2.1, o – experimental data for j_θ and ◆ for i_θ; 6- h_1=0.004, $k_p = \pi$, o –experimental data for i_θ; 7- h_1=0.004, k_p =3.83; 8- h_1=0.004, k_p =4.94. Fig. 4-13d shows calculated transmission ratio $i_\theta(b)$ and stiffness $j_\theta(b)$ functions with specified parameter λ.

It is clear that most of the relationships in Fig. 4-13c , 4-14 have a relative maximum or minimum. This prevented an unambiguous conclusion about the influence of certain parameter (except the thickness) upon the transmission ratio or the longitudinal stiffness. The form of the microstrip's cross section has an influence on the transmission ratio and especially on the longitudinal stiffness of the pretwisted microstrip mechanism (*see* Chap. 2). This influence on i_θ and j_θ for the solid microstrips with identical dimensions and material is not more than 10%, but the longitudinal stiffness of the perforated strips differs significantly.

The calculating relationships shown in Table 4-8 correspond well to the experimental data. The difference between calculated and experimental data obtained by many researchers (H. Abramson, B. A. Skolozubov, E. Ratajczyk, A. N.Grishin, and the author of this book) on a wide variety of twisted strips is not more than 10% for transmission ratio i_θ and is less than 15% for longitudinal stiffness j_θ. E. Ratajczyk (1968) and V.A. Shemetillo (1972) performed special research on the accuracy of the relationships proposed by the author of this book for the pretwisted double-helicoid strips shown in Table 4-8. The conclusions in all cases were positive.

Untwist limit. The limit of the elastic untwist for the pretwisted strip's crosspiece can be estimated on the basis of the ultimate stress state theory with the maximum shear stress as follows: $\theta_{lim} = \tau_e/[\pi(\delta\sigma_1 - \delta\sigma_3)]$, (4-35)

where $\tau_e = \sigma_e/(1.73 - 2)$ is the shear stress that corresponds to the yield stress of material; $\delta\sigma_1$and $\delta\sigma_3$ are the corresponding additions to the largest and the smallest principal stresses in the stress tensor with the extension of the pretwisted strip and rotation of its crosspiece to the unit angle of 1 rad. In accordance with (2-75) for the critical point of the strip's cross section, we have

$$(\delta\sigma_1 - \delta\sigma_3) = (4G\lambda / \kappa_p L)(1 + 0.379\,\chi_{kp}) \qquad (4\text{-}36)$$

because $\sigma^{33}_0 = Q/A$ and force $Q = 4G\lambda A\, C_\sigma \theta/(k_p L)$.

By substituting (36) into (35), we obtain the relationship for the limit angle of the strip's untwist at $\theta_{lim} < Lk_{p0}/4\pi$ as

$$\theta_{lim} = \frac{\sigma_e \kappa_p L}{8\pi\, G\lambda\,(1 + 0.379\chi_{kp})}. \qquad (4\text{-}37)$$

The coefficient 0.379 insignificantly differs from the corresponding coefficient in the expression for the transmission ratio $v_{if} = 0.345$. Hence, we can approximately express the limit of the untwist angle through the strip's transmission ratio:

$$\theta_{lim} \approx \frac{\sigma_e L i_\theta}{2 \times 0.0286\pi E}\,. \qquad (4\text{-}38)$$

This expression is effective for the pretwisted strips with elliptic and rectangular cross sections.

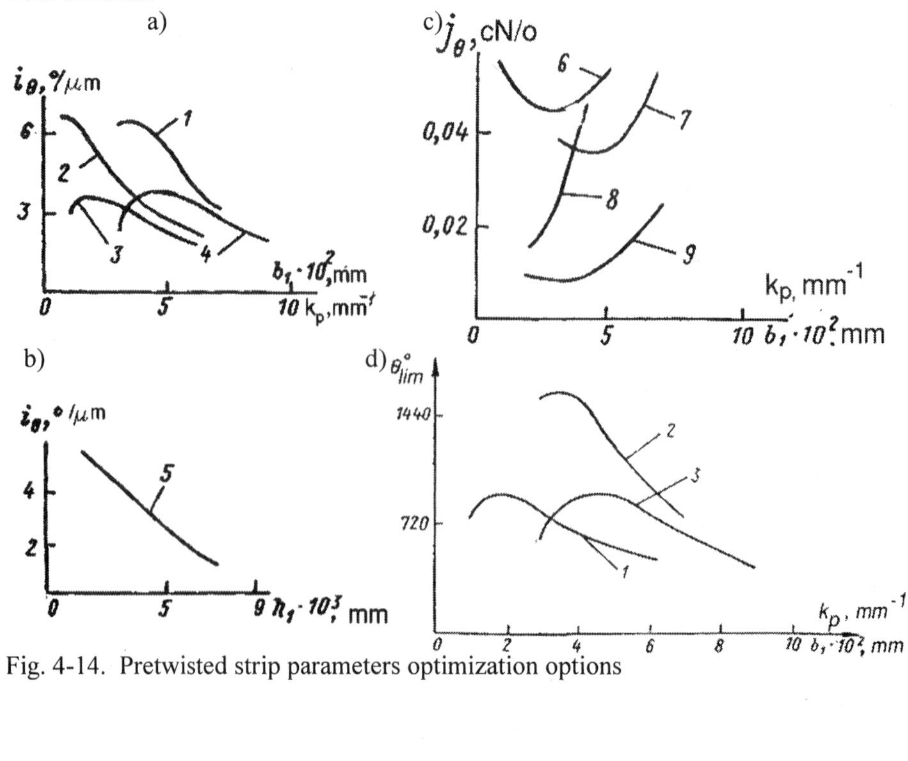

Fig. 4-14. Pretwisted strip parameters optimization options

Optimization. Optimal parameters of the pretwisted strips can be identified from the relationships shown in Table 4-8. From this we can find the partial effective relationships of the transmission ratio and the longitudinal force with thickness $2h_1$, width $2b_1$, helix twist parameter κ_p of the strip, and elastic moduli of material E and G. Functions $i_\theta = f_1(E, G)$ and $j_\theta = f_2(E, G)$ assumed to be linear. We defined the position of the extreme points (*see* Fig. 4-14a, c) on curves $i_\theta(b_1)$, $i_\theta(\kappa_p)$, $j_\theta(b_1)$, and $j_\theta(\kappa_p)$ through the partial differentiation of (33) and (34) at $L_{str} = 0$.
Fig. 4-14 shows calculated relationships of a) $i_\theta(b_1, k_p)$; b) $i_\theta(h_1)$; c) $j_\theta(b_1, k_p)$ for double pretwisted strips with parameters 1-h_1=0.002 mm, k_p=π mm^{-1}; 2- h_1=0.002, b_1=0.06 mm; 3- h_1=0.004, b_1=0.06; 4- h_1=0.004, k_p=π; 5- b_1=0.05, k_p=π; 6- h_1=0.004, b_1=0.06; 7- h_1=0.004, k_p=π; 8- b_1=0.06, k_p=π; 9- h_1=0.0022, k_p=π; d) $\theta^o{}_{\lim}$ as a function of pretwist parameter k_p for strip 1with b=0.12 mm, h=0.008 mm; as a function of the width b_1=1/2 b of the strip's cross section for strip 2 with k_p=π, h=0.004; and for strip 3 with k_p=π, h=0.008.

For the strips with elliptical cross sections, the relationship for the transmission ratio of width $2b_1$ has its maximum at $b_1 = 1.31\sqrt{h_1/\kappa_p}$, the same relationship of the helix twist parameter κ_p has its maximum at $\kappa_p = 1.70h_1/b_1^2$. The relationship for the longitudinal stiffness (Fig. 4-14c) of the strip's width has a minimum at $b_1 = 1.17\sqrt{h_1/\kappa_p}$ and a minimum at the helix parameter $\kappa_p = 2.36h_1/b_1^2$. The relationships for the transmission ratio and the longitudinal stiffness of the microstrip's thickness do not have extreme points that follow from the partial derivatives of these relationships (*see* Fig. 4-14b). Hence, with the change in the microstrip's geometrical and elastic parameters, one can choose during the structural synthesis the microstrip that effectively provides the specified transmission ratio and/or longitudinal stiffness. Moreover, the same transmission ratio or stiffness can be obtained with different combinations of parameters b_1, h_1, k_p. If the strip's cross-section thickness ultimately diminishes to zero, the transmission ratio theoretically approaches infinity, but the longitudinal stiffness approaches zero that excludes transformation options. The pretwisted strip's transmission ratio, indeed, can be obtained in the wide but limited range of $0.8 < i_\theta < 10°/\mu m$, and the longitudinal stiffness can practically be in the range of j_θ = 0.003 to 0.6 cN / ° with cross-section thickness in the range of h = 0.002-0.20 mm, width b = 0.060-0.20 mm, and ratio $b/h = 1/\sqrt{\lambda}$ = 15-25.

As it follows from (38) for θ_{lim}, the number of the crosspiece's revolutions in the elastic range of the strip's deformation is proportional to the strip's length L, its transmission ratio i_θ, and yield stress σ_e of the material, but inversely proportional to Young's modulus of elasticity E. Graphs of $\theta^o{}_{lim}$ in angular degrees that depend on the width, thickness, and parameter of the strip's pretwist κ_p are shown in Fig.4-14d for micro strips of 4-*Tin* 3-*Zinc* bronze with L =18mm. In this case, curve 3 corresponds to the strips with κ_p = π mm^{-1}; h = 0.008 mm; curve 2 corresponds to the strips with κ_p = π mm^{-1}; h=0.004 mm, and curve 1 corresponds to the strips with b=0.12 mm; h=0.008 mm. The position of the maximum-value points on these curves is found through the partial derivatives of (37):

$$\partial\theta_{\lim}/\partial h_1 = -\frac{\sigma_e h_1 k_p L / b_1^2}{4\pi G[(h_1^2/b_1^2) + 0.379\, b_1^2 k_p^2]^2}; \quad \partial\theta_{\lim}/\partial b_1 = -\frac{\sigma_e k_p L\left(-\frac{2h_1^2}{b_1^3} + 170 b_1 k_p^2\right)}{8\pi G(h_1^2/b_1^2 + 0.379 b_1^2 k_p^2)^2};$$

$$\partial\theta_{\lim}\Big/\partial\kappa_p = \frac{\sigma_e\, L(h_1^2/b_1^2 - 0.379 b_1^2 \kappa_p^2)}{8\pi\, G(h_1^2/b_1^2 + 0.379 b_1^2 \kappa_p^2)^2} \qquad (4\text{-}39)$$

By equalizing these expressions to zero, we can find that the relationship between θ_{lim} and the cross-section width has a maximum at $b_1 = 1.28\sqrt{h_1/\kappa_p}$, and the relationship between θ_{lim} and pretwist parameter κ_p has a maximum at $\kappa_p = 1.623\, h_1/b_1^2$. Relationship between θ_{lim} and the cross-section thickness $2h_1$ does not have extreme points. The limiting number of the working elastic untwist turns θ_{lim} of the microstrip's crosspiece does not continuously increase with the increase of the number of initial turns on the strip's twisted segment, as one would suppose. Conversely, the limiting number of elastic turns θ_{lim} and the transmission ratio start to decrease after the number of initial turns exceeds a value corresponding to the extreme point. The optimal value of the pretwisted turns on the strip's segment is equal to:

$$n_{v.opt} = \frac{1.8}{2\pi}\frac{h_1}{b_1^2}L + 3. \qquad (4\text{-}40)$$

In this case, of course, the calculated optimal number $n_{v.opt}$ should be larger than $\theta_{lim}/(2\pi)$, where θ_{lim} is expressed in rad. The increase of the pretwisted turns on the strip's segments can increase the linear segment II (*see* Fig. 4-13a,b) of the transmission ratio's function, but this can slightly simultaneously increase the initial nonlinear segment I. As a result, part of segment II can fall in the elastic-plastic zone of the strip's deformation because the large increase of pretwisted turns n_v will decrease θ_{lim}.

To diminish the driving force rate at the specified transmission ratio, one should use relatively thinner strips, i.e., strips with smaller λ at smaller absolute thickness h. These strips have a smaller longitudinal stiffness. For example, the transmission ratio is approximately equal to the strips with cross sections 0.006 x 0.122 mm and 0.008 x 0.12 mm. The longitudinal stiffness for the latter is 1.6 times larger.

Nonlinear transmission. Charts with initial I and quasi-linear II regions for calculated average $i_{\theta av}$ transmission ratios in accordance with nonlinear functions (2-56) and (2-57) for a set of the pretwisted strips with parameters of thickness×width×helix k_p and length of each pretwisted segment 18 mm are shown in Fig. 4-15a. Dash-dot lines correspond to function (2-59) and dashed lines correspond to the strip's elastic-plastic stress state in region III of transformation function. Let us discuss the influence of the driven element's mass clamped to the pretwisted strip's crosspiece on its transmission function. The load Q_N (Fig. 4-15b) from the mass clamped to the crosspiece at the horizontal strip's position is perpendicular to the strip's axis and causes sag s_{sa}, as in any thin string.

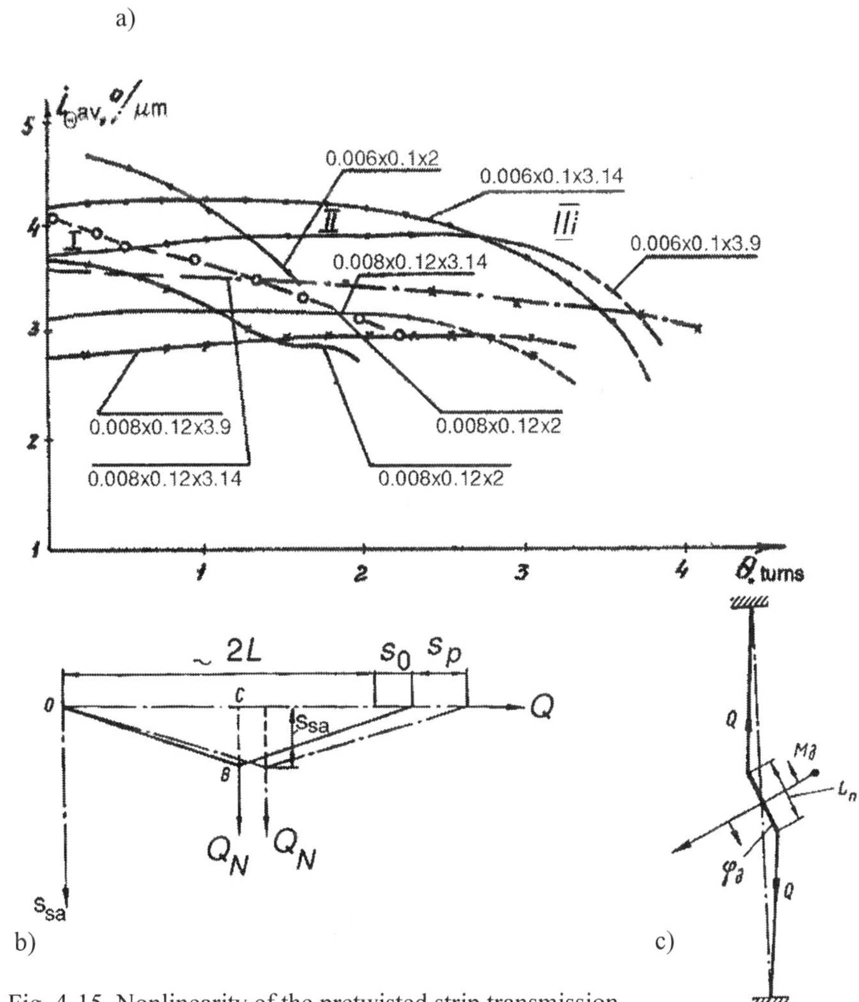

Fig. 4-15. Nonlinearity of the pretwisted strip transmission.

This sag can be decreased by the longitudinal preliminary stretch of the strip. The mechanism of the pretwisted strip will not have a constant transmission ratio with linear characteristics if the sag s_{sa} is large. This can be explained with the relationship that follows from the schematic in Fig.4-15b:

$$Q_N = \frac{2Qs_{sa}}{(L+s/2)} + j(L+s/2)\frac{s_{sa}^3}{(L+s/2)^3};\; j = Q/s_f = j_\theta i_\theta, \tag{4-41}$$

where $s=s_0+s_p$ is the change of the distance between the strip's ends on the straight line connecting them; s_f is the actual elongation of the strip. The relative additional strip's deformation in accordance with the triangle *OBC* is equal to:

$$k_\partial = \frac{s_{sa}^2}{2(L+s/2)^2}, \text{ but } k_\partial = \frac{s_f - s}{2(L+s/2)}.$$

Thus we have $Q_N = 2j_\theta i_\theta s\sqrt{2k_\partial} + j_\theta i_\theta (L+s/2)(2k_\partial)^{3/2}$. (4-42)

The preliminary longitudinal extension (s_0) of the distance between strip's ends should satisfy the condition $2L(k_{\partial 0} - k_\partial) = (s_f - s_p) \leq \{\delta s_n\}$, (4-43)

where $\{\delta s_n\}$ is the permissible additional nonlinearity of the transmission caused by the strip's sag; $k_{\partial 0} = \dfrac{s_f - s_0}{2(L + s_0/2)}$; and s_p is the working relative displacement of the strip's ends. The limit of the allowable transverse load on the strip's crosspiece can be found from (42) with condition $2Lk_\partial + s_0 + s_p < \theta_{lim} / i_\theta$. (4-44)
The torsional unbalance ΔM_∂ caused by the unbalanced load (ΔQ_g) clamped to the crosspiece at the distance r_{un} yields nonlinear deflection of the transmission function in the horizontal strip as a result of acting additional torque proportional to the sine of the crosspiece's rotation angle θ. This leads to additional rotation angle $\delta\theta$ of the crosspiece, which equals $\delta\theta = \pm \Delta M_\partial L / 2 j_{\tau\tau}$, (4-45a)
where $\Delta M_\partial = \Delta Q_g r_{un} \sin \theta$. (4-45b)
This component of the transmission nonlinearity is practically independent from the preliminary extension of the strip. The torque generated on the crosspiece of the microstrip is equal to: $M = j_{\tau k} s_\gamma - i_{\tau\tau}\theta_\gamma$.
The well-balanced mechanism has the torque on a crosspiece, which is close to zero. An additional error in the transmission can be caused by the longitudinal unbalanced load M_∂ (Fig. 4-15c) that is clamped to the crosspiece. We can show this by the following approximation. The strip's equilibrium yields the equality of the moment from the pointer's longitudinal unbalance and the moment from the strip's tension that are acting on the crosspiece with length L_n. This segment of the strip can be considered rigid as a result of clamping a rigid element to it or at a length less than $6b_1$. The equilibrium equation is $M_\partial = j_\theta \theta L_n \sin \varphi_\partial$, from which $\sin \varphi_\partial \approx \varphi_\partial = M_\partial / (j_\theta \theta L_n)$. This rotation of the crosspiece causes additional deformation of the strip's pretwisted segments whose value is equal to $\Delta_\partial = L_n (1 - \cos \varphi_\partial) \approx 0.5\, L_n \varphi_\partial^2$, or after substituting the expression for φ_∂ through the moment of unbalance, we get $\Delta_\partial = 0.5\, L_n^{-1} M_\partial^2 / (j_\theta^2 \theta^2)$. (4-45c)

Fatigue options. In real conditions, the angle of the crosspiece's rotation consists of the constant component θ_0, which corresponds to the preliminary extension of the strip, and the variable component θ_d, which corresponds to the working angle of the crosspiece rotation. In accordance with the theory of an asymmetric cycle loading for metals, the influence of the constant and variable components is not the same. As shown above, the pretwisted strip has a complex stress state with dominating extension ($\sigma_Q = \sigma_Q^{33}$) and torsion ($\tau_Q = \sigma_Q^{13}$) stresses. For the fatigue limit calculation, we will use the following equation:

$$\{\tau_{en}\}^2 = (A_\tau/2) + (B_\tau/2) + \frac{1}{2}\sqrt{(B_\tau - A_\tau)^2 + C_\tau^2}\,, \tag{4-46}$$

where $\{\tau_{en}\}$ is the endurance limit for the material's torsion with the symmetric loading cycle; $A_\tau = 0.125(\sigma_{Q1}^2 - \sigma_{Q1}\sigma_{Q2})$; $B_\tau = 0.125\,(\tau_{Q1}^2 - \tau_{Q1}\tau_{Q2})$;

$C_\tau = 0.25(2\sigma_{Q1}\tau_{Q1} - \tau_{Q1}\tau_{Q2} - \sigma_{Q1}\tau_{Q2})$, σ_{Q1} and τ_{Q1} are the largest working stresses of extension and rotation, respectively; σ_{Q2} and τ_{Q2} are the corresponding smallest stresses. For the pretwisted strip with an elliptical section, we approximately have

$\sigma_{Q1} = (1+0.308\chi_{kp})4G\lambda\theta_1/(\kappa_p L) = U_\sigma\theta_1$; $\sigma_{Q2} = U_\sigma\theta_2$;

$\tau_{Q1} = 2Gh_1\theta_1/L = U_\tau\theta_1$; $\tau_{Q2} = U_\tau\theta_2$, and $\theta_1 = \theta_0 + \theta_d$; $\theta_2 = \theta_0$.

Hence,

$$A_\tau = 0.125\,U_\sigma^2(\theta_1^2 - \theta_1\theta_2) = 0.125\,U_\sigma^2(\theta_d^2 + \theta_0\theta_d); \tag{4-47}$$

$$B_\tau = 0.125\,U_\tau^2(\theta_1^2 - \theta_1\theta_2) = 0.125\,U_\tau^2(\theta_d^2 + \theta_0\theta_d); \tag{4-48}$$

$$C_\tau = 0.5\,U_\sigma U_\tau\,(\theta_1^2 - \theta_1\theta_2) = 0.5\,U_\sigma U_\tau\,(\theta_d^2 + \theta_0\theta_d). \tag{4-49}$$

Substituting the expression of these coefficients into (46), we finally obtain

$$\{\tau_{en}\}^2 = 0.063\,(\theta_d^2 + \theta_0\theta_d)(U_\sigma^2 + U_\tau^2 + \sqrt{U_\sigma^4 + U_\tau^4}). \tag{4-50}$$

Angle $\theta_{eq} = \sqrt{\theta_d^2 + \theta_0\theta_d}$ from (50) may be called the equivalent fatigue angle of the strip's crosspiece untwist. In the real mechanism, the condition $\theta_{lim} > \theta_{eq}$ is required. The torsion stress endurance $\{\tau_{en}\}$ depends on the used material and the technology of the strip's treatment (see Sections 1-3, 2-3).

Any torsion endurance value $\{\tau_{en}\}$ for the pretwisted strip from bronze is connected to the duration of its finite lifetime. This process is quasi-accidental because of the random relationships between the endurance stress (σ_{en} or τ_{en}) and the number N_c of cycles, random values of angles θ_d and θ_0, and may be classified in accordance with the "relaxation model". The frequency functions $\varphi(\tau_{en}, N_c)$ of the conditional distribution for the fatigue curve section at N_c = constant are usually considered Weibull's distribution. The value of the equivalent stress τ_{eq} in the pretwisted strip at the cyclic loading is a random one because it depends on the variable random extension stress τ_u in each cycle and the random stress value τ_0 from the preliminary tension in the mechanism's strip. Therefore, with an equation analogous to the equivalent fatigue angle, we assume that

$$\tau_{eq} = \sqrt{\tau_u^2 + \tau_u\tau_0}\,. \tag{4-51}$$

Values τ_u and τ_0 depend on the adjustment of the structure and its working regime. If the frequency function of the equivalent stress distribution equals $\varphi(\tau_{eq})$, then the probability of an absence of the strip's break $P(N_c)$ at the specified number N_c of loading cycles is approximately equal to:

$$P(N_c) = 1 - A_1A_2, \tag{4-52}$$

where A_1 and A_2 are the hatched areas shown in Fig. 4-16a.

Table 4-10 shows the results of the elastic limit's estimation for strips with different parameters.

We can find the values for the limit of the strip's elastic deformation close to experimental data if we evaluate the reach of plasticity in a whole cross section of the strip under extension by the approximate formula

$$\theta_{lim}^{\circ} \approx \{\sigma_e\}\,A/j_\theta. \tag{4-53}$$

For responsible technical calculations of the elastic limit in the strip's work, one should certainly use expressions (37) and (50).

The pretwisted strip in the mechanism always works in the asymmetric cyclic regime, therefore the author of this book and other researchers have conducted a series of experiments with pretwisted strips installed in the special driving cycling fixtures.

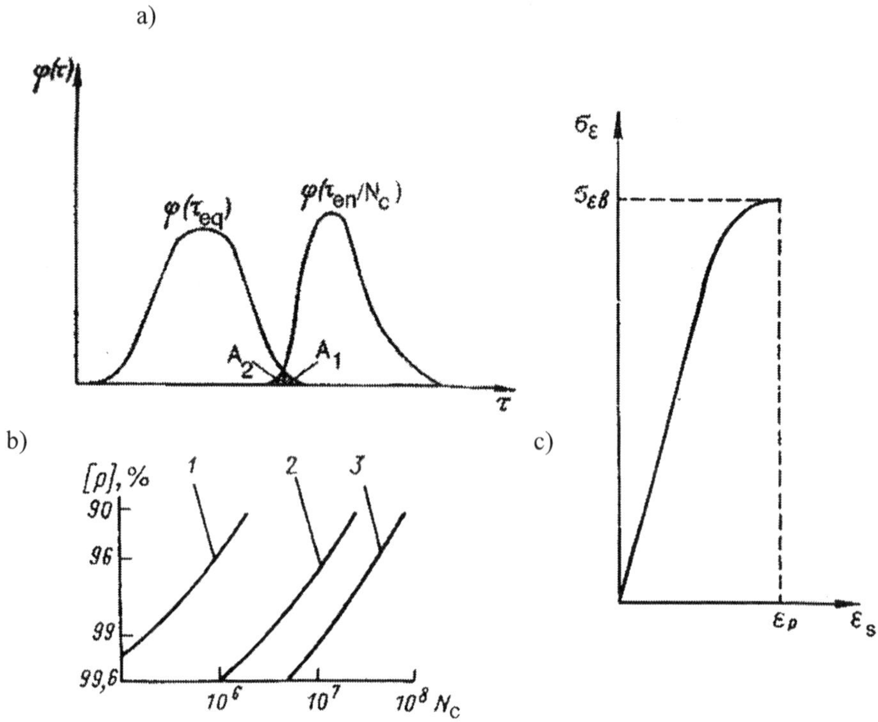

Fig. 4-16. Helicoid's fatigue probability and strain-stress diagram for its blank.

Table 4-10. The limiting and equivalent ultimate crosspiece untwist

Parameters of the strip				Working regime				
				Quasi-static		Cyclic, $N_c = 0.5\times10^7$		
$2b_1$	$2h_1$	L	n_v	$\theta^{\circ}_{\lim}$		θ°_{0}	θ°_{d}	θ°_{eq}
mm				Calculated with(4-37)	Experimental			
0.1	0.006	18	8	1343	1440	270	1080	1205
0.12	0.008	18	9	971	1080	270	830	880
0.20	0.016	18	8	479	486	180	360	405

It was found on this basis that the torsion endurance limit for the strips from zinc bronze is equal to τ_{en}= 147- 176 MPa (*see* Y.M. Tseytlin, 1972). In these tests, we also verified the strips' transmission errors before the cycling procedure and after it. No difference in the transmission errors was found if the strip's crosspiece equivalent rotation angle was less than the limiting angle θ_{lim}. A. N. Grishin in his tests (VNIIMP, Moscow, 1969) obtained endurance stress data σ_{en} = 333- 490 MPa for strips of beryl bronze that corresponds to τ_{en}= 167 – 245 MPa for torsion endurance stress. Fig. 4-16b shows endurance probability [p] experimental data for the micro-strip of beryl bronze *BrB*2 at the stress level 1with stress σ_{en}=450 MPa, 2 withσ_{en}=290 MPa, 3 withσ_{en}=245 MPa (after A. N. Grishin). Moreover, the author of the book has found from the mass production practice that the pretwisted strips have a break after 60 $\times10^3$ cycles or even less if the equivalent number θ_{eq} of the crosspiece turns exceeds the limiting calculated number θ_{lim} for the strip's crosspiece rotation.

Speed break options. Fast deformations of elastic elements are followed by the elastic and plastic waves in them. The simplest differential equation for elastic and plastic wave propagation along the beams is as follows

$$\rho_\gamma \partial^2 s/\partial t^2 = E_s\, \partial^2 s/\partial x^2, \qquad (4\text{-}54)$$

where $E_s = \partial\sigma_g/\partial\varepsilon$ is the modulus of the longitudinal deformation from the diagram of the material extension (Fig. 4-16c); ρ_γ is the density of the material, s is the elastic deformation, and x is a current coordinate for points in the beam's length along with the beam's axis. It is natural that for elastic deformations of material with small internal friction E_s=E, where the latter is the Young's modulus. With the fast displacement of one of the beam's ends while its other end is fixed and immovable, the deformations and stresses in the material can be much larger than that for the equal quasi-static displacement. Speed V_1 of the movable end displacement and the strain of the material are connected by the following relation:

$$V_1 = \int_0^{\varepsilon} s(E_s/\rho_\gamma)^{0.5} d\varepsilon\,. \qquad (4\text{-}55)$$

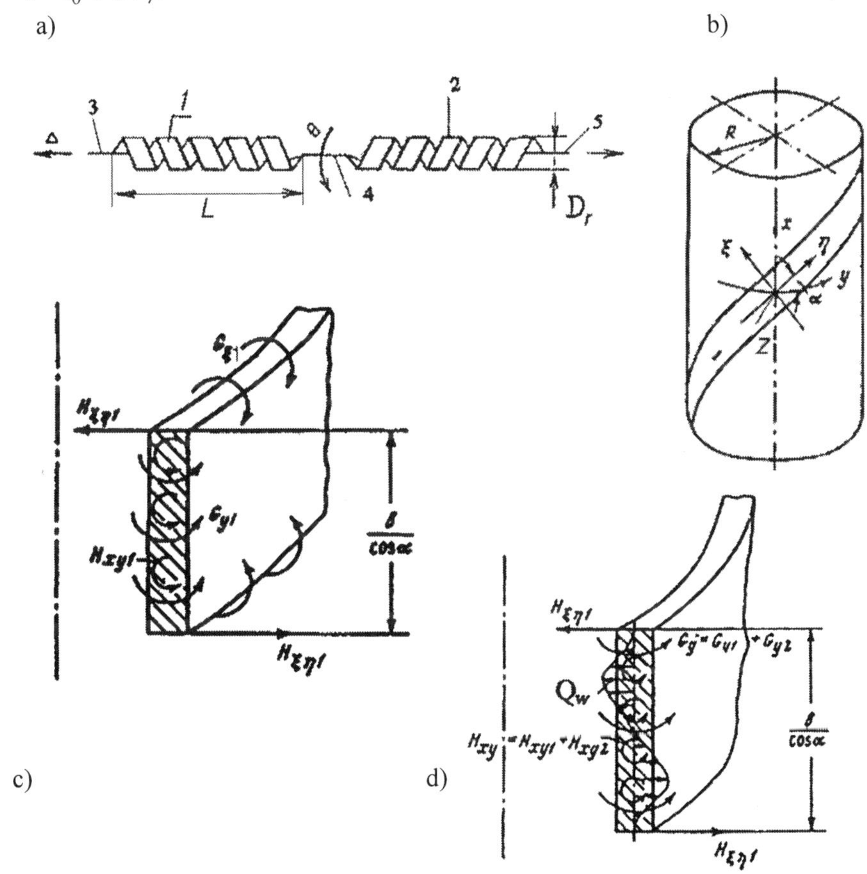

Fig. 4-17. Coiled strip schematic, middle surface, and two stages of deformation (depicted in accordance with V.L. Biderman and V.N. Shitikov, 1963, 1970)

A coiled ribbon spring mechanism.

History. The mechanism was first proposed by W. E. Ayrton and J. Perry (1884) and may be called after their names. This spring is usually considered a helical bar. But this model for double-helicoid (Table 4-8, ELKS of type SM-VII) wide coiled ribbon strips produces incorrect relationships between extension $s=\Delta$ and angular untwist θ of the crosspiece.

Parameters and features. The double-coiled helicoid strip has two parts 1 and 2 (Fig. 4-17a) coiled in opposite directions, straight ends 3 and 5, and the crosspiece 4 which connects the coiled parts and serves as the output device. These helicoids have a large helix angle α ranging from 40 to 70°, and a large ratio between width b and height h of the strip's cross section, such that $b/h = 30$ to 70. From a theoretical standpoint, these coiled strips are studied as cylindrical shells (V.L.Biderman, 1963) that provide more correct formulae for the transmission function in two stages of deformation.

Let us first look at the initial stage of deformation in one part of the coiled strip. Prior to deformation, the middle surface of the strip coincides with the surface of a cylinder of radius R_0, and the edges of the ribbon form the angle α_o with respect to the plane of the cylinder's parallel circle. Fig. 4-17 shows coiled strip schematic (a); middle surface presentation (b), fragments of internal load conditions after the first stage of the ribbon deformation (c), and after the second stage of the ribbon deformation (d). Following deformation, the radius of the cylinder and the angle of the helical line take the values R and α. Coordinate system x, y, z (Fig. 4-17b) is connected to a point of the shell's middle surface, and $y= R\varphi$,, where φ is a central angle. Along with these coordinate axes, orthogonal axes η and ξ are used which are tangent to the helix line and directed as a binormal to it, respectively. The components of displacement for the points on the middle surface in directions x, y, z are denoted as u, v, w. The whole deformation of the spring-shell, in accordance with V. L. Biderman's proposal (1963), has two constituents: 1) deformation without change to the cylindrical shell and without extension of the middle surface; and 2) deformation without change to the pitch and helix angle, but with extension of the middle surface. Thus, $u = u_1 + u_2$; $v = v_1 + v_2$; and $w = w_1 + w_2$. The values of R and α (in accordance with V.L. Biderman *et al.*, 1970) are associated with the spring longitudinal extension s and the mutual rotation angle θ of the ends through relationships that follow from the condition that the developed length L_r of the strip remains constant

$$s= L_r (\sin\alpha - \sin\alpha_0), \tag{4-56}$$

$$\theta = - L_r [(\cos\alpha/R) - (\cos\alpha_0/R_0)], \tag{4-57}$$

where $L_r = L_{r0}=2\pi R_0 \, n_{r0} / \cos\alpha_0$ (4-58)

$n_{r0} = L_r / S_{av}$ is the number of coils in one part of the coiled strip with certain direction of helix lines and average pitch S_{av}. In the first type of deformation, the initial curvatures of the lines x and y are $\kappa_1 = \sin^2 (\alpha - \alpha_0) / R_0$, $\kappa_2 = \cos^2(\alpha - \alpha_0) / R_0$. In the final position, the line x is a straight one, while line y has a curvature $1/R$. The change in curvatures therefore is $\Delta\kappa_1 = -\sin^2 (\alpha - \alpha_0) / R_0$ and $\Delta\kappa_2 = (1/R) - \cos^2(\alpha - \alpha_0) / R_0$. The change in the surface twist, referring to the same coordinates, is $\tau = -\sin (\alpha - \alpha_0)\times \cos (\alpha - \alpha_0) / R_0$. As a result, internal moments H_{xy1}, $G_{\xi 1}$, $G_{\eta 1}$, and $H_{\xi\eta 1}$ (Fig. 4-17c) are induced. Moment $G_{\xi 1}$ should be eliminated from the free sides of the strip by adding the second type of deformation. In this case, the conditions on the free ends of the ribbon strip (Fig. 4-17d) should be as follows

$$G_\xi = G_{\xi 1} + G_{\xi 2} = 0,\ \ - G_{\xi 1} = D\, w_2'' , \tag{4-59}$$

and the transverse force $Q_w = (d/\,d\xi)G_\xi = D\, w_2''' = 0$. (4-60)

In accordance with Lurie's theory for deformation of the thin cylindrical shells (*see* A.I. Lurie, 1947), displacement w_2 on the middle surface of the coiled spring satisfies the following differential relationship

$$w_2^{IV} + 4m_w^{\,4} w_2 = 0, \tag{4-60a}$$

where $m_w = [3(1-\nu_p^{\,2}) / R^2h^2]^{1/4}\cos\alpha$.

Equation (60a) has a similar form with the equation of the axis-symmetrical deformation of the thin cylindrical shells whose solution may be expressed through A.N. Krylov's (Table 4-11) beam functions $Y_1(m_w\xi)$, $Y_2(m_w\xi)$, $Y_3(m_w\xi)$, $Y_4(m_w\xi)$. These functions are connected through the following differential relations $Y_1{}' = -4\, m_w Y_4$, $Y_2{}' = m_w Y_1$, $Y_3{}' = m_w\, Y_2$, $Y_4{}' = m_w\, Y_3$.

Table 4-11. A.N. Krylov's beam functions (*see* A.N.Krylov, 1936)

Type	Formula	Type	Formula
$S(\alpha x)$	$\frac{1}{2}(\cosh\alpha x + \cos\alpha x)$	$T(\alpha x)$	$\frac{1}{2}(\sinh\alpha x + \sin\alpha x)$
$U(\alpha x)$	$\frac{1}{2}(\cosh\alpha x - \cos\alpha x)$	$V(\alpha x)$	$\frac{1}{2}(\sinh\alpha x - \sin\alpha x)$
$Y_1(m_w\xi)$	$S^2(\alpha x) - U^2(\alpha x) = \cosh m_w\xi \cos m_w\xi$;	$Y_2(m_w\xi)$	$S(\alpha x)T(\alpha x) - U(\alpha x)V(\alpha x) = \frac{1}{2}[\cosh m_w\xi \sin m_w\xi + \sinh m_w\xi \cos m_w\xi]$;
$Y_3(m_w\xi)$	$[T^2(\alpha x) - V^2(\alpha x)]/2 = \frac{1}{2}\sinh m_w\xi \sin m_w\xi$;	$Y_4(m_w\xi)$	$U(\alpha x)V(\alpha x) + U(\alpha x)T(\alpha x) = \frac{1}{4}[\cosh m_w\xi \sin m_w\xi - \sinh m_w\xi \cos m_w\xi]$;

Note: here $\alpha x = m_w\xi$; A.N. Krylov's functions may be presented through exponential functions as well because $\sinh(z) = \frac{1}{2}(e^z - e^{-z})$, $\cosh(z) = \frac{1}{2}(e^z + e^{-z})$, $\sin(z) = (1/2i)(e^{iz} - e^{-iz})$, and $\cos(z) = \frac{1}{2}(e^{iz} + e^{-iz})$.

The displacement $w_2(\xi)$ is symmetrical with relation to the middle of the coil's turn $\xi = 0$, therefore it was represented (V.L. Biderman, 1963) by the formula

$$w_2 = A_w Y_1(m_w\xi) + B_w Y_3(m_w\xi). \tag{4-60b}$$

Coefficients A_w and B_w were defined in order to satisfy the boundary conditions (4-60). Hence, on the boundaries of the strip at $\xi = \pm \frac{1}{2} b$ there should be the conditions $w_2'' = - G_{\xi 1}/ D$, $w_2''' = 0$, and coefficients A_w and B_w are connected through the system of equations

$-4\, A_w\, Y_3(\lambda_{co}) + B_w Y_1(\lambda_{co}) = - G_{\xi 1}/ D\, m_w^{\,2}$, $A_w Y_2(\lambda_{co}) + B_w Y_4(\lambda_{co}) = 0$, where $\lambda_{co} = \frac{1}{2} m_w b$; $G_{\xi 1} = -D\{(\Delta k_1 + \nu_p \Delta k_2)\cos^2\alpha + (\Delta k_2 + \nu_p \Delta k_1)\sin^2\alpha - (1-\nu_p)[\sin(\alpha-\alpha_0)\cos(\alpha-\alpha_0)/R_0]\}$. Substituting the expressions for coefficients A_w and B_w obtained from these system of simultaneous equations into (60b) allowed (V.L. Biderman, 1963) to yield the following expressions for the displacement w_2 and its second derivative w_2'' as $w_2 = (G_{\xi 1}/D m_w^{\,2})\{[Y_1(m_w\xi)Y_4(\lambda_{co}) - Y_2(\lambda_{co})Y_3(m_w\xi)] / [Y_1(\lambda_{co})Y_2(\lambda_{co}) + 4Y_3(\lambda_{co})Y_4(\lambda_{co})]\}$, $w_2'' = \dfrac{G_{\xi 1}}{D m_w^2} F(\xi)$, where $F(\xi) = \dfrac{Y_1(m_w\xi)\, Y_2(\lambda_{co}) + 4Y_3(m_w\xi)\, Y_4(\lambda_{co})}{Y_1(\lambda_{co}) Y_2(\lambda_{co}) + 4Y_3(\lambda_{co}) Y_4(\lambda_{co})}$

and $D = Eh^3 / 12(1-\nu_p^{\,2})$ is a flexural cylindrical rigidity. The interior forces in the axial section of the ribbon strip reduce (V.L. Biderman *et al.*, 1970) to a "twisting" moment M^t and a "flexural" moment M^f that are equal to:

$M^f = \int_{-1/2b}^{1/2b} (G_{y1}+G_{y2})\frac{d\xi}{\cos\alpha}$, and

$M^t = H_{\xi\eta 1}\frac{b}{\cos\alpha} + \int_{-1/2b}^{1/2b} (H_{xy1}+H_{xy2})\frac{d\xi}{\cos\alpha} + \int_{-1/2b}^{1/2b} Q\frac{\xi}{\cos\alpha}\frac{d\xi}{\cos\alpha}$. The integrals occurring in the expressions for M^t and M^f are equal to

$\int_{-1/2b}^{1/2b} F(\xi)d\xi = K_\xi b, \quad \int_{-1/2b}^{1/2b} \frac{dF(\xi)}{d\xi}\xi\, d\xi = (1-K_\xi)b$,

where $K_\xi = \lambda_{co}^{-1}(\cosh 2\lambda_{co} - \cos 2\lambda_{co}) / (\sinh 2\lambda_{co} + \sin 2\lambda_{co})$ and $\lambda_{co} = [3(1-\nu_p^2)]^{1/4}\frac{b\cos\alpha}{2\sqrt{Rh}}$. Finally, the "twisting" M^t and "flexural" M^f moments are equal to: $M^f = (Db/\cos\alpha)\{(1/R_0 - 1/R)[1-K_\xi(\sin^2\alpha + \nu_p\cos^2\alpha)^2]$

$- (1-\nu_p)R_0^{-1}\sin(\alpha-\alpha_0)[\sin(\alpha-\alpha_0) - K_\xi(\sin^2\alpha + \nu_p\cos^2\alpha)\sin(\alpha+\alpha_0)]\}$, (4-61a)

$M^t = (Db/\cos^2\alpha)\{(1/R - 1/R_0)\{1-K_\xi[1-(1-\nu_p)^2\cos^4\alpha]\}\sin\alpha + (1-\nu_p)R_0^{-1}\sin(\alpha-\alpha_0)\{2\cos\alpha\cos(\alpha-\alpha_0) + \sin\alpha\sin(\alpha-\alpha_0) - K_\xi\sin\alpha\sin(\alpha+\alpha_0)[1+(1-\nu_p)\cos^2\alpha]\}\}$. (4-61b)

Besides that, $M^t = Q_p R$, where Q_p is the longitudinal stretching force. If flexural moment $M^f = 0$ under spring tension with free rotation of ends, then one can find that

$$\frac{1}{R} = \frac{1}{R_0} + \frac{1-\nu_p}{R_0}\sin(\alpha-\alpha_0)\frac{K_\xi(\sin^2\alpha + \nu_p\cos^2\alpha)\sin(\alpha+\alpha_0) - \sin(\alpha-\alpha_0)}{1-K_\xi(\sin^2\alpha+\nu_p\cos^2\alpha)^2}. \quad (4\text{-}61c)$$

Substituting this expression into (59b), we finally derive sought formula with variable parameter α for untwist angle θ of the coiled strip under large extension as follows

$$\theta = -L_r\left[\frac{\cos\alpha - \cos\alpha_0}{R_0} + \frac{1-\nu_p}{R_0}\cos\alpha\sin(\alpha-\alpha_0)\frac{K_\xi(\sin^2\alpha+\nu_p\cos^2\alpha)\sin(\alpha+\alpha_0) - \sin(\alpha-\alpha_0)}{1-K_\xi(\sin^2\alpha+\nu_p\cos^2\alpha)^2}\right],$$

where $\sin\alpha = \frac{S}{L_r} + \sin\alpha_0$ and $\alpha = \sin^{-1}\left(\frac{S}{L_r} + \sin\alpha_0\right)$. (4-61d)

If quantity λ_{co} is enough large, i.e., the corresponding expression $b_{co}\alpha/\sqrt{Rh}$ is also large, then $K_\xi \sim 1/\lambda_{co} \to 0$ and (61d) reduces to

$$\theta = -L_r\left[\frac{\cos\alpha - \cos\alpha_0}{R_0} - \frac{1-\nu_p}{R_0}\cos\alpha\sin^2(\alpha-\alpha_0)\right]. \quad (4\text{-}61e)$$

Now we assume that with a small longitudinal displacement s the change $\Delta\alpha$ in helix angle α is also small. Therefore, $\alpha = \alpha_0 + \Delta\alpha$ and, in accordance with (59a), $s = L_r[\sin(\alpha_0 + \Delta\alpha) - \sin\alpha_0]$, where $\Delta\alpha \ll 1$.

Hence, $s = L_r\Delta\alpha\cos\alpha_0$ and $\Delta\alpha = \frac{s}{L_r\cos\alpha_0}$. (4-61f)

In addition, $(\cos\alpha - \cos\alpha_0) = -\Delta\alpha\sin\alpha_0$ and $\sin^2(\alpha-\alpha_0) = (\Delta\alpha)^2$. Substituting these expressions into (61e), we obtain the following formula for untwist angle θ at a small longitudinal stretch of the coiled ribbon strip $\theta = \frac{s \sin\alpha_0}{R_0 \cos\alpha_0} = \frac{s}{R_0}\tan\alpha_0$, because the limit of the second member in brackets of (61e) has a very small value, $\lim_{\Delta\alpha\to 0} \frac{1-\nu_p}{R_0}(\Delta\alpha)^3 \sin\alpha_0^2 = 0$. Let us also recall that tan $\alpha_0 = \frac{S_{av}}{\pi D_{ro}}$, where S_{av} is the average helix pitch and $D_{r0}=2R_0$ is the initial diameter of the helix middle surface. Hence, we derive a very simple formula for the transmission ratio under a small extension s of the coiled ribbon strip $i_\theta = \frac{\theta}{s} = -\frac{2S_{av}}{\pi D_{r0}^2}$, which entirely corresponds to the similar formula derived by V.L.Biderman in 1963 with stipulation that the radius of helix surface does not change under small extension, i.e., $R-R_0=0$. For double-coiled ribbon strip this formula transforms to: $i_\theta = -\frac{S_{av}}{\pi D_{r0}^2}$. (4-62)

This expression well corresponds to experimental data at the beginning of extension of the spring. However, the transformation function at a larger extension becomes a nonlinear one (Fig. 4-18a), which corresponds (K.N. Manuilov, 1959) to the quasi-elliptical expression

$$[\theta^2/(2L_r/D_r)^2] + [(L+s)^2/2L_r^2] = 1, \tag{4-63a}$$

where D_r is the helicoid's outside diameter; and L is the length of each coiled part.

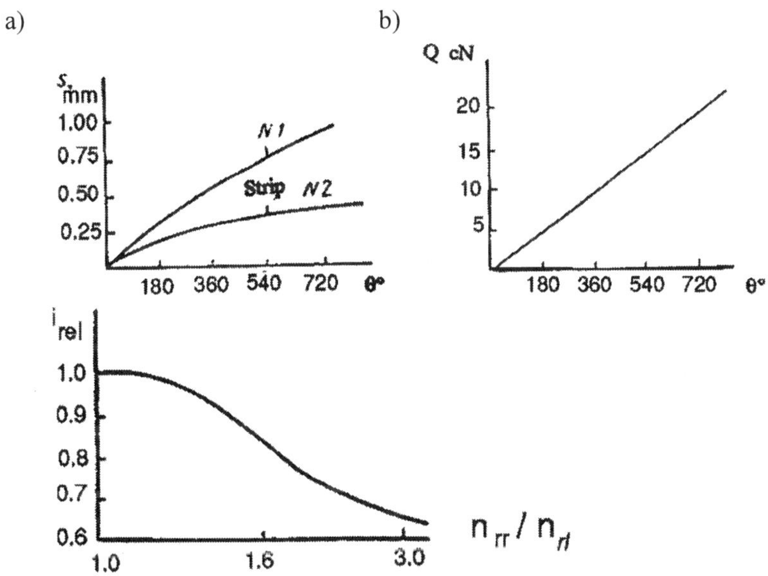

c) Fig. 4-18. Coiled strip kinematics and force functions

Expression (63a) is the canonical equation of an ellipse with semi-axes $\sqrt{2}\, L_r$ and $2L_r/D_r$ in the Cartesian coordinates. Charts in Fig. 4-18a reflect features of strip №1 with D_r=0.2 mm, S_{av}=0.9 mm, b=0.5 mm, h=0.01 mm, and n_r=76, and strip №2 with D_r=0.1 mm, S_{av}=0.7 mm, b=0.3 mm, h=0.007 mm, and n_r=40. Material of the strips is the iron-nickel-chrome-titanium alloy. At the beginning of the crosspiece rotation, the mean transmission ratio (in arc degrees per micrometer) is equal to $i_{\theta m} \approx - 0.0573 S_{av}/\pi D_r^2$, which corresponds well to the above shown theoretical formula (62). Fig. 4-18b shows the relationship between tension force Q and angle of rotation θ°for the double-coiled helicoid which is close to linear with stiffness j_θ (Table 4-8, in cN per angular degree). The limit of the crosspiece's elastic untwist in accordance with K.N. Manuilov (1959) for the strip of iron-nickel-chrome-titanium (*Ni41ChrTi*) corresponds to the empirical expression $\theta_{\lim} = n_r$ (0.15 to 0.33). The transmission ratio of the double-coiled helicoid becomes more linear with an increase in the number n_r of coils. If the number of coils in the part with the right-hand helix n_{rr} and in the part with the left-hand helix n_{rl} is different, then the mean transmission ratio decreases as $i_{\theta m} = - 4\sqrt{n_{rr} n_{rl}}\ S_0/[(n_{rr} + n_{rl})\,\pi D_r^2]$. (4-63b)

Fig. 4-18c shows the relative transmission ratio of double-coiled strip with the different ratio of the coils on the left- and right-hand sides, n_{rr}/n_{rl}. The largest transmission ratio has a double-coiled strip with $n_{rr} = n_{rl}$. Small deflections from 1 in the ratio n_{rr}/n_{rl} changes the coiled strip's transmission ratio insignificantly. For example, if n_{rr}/n_{rl} =0.8, the transmission ratio decreases only by 0.8%. This frees us from the necessity to keep the exact equality between the number of coils with left-hand and right-hand helices. The combined coiled helicoid consists of a strip coiled in one direction and a flat strip attached to its end. The transmission ratio for this system can be expressed as

$$i_{\theta m} \approx -(2\, S_{om}/\pi D_r^2)/(1+j_{k2}/j_{k1}), \tag{4-64}$$

where $j_{k1} = \dfrac{Ebh^3 \cos\alpha}{12\pi(1+\nu_p)Rn_r}\,\dfrac{1}{1+(1-2\nu_p)\cos^2\alpha}$ at K_ξ=0 and $j_{k2} = Gbh^3/3L$ are the torsion stiffness coefficients for the coiled and flat strips, respectively. If ratio j_{k2}/j_{k1} decreases, the absolute value of the transmission ratio increases. Its upper limit value in this case is equal to the transmission ratio of the double-coiled strip:

$$\lim_{j_{k2}/j_{k1}\to 1} i_{\theta m} = -S_{om}/\pi D_r^2\,. \tag{4-64a}$$

The spatial cable system SM-VIII. This mechanism has a quasi-helicoidal form of motion and can also be used as a transmission mechanism. The elastic elements of the system are in the form of thin wires (cables) with diameter d_w< 0.5 mm that are crossed in relation to the axis of rotation for the driven element. Three versions of this system are known. In all versions, one link between driven element 1 (Fig. 4-19a, b, c) and driving slide 3 or stand 4 consists of the crossed wires that convert the translational motion into a rotational one. The link with the rigid non-rotating element can be different: (i) – two wires 2' are crossed (Fig. 4-19a), (ii) two wires 2' are parallel (Fig. 4-19b) in initial position, and (iii) one wire 2' has a central position (Fig. 4-19c).

The transmission ratio of this mechanism is nonlinear (Fig. 4-19d), but its force characteristic is close to linear (Fig. 4-19e). The parameters of this ELKS corresponding to charts I-IV in Fig. 4-19f are as follows: d_w =0.1 – 0.2 mm; X_0/ L_aL_b = 1.95 (chart I); 4.16 (II); 8.32(III); 16.64 (IV).

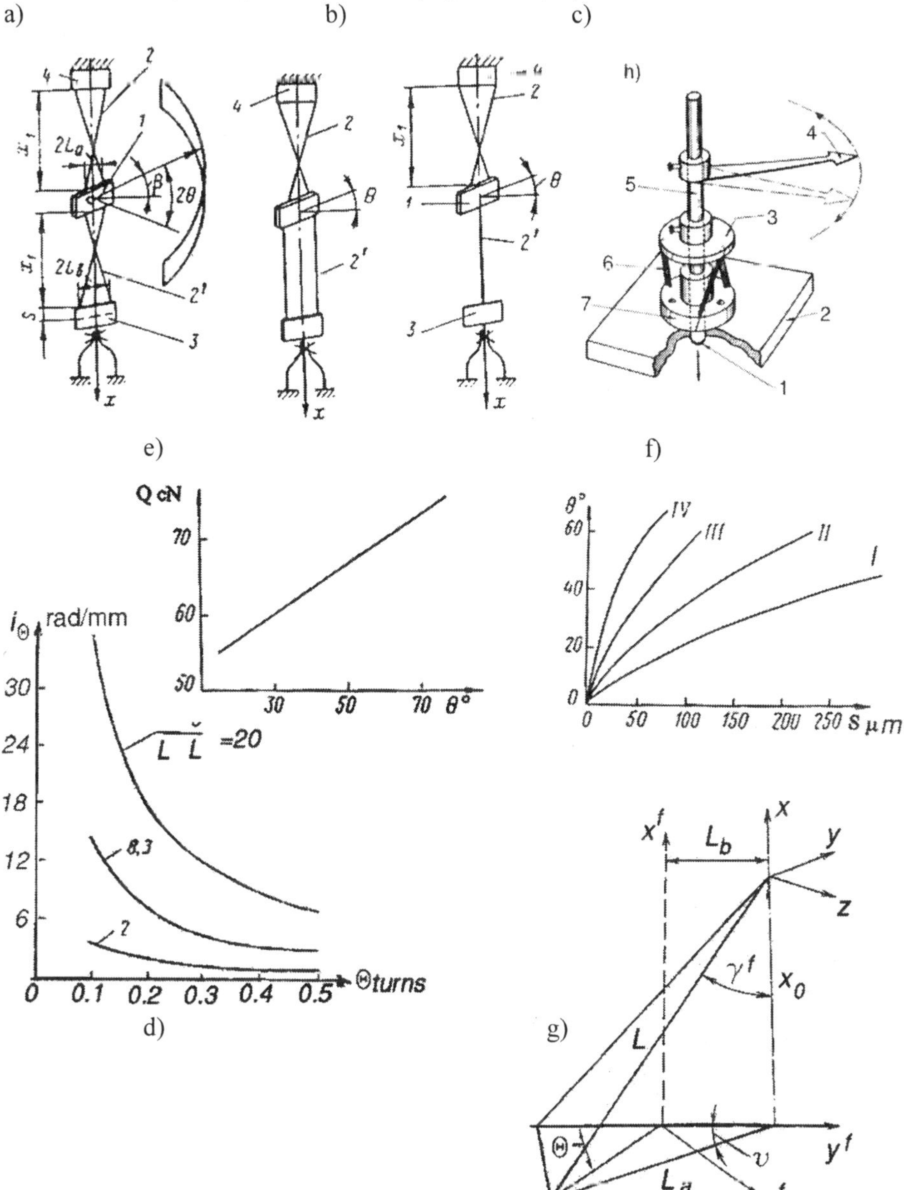

Fig. 4-19. Spatial cable system schemes with functional features and strip's recorder diagram (the latter is after N.P.L., 1956).

To estimate a transmission and force function for this ELKS, let us use the transformation matrix A_{mn} between moving x, y, z coordinates and fixed x^f, y^f, z^f coordinates (Fig. 4-19g). To find the displacement S^f_κ of the rotational element, the coordinate-transformation matrix A_{mn} should be multiplied from the left-hand by the matrix-row S_κ =[$L\cos\gamma^f$; $L\sin\gamma^f$; 0; $\cos\theta$; 0; $\sin\theta$], and from the right-hand side by the column-matrix of the unit-load force $Q_{\kappa e}$ =[000100]. On the basis of the rule for the multiplication of the three fore-mentioned matrices, we have:

$$S^f_\kappa = S_\kappa A_{mn} Q_{\kappa e} = \sum_{j=1}^{m} \sum_{i=1}^{n} S_i a_{ij} Q_j. \tag{4-65}$$

Matrix $Q_{\kappa e}$ has only one member that is not equal to zero $Q_{j=4} \neq 0$, and matrix S_κ has two members that are equal to zero $S_{\kappa 3} = S_{\kappa 5} = 0$. Hence, it is necessary to define only the following members of matrix A_{mn}: a_{14}; a_{24}; a_{44}; a_{64}. In this case, $a_{14} = \cos\theta$; $a_{24} = 0$; $a_{44} = L_b \sin\theta\cos\vartheta$; and $a_{64} = L_b \cos\theta\cos\vartheta$. Thus, by neglecting the small members with s^2 and $(\Delta L/L_0)^2$, we derive the equation for connection of the longitudinal displacement s and the angle θ of rotation for the movable element:

$$s[L_0^2 - (L_a + L_b)^2]^{1/2} + 2L_aL_b(1-\cos\theta) + 2\,L_0\Delta L = 0, \tag{4-66}$$

where $L = L_0 + \Delta L$ is the length of the wire (cord) with accounting for its elastic elongation ΔL under the acting forces. The kinematic transformation function can be obtained as the ratio of the partial derivatives from expression (66) with respect to parameters s and θ (differentiation of the implicit function). If the wires are supposed to be non-extensible, this differentiation yields the formula for the transmission ratio:

$$i_\theta = (X_0 / 2L_aL_b)\,\theta / (1-\cos\theta), \tag{4-67}$$

where $X^2_0 = L_0^2 - (L_a + L_b)^2$ at $\theta = 0$. We derive the longitudinal driving force Q_w for the round four wires at $\theta << 1$ from the principle of virtual displacements of element 1. For the sake of simplification, we consider the wires to be non-stretchable, and we add together the three virtual works for: 1) driving forces of element 3 (Fig. 4-19a) δR_{vQ}; 2) twisting four wires δR_{vtw}, and 3) transverse bending of four wires δR_{vben}, such that $\delta R_{vQ} + \delta R_{vtw} + \delta R_{vben} =$

$$\{Q_s + [\,0.2G\, d_w^4\, \theta X_o /(L\, L_aL_b \sin\theta)] + 1.2\, EL_a\, d_w^4 \theta X_o / L^3 L_b\}\, \delta S. \tag{4-68}$$

$$|Q_s| = [0.2\, G\, d_w^4 \theta X_0 /(L\, L_0 L_b \sin\theta)] + 1.2\, EL_a\, d_w^4 \theta X_0 / L^3 L_b, \tag{4-69}$$

or if $\theta << 1$ and $\sin\theta \approx \theta$, then

$$|Q_s| = [0.2\, G\, d_w^4\, X_0 /(L\, L_aL_b)] + 1.2\, EL_a\, d_w^4 \theta X_0 /(\, L^3 L_b), \tag{4-70}$$

and

$$Q_w = Q_s / 2\cos\gamma^f, \tag{4-71}$$

where γ^f is the angle at which the wires cross.

The derivative of (70) yields the expression for driving rigidity of this mechanical transformation mechanism j_θ (*see* Table 4-8). Another spatial mechanism with quasi-helicoid motion is based on the system of three oblique spring-strips 6 (Fig. 4-19h) clamped to disk 7 which is bolted to base plate 2. Other ends of strips 6 are clamped to disk 3. Operating spindle 5 with anvil 1 moves together with clamped to it disk3 and pointer 4, which proportionally rotates. Mechanical amplification of 50 times of the small axial movement of the spindle was easily obtained with strips 0.063 mm thick, 3.2 mm wide, and 15.1 mm long mounted at an inclination of 20 ° in disks of 19 mm diameter. The separation of the inside faces of the disks 3 and 7 equals 9.5 mm. This mechanism was developed by G.A. Tomlinson in N.P.L., UK (1956) for waviness recorder.

Transformability.

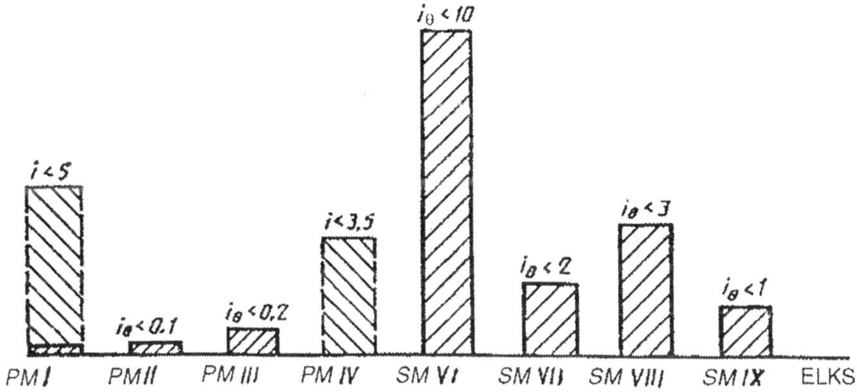

Fig. 4-20. Typical upper limits of transmission ratio in plane and spatial ELKS.

Comparison of the realized transmission ratios for different planar *PM* (Table 4-5) and spatial *SM* (Table 4-8) elastic mechanisms (Fig. 4-20) proves the conclusion made in Chap.2 that the greatest transmission ratio (i_θ in angular degree per μm or dimensionless i) in the straight chain of linear motion transformation has a thin pretwisted double-helicoid in ELKS of type SM-VI which correspondingly has the greatest sensitivity to displacements and $i = s_{out}/ s_{in}$ (*see* Chap. 8).
The PM-III mechanism has $i=(f_k-f_{k0})/s_1$ ratio up to 200 (*see* Table 4-5 with reference to M.T.A. Saif *et. al.*, 1998) and minimal structural stiffness as force corrector $j = \delta F/(f_k - f_{k0}) \approx 0.018 \times 10^{-3}$ N/m, where f_{k0} is the initial sag.

4-4. Couplings, Correctors, and Compensators

Couplings, correctors, and compensators are an important part of precision elastic structures. They are necessary for adjustment of initial and working parameters with the aim to minimize bias and random uncertainties and eliminate to the possible extent negative action of the influence quantities. The difference between correctors and compensators is conditional. The former are supposed to be controllable in the running fashion. The latter are working automatically after some adjustment at calibration. Elastic couplings represent one type of the kinematic compensators which are necessary to compensate the movable elements disaliment. Table 4-12 shows the main principles on which the work of correctors and compensators is based.

Kinematical correctors. KC-II Cover plate corrector. The cover plates that exclude a part of the flexure's body from the work are more specific for ELKS. In this corrector, elastic strip 2 (Fig. 4-21a) is tightened between the side of the slot in bracket 1 and rigid insert 3, which is pushed by a screw from the opposite side of the slot. If the clamping device is positioned in the middle of a strip near the strip's inflection point, its action is not significant. Conversely, the clamping device has a significant effect on the trajectory of the guiding carriage if it is positioned near the strip's ends where the curvature of the strip is the largest. This adjustment is fast, reversible, and practically continuous because it does not inflect any damage on the strip. It allows one to reduce pitch angle ϑ (angular seconds) of the trajectory in the spring parallelogram by 5 – 50 times (*see* R.V. Jones, 1988) that is seen from the comparison of curves I (with corrector) and II (without corrector).

Table 4-12. Couplings, correctors, and compensators base principles

Correction/compensation target	Structural specifics, examples
Kinematics	Adjustable cantilevers, plates, diaphragms, bellows, cables (Sect. 4-4)
Force compensation	Flat and curved tapes, levers, spiral and sine springs (Sect. 3-1, 4-4,Chap.8)
Orientation in gravitational field	Weight balances (Chaps. 4, 8)
Temperature	Bimetallic and shape memory material (Chap. 7)
Forced vibration, speed variation	Dampers, isolators, magnetic stabilizers (Sect. 4-5, Chap. 7, 8)
Air pressure	Differential function (Chap. 3, 7)
Air velocity, humidity, dust	Polymers, coatings, thin strips deflection (Chap. 7)
Electromagnetic field	Shields (Chap. 7)
*Electrostatic	Ionizers with high voltage and low radiation
Radiation	Shields, neutralizers (Chap. 7)
*Compound random ambient influence	Parallel structures, cantilevers, differential system
Energy supply	Stabilizers (*see* P.V. Novitskii *et al.*, 1990)

* *See* details in Chaps. 7and 8.

Changeable compliance. The work of the elastic correctors is based on the reduction of motion owing to the internal compliance feedback. They usually have the structure of cantilever strips 1 (Fig. 4-21b,c), which allow us change their compliance smoothly by screw 2 or by steps that correspond to the distance (Fig. 4-21c) between stop screws (multi-stop option). The requirements that are specified for these systems concern with the necessary sensitivity and the range of adjustment of the mechanism's transmission ratio *i*. These requirements for the adjustment sensitivity of a cantilever strip with adjustable length can be represented as follows:

$$(\partial i \,/\, \partial L_1)\, \Delta\, L_1 < \{\Delta i\}, \quad (4\text{-}72a)$$

where $\{\Delta i\}$ and ΔL_1 are the small values that characterize the necessary threshold of the transmission adjustment and the minimum possible change in the cantilever's

length, respectively. We have a requirement for the necessary limit of the transmission ratio adjustment δi: $(\partial i / \partial L_1) L_1 \geq \delta i$. (4-72b)
The elastic corrector can be linear or nonlinear. A cantilever strip may be considered linear when the displacement of its end is within $\pm$ 0.1 L_1. The nonlinear corrector may be based on the coiled strip (Table 1-3, ELKS of type SM-VII), spring parallelogram (Table 1-2, ELKS of type TG-I), elastic-lever mechanisms (Table 1-3, ELKS of type PM-I*el*), bowed beam (*see* notes about its dynamics in section 4-5), etc.

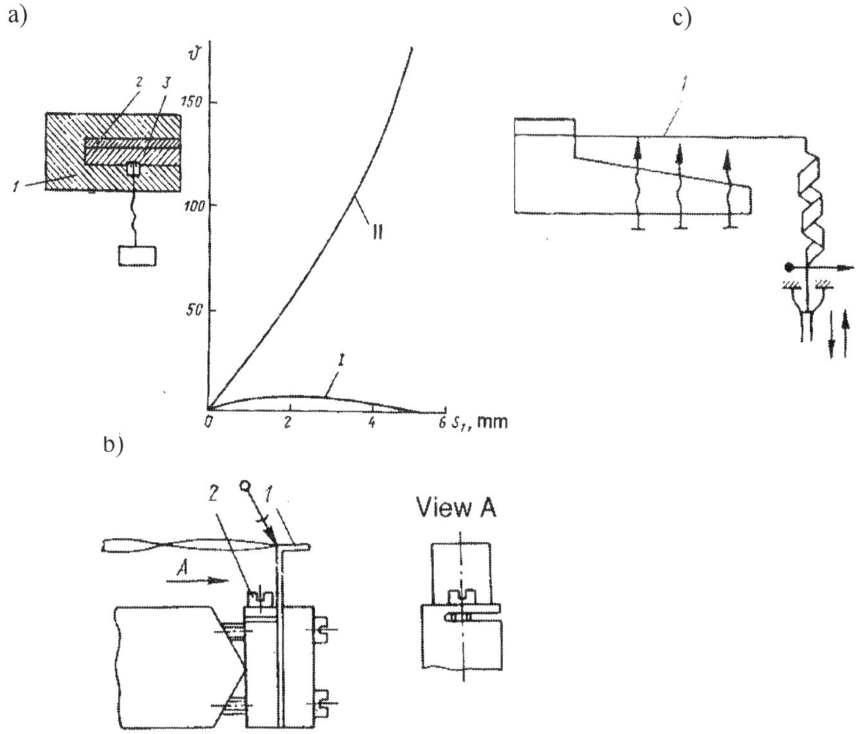

Fig. 4-21. Kinematical correctors and compensators.

Force correctors

Fig. 4-22e shows an example of the simplest force corrector providing two swichable measurement force values on the measuring spindle 1. Additional spring 2 acts through flexible link 6 and lever 4, which is connected with cam switch 5. A lesser measuring force is provided by spring 3 connected to the spindle permanently.

FC-Ipl The structural schematic of the *force pull-spring-lever* compensator with negative rigidity ($Q_g < 0$; $dQ_g > 0$; $j = dQ_g / ds < 0$) is shown in Fig. 4-22a. Lever 2 is installed on the elastic hinge 3 and kinematically connected with the translational driving element 1. Cylindrical extension spring 5 and counterbalance 4 with mass *m* are clamped to the lever's arms. The function of this corrector is based on the automatic change of arm a_{nz} of spring 5 action on lever 2 at its rotation. As a result, the reaction force Q_R on driving element 1 and its change δQ_R can be made small enough ($Q_R \to 0$; $\delta Q_R \to 0$). If the spatial position of ELKS changes, the compensation of the change in the components' gravity force occurs automatically due to changes in the counterbalance 4's moment acting on lever 2. Let us write the calculation relationships after replacing the single-strip hinge by a point-spring element with flexible center C_z. In this case, we have

$\theta_{c_z} = \sin^{-1}[(r_{z1} \sin\vartheta_{1z} + s) / r_{z1}]$;

$$\Phi_{Az} = 180° - (\psi_{Az} + \upsilon_{1z} + \theta_{c_z});\ a_{nz} = \frac{r_z r_{z2} \sin\Phi_{Az}}{\sqrt{r_z^2 + r_{z2}^2 + 2 r_z r_{z2} \cos\Phi_{Az}}};\ \Delta Q_g = \frac{Q_{spr}(a_{nz} - a_{nz0})}{r_{z1} \cos(\vartheta_{1z} + \theta_{c_z})},$$

where θ_{c_z} is the angle of lever 2's rotation around the flexible center at displacement s of driving element 1; ϑ_{1z} is the angle of the lever's initial position; a_{nz} and a_{nz0} are the end and initial value of the spring force Q_{spr} action arm on lever 2; ΔQ_g is change in the force with which the lever acts on the driving element. Index z marks the angles and the radius-vectors relative to flexible center C_z. We can find the flexible center's current coordinates and the pole of rotation as well using the formulae in Table 3-4 and (3-37) as the functions of the hinge parameters.

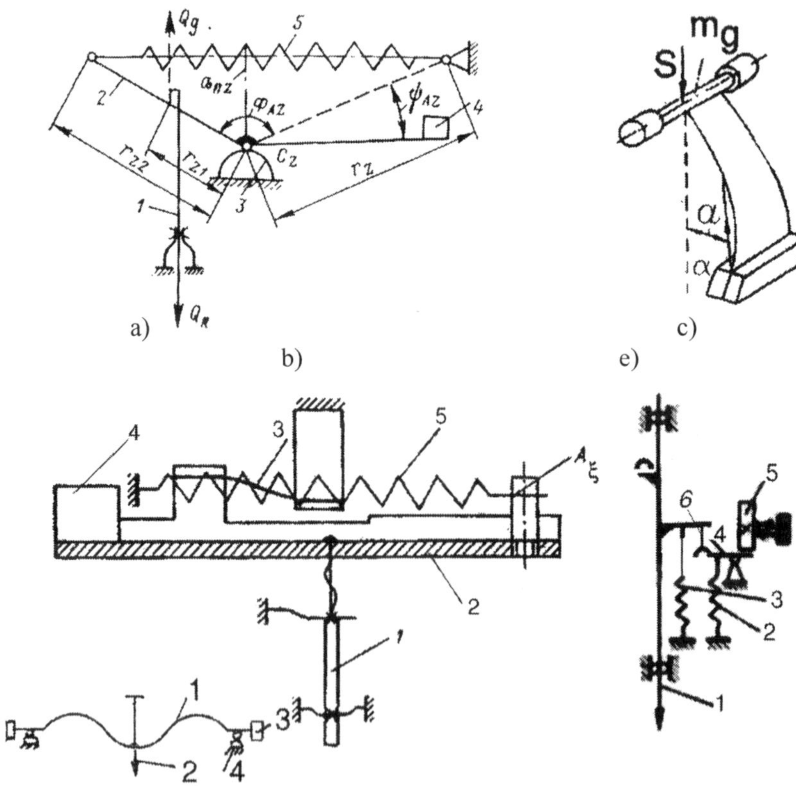

d) Fig. 4-22. Typical force compensators and correctors.

The position of the lever's axis of rotation is assumed to be unchangeable in the case of a mechanism with the rigid elements of kinematic pairs (K. Mahr. Patent Germany 802545), but all of the above shown relationships are applicable to the rigid kinematic structure also.

The pivot of the corrector's lever should provide reliable and stable work under relatively large values of force Q_{spr} because by increasing this force we can compensate for the small difference in the arms of its action ($a_{nz} - a_{nz\,0}$). The latter is typical at a small displacement s. If the pivot has external Coulomb friction, the rotational angle of the lever should be larger than the angle of friction in its pivot ($\theta_z > 0.07$). This condition may be avoided in the precision rigid structures with $\theta_z < 0.02$. In addition, it is necessary to decrease the initial component $a_{nz0}Q_{spr}$ of the compensating moment that can be achieved if the axis of the spring in the initial position is close to the axis of the lever's pivot ($a_{nz0} \rightarrow 0$). This allows us to provide a small nonlinearity for the compensating moment with respect to the angle of rotation because the position is close to an extreme one.

Structure. Fig. 4-22b shows the structure of the discussed corrector that is close to the above described requirements. The mutual position of point A_ξ in which spring 5 is clamped to the lever and its pivot 3 provide the longitudinal extension of the strip in the elastic pivot (ELKS of type RG-I), which guarantees its working stability. The thickness of the strips in this pivot may be minimal with respect to the extension and flexure strength. The resistance to rod 1's motion in the kinematic chain usually increases in proportion to displacement s, $Q_r = Q_{r0} + j_m s$, where j_m is the stiffness of the mechanism related to the driving element. But arm of the acting force in the corrector also increases with the following rotation of the lever 2. As a result, the lifting force $Q_g = Q_{g0} + \Delta Q_g$ increases, and thus diminishes the acting force on the rod, where the resulting force $Q = Q_r - Q_g$. Balance 4 clamped to the lever provides compensation at changeable orientation of the structure in the gravitational field. At the adjustment of the described corrector, it is necessary first to apply the longitudinal extension force to the pivot's strip. This sequence is important because the system is not linear and does not follow the principle of superposition.

FC-Ipu The analogous system may be built on the principle of the loading force-push corrector with negative stiffness ($Q_g > 0$; $dQ_g < 0$; $j = dQ_g / ds < 0$). However, in the latter case it is only possible to minimize the increment of the acting on the rod force dQ_r, but the value of the acting force is not diminished.

FC-II, III. The approximately constant force of deformation ($Q \approx$ constant) has a *buckled spring* and a *thin buckled strip* with the changeable arm a (Fig. 4-22c) of the force action at displacement s. However, in this system, in accordance with M. Born's (1956) research, there are possible bifurcations in the strip's form, which depends on the weight m_g, angle of clamping α, rigidity, and the length of the strip. It is necessary to keep strip's clamp in a definite orientation in the gravitational field to eliminate ambiguity of this corrector's function.

FC-V. Sine-strip corrector may provide force stabilization. Strip 1 (Fig. 4-22d) that is flat in the initial state gets sinusoid form at its clamping in the structure. It is possible to provide positive, negative, or close to zero stiffness of this system in the direction of rod 2 displacement by changing of the strip's ends angular position within 3-5° that causes the change of the form of strip's equilibrium. The version of this structure has two (Patent GB 617076) and three (Patent SU 262448) semi-waives with equal curvature.

The principle of tension and bending moment dependence was used by F.S. Eastman (1935) in flexure pivots with four parallel strips for achieving self-compensating restoring moment by preliminary adjusted tension which is increasing in two and decreasing in another two strips after application of external load.

4-5.Dynamic Functions of the Amplitude Transducers

Dynamic properties. The dynamic functions of the amplitude transducers can be linear and nonlinear. The former once may be characterized by more simple methods through transient and transfer functions.

Acceleration. If the movable carriage is in planar transversal motion, then we can find an instantaneous center of acceleration (*ICA*) on the moving plane with the carriage. This center O_a is positioned at the point of intersection between rotational circle $d = \overline{P_1P_2}$ and the straight line which passes through pole P_1 under angle α_a such that $\tan \alpha_a = |\dot{\Omega}|/\Omega^2$, where Ω *and* $\dot{\Omega}$ are the angular speed and angular acceleration of the carriage, respectively. The *ICA* has neither a tangent ($w_t = 0$) nor a normal ($w_n = 0$) acceleration. The acceleration in any point B_a on the moving body is proportional to the point's distance $R_{OB} = \overline{O_aB_a}$ from the *ICA*,

$$w_B = R_{OB}\sqrt{\dot{\Omega}^2 + \Omega^4} \ . \tag{4-73}$$

Dynamic characteristics. The linear system's transient response $\theta_e(t)$ can be used for the estimation of motion s_{out} of the driven output element by convolution if motion s_{in} of the input element is specified: in the operational imaging form,

$$s_{out}(\lambda) = \lambda \ \theta_e(\lambda)\, s_{in}(\lambda), \tag{4-74a}$$

where $s_{out}(\lambda)$, $s_{in}(\lambda)$, and $\theta_e(\lambda)$ are the output, input, and transient function's Laplace transforms with argument λ, respectively; and in the regular form of Duhamel's superposition integral, $s_{out}(t) = \int_0^t \ddot{\theta}_e(t - t_{st})\, s_{in}(t_{st})\, dt_{st}$. (4-74b)

Periodic forcing function. The bowed beam corrector is similar to the bridge mechanism. Its transmission ratio increases with the decrease of beam's sag. However, the decrease of beam's sag has some limitations. This follows with the increase of the driving force, larger compression strain, and possible bifurcation and loss of the motion stability if the driving force is a periodical one. The latter phenomenon is similar to Mathieu's model (*see* N.W. McLahlan, 1964) shown in Fig. 4-23a.

The dynamic differential equation for this beam (strip) structure,

$m\ddot{y} + C_v\dot{y} + jy = (F_0 \cos 2\omega t)y$, reflects an option for instability at certain parameters of the beam- strip spring rigidity j, viscose resistance C_v per unit velocity of the transverse beam's motion, circular frequency ω, and amplitude F_0 of driving force. This equation corresponds to the standard Mathieu's equation with stable and unstable regions $\ddot{y} + 2k_z\dot{y} + (\bar{a}_z - 2q_z \cos 2z)y = 0$, where $k_z = C_v / 2\omega t$, $\bar{a}_z = j/\omega^2 m$, $z = \omega t$ is the modulation, and $2q_z = F_0/\omega^2 m$.

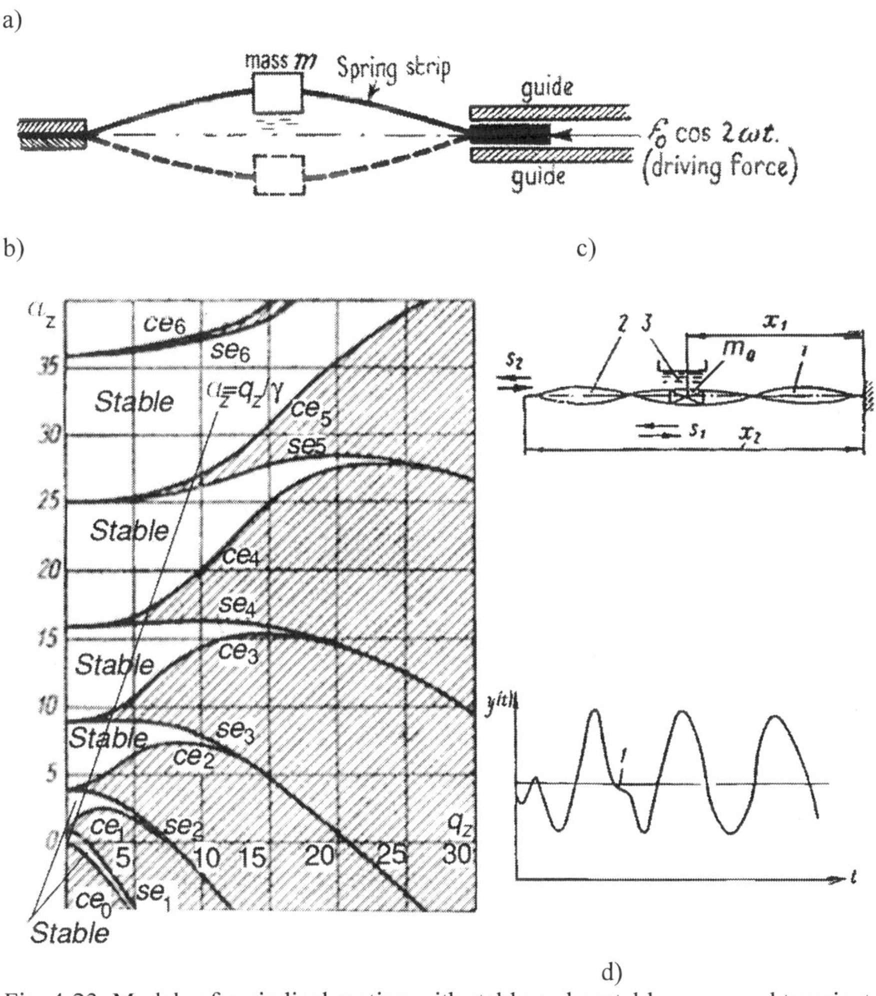

Fig. 4-23. Models of periodical motion with stable and unstable zones and transients in dynamics.

Fig. 4-23b shows an example of the stability Strutt diagram (M.J.O. Strutt , 1932) at C_v=0 with Mathieu's functions ce_n and se_n of the n-th order (after N.W. McLachlan, 1964). The hatched zones in Fig. 4-23b represent the unstable regions for the

equation with $q_z = \bar{a}_z\ \gamma_z$, where $\gamma_z = F_0/2j$. We assume that viscose damping k_z in the bowed beam application is very small. The influence of viscose resistance on the dynamic stability in Mathieu's model is shown in Chap. 5.
SM-VI dynamics. The most sensitive pretwisted strip mechanisms have been used in amplitude transducers for the measurement of length, temperature, force, and other parameters. The connection between the deformation and developing forces

$$Q_1 = j_{kk}\frac{s_1}{L} - j_{\tau k}\frac{\theta}{L};\qquad (4\text{-}75a)$$

$$Q_2 = j_{kk}\frac{s_2 - s_1}{L} - j_{\tau k}\frac{\theta}{L};\qquad (4\text{-}75b)$$

$$M_1 = j_{k\tau}\frac{s_1}{L} - j_{\tau\tau}\frac{\theta}{L};\qquad (40\text{-}75c)$$

$$M_2 = j_{\tau k}\frac{s_2 - s_1}{L} - j_{\tau\tau}\frac{\theta}{L},\qquad (4\text{-}75d)$$

where Q_1 and Q_2 are the longitudinal forces; M_1 and M_2 are the torsion moments developing in segments 1 and 2 of the strip, respectively; s_1 is the longitudinal displacement of mass m_Q on the crosspiece which is equal to the elastic deformation of the first strip's segment; and s_2 is the similar deformation for the second strip's segment connected to the driver. The rigidities of both segments are considered equal. The strip's mass m_s is practically small in comparison with mass m_Q clamped on it. Therefore, we neglect the influence of mass m_s in this discussion. This mass is effectively used in Chap.5 for research of the multivibrator functions. Let us express the differential equations for longitudinal-torsion displacements of mass m_Q in the following form:

$$m_Q\ddot{s}_1 = j_{kk}\frac{s_2}{L} - 2j_{kk}\frac{s_1}{L} - C_s\dot{s}_1,\qquad (4\text{-}76a)$$

$$J_Q\ddot{\theta} = 2j_{\tau\tau}\frac{\theta}{L} + j_{k\tau}\frac{s_2}{L} - C_\theta\dot{\theta},\qquad (4\text{-}76b)$$

where J_Q is the moment of inertia of the movable mass; C_s is the coefficient of viscose resistance to the translational motion in damper 3; C_θ is a similar coefficient in the rotational motion in the same damper. These are the linear differential equations of the second order with separated variables.

Transient function. This separation of variables allows us to find the transient function $\theta_e(t)$ by the solution of differential equation (76b) at $s_2(t)=sU(t)$, where unit-step $U(t) = \int_{-\infty}^{+\infty}\delta(t)dt = 1$.

Therefore, $J_Q\ddot{\theta} + \frac{2j_{\tau\tau}}{L}\theta + C_\theta\dot{\theta} = \frac{j_{k\tau}}{L}sU(t)$. (4-77)

The solution of this equation can be derived by the Laplace-transform method with initial zero values $\theta(0) = 0; \dot{\theta}(0) = 0$ as follows

$$\pounds\theta_e(t) = \theta_e(\lambda) = \frac{j_{k\tau}}{L}s[\lambda(J_Q\lambda^2 + C_\theta\lambda + \frac{2j_{\tau\tau}}{L})]^{-1}.\qquad (4\text{-}78)$$

The roots are correspondingly equal to:

$\lambda_1 = 0$; $\lambda_{2;3} = -[C_\theta/(2J_Q)] \pm \sqrt{\left(\frac{C_\theta}{2J_Q}\right)^2 - \frac{2j_{\tau\tau}}{J_Q L}}$. The following partial cases influence the system's transfer function, which depends on the relationship between the values of mass moment of inertia, specific torsion stiffness ($j_{\tau\tau}$/L) and coefficient C_θ of the viscose resistance:

1. $(C_\theta/2J_Q)^2 > 2j_{\tau\tau}/J_Q L$ where transfer function is with all real roots;

2. $(C_\theta/2J_Q)^2 < 2j_{\tau\tau}/J_Q L$ with two imaginary roots;

3. $(C_\theta/2J_Q)^2 = 2j_{\tau\tau}/J_Q L$ with two equal real roots. The corresponding transient functions see in Chap. 8.

Frequency response. When we look for the solution of a differential equation similar to (77), but with frequency forcing function

$$A_\Omega \ddot{y} + B_\Omega \dot{y} + C_\Omega y = F_m \cos(\Omega_1 t + \psi_\Omega), \tag{4-78a}$$

we find that this solution has two parts:

$$y(t) = y_{st}(t) + y_{fr}(t), \tag{4-78b}$$

where $y_{st}(t)$ is a stable part of the solution; $y_{fr}(t)$ is a free transient part of the solution. Chart $y(t)$ corresponding to (78a,b) is shown in Fig. 4-23d. One can see step 1 on the descending segment of the first semi-wave. This was noted in the experiments with the impulse transducers based on the pretwisted strip mechanisms (see Chap.8).

SM-VI versus SM-VII in dynamics. The stiffness of the double-coiled strips (ELKS of type SM-VII) and their stability under dynamic loading depends on the relationship between the angle of crosspiece rotation and the induced on it torque. K N. Manuilov used a conditional radius for the comparison of different strips' performance in dynamics. This radius is equal to the ratio between the induced torque and the applied extension longitudinal force when the crosspiece rotates on the specified angular value. The conditional radius ρ_c can be used to compare ELKS of type SM-VI and ELKS of type SM-VII in their ability to convert certain part of the longitudinal force to the torque. The conditional radius for the pretwisted strips can be calculated by the following formula (Y.M. Tseytlin, 1972)

$$\rho_c = \frac{j_{k\tau} - j_{\tau\tau} i_\theta}{j_{kk} - j_{\tau k} i_\theta}. \tag{4-79}$$

We obtain it by dividing corresponding expressions of torque M and stretching force F_Q presented through the three partial rigidity parameters $j_{k\tau} = j_{\tau k}$, $j_{\tau\tau}$, and j_{kk}, on the basis of (2-35) and (2-36) as follows: $M/F_Q = [j_{k\tau} s_\gamma - j_{\tau\tau}\theta_\gamma]/[j_{kk} s_\gamma - j_{\tau k}\theta_\gamma]$.
Let us also recall that $\theta_\gamma/s_\gamma = i_\theta$ which represents the strip's transmission ratio. Samples of the pretwisted strips regular used in precision instrumentation have ρ_c in the order of 0.001-0.004 mm. The double-coiled strips have ρ_c within the range of

$\dfrac{M^t}{Q_{pM}} \approx R$=0.05 to 0.1 mm as showed their tests. This means that the pretwisted strips (ELKS of type SM-VI) under extension transform longitudinal force into torque on the crosspiece ten times less than the double-coiled strips (ELKS of type SM-VII) under the same tension. Despite of that, tests of the ELKS of type SM-VI installed in the amplitude transducers show that they are enough stable in dynamics of high rate automatic measuring machines (up to 10 measurement per second) if their structural synthesis is performed correctly (*see* Chap.8).

5.

Vibrating Transducers

Remote measurements of small displacement and force taken with frequency methods are usually produced by the string transducers that are simple, reliable, and accurate. Two types of string transducers were widely known and used for a long time: (1) with a round string and (2) with an initially flat string. Later the author of this book developed a new type of string transducer with helicoidal pretwisted strips. Other types of frequency transducers are the vibrating cantilevers and diaphragms. Vibrating miniature cantilevers with different shapes and coatings are essential as precision sensors for the Atomic Force Microscopy (*AFM*).

5-1. Twisting Tape Vibrators

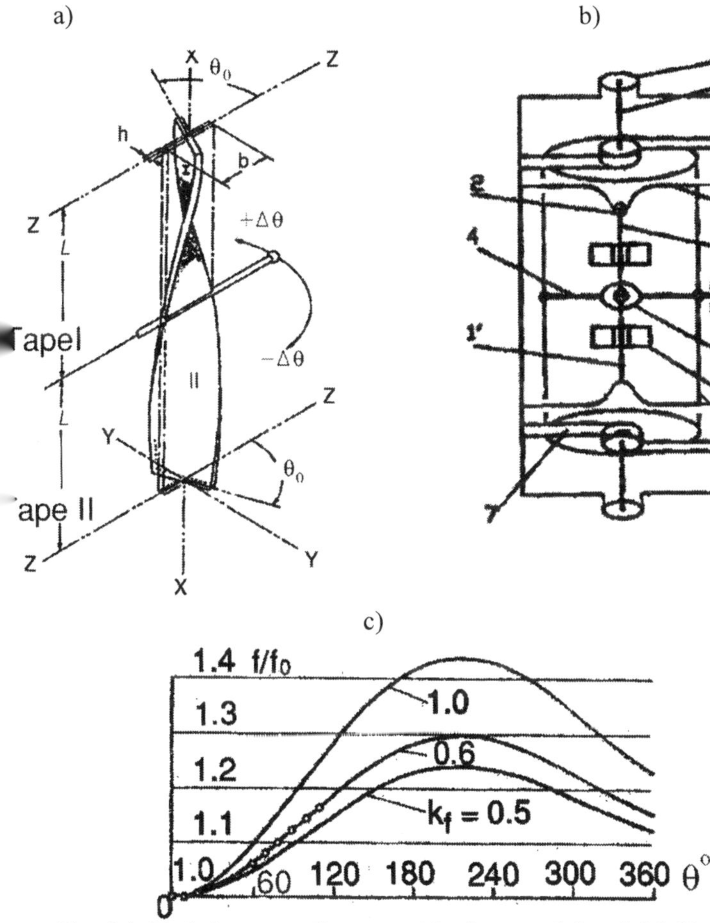

Fig. 5-1. Twisting tape vibrator and its features (after A.M. Voutsas, 1963).

FR-II. These vibrators have certain design and functional specifics. A miniature electromechanical device developed by A. M. Voutsas (1963) measures input angles within a fraction of a second of arc through conversion from a mechanical angular displacement $\Delta\theta$ to a difference in frequency Δf, which function is expressed as $\Delta\theta = B_f \Delta f$. The input angle $\Delta\theta$ (Fig. 5-1a) is shown to be linearly dependent on the difference in transverse natural frequencies of two equally initially twisted flat tapes (I and II). The tapes with width b and thickness h are identical in shape and are connected to each other at a common junction with their outer ends locked to a fixed reference base at a predetermined twist angle θ_0. The vibrating tapes are also shown in the assembly schematic (Fig. 5-1b) under item 1, 1' fastened to the input arm 6 through diaphragm 4 and to end clamps 2. The stationary end of the tape is clamped to the housing of instrument 9 through a leaf isolation spring assembly 5 that maintains constant tension of the tape during the operation. Cross support tapes 7connected with internal rotating cylinder 12 and torsion wires 8 clamped to covers 10 maintain stability of the rotating assembly with respect to the housing of the instrument 9 while the internal assembly maintains free rotation. The tapes 1, 1'are attached to each other in a series through a spring coupling 3. Rotation of the input arm 6 to the desired input angle will twist both tapes simultaneously. These metallic tapes are located in the transverse magnetic field between the poles of a pair of permanent magnets 11. A separate oscillator-amplifier is connected to the extreme ends of each tape, which serves as a transducer of the oscillator-amplifier circuit. The tapes are energized by the output of the amplifier, which causes them to move in the magnetic field imposed by the permanent magnets. The motion of each tape induces a voltage across itself, which is fed back to the oscillator-amplifier, and thus tape resonance is sustained. This resonant frequency is fed back to the oscillator–amplifier circuit, which may be measured by a frequency meter. An input angle $\Delta\theta$ will twist one tape to a total angle $\theta_o + \Delta\theta$ and untwist the second tape to an angle of $\theta_o - \Delta\theta$. It is shown below that a change in twist angle changes the natural frequency of the tape and that the measurement of the difference in frequency between the two aforementioned tapes, f_1-f_2, is a measure of the input angle $\Delta\theta$. The input angle may also represent a linear displacement, such as the motion of a diaphragm, when translated to an angular motion. This electromechanical transducer is miniature device (about the size of a small matchbox) with total power consumption in the order of $0.04\text{x}\ 10^{-6}$ W.

The derivation of the relation between input angle $\Delta\theta$ and transverse vibration frequency difference f_1-f_2 of two identical tapes twisted to angles $\theta_o + \Delta\theta$ and $\theta_o - \Delta\theta$, respectively, considers the resonating system of a beam with its ends clamped. Each tape is represented simply by a beam with varying moments of inertia under a constant axial tension, given by

$$J = J_1 + (J_2 - J_1)\sin^2 \vartheta_x , \tag{5-1}$$

where J_1 is the moment of inertia of cross-sectional area with respect to Z-Z coordinate axis at zero preliminary twist angle (θ=0°), J_2 is the moment of inertia of cross-sectional area with respect to Y-Y coordinate axis at zero twist (θ=0°), $\vartheta_x = \theta dx / L$ is a local twist angle of element dx along the beam length L, and θ is the total angle of tape twist. The wave shape of the vibrating beam clamped at both ends is assumed to be represented by the equation

$$Y_x = Y_0[1 + \cos(2\pi x/L)]\ e^{i\omega_v t} .$$

It was found in practice that ideal clamping conditions are not obtained to isolate total vibratory energy in the tape, and the error introduced by this uncertainty exceeds the error introduced by assuming the aforementioned wave shape. The natural frequency of a tape at zero twist is given (A.M. Voutsas, 1963) as

$$f_0 = (1/1.015)\ \{(\frac{1}{3m_sL})[T_Q + (4\pi^2 EJ_1/L^2)]\}^{1/2} =$$

$$0.57\ \{(1/m_sL)\ [T_Q + (4\pi^2 EJ_1/L^2)]\}^{1/2}, \qquad (5\text{-}2)$$

where T_Q is an axial tension, E is Young's modulus of elasticity, and m_s is the tape mass. The natural frequency of a twisted tape is given by A. M. Voutsas as follows

$$f = f_0(1 + K_f A_{f\theta})^{1/2} = f_0\{1 + K_f[1\text{-}(\sin\theta/\theta) + (\theta\sin\theta/(4\pi^2\text{-}\theta^2))]\}^{1/2}, \qquad (5\text{-}3)$$

where $K_f = (J_2 - J_1) / [(L^2\ T_Q/2\pi^2E) + 2J_1]$, (5-4)

$$A_{f\theta} = 1 - (\sin\theta/\theta) + [\theta\sin\theta/(4\pi^2 - \theta^2)]. \qquad (5\text{-}5)$$

A. M. Voutsas recommended multiplying the tape length L used to calculate (2) by a numerical constant, which is larger than one, for an equivalent adjusted length to compensate for kinematic energy absorbed by the end clamps. He claims that this constant is best obtained experimentally. It should be mentioned, however, that losses of energy in attachments of high quality oscillators usually cause the decrease of their Q-factor with much smaller change of frequency (*see* Chap. 8). The natural frequency of a thin twisted strip is lower than that for a straight strip of equal length, tension, and of the same material (*see* Section 5-4). Expanding the transverse frequency solution for each tape in terms of Taylor series, the transverse frequency difference of the two tapes is given by the polynomial of the third order

$$f_1\text{-}f_2 = 2(df/d\theta)\Delta\theta + 1/3\ (d^3f/d\theta^3)\Delta\theta^3 + \ldots \qquad (5\text{-}6)$$

This relationship provides the calibration formula, since f_1-f_2 is the direct quantity and $\Delta\theta$ is the indirectly measured input angle. We can use now a unitless variable

$\tilde{\Delta\theta} = \Delta\theta / \Delta\theta_{\max}$ in the range ±1, express through it the input angle

$\Delta\theta = \Delta\theta_{\max}\tilde{\Delta\theta}$, and recall that Chebyshev's polynomial $P_3(Z)$ of the third order

(*see* Table 1-7) allows us by substituting into (6) $\tilde{\Delta\theta} = (3/4)\tilde{\Delta\theta}$ obtain with the minimal possible uncertainty the calibration equation of the first order such as

follows $f_1\text{-}f_2 = [2(df/d\theta)\Delta\theta_{\max} + 1/4\ (d^3f/d\theta^3)\Delta\theta^3_{\max}]\ \tilde{\Delta\theta}$. (5-7)

This transformation causes in the whole measurement range an additional

uncertainty, which is equal to: $1/3\ (d^3f/d\theta^3_{\max})\tilde{\gamma} = 1/12\ (d^3f/d\theta^3_{\max})$. (5-8)

Therefore one can have the calibration formula of the first order instead of that of the third order, which is practically more convenient if this additional conversion uncertainty is small enough and acceptable. Here $df/d\theta$ and $d^3f/d\theta^3$ are represented in cps / rad and cps/ rad^3, respectively, where cps is a cycles per second (Hz). The ratio of the obtained frequency f of a single tape twisted to angle θ over its zero twist frequency f_0 is shown in Fig. 5-1c. Experiments conducted from 0° to 360° of the total twist angle confirm the relationship presented by (3). Additional experiments were conducted to evaluate the sensitivity, nonlinearity, and accuracy of (6). The total range of input angle tested was 45°, from 42.5° to 87.5° of total twist on each tape, starting with an initial twist angle of 65°. The measured frequency difference of tapes having K_f equal to 1.06 is given by the following expression as a function of the input twist angle $f_1\text{-}f_2 = 24.4984(1\text{-}0.000059\Delta\theta^2)\Delta\theta.$ (5-9)

The maximum deviation from this relation during the entire range of the input angle of 45° was within an uncertainty of 0.02 Hz, equivalent to 3.2 second of arc. This uncertainty was within the accuracy of the input angle turntable on which the device was mounted, taking into account the added error shift of the frequency counter. The transformation of (9) to the relative expression similar to (7) with the first order unitless variable causes an additional uncertainty which equals $0.000015\,\Delta\theta^3{}_{max} = 7.26\times10^{-6}$. This uncertainty may be considered negligible because it corresponds to 1.2 arc second in the 45° range and in most applications is ignored by using the following simplified expression

$f_1\text{-}f_2 = (2df/d\theta)\Delta\theta.$ (5-10)

Figure 5-1c shows that the theoretical maximum frequency on one tape is reached at approximately θ^{o}=210° of the total twist angle. However, in practice, absolute maximum frequency is reached at lower angles because, as mentioned earlier, part of the vibratory energy is transmitted to the end clamp assembly and effective tape length is seemed to be longer than the theoretical tape length. We consider that part of these results is induced by the model corresponding to expression (2). The twisting tape transducer is a laboratory proven device that lends itself to a wide use in instrumentation, space guidance components, and telecommunications. Its inherent advantages are the ability to convert a mechanical motion directly to a difference in frequency output, small threshold, which is limited only by the readout equipment, high accuracy, and miniature size. These characteristics present advantages for transmitting the frequency difference over long distances with repeated amplification and without signal distortion. However, the discussed vibrator has a limited angular measurement range, which is less than one revolution.

5-2. Helicoidal Multivibrators

In this section, we discuss the specifics of bifurcation in the helicoidal multivibrators, small and large impedance relations in vibrating contours, zigzag deflections in their frequency functions, three ways of the frequency function linearization, internal friction problems, and differential transducers' advantage. We also consider the options of these transducers' integration into an intelligent computerized systems, and expansion their range of motion with the rolamite connection.

In Chaps 2 and 4 we discussed the helicoidal amplifier that is built on the basis of the thin helicoidal strip and found that this mechanical system has the largest transmission ratio in the straight chain of motion transformation from a translational to a rotational one.

FR-III. The author (Y.M. Tseytlin, 1978a, Patent SU 706677) has developed a string system emitting a frequency signal in which the helicoidal thin strip in addition to its usual transformation function of a spring multiplier functions as a string-type frequency transducer. As a result, this system, which is called a helicoidal multivibrator (multiplier vibrating), has new specific features.

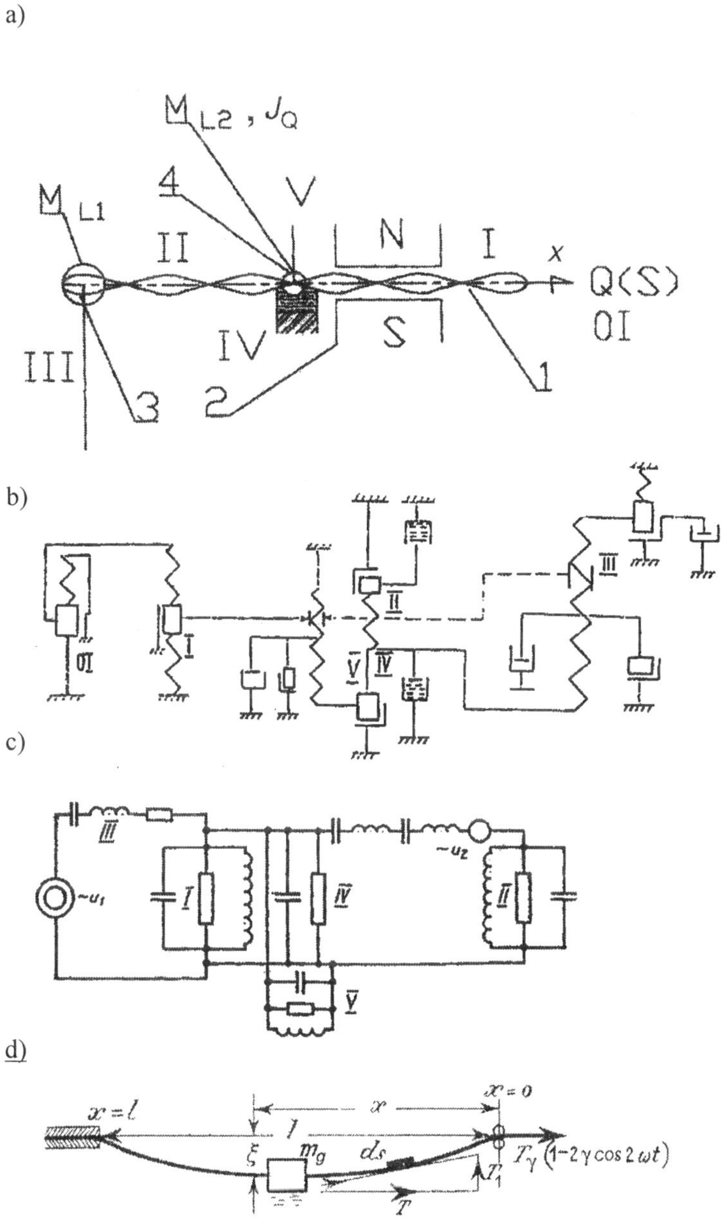

Fig. 5-2. Schematic and oscillators of a helicoidal multivibrator and model for longitudinal-transverse motion stability of string (latter after N.W. McLachlan).

The helicoidal multivibrator schematic is shown in Fig. 5-2a. The transducer is a bifurcating one because it has two output signals at one input of strip 1tension: (i) an amplitude-type signal through the angle of crosspiece 4 rotation and (ii) a frequency-type signal from the helicoidal strip transverse vibration in the magnetic field 2 (permanent *N-S* or *ac* type). The bifurcating transducer with amplitude and frequency outputs automatically satisfies well-known Abbe's principle for alignment of the compared arrangements to avoid an additional uncertainty in the measuring chain.

When the electrical pulses from a self-excited oscillator are applied to the helicoidal string lying within the magnetic field, Lorentz force acts on the string in a transverse direction, as a result of which the string moves out of equilibrium and starts to vibrate. These vibrations are maintained by the self-oscillator in the positive feedback chain and can be measured with the standard digital or analog frequency (period) meters that are connected to the oscillator on one side and to any intelligent calculating or servo system on the other side. The frequency output is useful for the large distance transmission of the signal without disturbances. But the period output is faster because the frequency (period) meter has its own quartz oscillator with a natural frequency in MHz that is approximately 1000 times larger than the output frequency of the helicoidal multivibrator which is in the range of kHz (*see* below and sec. 5.4).

The helicoidal strip-string in the transverse vibration regime can be considered a string with the specific longitudinal stiffness at the quasi-static tension, $j_\theta\theta = j_\theta\, i_{\theta}s = js$. However, in the dynamics its longitudinal stiffness is approximately equal to EA/L as for regular round or flat strings with Young's modulus E, cross-section area A, and length L. Fig. 5-2b shows mechanical model of the multivibrator system. The double-helicoid multivibrator's system has indeed a set of interconnected oscillators (Fig.5-2a, b): (I) of the helicoidal strip's segment L located in the transverse magnetic field, (II) the symmetrical part of the helicoidal strip, which is not connected directly with the magnetic field, (III) of a mass M_{L1} on the elastic support; and (IV) of a mass M_{L2} with the mass moment of inertia J_Q on the strip's crosspiece. Oscillators I – III have air damping but oscillator IV may have liquid-type viscose damping. The natural frequencies of oscillators I and II are equal ($f_{\mathrm{I}} = f_{\mathrm{II}}$) and depend on the helicoidal strip's tension. Oscillator III usually has a higher natural frequency ($f_{\mathrm{III}} > f_{\mathrm{I,II}}$) that is practically independent from the strip tension. Oscillator IV has a comparatively low natural frequency ($f_{\mathrm{IV}} << f_{\mathrm{I,II}}$) that increases with the extension of the strip. The parametric resonant occurs at the frequency of the strip's longitudinal vibration that is equal to the multiple of the natural frequency for the transverse vibration of oscillator IV. As a result of the action of the liquid viscose damping, oscillator IV can be in the non-periodic vibration regime especially at the weak tension of the helicoidal strip. In this case the system of oscillators I, IV, and II is similar to the electrical wave band filter (Fig 5-2c). In addition to the above mentioned oscillators, in certain realizations input oscillator 0I can be found with the natural frequency $f_{0\mathrm{I}}$ which is equal to the natural frequency of the input sensing element (measuring spindle), and the output oscillator V with natural frequency f_{V}. The analog output oscillator V can be in the form of an arrow-like pointer, a light mirror, Hall's transducer plate, a small thin plastic or metallic non-transparent shield on the double-helicoid crosspiece. Analog output may be performed even without adding anything to the crosspiece by indicating its angle of rotation with the laser beam diffraction (*see* Chap.8).

The general relations.

Mechanical oscillators. The typical differential equations for the oscillators 0I, III, IV, and V at the small natural vibrations can be represented as follows:

for oscillator 0I $m_1\ddot{x}+j_1 x=0,\ m_1\ddot{x}+j_1 x=Q\delta(x-a)$, (5-11a)

for oscillator III $M_{L1}\ddot{y}+j_3\ y=0$, (5-11b)

for oscillator IV $M_{L2}\ddot{z}+C_x z=2L^{-1}Q(z-z_1)+2j_\theta i_\theta L^{-2}(z-z_1)^3$, (5-11c)

$$J_Q\ddot{\theta}+C_\theta\theta=-2(j_{\tau\tau}\theta/L)+j_{k\tau}z^2/L^2, \quad (5\text{-}11d)$$

and for the oscillator V with cantilever pointer whose two segments supported in one intermediate point $(EJ_{1,2}m^{-1}{}_{1,2}\,\partial^4 v/\partial x^4)+\partial^2 v/\partial t^2=0$. (5-11e)

Solution to the last equation is shown in Chap. 7. In this section we will study oscillators I, II, and IV, which have not been previously studied in all options and are essential for the helicoidal multivibrators. Let us discuss this problem in more detail.
Pretwisted string. The helicoidal (initially pretwisted) beam-strip with fixed ends has flexural (in two planes), longitudinal, and torsional vibrations that are mutually related (*see* B.F. Shorr, 1961). In the linear problem for quasi-constant relations at the vibration of the pretwisted beam-strip with double-symmetrical profile under longitudinal tension, this system splits into two independent systems: the flexural and the longitudinal-torsional vibrations. Let us note that the inductance voltage in the helicoidal multivibrator is generated only by its transverse motion which is perpendicular to the magnetic lines, i.e., by the flexural vibration in one plane. Hence, in the linear problem with the elimination of non-used expressions for torsional vibrations, the transverse vibration of the helicoidal strip-string can be described by the partial differential equation with the periodic coefficients such as follows

$$Q_0(\partial^2 v/\partial x^2)-EA\,\partial^2/\partial x^2\,[\kappa_j^2(x)\partial^2 v/\partial x^2]=\rho_\gamma A(\partial^2/\partial t^2), \quad (5\text{-}12)$$

where $\kappa_j^2(x)=(1/A)\,[J_\eta\sin^2\varphi(x)+J_\xi\cos^2\varphi(x)]$ is the equivalent of radius of gyration for the helicoidal strip's cross section, Q_0 is the longitudinal pretension of the strip, ρ_γ is the density of the material, A is an area of the strip's cross section, J_η, J_ξ are the section's moments of inertia, $\varphi=(2\pi/S_0)x$, and S_0 is the pitch of the pretwisted helix. Note that function $\kappa_j^2(x)$ is a periodical one. Let us evaluate the mean value $\bar{k}_j^2(x)$ in the limits of one helix pitch S_0 on the pretwisted strip

$$\bar{k}_j^2(x)=[1/(S_0 A)]\int_0^{S_0}dx\ [J_\eta\sin^2(2\pi/S_0)x+J_\xi\cos^2(2\pi/S_0)x]=$$

$$\tfrac{1}{2}(J_\xi/A)+\tfrac{1}{2}(J_\eta/A)=\tfrac{1}{2}J_p/A. \quad (5\text{-}13)$$

Furthermore, if the radius of gyration is constant and equal to the found $\overline{\kappa_j(x)}$, we have from (12) a solution similar to that for the rigid flat string with $\kappa_j^2(x)=\bar{k}_j^2(x)$ and at the fixed ends $Y(0)=Y(L)=Y'(0)=Y'(L)=0$, which yields

$$f_n\approx(n_f/2L)\sqrt{Q/\rho_\gamma A}[1+(2/L)\sqrt{EJ_p/2Q}+(4+\frac{n_f^2\pi^2}{2})EJ_p/(2QL^2)], \quad (5\text{-}14)$$

where f_n is the natural frequency of vibration for the elastic helicoidal multivibrator at $n_f \ll L/S_o$. So the number of harmonics n_f should be not more than 3 because in the typical multivibrator $L/S_o = 9$. The difference between (14) and the known analogous relationship for the flat string is in the substitution of the radius of gyration for the parameter $\sqrt{(1/2)J_p/A}$. At $J_p \rightarrow 0$ expression (14) corresponds to the formula for the vibration of the regular flat string under tension. By substituting the expression $v = Y(x)e^{-2\pi ift}$ into (12), we obtain the ordinary quartic differential equation of the fourth order with periodic coefficients:

$$-QY'' + [f(x)Y'']'' + \chi g_x Y = 0, \quad (5\text{-}15)$$

where $f(x) = \kappa_j^2\,[(2\pi/S_o)x]$, $g(x) = 4\pi^2\rho_\gamma A$; $\chi = f^2$. (5-16)

This conjugate equation with the conjugate boundary-value conditions $Y(0) = Y(L) = Y'(0) = Y'(L) = 0$ has eigenvalues with multiplicity not more than two. Equation (15) fully corresponds to the problem for vibration of the rigid string with the section radius of gyration equal to the mean value of the radius of gyration for the initially pretwisted helicoidal string $k_j = \bar{k}_j = \sqrt{J_p/2A}$. (5-17)

Thus, we again come to the same expression (14) for the natural frequency of the helicoidal multivibrator, which was established earlier by the physical approach to the problem.

Uncertainty of the derived equation. The approximate expression similar to (14) for the rigid uniform flat or round string is an effective one at $EA\kappa_j^2 < L^2Q_0$. But the differential equation for vibration of the rigid string in general can be solved only by numerical methods. In Table 5-1 the uncertainty of the solution for (14) is shown in comparison with the numerical solution to a similar flat string.

Table 5-1. Estimation of an uncertainty in the application of (14)

$\frac{Q_0L^2}{AE\bar{k}_j^2}$	$\sqrt{\frac{\rho_\gamma L^4}{Ek_j^2}}\,\omega_0$ for (12)	$\omega\sqrt{\frac{\rho_\gamma L^4}{E\bar{k}_j^2}}$ from (14)	$\frac{\omega-\omega_0}{\omega_0}$ %	(i) Q_0, cN	θ^o	(ii) Q_0, cN	θ^o	(iii) Q_0, cN	θ^o
106.5	41.9	41.42	-1.10	2.17	48.4	-	-	-	-
243.2	57.3	57.08	-0.39	4.97	110.5	2.15	86.3	0.74	73.7
564.3	82.2	82.07	-0.16	11.54	256.4	5.00	200	1.71	170.9
1358.0	122.9	122.8	-0.08	27.76	617.0	12.05	500	4.11	411.3

Note: $\omega_0 = 2\pi f_0$; $\omega = 2\pi f$; Helicoidal strip of zinc bronze with dimensions (i) 0.008x 0.12x π x 18; (ii) 0.006x 0.10x π x 18; (iii) 0.004x 0.08x π x 18.

The uncertainty of natural frequency calculation in the case of the helicoidal multivibrator with preliminary untwist $\theta = 90°$ and $Q_0L^2/(AE\bar{k}_j^2) = 243.2$ is less than 0.4%, and an uncertainty is less than 0.16% for the more typical adjustment with $\theta \geq 200°$ and $QL^2/(AE\bar{k}_j^2) \geq 564.3$. However, the influence of the section rigidity on the frequency value ω for the helicoidal multivibrator is significant in accordance with the formula

$$\delta(\bar{k}_j^2, Q_0) = 2(EA\bar{k}_j^2/Q_0L^2)^{1/2} + (4+\pi^2/2)\,EA\bar{k}_j^2/(Q_0L^2), \quad (5\text{-}18)$$

which predicts the following values of this influence, in %,

$\delta(Q_0L^2/\,EA\,\bar{k}_j^2)$=27.77(106.5), 16.50(243.2), 10.00(564.3), 6.08 (1358,0) in

comparison to the stretched string with negligible flexural rigidity.
One can use a thinner strip with a smaller sectional polar moment J_p in order to decrease the influence of the section flexural rigidity at equal preliminary tension of the helicoid.

Function of impedance.

The compliance of the end of helicoid with the pointer's mass $M_{L2} = M_Q$ on the crosspiece (contour IV) is reflected in the differential equation of its transverse vibrations

$$M_Q\,\ddot{\eta}_y + C_y\dot{\eta}_y - j_y\,\eta_y = 2Q(\partial y/\partial x)_{x=L}\cos(\lambda_a at - \varepsilon_\lambda)\,, \qquad (5\text{-}19)$$

where C_y is a coefficient of the resistance to a motion in the liquid damper; j_y =2Q/L is an equivalent rigidity of the transverse motion for the mass M_Q, and $y=\varphi(x)$ is the eigen form of the helicoid's segment natural vibration.

The natural frequency of vibration for the pointer (indicator) on the strip is approximately equal to: $f_{IV} \approx (1/2\pi)\sqrt{(2Q/LM_Q)-(C_y/2)^2}$. (5-20)

One can neglect the transverse motion of the pointer if it is less than the amplitude of the string's vibration, $\eta_y << y_{x=L}$. However, if the frequency of helicoid vibration is close to $f=Nf_{IV}$ with N as an even integer, the parametric resonant influence effect can emerge. Let us discuss this in detail. Assume that the function of motion (H. Jeffreys and B. Swirles, 1956) for the string is

$y(x) = (\alpha_\lambda \sin\lambda_a x + \beta_\lambda \cos\lambda_a x)\cos(\lambda_a at - \varepsilon_\lambda)$

with the following conditions at the ends of the helicod:

a) $x=0 \Rightarrow Y=0$; b) $x=L \Rightarrow M_Q\ddot{y} + C_y\dot{y} + j_y y = -2Q(\partial y/\partial x)_{x=L}$.

Condition a) yields $\beta_\lambda=0$, $y=\alpha_\lambda \sin\lambda_a x\cos(\lambda_a at-\varepsilon_\lambda)$. Condition b) after substitutions for y, $\dot{y}$, $\ddot{y}$, and $(\partial y/\partial x)_{x=L}$ yields

$$\tan(\lambda_a L) = 2Q\,\lambda_a/(M_Q a^2\lambda_a^2 - j_y) = 2/\nu_a \approx 2\nu_a/(1-\nu_a^2), \qquad (5\text{-}21)$$

where $\nu_a = (M_Q a^2\lambda_a^2 - j_y)Q^{-1}\lambda_a$.

An analogous relationship can be obtained from the fundamental form for the elastic line of a string with the fixed ends and the vibrating middle point of mass M_Q that leads to the same result as in (21). Note that (21) is analogous to Rayleigh's relationship derived from investigation of vibrations for the string on the symmetrical compliant supports. An influence of the mass on the crosspiece transverse displacements can also be evaluated through the impedance of its motion. *Impedance.* The corresponding boundary-value conditions for the left-hand and right-hand end sections of the vibrating elastic helicoid are as follows $y(0)=0$; $y(L)=(Q/i\omega Z_L)\,\partial y/\partial x|_L$. The application of the fundamental functions (*see* Ph. M. Morse, 1948) in the form $\psi_n(x,\omega)=\sin[(\omega_n/\rho)(x-a_n)]$ or

$\psi_n = \sinh(i\omega_n/C_\rho)(x-a_n)/\sinh(i\omega_n/C_\rho)$ yields

$a_n = \pi n_f C_\rho/\omega_n$, $C_\rho = \sqrt{Q/\rho_\gamma A}$, and

$$\tan[(\omega_n/C_\rho)(L-\pi n_f C_\rho/\omega_n)] = (Q/i\omega Z_L)\,\omega_n/C_\rho\,, \qquad (5\text{-}22)$$

where Z_L is an impedance.

Small ratio $Q/(\omega Z_L)$ with large impedance. At this suggestion, we have an expression for the circular frequency ω_n of vibration

$$\omega_n = (n_f \pi / L)\ \sqrt{Q/\rho_\gamma A}\,(1+\frac{Q}{i\omega_n L Z_L}), \tag{5-23}$$

where impedance Z_L is equal to:

$Z_L = C_y - i\,(\omega_n M_{L2} - j_y/\omega_n)$ with $j_y = 2Q^{-1}N^2$ and N=3, 4,5, 6…

Substituting the impedance expression into (23) and assuming C_y=0, we derive after elementary transformations the cubic equation in relation to the circular frequency

$$(LM_{L2})\ \omega_n^3 - \pi\, n_f\, M_{L2}\ \sqrt{Q/\rho_\gamma A}\ \omega_n^2 - 2Q\omega_n N^2$$

$$+\ 2\pi\, n_f L^{-1}\ \sqrt{Q/\rho_\gamma A}\ Q\,(N^2 - ½) = 0 \text{ with } n=1,2\ldots \tag{5-24}$$

The cubic equation in relation to the linear frequency is a similar one

$$4(LM_{L2})\pi^2 f_n^3 - 2\pi^2 n_f M_{L2}\sqrt{Q/\rho_\gamma A}\ f_n^2 - 2Qf_n N^2$$

$$+\ n_f L^{-1}\ \sqrt{Q/\rho_\gamma A}\ \ Q(N^2 - ½) = 0, \tag{5-25}$$

where N= 3,4,5,6… If M_{L2}= 0, both parts of the double-helicoid are vibrating together and $f_n = (n_f/2)(1/2L)\ \sqrt{Q/\rho_\gamma A}$. Because the decrement Δ_M of (25) is less than zero [$\Delta_M = -24(LM_{L2})Q\pi^2N^2 - 4\pi^4 n_f^2 M_{L2}^2 Q/(\rho_\gamma A) < 0$], (25) corresponds to a cubical parabola with three real roots. Its analysis shows that the two roots with plus and minus signs have an approximately equal module, and the third root corresponds to the natural vibrations of the string with fixed ends, which is equal to: $f_n \approx (1/2\mathrm{L})\ \sqrt{Q/\rho_\gamma A}$.

Let us recall that (25) was derived with the assumption of a small ratio $Q/\omega Z_L$,or a large enough impedance Z_L. But the system under consideration can have a small impedance, which, at resonance of mass M_{L2} (oscillator IV), can be close to zero. In this case, the discussed ratio cannot be considered a small one.

Large ratio $Q/\omega Z_L$ with small impedance. It should be noted that the problem of string vibration in the whole field of the impedance values was not previously investigated. So we try to make up a deficiency in the solution to the problem (Y.M. Tseytlin *et al.*, 1989) with only one assumption: that an influence of the liquid damper C_y is small. Now let us transform the transcendent equation (22) as follows:

$\tan[\,(\omega_{vn}/C_\rho)(L - \pi n_f C_\rho/\omega_{vn})] = Q\omega_{vn}/[\,C_\rho\,(\omega_{vn}^2 M_{L2} - 2QN^2/L)]$.

Then, we transform its right-hand side with the assumption that

$$(\sqrt{\rho_\gamma A L}/N\sqrt{2M_{L2}})\,[\kappa_\omega/(\kappa_\omega^2 - 1)] = q_N(\kappa_\omega/(\kappa_\omega^2 - 1) \approx (\Delta\kappa_\omega)^{-1} q_N/2,$$

where $0 \le \kappa_\omega \le 1$; $\Delta\kappa_\omega = 1 - \kappa_\omega$. As a result, we obtain the following simplified expression

$$\tan[\,(\omega_{vn}/C_\rho)(L - \pi n_f C_\rho/\omega_{vn})] = q_N\,[\kappa_\omega/(\kappa_\omega^2 - 1)] \approx (\Delta\kappa_\omega)^{-1} q_N/2 \tag{5-26}$$

that at n_f=1 corresponds to $\omega_1 = (1/2)\ \pi C_\rho\ /\ L$. Therefore, we find from (26) that the circular frequency of oscillator I is

$$\omega_{vn} = (\pi n C_\rho/L)\ \{\ 1 - [(1/(\pi n)|\ \tan^{-1} q_N(\kappa_\omega/(\kappa_\omega^2 - 1)|]\ \},$$

$$\omega_{vn} = (N \pm k_\omega \mp 1)\sqrt{2Q_g/(LM_{L2})}\ ,\ \text{and}\ \ f_{\mathrm{I}} = \frac{\omega_{vn}}{2\pi}\ . \tag{5-27}$$

The chart of this function (Fig. 5-3a) has typical zigzag deflections (deviations) near the frequencies $f_I = Nf_{IV}$.

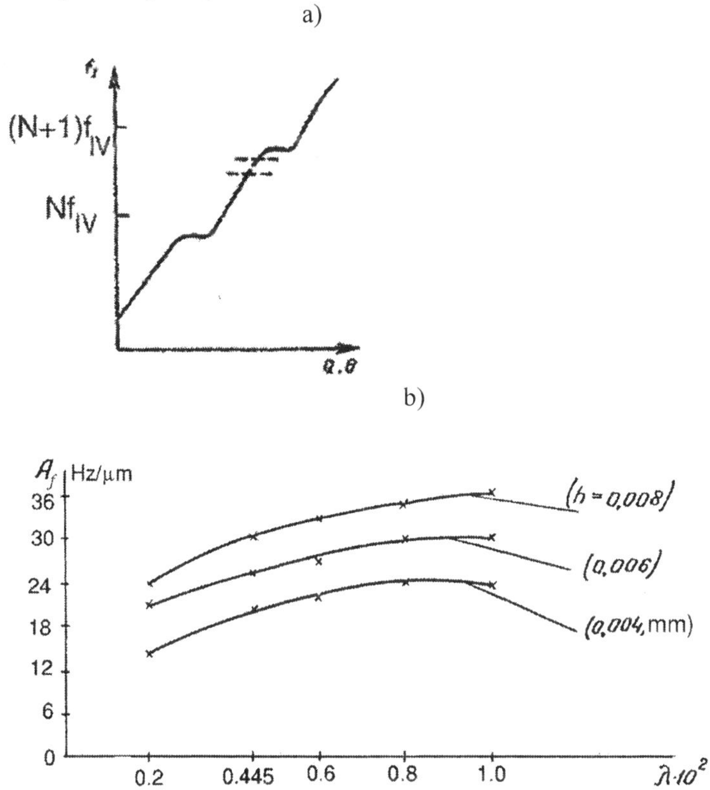

Fig, 5-3. Frequency functions of the double helicoidal multivibrator

There is a parametric resonance at the even values of integer N with a break in the transformation function (shown by dashed lines).

Following from (27), the sensitivity of a force to the frequency transformation and also a displacement to the frequency transformation is lower at the frequency value which is a multiple of the natural frequency for the mass M_{L2} vibration, i.e., when $\Delta\kappa_\omega \to 0$. The analysis of (27) shows that the influence of mass M_{L2}'s motion is small when $\Delta\kappa_\omega > 0.05$, and the frequency characteristic approaches the usual function for a string transducer. Consequently, the operating range of the characteristic of a multivibrator without special scaling or correcting system does not exceed $\Delta f_I = 0.9 f_{IV}$. The operating range is selected during adjustment of the measuring transducer by varying the longitudinal tension in the strip.

Effective parameters. Tension Q_o, corresponding to $Q_o L^2/AE\bar{k}_j^2 > 560$, yields rotation of the crosspiece by an angle more than170°. The relationship between the increment in frequency of oscillation and the tension in the pretwisted strip $\Delta f = ½\, \Delta Q\, (f_o/Q_o)$ under these conditions is a quasi-linear, within the limits of operating range. If we substitute the elastic and geometrical parameters of the strip into the latter expression, then

$$\Delta f = 1.99E(4 + v_{if}h^2\kappa_p^2/\lambda^2)^{-1}(1 + ¼\, v_{jf}h^2\kappa_p^2/\lambda^2)h\,(Q_o\rho_\gamma)^{-1/2}(4L^2\lambda^{1/4})^{-1}\Delta s, \qquad (5\text{-}28)$$

where Δs is the corresponding variation in the distance between the ends of the strip.

When choosing the parameters of the pretwisted strip, we should attempt to achieve a good frequency conversion $A_f = \Delta f / \Delta s = ½\, j_\theta\, i_\theta\,(f_0/Q_0)$ and visual amplitude conversion ($B_f = i_\theta = \Delta\theta / \Delta s$) at the same time, i.e., to ensure that the ratio of A_f / B_f is a representative enough or to achieve the greatest frequency conversion A_f with an adequate amplitude conversion B_f. Figure 5-3b shows charts of $A_f = \sqrt[4]{\lambda}h^{-1}i_\theta j_\theta(Q_0\rho_\gamma)^{-1/2}/4L$ for pretwisted strips with the rectangular cross section, ρ_γ=8.8 g/cm^3, $k_p = \pi$ mm^{-1}, L=18 mm, and preliminary tension Q_0= 7.2 cN. It is clear that to obtain a larger value for A_f we should use a strip with larger thickness h. However, we need to have a thinner strip that is optimized with respect to λ ratio in order to obtain a higher value $B_f = i_\theta$ of the amplitude transformation (*see* Fig. 4-13c).

The elastic longitudinal links.

The dynamically stable fields of the helicoidal multivibrator at the longitudinal-transverse vibrations can be estimated with the help of Mathieu's model [*see* N.W. McLachlan ,1964 and Fig. 5-2d with the exciting force T_γ $(1\text{-}2\gamma\cos 2\omega t)$] because our corresponding model has similar periodic coefficients in the differential equation

$$\ddot{S}_u + 2n_u\dot{S}_u + \omega_{\nu\Omega}^2\,[1 - h_\nu\cos 2\omega_\nu t]\,S_u = 0\,, \qquad (5\text{-}29)$$

where $S_u = \xi$ is the possible transverse displacement of the mass M_{L2}=m_g that is clamped to the string's crosspiece, $h_\nu = 2\gamma$ is the depth of the periodical longitudinal modulation of the string's tension F_{Q0}=T_γ, $\omega_{\nu\Omega} = (2Q_o/LM_{L2})^{0.5}$, and n_u is a coefficient of the liquid damper resistance to the transverse vibrations at the middle point of the string. The transverse vibrations are accompanied by the secondary longitudinal deformation of the string and by the corresponding longitudinal vibrations. It is proved (V.V. Bolotin, 1956) that for the main region of instability in the case of model similar to (29) the boundary value is

$$\omega_{\nu^*} = 2\omega_{\nu\Omega}\sqrt{1 \pm \sqrt{\gamma^2 - (\Delta_\nu/\pi)^2}}\ , \text{ where } \Delta_\nu = 2\pi n_u/\omega_{\nu\Omega}\,.$$

We can see from this expression that instability in the region near ω_ν=$\omega_{\nu\Omega}$ is possible if $\gamma > \gamma_*$ where $\gamma_* = (\Delta_\nu/\pi)$ is a critical value of the excitement's coefficient. For n-th region of the parametric excitement, the critical value of $\gamma_* \sim (\Delta_\nu/\pi)^{1/n}$. If the clamped mass and viscose damping are small, $m_g \to 0$, $n_u \to 0$, and the string's mass per unit length is equal m, then the corresponding differential equation for transverse motion $\xi = f_\xi(t)\sin b_\xi x$ of the string may be expressed as standard Mathieu's equation with possible stable and unstable zones

$$\frac{d^2 f_\xi}{dz^2} + (a_z - 2q_z\cos 2z)f_\xi = 0\,,$$ where $a_z = b_\xi^2 T_\gamma/\omega_\nu^2 m$, $q_z = \gamma a_z$, and $z = \omega_\nu t$. Critical values of a_z and q_z can be found on the stability diagram (*see* Fig. 4-23b). In practice owing to damping, the motion once started could die away with increase in time.

Nonlinear elasticity and inertia. We recall that the differential equation for the natural vibration of a thin string has the following form

$$\partial^2 v/\partial t^2 - a_f^2 \partial^2 v/\partial x^2 = 0, \qquad (5\text{-}30)$$

where $a_f^2 = Q/\rho_{\gamma L}$ and $\rho_{\gamma L}$ is the string's mass per unit length.

The longitudinal motion w and tension force Q in the system with a pretwisted string that has longitudinal stiffness j_1, masses M_{L1}, M_{L2}, and the coefficient C_ε of the liquid damper resistance corresponds in vibration to the differential equation which describes the influence of the longitudinal w vibration

$$Q = Q_o + j_\theta \Delta\theta + (EA/2L)\, w\, [1 - (EA/L)(j_1 + EA/L)^{-1}] + j_1 w (EA/2L)(j_1 + EA/L)^{-1}$$

$$-(EA/L)\, M_{L1}\, \ddot{w}\, (j_1 + EA/L)^{-1} - M_{L2}\, \ddot{w}\, [1 - (EA/L)(j_1 + EA/L)^{-1}] - C_\varepsilon \dot{w}, \qquad (5\text{-}31)$$

where Q_o is the longitudinal pretension and $\Delta\theta$ is an angle of the crosspiece's untwist corresponding to the strip's extension. One can assume that as a result of the negligible flexural rigidity of the strip its transverse motion can be described by a sinusoidal function

$$v = \varphi_t(t)\sin(\pi x/L) \text{ and } w = L - \int_0^L \sqrt{1-(dv/dx)^2} = \frac{\pi^2 \varphi_t^{\,2}}{4L} + \ldots,$$

$$\dot{w} = \pi^2 \varphi_t \dot{\varphi}_t / 2L, \quad \ddot{w} = \frac{\pi^2}{2L}[\varphi_t \ddot{\varphi}_t + (\dot{\varphi}_t)^2] + \ldots$$

Substitution of these constituents into (31) leads to the differential equation with the functions $\varphi_t, \dot{\varphi}_t, \ddot{\varphi}_t$

$$\ddot{\varphi}_t + \omega_v^2 \varphi_t + \psi(\varphi_t, \dot{\varphi}_t, \ddot{\varphi}) = 0, \qquad (5\text{-}32)$$

where $\psi(\varphi_t, \dot{\varphi}_t, \ddot{\varphi}_t) = \gamma_\varphi \varphi_t^3 + 2\,\varepsilon_L \varphi_t^{\,2} \dot{\varphi}_t + 2\,\aleph_\varphi\, \varphi_t[\varphi_t \ddot{\varphi}_t + (\dot{\varphi}_t)^2]$,

$$\omega = (\pi/L)\sqrt{\frac{Q_0 + j_\theta \Delta\theta}{\rho_{\gamma L}}}, \quad 2\gamma_\varphi = \omega^2 \pi^4\, [(EA/L) + q_\pi]/[4L^3 (Q_o + j_\theta \Delta\theta)],$$

$$q_\pi = (EA/L)\,(j_1 - EA/L)/(j_1 + EA/L),$$

$$\aleph_\varphi = \Phi\,(M_{L1}, M_{L2}, \rho_{\gamma L}, EA, j_1, L), \text{ and } \varepsilon_L = \pi^4 C_\varepsilon/(2\rho_{\gamma L} L^3).$$

It is assumed that $EA/(L j_1) \ll 1$. The type of differential equation similar to (32) with members of the nonlinear elasticity $\gamma_\varphi \varphi_t^{\,3}$ and nonlinear inertia $2\aleph_\varphi\, \varphi_t\, [\varphi_t \ddot{\varphi}_t + (\dot{\varphi}_t)^2]$ are carefully studied with the van der Pol method of slowly changing amplitudes by V.V. Bolotin, 1956, who found the relationship between the amplitude a_ω and the frequency ϖ of oscillations for the corresponding nonlinear system as follows $\varpi = \omega\sqrt{[1 + \frac{3}{4}(\gamma_\varphi/\omega^2)a_\omega^2]/(1 + \aleph_\varphi a_\omega^2)}$

or with respect to the negligibility of nonlinear corrections

$$\varpi = \omega\{1 + (a_\omega^{\,2}/2)[3/4\,(\gamma_\varphi/\omega^2) - \aleph_\varphi]\}. \qquad (5\text{-}33)$$

It is important that in the case of the helicoidal multivibrator the value of γ_φ/ω^2 decreases with the increase of untwist, i.e., with the increase of the tension force $(Q_o + j_\theta \Delta\theta)$. Hence, the initial elastic nonlinearity decreases with the increase of the stretch and untwist. Then, after passing through the state of isochronic vibrations, we automatically come to the system with predominant nonlinear inertia.

Let us analyze function $\varpi(\theta)$. If the initial value of the argument's origin $\theta = \theta_\kappa$ at $\aleph_\varphi = ¾\,(\gamma_\varphi/\omega^2)$, then

$$\varpi = \omega\{1 \mp (a_\omega^2/2)(3\pi^4 j_\theta \theta_\kappa/32L^3)[(EA/L)+q_\pi]\}$$

and the nonlinear change of frequency in relation to the new origin is equal to:

$$|\Delta\varpi| = \omega\,(a_\omega^2/2)\,)(3\pi^4 j_\theta \theta_\kappa/32L^3)[(EA/L)+q_\pi]. \tag{5-34}$$

It is clear that this nonlinear function has convexity in the pre-isochronic state of vibration and concavity in the post-isochronic state of vibration (Y.M. Tseytlin, 1982b). This conclusion corresponds to the above shown zigzag frequency function (Fig. 5-3a) for the helicoidal multivibrator at the small impedance. The same type of nonlinearity we can see from the experimental plots in Fig. 5-6.

Internal friction (damping) in a helicoidal multivibrator.

Experimental data. The damping variations of the longitudinal displacements s of the ends in the helicoidal multivibrator corresponding to the certain crosspiece rotation θ in quasi-static deformations is not exceed $K_{\theta e} = 0.13\%$ for the tin-zinc bronze or $K_{\theta e} = 0.05\%$ for the platinum-silver alloy (*see* Table 1-13). However, it has been found from our experiments executed in 1960s (*see* Y.M. Tseytlin, 1972) that there is a stable difference $\delta\theta_\kappa$ in the angular displacements of the crosspiece with its indicator (pointer) when equal longitudinal distances between the ends of the initially pretwisted strip have been established by fast and slow displacements. We can assume that $\delta\theta_\kappa = \psi_{\Delta\kappa}\Delta\theta_d$, where $\psi_{\Delta\kappa}$ is the coefficient of the elastic energy absorption due to internal friction, $\Delta\theta_d$ is the dynamic lag of crosspiece rotation behind the quasi-static transmission ratio, and κ is an index that characterizes the direction of the longitudinal deformation: $\kappa = t$ for tension and $\kappa = r$ for relaxation. According to experimental data $\psi_{\Delta t} > \psi_{\Delta r}$ and $\psi_{\Delta t}/\psi_{\Delta r} = \delta_{tr}^2 >> 1$. It has been established experimentally that similar effects also arise in difference for slow longitudinal deformations of an elastic helicoid with and without forcible immobilization (temporary arrest) of the crosspiece rotation (*see* Table 5-2).
In the Table, helicoid No 1 has dimensions 0.006 x 0.1 x π x 18; helicoid No 2 has dimensions 0.008 x 0.12 x π ×18 with the quasi-static transmission ratio 5.2 and 3.6 °/ μm, respectively. The material is 4tin-3zinc bronze in both cases.

Table 5-2. Transmission uncertainty in helicoid strips at a different working regime

Helicoid No	Nominal untwist angle θ^o	$(\theta^o_{ts} - \theta^o_{tf})$	$(\theta^o_{rs} - \theta^o_{rf})$	$(\theta^o_{ts} - \theta^o_{tf\Phi})$	$(\theta^o_{rs} - \theta^o_{tf\Phi})$	$(\theta^o_{ts\Phi} - \theta^o_{ts})$
1	180	8.2-11.6	-	-	-	-
	360	17.5	-	-	-	-
	720	13.9 –19.7	0.7	13.9-27.3	0.6	-
2	180	5.4	0.0	6.6	0.6	-
	360	12.8	0.08	13.9	1.1	-
	440	-	0.4	16	1.9	12.6-15.9
	720	14.2	1.2	23	4.2	-

Note: θ^o_{ts} is an angle position of the crosspiece relative to its theoretical angular position θ^o after a slow tension of the helicoid; θ^o_{tf} is the same after a fast tension of the helicoid; θ^o_{rs} and θ^o_{rf} are the same after a slow and fast relaxation of the helicoid's tension. Subscript Φ marks the angular position of the crosspiece after the corresponding longitudinal deformation of the helicoid with forcible immobilization of the crosspiece rotation

Experimental data of these transmission uncertainties ΔΘ° at slow and fast deformation in the double-helicoid can be presented as a regression (Fig. 5-4a), where chart $N1(\Delta\theta_t)$ is for tension and chart $N2$ $(\Delta\theta_r)$ is for relaxation. The experimental points are shown by the crosses for extension and by dots for relaxation.

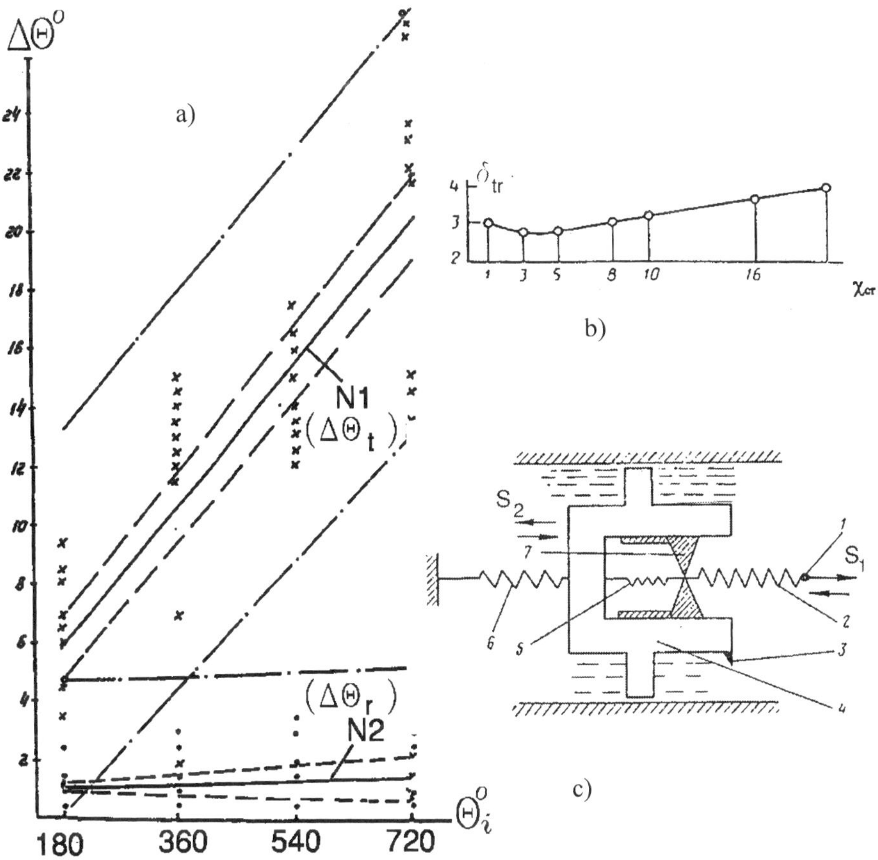

Fig. 5-4. Internal friction asymmetry in pretwisted microstrip and its macroscopic model

The boundaries of the distribution of ΔΘ correspond to dash-dot lines and correlation limits are shown by dashed lines. These latter limits are defined with the known statistical formula

$\Delta\theta(\bar{\theta}) = \bar{\Delta\theta} \pm t_{q,n-2}(n-2)^{-1}\sqrt{1+n(\theta-\bar{\theta})^2/[\theta\theta]}$, where $n = 128$ is the number of experimental data, $t_{q,n-2} = 2.241$ is the 2.5 percent point of Student's distribution at the level of significance $q = 2.5\%$; $\bar{\Delta\theta}$ and $\bar{\theta}$ are the corresponding mean values for uncertainty (deflections) and rotation angles of the pointer. Calculations show that for tension $\bar{\Delta\theta}_t\,(180°) = [5.8 \pm 0.9]°$, for relaxation $\bar{\Delta\theta}_r\,(180°) = [1.0 \pm 0.2]°$, $\bar{\Delta\theta}_t$ $(720°) = [20.0 \pm 1.3]°$, and $\bar{\Delta\theta}_r$ $(720°) = [1.5 \pm 0.5]°$. Described phenomenon in some aspects reminds, largely increased by the transmission ratio in the pretwisted strip, the so-called microBaushinger effect of microplasticity found by R.D. Carnahan in 1964 who showed that the stress required to operate the dislocation in materials depends on the shear modulus, Burgers vector (denotes the amount and direction of atomic displacement under the motion of dislocation), and the length of dislocation. Besides that, the shear strain produced by the motion of a dislocation depends on the density of moving dislocations and their pinning (*see* Chap. 1).

Let us look in the specifics of state of stress in the pretwisted strip under dynamic deformation.

State of stress. In this study, relationships from elasticity and plasticity theory are used to derive the theoretical ratio $\delta_{tr} = (\psi_{\Delta t} / \psi_{\Delta r})^{0.5}$ as a function of the elastic and geometric parameters of a helicoidal displacement transducer (with multivibrator or amplitude version). Generally, the coefficient of absorption $\psi_{\Delta\kappa}$ in the elastic helicoids depends on the conditions, limits, and type of deformation as well as the inelastic properties of the material. The stressed state of an initially pretwisted strip under elastic linear longitudinal-torsional deformations is described by a tensor $\hat{\sigma}^{\xi\eta}$ with the components

$$\hat{\sigma}^{\xi\eta} = \sigma_Q^{\xi\eta} S_\gamma + \sigma_M^{\xi\eta} \theta_\gamma , \tag{5-35}$$

where $\sigma_Q^{\xi\eta}$, $\sigma_M^{\xi\eta}$ are the components of the stress tensor in the respective cases: (i) longitudinal tension in the absence of the torque (M=0; $Q\neq 0$) and (ii) with torque M in the absence of tensile force Q (Q=0; $M\neq 0$) , S_γ is the longitudinal deformation per unit length, θ_γ is the torsional deformation per unit length of the helicoid, and ξ, η are the following coordinates of the cross section (*see* Chap. 2). Since the transverse-vibration amplitude A_{tv} of a helicoidal multivibrator crosspiece is a small one in the normal operation ($A_{tv} / L < 2.10^{-3}$), these vibrations practically have a small influence on the longitudinal force in the helicoid. This is evident because the ends of the helicoidal multivibrator are mounted to an elastic support, which provides a quasi-permanent tension force. During slow quasi-static longitudinal elastic deformation of the helicoid with M=0, $Q \neq 0$, the stress tensor has a following form at the most heavily loaded point of the strip's cross section

$$\hat{\sigma}_{1T} = \begin{bmatrix} \sigma_Q^{11} & 0 & \sigma_Q^{13} \\ 0 & \sigma_Q^{22} & 0 \\ \sigma_Q^{31} & 0 & \sigma_Q^{33} \end{bmatrix}, \tag{5-36}$$

where $\sigma_Q^{11} \approx \sigma_Q^{22} = -(18/224)\, \chi_{kp}\, C_\sigma^{-1} \sigma_0^{33}$, $\sigma_Q^{13} = h_1 \kappa_p C_\sigma^{-1} \sigma_0^{33} / (2\lambda)$, $\sigma_Q^{33} = (1 + 69\chi_{kp}/224)\, C_\sigma^{-1} \sigma_0^{33}$; $C_\sigma = 1 + \nu_{jf}\chi_{kp}$, ν_{jf} is the force factor for the certain form of the strip's cross section, $\sigma_0^{33} = Q/EA$; $\chi_{kp} = b_1^2 \kappa_p^2 \lambda^{-1}$, $\lambda = (h/b)^2 \in [\,0.16 \cdot 10^{-2} - 0.45 \cdot 10^{-2}]$ characterizes the relative thickness of the strip's cross section, $h = 2h_1 \in [\,0.002;\ 0.02]$ is the thickness (mm), b=$2b_1 \in [0.08;0.2]$ is the width of the cross section (mm), and $\kappa_p = 2\pi/ S_0$ is the initial pretwist with a helix pitch S_0. When tension is quickly applied to the pretwisted strip, the longitudinal stress σ^{33} temporarily increases more than at slow tension because M≠ 0 and

$$\sigma^{33} \approx (j_\theta\, \theta/A) + [\, I_Q \partial^2[\theta_e]_t / \partial t^2 + C_\theta \partial[\theta_e]_t / \partial t] / \rho_c A\,, \tag{5-37}$$

where I_Q is the mass moment of inertia of the driven pointer (indicator) on the crosspiece between the helicoids with left-hand and right-hand helices;

$\partial[\theta_e]_t/\partial t$, $\partial^2[\theta_e]_t/\partial t^2$ are the angular velocity and acceleration of its motion, respectively, C_θ is the resistance of the liquid damper, $t = t_{max}$ is the time corresponding to the largest value of the derivatives of the displacement transducer's transfer function $[\theta_e]$, ρ_c, j_{kk}; $j_{k\tau}$, and $j_{\tau\tau}$ see in Chaps 2 and 4 [(2-35), (2-36), and (4-79)]. With a sudden sharp extension of the pre-loaded helicoid, $\sigma^{33} = (j_\theta \theta_0 / A) + ES_\gamma$ where θ_0 is the initial untwist angle of the crosspiece. On the other hand, if the tension in a pretwisted strip is relaxed quickly, the longitudinal tension is initially lower than under equivalent slow relaxation.

The state of stress in the most critical point of the cross section of the initially pretwisted strip can be approximately described under quasi-dynamic longitudinal tension by a tensor $\hat{\sigma}_{2T}$ with the components

$\sigma_Q^{11} \approx \sigma_Q^{22} = -(18/224)\upsilon_\kappa^{11}\, \chi_{cr}\, C_\sigma^{-1} \sigma_0^{33}$; $\sigma_Q^{13} = \upsilon_\kappa^{13}\, h_1 \kappa_p C_\sigma^{-1} \sigma_0^{33}/(2\lambda)$;

$\sigma_Q^{33} = \upsilon_\kappa^{33}\,(1 + 0.308\chi_{cr})\, C_\sigma^{-1} \sigma_0^{33}$; $\chi_{cr} = \chi_{kp}$.

Coefficients υ_κ^{ij} have different values under slow and fast longitudinal deformation. Under slow quasi-static deformation, $\upsilon_k^{11} = \upsilon_k^{13} = \upsilon_k^{33} = 1$ and the tensor $\hat{\sigma}_{2T}$ becomes equal to tensor $\hat{\sigma}_{1T}$. Initially, with a fast tensile loading of the helicoid, the stress coefficients are

$\upsilon_t^{11} \le 1$, $\upsilon_t^{13} = 0$; $1 < \upsilon_t^{33} \le EAL^{-1} i_\theta^{-1} j_\theta^{-1} (1 + 0.308\,\chi_{cr})^{-1}\, C_\sigma$,

and with a fast unloading (relaxation) of the helicoid their initial values are

$\upsilon_r^{11} \le 1$; $\upsilon_r^{13} = 1$; $0 \le \upsilon_r^{33} < 1$.

Microplastic shifts. The microplastic shifts e_s in the continuum media are proportional to the stress intensity σ_s, which for a fast relaxation of the helicoid has the value

$$\sigma_{sfr}(\hat{\sigma}_{2T},\, \upsilon_r^{33} = 0) = [I_2 \operatorname{dev} \hat{\sigma}_{2T}\ (\upsilon_r^{33} = 0)]^{0.5} \approx \sigma^{13}_Q\,(\upsilon_r^{13} = 1)\,, \tag{5-38}$$

where $I_2 \operatorname{dev} \hat{\sigma}_{2T}$ is the second invariant of the deviator of tensor $\hat{\sigma}_{2T}$. The angle factor of the stress tensor at the fast relaxation of the strip equals

$$\mu_{fr} = (\sqrt{3}) \cot [\, 0.33 \cos^{-1}(3/2)(\sqrt{3})\, \sigma_{fr}^{-3}\, I_3 \operatorname{dev} \hat{\sigma}_{2T}\ (\upsilon_r^{33} = 0)] = 0, \tag{5-39}$$

which corresponds to a pure torsion. Here $I_3 \operatorname{dev} \hat{\sigma}_{2T}\ (\upsilon_r^{33} = 0)$ is the third invariant of the deviator of tensor $\hat{\sigma}_{2T}$. The stress intensity during a fast stretching of the helicoid is initially equal to:

$$\sigma_{sft}[\hat{\sigma}_{2T},\, 1 < \upsilon_t^{33} \le EAL^{-1} j_\theta^{-1} i_\theta^{-1} (1 + 0.308\,\chi_{cr})^{-1}\, C_\sigma] = EA\sigma_0^{33} L^{-1} i_\theta^{-1} j_\theta^{-1} / \sqrt{3}\,. \tag{5-40}$$

Then the angle factor of the stress tensor $\mu_{ft} = -1$ which corresponds to a pure tension.

Dissipation. The loss of the energy as a result of inelastic properties of the material with the coefficient of absorption $\psi_{\Delta\kappa}$ can be expressed by the formula

$$R_{\nu\kappa} = (1/L)\Delta\theta^2_{max}\, \psi_{\Delta\kappa}\, j_{\tau\tau}\,, \tag{5-41}$$

but $\partial R_{\nu t}/\partial\Delta\theta_{max} = M_t = 2\Delta\theta_{max}\, \psi_{\Delta t}\, j_{\tau\tau}/L$. Hence, the uncertainty of the angle position for the helicoid's crosspiece is

$$\delta\theta = M_t\, L/2\, j_{\tau\tau} = \psi_{\Delta\kappa}\Delta\theta_{max}. \tag{5-42}$$

Assuming that the absorption coefficients are proportional to the work $R_{\nu\kappa}$ of the microplastic shifts which equals for tension $R_{\nu t}=\sigma_{st}\, e_{st}$ and for relaxation $R_{\nu r}=\sigma_{sr} e_{sr}$, respectively, we derive the sought relationship for the absorption coefficients ratio δ_{tr}

$$\delta_{tr}^{2} = \psi_{\Delta t} / \psi_{\Delta r} = R_{\nu t} / R_{\nu r} = \sigma_{st}\, e_{st} / (\sigma_{sr} e_{sr}) \approx \sigma_{st}^{2} / \sigma_{sr}^{2}\,, \tag{5-43}$$

where e_{st} and e_{sr} are the microplastic shifts under the fast tension and the fast relaxation, respectively, which are proportional to the corresponding stress intensities σ_{st} under tension and σ_{sr} under relaxation. Then, after substitutions and simple rearrangements, we finally obtain the following formula for the absorption coefficients ratio

$$\delta_{tr} = (\psi_{\Delta t} / \psi_{\Delta r})^{0.5} = \sigma_{st} / \sigma_{sr} = 2.32\, (\chi_{cr}^{-0.5} + \nu_{if} \chi_{cr}^{0.5})\,, \tag{5-44}$$

where ν_{if} is the kinematic form factor for the cross section of the elastic helicoid strip.

Analysis of δ_{tr} curve. It is seen from the plot of internal friction asymmetry δ_{tr} (Fig. 5-4b) that this function has a weak minimum $\delta_{tr\min}$ at $\chi_{cr} = \nu_{if}^{-1}$ since the first derivative of the function (44) is equal to zero and the second one is greater than zero, such that $\frac{d\delta_{tr}}{d\chi_{cr}}(\chi_{cr} = \nu_{if}^{-1}) = 0$ and $\frac{d^2\delta_{tr}}{d\chi_{cr}^2}(\chi_{cr} = \nu_{if}^{-1}) > 0$.

The typical kinematic cross-sectional form factor of the initially pretwisted thin helicoids is in the range of $\nu_{if} = 0.32$- 0.39 and the parameter χ_{cr} is in the range of $\chi_{cr} \in [3;12]$. Let us recall that for the elliptic cross section $\nu_{if} = 0.345$.
Hence, $\delta_{tr} = \sigma_{st} / \sigma_{sr} = 2.32\, (\chi_{cr}^{-0.5} + \nu_{if} \chi_{cr}^{0.5}) = 2.6 - 3.5 \in \chi_{cr} [1;\, 12]$, $\delta_{tr\min}(\nu_{if} = 0.345) = 2.72$, and the mean value of the absorption coefficients ratio $<\psi_{\Delta t} / \psi_{\Delta r}> = <\delta_{tr}^{2}> = 9$. The average value of $<\delta_{tr}> = 3$ corresponds to the ratio $\psi_{\Delta t} / \psi_{\Delta} \approx 6 - 12$ observed in the experiments with the average value of the absorption coefficients ratio $<\psi_{\Delta t} / \psi_{\Delta r}> = 9$.

Macromodel. The described effects can also be expressed through the macroscopic model with the elastic-viscose-plastic elements (Fig. 5-4c). The displacement of point 1 corresponds to the input displacement S_1 (longitudinal deformation), and the displacement of index 3 clamped to inertial body 4 corresponds to the output (driven) displacement S_2. The ideal elastic links are represented by elements 2, 5, and 6. The plastic deformations are modeled by the dry friction link between driving body 7, and driven body 4. In addition, the viscose resistance acts on body 4. At the slow displacement of point 1 in S_1 direction, the motion through elastic link 2 transmits to body 7 and connected to it body 4 which at slow speeds and acceleration meets a small viscose resistance and has negligible inertia forces. At the sharp fast displacement of point 1 in S_1 direction, inertial body 4 lags behind body 7 and the sliding in the friction pair 7-4 occurs. In this case, at the same fast displacement S_1 the value of the output displacement S_2 will be less than at the corresponding slow displacement S_1. The sequential relaxation of the helicoid's tension causes the removal of the microplastic displacements by the backward forces (the backward displacement of the dislocation). The inertial body for the double-helicoid system corresponds to an element clamped to the crosspiece. The less the mass and the moment of inertia of this element the less the uncertainty of the transmission from the instability of the speed in longitudinal extension of the helicoid. If the moment of inertia for the body on the crosspiece is more than the critical one,

$J_Q > (C_\theta)^2 L/ 8j_{\tau\tau}$, then the system of the helicoid can become sensitive to the instability of the relaxation speed as well.

Optimization of the FR-III transformation function.

Function (14) for the transformation of the longitudinal displacements into the frequency is a nonlinear one. There is a necessity of its linearization for the linear measurement systems (*see* Chap.8). This can be executed by the following methods:
1) Connection of the calculator with pre-adjusted polynomial $R_n(Z)$ coefficients to the output of the helicoidal multivibrator for the appropriate approximation of the transformation function;
2) Installation of the differential-type multivibrator structure with the optimized ratio of the frequency outputs;
3) Certain adjustment of the transducer' s parameters.

i) In the first method with thin string, we have the following relationship for the output with the average period T_s

$$(A_s / \sqrt{S_{0A}+\Delta S}) - (A_s/\sqrt{S_{0A}}) = T_s - T_{s0} = \Delta T_s, \qquad (5\text{-}45)$$

where A_s is a constant of the transformation, S_{0A} and T_{s0} are the extension and the corresponding period of vibration for the original (zero) position of the analog pointer (indicator, analog output) of the helicoidal multivibrator; ΔS, ΔT_s are the change of the elongation and corresponding change of the period of vibration in the helicoidal multivibrator with respect to these parameters at the zero position of the analog indicator.

On this basis, we derive from (45) the following calibration expression

$$\Delta S = S_{0A}\{[(\Delta T_s / A_s) + 1]^{-2} - 1\}. \qquad (5\text{-}46)$$

One can find constants A_s, and S_{0A} by using two equations of the type which is similar to (46) and corresponds to two symmetrical relative to the zero analog output pairs of values ΔS_1; ΔT_{s1} and $\Delta S_2 = -\Delta S_1$; $\Delta T_{s2} = -\Delta T_{s1}$. As a result, we have the following equality $\{[(\Delta T_{s1}/ A_s) + 1]^{-2} - 1\} = - \{[(\Delta T_{s2}/ A_s) + 1]^{-2} - 1\}$, which corresponds to the cubic equation

$$A_s^3 + p_A A_s^2 + q_A A_s + r_A = 0, \qquad (5\text{-}47)$$

whose coefficients are equal to:

$$p_A = ½ (\Delta T_{s1}^2 + \Delta T_{s2}^2 + 8 \Delta T_{s1}\Delta T_{s2}) / (\Delta T_{s1} + \Delta T_{s2}),$$

$$q_A = 2\Delta T_{s1}\Delta T_{s2}, \text{ and } r_A = \Delta T_{s1}^2\Delta T_{s2}^2/ (\Delta T_{s1} + \Delta T_{s2}).$$

Constant A_s is equal to the appropriate root of (47). Constant S_{0A}' s value easily yields from (46) with known $\Delta S_{1(2)}$, $\Delta T_{s1(2)}$, and A_s.

ii) In the second method, frequency signals f_1, f_2 from two strings are linked with the preliminary tension Q_o, transmission ratio i to the string, longitudinal stiffness j of the string; and the density ρ_m of its material per unit length of the string as follows:

$$f_1/ f_2 = (L_2 / L_1) \sqrt{Q_{01}\rho_{m2} / Q_{02}\rho_{ml}} \sqrt{(1+j_1 i_1 S/Q_{01})(1-J_2 i_2 S/Q_{02})^{-1}} = 1+K_s S, \qquad (5\text{-}48)$$

where $L_1$and L_2 is the length of each string, and K_sS is the sought lincar function of transformation.

If we assume that in the initial state $(L_2 / L_1) \sqrt{Q_{01}\rho_{m2} / Q_{02}\rho_{ml}} = 1$, then

$$(1 + j_1 i_1 S/Q_{01})/(1 - j_2 i_2 S/Q_{02}) = 1+2K_s + K_s^2 S^2 \text{ and } S = S_{\sim} S_{max},$$

where $S_{\sim} \in [0; 1]$ is a unitless variable in the range from 0 to 1.

By using Chebysev's polynomial of the fourth degree $R_4 (S_{\sim})$ with the smallest deflection from zero in the range $S_{\sim} \in [0; 1]$, we obtain the fourth degree expression

$S_{\sim}^{4} = 1.9209\, S_{\sim}^{3} - 1.1135\, S_{\sim}^{2} + 0.1835\, S_{\sim}$, which after substitution into (48) leads to the relationship of the third degree relative to $S_{\sim}$

$$[1.9209 + (K_s S_{max})^{-1}(1-S_{02}K_s/i_2)]S_{\sim}^{3}$$
$$-\{1.1135 + (K_s^{2}S^{2}_{max})^{-1}[1-(S_{02}i_1/S_{01}i_2) + S_{02}K_s/i_2]\}S_{\sim}^{2}$$
$$+ \{0.1835 - [(i_1S_{02}/ K_s^{3}i_2S_{01}) - (2S_{02}/K_s^{2}i_2) + K_s^{-3}]/S^{3}_{max}\}S_{\sim} = A_3R_3(S_{\sim}). \qquad (5\text{-}49)$$

Let us recall that Chebyshev's polynomial of the third degree is equal to:

$$R_3 (S_{\sim}) = S^{3}_{\sim} - 1.3923\, S^{2}_{\sim} + 0.4308\, S_{\sim}. \qquad (5\text{-}50)$$

After equalizing the left-hand side coefficients of (49) divided by the coefficient at $S_{\sim}^{3}$ and right-hand side of polynomial $R_3 (S_{\sim})$, we derive the following optimal relationships

$$1.3923 = \frac{1.1135 + (K_s^2 S_{max}^2)^{-1}(1 - X_s + Y_s)}{1.9209 + (K_s S_{max})^{-1}(1 - Y_s)}, \quad 0.4308 = \frac{0.1835 - (K_s^3 S_{max}^3)^{-1}(X_s - Y_s + 1)}{1.9209 + (K_s S_{max})^{-1}(1 - Y_s)} \qquad (5\text{-}51)$$

where $X_s = S_{02}i_1 / S_{01}i_2$ and $Y_s = S_{02}K_s / i_2$. Thus,

$$Y_s = \frac{-0.644\, K_s^3 S_{max}^3 + 1.1302 K_s^2 S_{max}^2 + 1.3923 K_s S_{max} - 2}{-0.4308 K_s^2 S_{max}^2 + 1.3923 K_s S_{max} - 1}, \qquad (5\text{-}52)$$

$$X_s = -1.5610 K_s^{2}S^{2}_{max} - 1.3923\, K_s S_{max} + (1.3923 K_s S_{max} + 1)Y_s + 1. \qquad (5\text{-}53)$$

If $S_{02} = S_{01}$, $Y_s = 2$, then the optimal relations between the kinematic transmission ratios for each string should be $X_s = i_1/ i_2 = 3$. (5-54)

In this case, the uncertainty of the frequency approximation Δ_f corresponds to the largest deflection from zero $\tilde{\gamma}_{p3}$ for Chebyshev's polynomial of the third degree, i.e.,

$$\Delta_f = \tilde{\gamma}_{p3}\, [1.9209 - K_s^{-1}\, S^{-1}_{max}]i_2 i_2 K_s^{3} S^{4}_{max}/(4Q_{02})\,, \qquad (5\text{-}55)$$

where $\tilde{\gamma}_{p3} = 0.0385$ for the Chebyshev's polynomial $R_n(Z)$of the third degree with zero point (*see* Table 1-8). The string transducer with $i_1/i_2 = 3$ may be called an asymmetric one. At the same requirement $X_s = 3$, the optimal symmetric string transducer can be built with $i_1 = i_2 = i$, but at $S_{02} = 3\, S_{01}$. In the latter case, however, we have an increased initial tension in one of strings. The frequency transducer with the two vibrating strings is a differential one. If both strings have the same flat type and equal sections, their deformation work for symmetric version is equal to $W_{\gamma s} = 2EA(iS)^2 / L$ and for asymmetric version the work equals $W_{\gamma a} = 10EA(i_1S)^2 / 9L$ (*see* Y.M. Tseytlin *et al.*, 1989). Therefore $W_{\gamma a} < W_{\gamma s}$.

Let us recall that the cross- sectional equal longitudinal rigidity for one flat string and one helicoidal *multivibrator will occur if* $EA_{fl} = L_{fl}\, i_\theta j_\theta$, *where* A_{fl}, L_{fl} are the cross-sectional area and the length for a string, respectively; i_θ and j_θ are the transmission ratio and the longitudinal rigidity for the helicoidal multivibrator related to the angular output θ.

iii) In the third method, the adjustment of the working range of the frequency transformation function and the necessary imbalance of the indicator on the crosspiece can be used. The latter option indeed is a limited to the stationary structures.

Experimental verification.

Models. On the basis of the analytical and experimental research, the author developed three models of the double-helicoidal multivibrators (Patent SU 706677, 721658, 1980): (i) mikrochator and michator (preferable pronunciation as "mikrotchator" and "mitchator", respectively) with the pointer-arrow 1 on the crosspiece (Fig. 5-5a), (ii) optichator ("optitchator") with the mirror-pointer 2 (Fig. 5-5b), and (iii) differential type of helicoid-cable version (Fig. 5-5c). They are built on the basis of well-known spring measuring heads: mikrokator, mikator (small size mikrokator), and optikator (*see* Chap. 8).

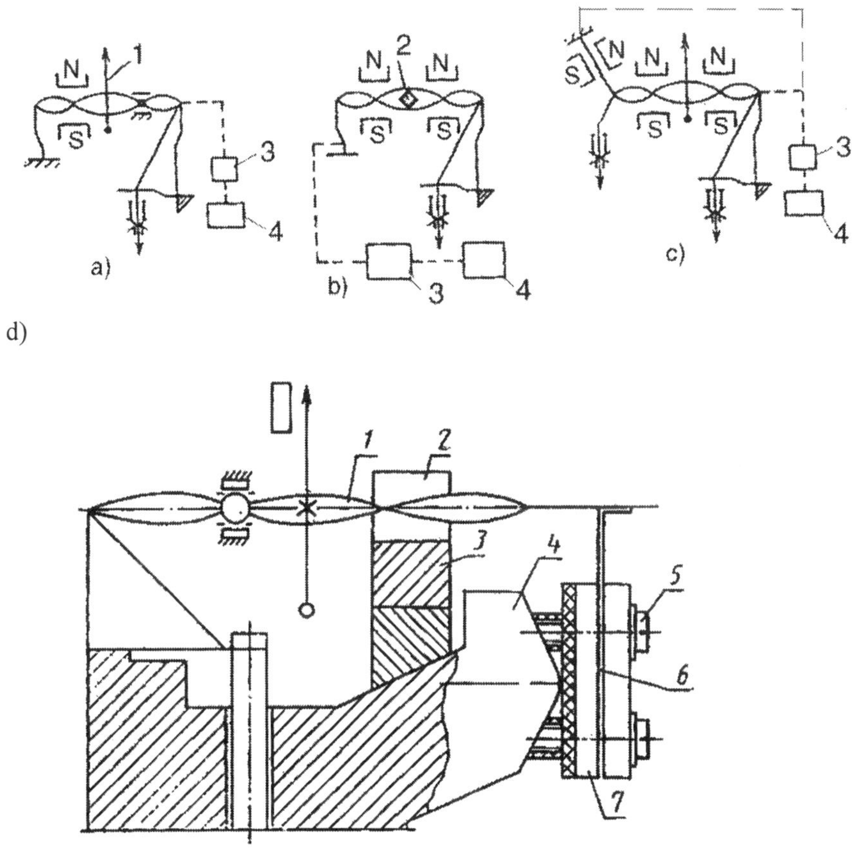

Fig. 5-5. Schematics and structural elements of helicoidal multivibrators.

In all cases, pretwisted strip passes through pole pieces of magnet *NS*. The self-oscillators 3 are built as the operational amplifiers with positive feedback to the pretwisted strip and connection to the frequency (oscillation period) meters 4.

The permanent magnet 2 with clamping fixture 3 (Fig. 5-5d) is installed on the front bracket 4 of the used spring measuring head. The magnet pole tips surround not less than 0.7*L* length of the front part of the double-helicoid 1 in the spring mechanism. The magnet inductance in the gap between the magnet pole tips in the experimental samples was equal to $B_M = 0.34$ T. One end of the double-helicoid strip is electrically isolated from the mechanism by glue or by isolator inserts together with the helicoid front support 5, 6, 7. The signal up to 20mV from the self-excited oscillator passes directly to this end through the elastic front support 6. Another end of the helicoid strip is electrically connected to the base of the mechanism and to the electrical "ground". At the measurement of a mean period of the helicoid vibration, its time intervals are filled with the megahertz periods of the quartz resonator in the frequency meter. As a result, the time for measurement of hundred periods is less than 0.2 second with the resolution up to 4 digits after the decimal point at the natural frequency of the multivibrator 3-5 kHz.

a) b)

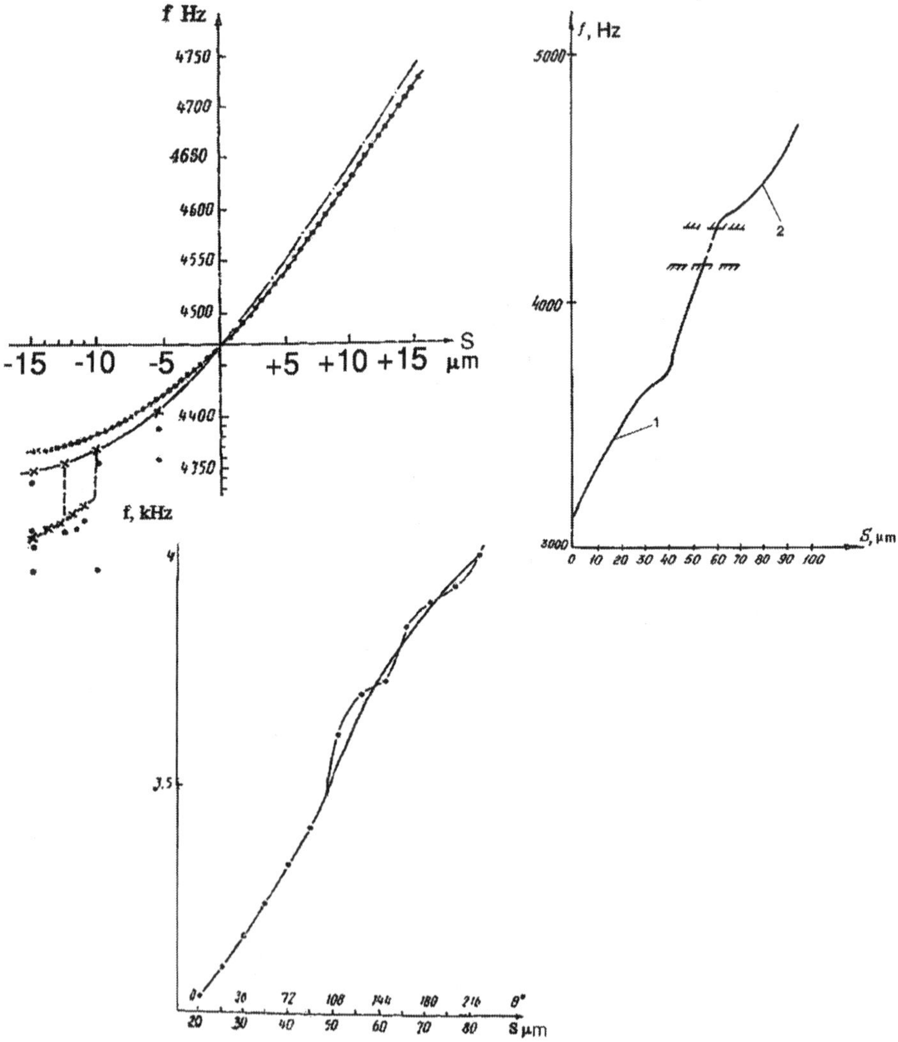

c)

Fig. 5-6. The experimental fuctions of mikrochator and the multi-revolution michator

Experimental data. The experimental characteristics of the helicoid multivibrator (strip 0.008 x 0.12 x π x 20) in the mikrochator (Fig. 5-6a,b) and multi-revolution michator (Fig. 5-6c) systems well correspond to the theoretical predictions. One can see from the chart in Fig. 5-6b that the type of transformation function inside of the range 3.1 to 4.7 kHz changes from a convex 1 (pre-isochronous) in the range 3.1 to 3.6 kHz to a concave 2 (post isochronous) in the range between 4.3 and 4.7 kHz. The break of the output frequency (black experimental points in Fig. 5-6a, dashed lines in Fig. 5-6b) occurs in the range 3.8 to 4.4 kHz as a result of the even integer ratio N between the natural frequencies for the oscillators I and IV (f_{IV} ≈700Hz) in the mikrochator, i.e., the ratio N is equal to 4200/700 = 6. The range 3.8-4.4 kHz of the output frequency was eliminated from the work by the preliminary adjustment of the tension in the pretwisted helicoidal strip. It was shown above that for the quasi-linear part of the transformation

$$\Delta f / S = 0.5(f_0 / Q_0)\, i_\theta j_\theta = \lambda^{1/4}\, h^{-1} i_\theta j_\theta /(4L \sqrt{Q_0 \rho_\gamma}\,). \quad (5\text{-}56)$$

Hence, the sensitivity of the transformation of the displacement into the frequency in the helicoidal multivibrator increases with the decrease of the longitudinal tension. For example, the sensitivity of transformation in the pretwisted strip with dimensions 0.008×0.12×π×18 of Tin-Zinc bronze is equal to 30 Hz /μm in the frequency range from 2.0 to 2.5 kHz, and at the same time the sensitivity is only 20 Hz /μm in the frequency range 3.1 to 3.6 kHz with increased preliminary tension. An uncertainty of the readout on the dial of the measuring head corresponds to 1-2 Hz that is equivalent to the uncertainty in the linear displacement of 0.05-0.1 μm. An uncertainty of the backward displacement is on the same level. The research of stability in the adjustment for the mikrochator system provided during 1.5 years showed a change in the frequency that is less than 4 Hz per month. This change can be compensated by the null-corrector of the mechanism.

The multi-revolution michator with the pretwisted strip 0.006×0.1×π× 18 and the measuring range of 200 μm has a frequency characteristic with the typical zigzag deflections. Part of this characteristic in the range 20-80 μm is shown in Fig. 5-6c. The differential mikrochator with the additional magnet 2 (Fig. 5-7a) and the auxiliary string 1(Y.Tseytlin, Patent SU 721658) works on the basis of the ratio between natural frequency of this string and the natural frequency of the helicoidal multivibrator 5. Note that element 3 is the transmission ratio's adjustment device and 4 is the auxiliary string elastic support. Both strings are electrically isolated from one end and connected to self-excited oscillators 6. Their transformation characteristics 1 for basic and 2 for auxiliary strings are shown in Fig. 5-7b. The additional string in the differential helicoid multivibrator can be of any possible kind: round, flat string, or shroud cable system with optimal displacement transformation ratio in accordance with (54). It is not necessary to build this string from the pretwisted elastic helicoid unless the second analogous output is specified.

The differential structure with helicoidal pretwisted strip multivibrator can have a schematic with "straight levers" I (Fig. 5-7c) at the measuring tips pairs (2-3), (1-4), (1-3), or (2-4), and a scheme with " intersecting levers" II with the analogous pairs of measuring tips.

The probable absorption coefficients ψ_Δ are different at simultaneous retraction and the release of both levers in accordance with the matrices: For scheme I

$$\begin{bmatrix}\psi_\Delta(1-2)\\ \psi_\Delta(4-3)\\ \psi_\Delta(1-3)\\ \psi_\Delta(2-4)\end{bmatrix}=\begin{bmatrix}0 & 1\\ 1 & 0\\ 1/2 & 1/2\\ 1/2 & 1/2\end{bmatrix}\begin{bmatrix}\psi_{\Delta t}\\ \psi_{\Delta r}\end{bmatrix} \quad (5\text{-}57) \qquad \begin{bmatrix}\psi_\Delta(1-2)\\ \psi_\Delta(4-3)\\ \psi_\Delta(1-3)\\ \psi_\Delta(2-4)\end{bmatrix}=\begin{bmatrix}1 & 0\\ 0 & 1\\ 1/2 & 1/2\\ 1/2 & 1/2\end{bmatrix}\begin{bmatrix}\psi_{\Delta t}\\ \psi_{\Delta r}\end{bmatrix} \quad (5\text{-}58)$$

for scheme II.

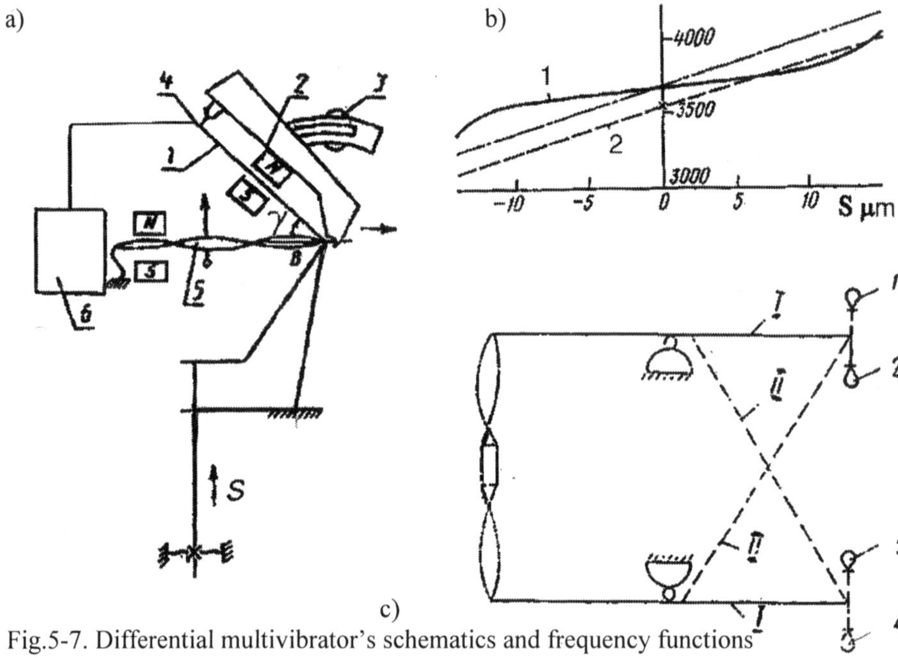

Fig.5-7. Differential multivibrator's schematics and frequency functions

The matrix component ½ shows that the realization probability of absorption coefficients with tension $\psi_{\Delta t}$ and relaxation $\psi_{\Delta r}$ in these cases are equal.

Theoretical 1 and experimental 2 frequency charts corresponding to certain preliminary tension with untwist angle θ_0 for the helicoidal multivibrator with the pretwisted strip 0.008x 0.12 x π x 18 from tin-zinc bronze and a small thin mirror on the crosspiece in the optichator version are shown in Fig. 5-8a. From these charts and regression coefficient m_σ^{-1}, μm/Hz, follows, as from the theoretical relationships, that the sensitivity for transformation of displacements s to the frequency f increases with the decrease of the preliminary tension (decrease of the frequency magnitude). The latter can be estimated through the angle of the preliminary untwist θ_0 of the helicoid. The difference between calculated 1(solid line) and experimental 2 (dashed line) curves is less than 5 %.

Aftereffect. The elastic aftereffect in strings corresponds to a stress variation described (*see* N.N. Davidenkov, 1933) by the empirical expression $\Delta\sigma_Q = N_c(\sigma_Q - \sigma_{Qo})(t^{0.23} - M_N)$. This variation of the stresses in the strip of a helicoidal multivibrator corresponds to a lower frequency at the zero mark on the dial as $\Delta f_e = \frac{1}{2}\Delta\sigma_Q(\sigma_Q - \sigma_{Qo})^{-1}(f^2 - f_0^2)f_0^{-1}$. (5-59)

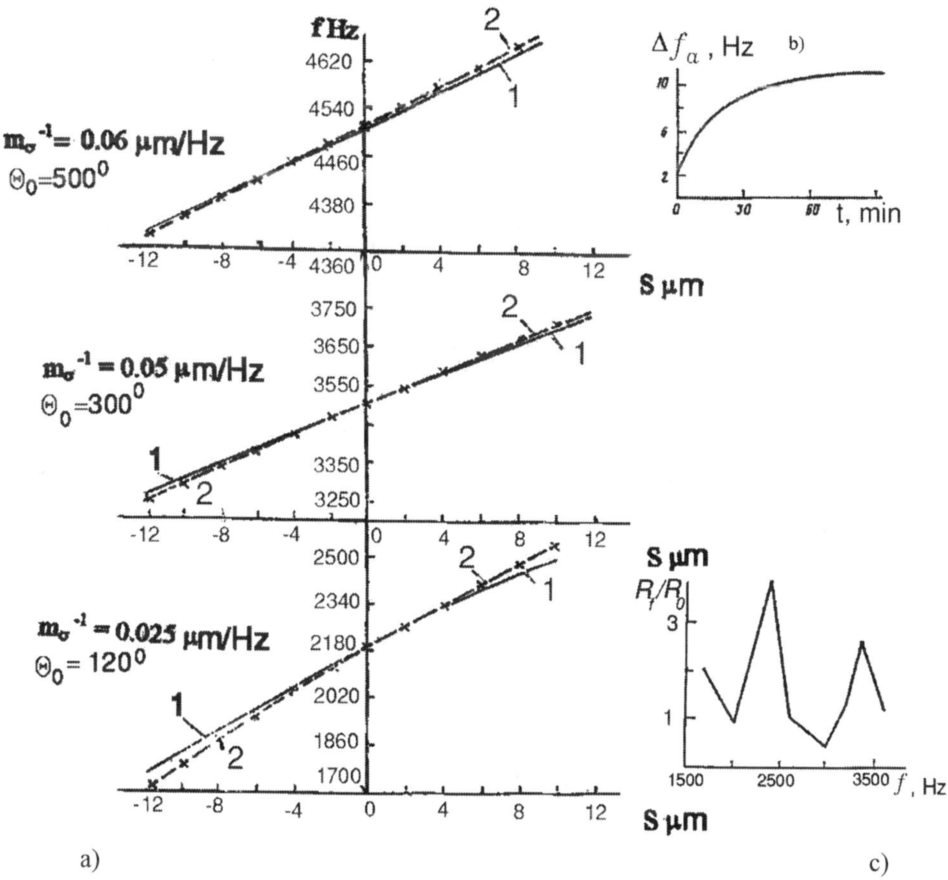

Fig. 5-8. Optichator's frequency functions, their aftereffect shift, and damping frequency dependence

It is assumed that $N_c = 2 \cdot 10^{-3}$, M_N =0.6 for artificially aged strings. Therefore, we obtain the variation which is equal after time t = 1 day to:
$\Delta f_e = 2 \cdot 10^{-3}$ (1-0.6)[(4200/3500) +1] × 700 =1.2 Hz for f= 4200 Hz at the extreme position of the readout indicator and f_0 = 3500 Hz at the zero mark. Similar values we observed in devices that allow limited rotation angles for the readout indicator and the helicoid crosspiece in the range $\theta = \pm 60°$. In optical devices with no limiters (stops)on the rotation of the reading mirror attached to the crosspiece, the angle of the mirror rotation from the zero mark may exceed 300° with the measuring rod (spindle) retraction. After holding the measuring rod in the lifted (arrested) position, the frequency at the zero mark may change by 8-12 Hz when the retractor is released and the rod come to the contact with the datum surface. This frequency shift cannot be observed with the usual brief retractions. If the elastic aftereffect in a helicoidal

multivibrator can be modeled as a non-relaxing Kelvin body, the relationship between the stress tensor deviator dev $\hat{\sigma}$ and the deviator of micro-displacements tensor e_s can be described by the formula

$$\text{dev}\ \hat{\sigma} = 2G\ \text{dev}\ e_s + 2\eta_{VR}\ \text{dev}\ \dot{e}_S, \tag{5-60}$$

where η_{VR} characterizes the "viscosity" (internal damping)of the rigid body deformation. The helicoidal multivibrator usually does not have a significant stress relaxation in time. This follows from the fact that instruments based on the initially pretwisted strip can be stored long time in the tensioned (arrested) state without loss of quality. Therefore, we can assume in the case of mikrochator that displacement derivative of e_s with the respect to time $\dot{e}_S = 0$ for the arrested state, therefore dev $\hat{\sigma}$ = constant is irrespective of the holding time t. For an optical device, derivative $\dot{e}_S \neq 0$ since the crosspiece of the double-helicoid does not have the rotational stops. This means that the frequency shift at the zero mark on the scale is an exponential function of the holding time in the multivibrator's with mirrors arrested state. Therefore,

$$\text{dev}\ \hat{\sigma} = 0 \text{ and } \text{dev}\ e_s = [\text{dev}\ e_{st=0}]\ e^{-Gt}/\eta_{VR}. \tag{5-61}$$

This type of elastic aftereffect is confirmed by experimental data (Fig. 5-8b), where $\Delta f_a \approx \Delta f_{emax}(1\text{-}e^{-t/20})$ corresponds to the frequency shift, t is the time in the arrested state, in minutes, and $\Delta f_{emax} \approx 10$ Hz. When a multivibrator is held arrested in the extreme tensioned position, the frequency adequate for return to the zero mark without resetting is lowered by $\Delta f_e \leq 10$ Hz. However, it increases by $\Delta f_e \leq 10$ Hz if the device is stored with completely relaxed tension. The experiment showed that the shift in frequency for the optichator is approximately equal to:

$$|\Delta f_{opt}| \leq 4(|\Delta\theta_{opt}|\ i_\theta^{-1} + (2/3)L|\varepsilon^{33}|\ )0.0013m_\sigma \tag{5-62}$$

and the mikrochator has a shift in frequency that equals

$$|\Delta f_{mic}| \leq 2(|\Delta\theta_{mic}|\ i_\theta^{-1} + (4/3)L|\varepsilon^{33}|\ )0.0013m_\sigma, \tag{5-63}$$

where $|\varepsilon^{33}|$ is the module of the longitudinal strain in the helicoid at its retraction, $\Delta\theta$ is the angle of untwist during the measuring rod retraction, and $|\Delta\theta_{mic}| < |\Delta\theta_{opt}|$. It should also be noted that a damper with fluid of comparatively low viscosity (10^{-2} m^2/s) was used in the experimental mikrochator devices. The damping fluid in the optical device was of higher viscosity (the order of 10×10^{-2} m^2/s), which might also influence the aftereffect. It would be helpful for reduction of the aftereffect in the latter case to limit the crosspiece rotation angle inside the dail (within ±45°) using, for example, a small auxiliary needle on the crosspiece and stops on the housing. It would also be advantageous to use a damping fluid of the lowest possible viscosity.

The amplitude Y_m of the vibration in the middle of the pretwisted helicoidal section of the multivibrator was approximately equal to 0.04 mm. The experimental estimation of this amplitude was provided with the double microscope, and its calculations were done with the known formula

$$Y_m = 2E_m B_{1M}^{-1}\ L^{-1}\omega_{vo}^{-1}, \tag{5-64}$$

where $E_m = 3.5\times10^{-3}$ V is the voltage of supplied electrical power, $B_{1M} = 0.34$ T is the magnet field inductance in the gap of the permanent magnet, $L = 0.02$ m is the length of the string segment interacting with the magnetic field, and ω_{vo} is the circular frequency of oscillation. In this case, the mean induced longitudinal displacement of the ends of the vibrating helicoid corresponds to the relationship

$$\overline{\Delta S} = (t^{-1}\int_0^t dt \int_0^L \sqrt{1+y'^2}\,dx)(1-i_\theta j_\theta) \approx Y_m{}^2\pi^2\,(1- i_\theta j_\theta/j_1)/8L, \tag{5-65}$$

where j_1 is the flexural stiffness of the front cantilever support of the helicoid.
It is possible for oil to drip out of the liquid damper at the large amplitude of these longitudinal oscillations. One can decrease the amplitude Y_m by increasing of the magnetic inductance B_{1M} in the magnet poles' gap and decreasing of the supplied voltage from the self-excited oscillator.
Internal friction's dependence on frequency. It was experimentally found (Y.M. Tseytlin, 1988) that internal friction may be lowered in a helicoidal multivibrator when it is excited with high-frequency vibrations in the range of 2.5 to 3.5 kHz. This effect is manifested in the uncertainty of readings from an appropriate spring-type head during multiple lifting (retraction) and dropping of the measurement rod whose tip rests in its working position on a fixed datum (e.g., gage block) of the column in which the head is mounted. The smaller this uncertainty range of reading the lower the internal friction in the precision mechanical system. Comparison of an uncertainty of readouts at zero exciting frequency R_o and R_f at various frequencies f of oscillations for an initially pretwisted helicoid indicates that an uncertainty $R_f < R_0$ in certain frequency bands (Fig. 5-8c), and that $R_f / R_o < ½$ for the frequency bandwidth of $f = 2800$-3200 Hz.

Expansion of FR-III measuring range.

Elastic kinematic systems including the helicoidal one have a small working range due to the limited elastic strains and stresses in real materials. Elastic elements with thin walls are used in order to increase deformation options and decrease the corresponding stresses. If it possible, bending is preferable over stretching and twisting. However, the latter is possible not in all cases including the helicoidal multivibrator.
Rolamite options. To increase the working range in the helicoidal multivibrator's system, it can be kinematically linked to the rolamite-type precision band mechanism (ELKS BL-IV, Table 1-4) that has a wide working range and minimal external friction among band mechanisms.

The rolamite has been invented by D. F. Wilkes (Patents USA 3,452,175; 3,471,668, 1969). Further developments of the rolamite are shown in USA Patent 3,730,007 (inventor C.K. Wellington, 1973). As we know, the helicoidal multivibrator is a very sensitive transducer with the threshold up to nanometers. But its measuring range is very limited. To improve this deficiency, we proposed to connect helicoidal multivibrator 11 with magnet 10 to a ring-like elastic element 1 (Fig. 5-9a) that is linked (through an elastic attenuation compensator not shown on the schematic) to band 2. This flexible band 2 is engaged with round roll 3 that is installed on the same axis and connected to polygonal roll 4. The latter one together with conjugate roll 5, flexible band 6, and slot with sides 7, 8 in the stand fulfill the rolamite system. The sequential turns and displacement by pitch b_n of the polygonal roll with radius R_m at the continuous motion of spindle 9 divide the working range to sub-ranges b_{nr}. These sub-ranges are equal to the side-size b_n of the polygonal roll and to the measurement range in use of the helicoidal multivibrator connected to self-excited oscillator 13 and frequency meter 14. Size b_n may be larger than sub-range b_{nr} if the elastic attenuator-compensator installed between band 2 and ring 1. This measuring system may be called "quasi-adaptive" to the environmental temperature (*see* Chap. 7) as a result of the temperature influence on flexible band 2. The more invariant version with shorter band 2 in the relation to the changes of ambient temperature is shown in Fig. 5-9b, where the elastic ring element together with the helicoidal multivibrator are movable. The second rolamite 12 with a punched tape can be connected to the spindle for the compensation of possible jumps in the measuring force.

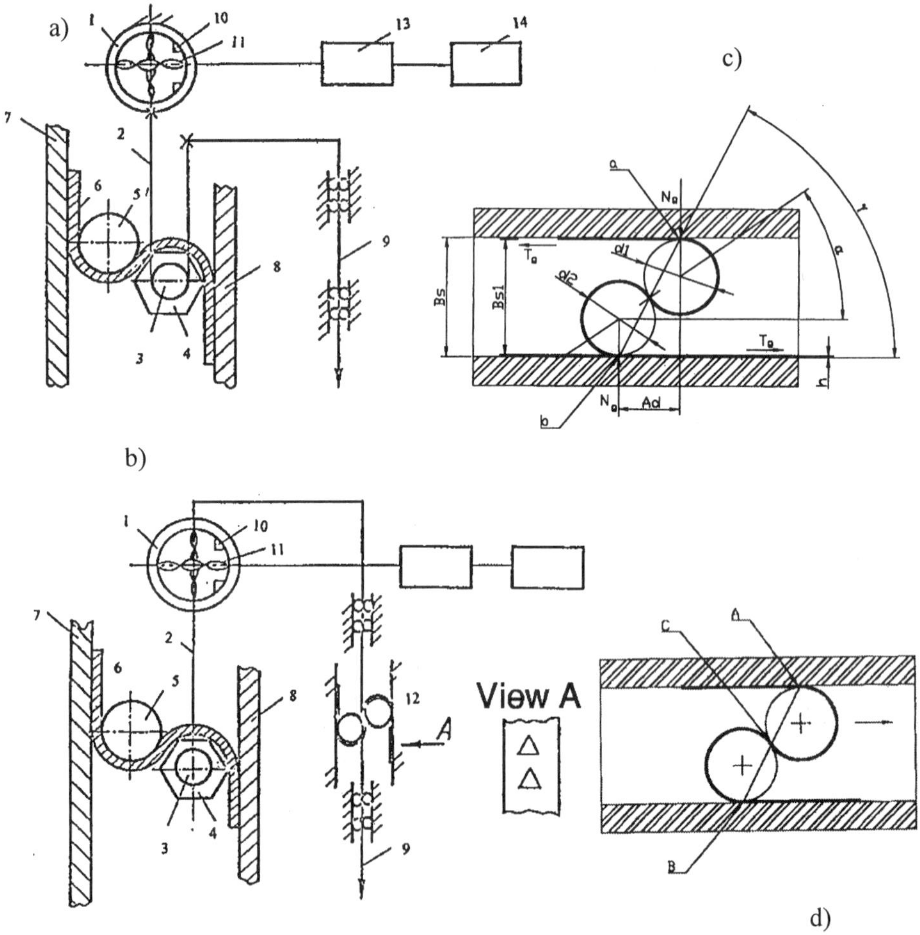

Fig. 5-9. Helicoidal multivibrator with rolamite (Patent SU 1620803, 1991).

BL-IV. Specifics of the rolamite (after D.F. Wilkes, 1968). A rolamite has some engineering constraints. In order to insure fully elastic behavior or some minimum endurance level for the band, the maximum strain induced by the combined effects of bending must be kept below allowed level to conform with the roller diameter and all other axial strains induced by initial tensioning, loading, or Poisson's ratio effects due to normal loading. Therefore, we consider all of the induced strains summed as follows:

$$\varepsilon^* = \varepsilon_{F\max} + \varepsilon_T + \varepsilon_L + \nu_p \varepsilon_N , \qquad (5\text{-}66)$$

where $\varepsilon_{F\max}$ is the maximum tensile fiber strain induced by the flexure around rollers, ε_T is the strain induced by band tension, ε_L is the tensile strain induced by loading, ε_N is the normal compressive strain in the band. In most situations, flexural strains are strongly dominant. If $\varepsilon_{F\max}$ is set at approximately 70% of allowed ε^* strain limit for the used material, a safe initial structure estimate will emerge without a detailed consideration of other terms that can be done later. In the helicoid multivibrator case, there is no large loading on the band except those induced by the preliminary and working small tension.

The constraints imposed by the free-rolling locked cluster (Fig. 5-9c) are

$$(B_s - 3h) > d_1 \text{ and } d_1 \geq d_2 , \qquad (5\text{-}67)$$

where h is the thickness of the band. The position of rolls corresponds to angles $\alpha$and γ. Additionally, $(B_s - 3h)$ must be smaller than the sum of the roller diameters, or the cluster will not lock. Thus, the following inequality should be satisfied

$$(B_s - 3h) < (d_1 + d_2). \qquad (5\text{-}68)$$

Cluster equilibrium requires that the moment induced by the nonaligned band tension T_Q counteracts by an equal and opposite moment induced by normal force N_Q. Hence, $T_Q\, B_{s1} = N_Q\, A_d$, (5-69)

where $B_{s1} = B_s - h$ and A_d is the distance between the axes of rolls.

An essentially tight geometry is not necessary. It occurs when the tangent lines between the band and guide surfaces and the band and rollers fall inside the contact interface between the rollers, band, guide and tension is of the order

$$T_Q \geq EJ_b / hR_d , \qquad (5\text{-}70)$$

where J_b is the band's cross-sectional moment of inertia, and R_d is its rolling radius. Most rolamites have geometry which exhibit some degree of looseness. The friction coefficients on various rolamite configurations without lubrication and with the rollers' diameters of 5.2 to 12.7 mm have fallen in the range of 0.0004 to 0.0016, which is significantly better than on the lubricated surfaces (D.F. Wilkes, 1968). One of the most useful features of the rolamite geometry is an ability of the band to generate varying forces of any desired character along the length of travel. The following equation is useful in predicting the axial force, which may be obtained or used for designing width changes in the band to produce a certain force-deflection characteristic, providing that the yield stress of the band is not exceeded,

$$T_{QX} = E(b-a)h^3 / 6d^2 , \qquad (5\text{-}71)$$

where a and b are the effective width of a band at points A and B (Fig. 5-9d), respectively, and d is the diameter of the conjugate roller. Application of this equation with different types of width-position functions substituted for a and b produces force-displacement curves of any desired character. The normal stress value in the rolamaite band can be estimated from the following expression

$$\sigma = (1/d)[Eh/(1-\nu_p^2)] + T_Q / (bh).$$

There are known rolamite versions with two bands that work in reciprocate parallel for more accurate kinematic transmission at the reverse motion. However, it is essential for further rolamite applications in precision systems to study in detail its dynamics including the critical speed of rolls rotation.

5-3. Multivibrators with Shells and Quasi-Helicoids

Helicoidal multivibrators can be built not only on the basis of the pretwisted thin solid or perforated strips, but also on the basis of coiled and pretwisted shells as well as elastic systems with quasi-helicoid motion. We have two suitable types of shells: (i) coiled strip and (ii) pretwisted tube. Type (i) indeed is a bar but from a theoretical standpoint it should be considered a shell. Type (ii) is a tube that can be simultaneously used as a helicoidal multivibrator and the pressure sensor with the effective internal kinematic transmission. These helicoidal multivibrators have special features and can be used appropriately in different ways than the thin helicoidal multivibrators studied in Section 5-2. The pretwisted tube features were discussed in Chap. 3 and coiled strip was studied in Chap.4. In this section we consider their options as multivibrators and also discuss another quasi-helicoid multivibrator that can be based on the cable shroud spatial mechanism.

The ribbon coiled spring's multivibrator.

FR-IIIqco. A helicoidal multivibrator can be built on the basis of the double-coiled and combined coiled strips. In the first case, at least one coiled part of the double-coiled spring is located in the *N/S* magnetic field (Fig. 5-10a). In the second case, the initially flat strip 1 (Fig. 5-10b) or the coiled part 2 (Fig.5-10c) may be located in the magnetic field *N/S*. However, electrical current in the thin helical nanospring may cause its compression and bending by generated magnetic force.

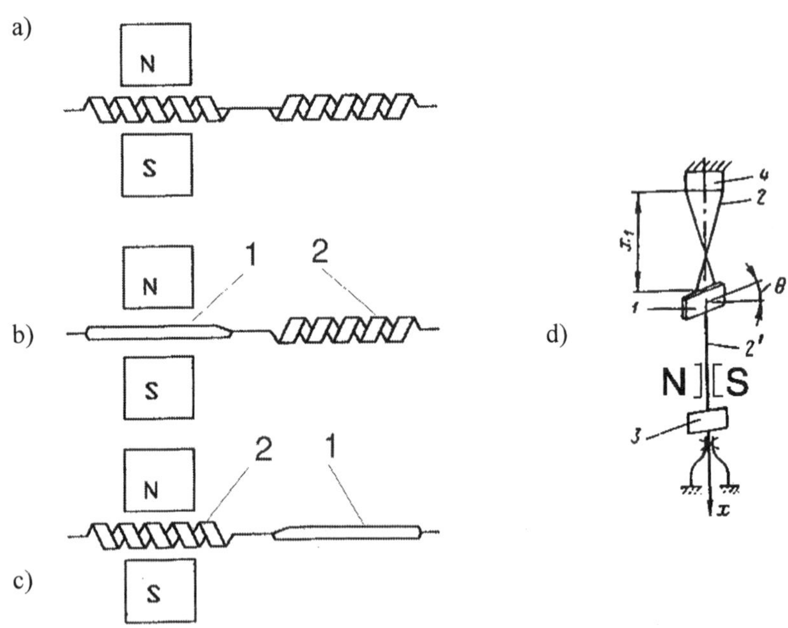

Fig. 5-10. Quasi-helicoidal multivibrators with the double coiled strip and shroud-cable system.

Natural frequency. Let us evaluate the natural frequency of the coiled strip-string. If we suppose that the thin coiled strip with a small diameter $D=2R$ and thickness h has negligible flexural rigidity, its natural frequency may approximately be calculated on the basis of the formula for a thin flexible string:

$f_{nc} = (n_f/2L)\sqrt{Q_p/\rho_\gamma A} = (n_f/2L)\sqrt{Q_p\cos\alpha/\rho_\gamma bh}$ (5-72)

because a cross-section area of the coiled ribbon strip is equal to $A = bh/\cos\alpha$, L is the length of a coiled part, Q_p is the force of longitudinal tension, and ρ_γ is the density of the strip's material. If the flexural moment M^f in the double-coiled strip equals zero, the corresponding longitudinal force derived from (4-61b) and (4-61c) is equal to: $Q_{p\,M^f=0} = M^t_{M^f=0}/R$, which after substituting $\frac{1}{R} = (\frac{\cos\alpha_0}{R_0} - \frac{\theta}{L_r})/\cos\alpha$ yields the following expression

$$Q_{p\,M^f=0} = 2(1-\nu_p)(Db/R_o)\left(\frac{\cos\alpha_0}{R_0} - \frac{\theta}{L_r}\right)[\sin(\alpha-\alpha_o)/\cos(\alpha-\alpha_o)^2]\text{x}$$

$$\{\cos(\alpha-\alpha_o)-K_\xi[\cos(\alpha-\alpha_o)-2(1-\nu_p)\cos^2\alpha\sin\alpha$$

$$+(1-\nu_p)^2\cos^2\alpha(2\alpha-\alpha_o)]\}\,/\,[1-K_\xi(\sin^2\alpha+\nu_p\cos^2\alpha)^2]. \quad (5\text{-}73)$$

Specifying the angle α of the deformed spring, we use (4-59a) to determine the longitudinal travel s and (4-61d,e) for estimation of the corresponding untwist angle θ. For small displacements, we assume that $(\alpha-\alpha_o)=\Delta\alpha$ is small and $K_\xi \to 0$.

Then $Q_{pM^f=0} = \dfrac{2(1-\nu_p)Db\Delta\alpha}{R_0^2\cos\alpha_0}$. (5-74)

Substituting expression for flexural cylindrical rigidity $D = \dfrac{Eh^3}{12(1-\nu_p^2)}$ and

$\Delta\alpha = \dfrac{sS_{av}}{2\pi R_0 L/\cos\alpha_0}$ from (4-61f) into (74), we finally have a linear expression for the tension force at small displacements s of double-coiled strip's ends as follows

$$Q_{pM^f=0} = \frac{Ebh^3 S_{av}}{12\pi R_0^3 L}s. \quad (5\text{-}75)$$

If we recall that in accordance with Table 4-8 $s = \dfrac{\pi D_{r0}^2}{0.0573 S_{av}}\theta$ and $D_{r0}=2R_0$, then the rigidity of the double-coiled strip's extension related to the angle θ of its crosspiece rotation is equal to: $j_\theta = \dfrac{Q_{pM^f=0}}{\theta} = C_Q\dfrac{Ebh^3}{D_{r0}L}$, (5-76)

where coefficient $C_Q \approx 5.8/(1+\nu_p)$ if j_θ is expressed in cN/(angular degree). Comparison of (5-76) with the analogue expression in Table 4-8 proves their structural similarity.

Combined spring calibration. In the version of a combined coiled spring with the initially flat strip in a magnetic field, the natural frequency of the latter strip can be estimated through the means similar to ones used in Section 5-1. The natural frequency of a strip at zero twist is given by the known expression

$f_0 = 0.5\text{L}\sqrt{1/(\rho_\gamma A)}\,[Q_{pM} + (4EJ_1/L^2)]^{1/2}$, (5-77)

and the natural frequency of a twisted strip is similar to (5-3) and equal to:

$f_{0t} = f_0 (1 + K_f A_{f\theta})^{1/2} = f_0\{1 + K_f[1-(\sin\theta/\theta) + (\theta\sin\theta/ (4\pi^2-\theta^2))]\}^{1/2}$, (5-78)
where $K_f = (J_2 - J_1) / [(L^2 Q_{pM} /2\pi^2 E) + 2J_1]$, J_1 is the cross-section moment of inertia at zero twist ($\theta = 0$) with respect to cross-section coordinate axes, Q_{pM} is the axial tension, and the untwist angle at the input displacement s in accordance with (4-64) is equal to: $\theta = i_{\theta m}\, s$. However, unlike in (5-3), where axial tension T_Q is assumed to be constant, axial tension Q_{pM} in accordance with (75) is a variable that makes expressions (77) and (78) more complicated for linearization. If the strength conditions and attainable frequencies allow, one can substitute the initially flat strip by a round string with equal torsional rigidity which calibration formula is simpler as for regular string where small stretch s and change of natural frequency Δf will have the following relationship: $\Delta f = ½\, \Delta Q_{pM}\, (f_0/Q_{pM0}) = \dfrac{1}{2}\dfrac{Ebh^3 S_{av}}{12\pi R_0^3 L}(f_0/Q_{pM0})s.$ (5-79)

Here the ratio between initial frequency and tension force (f_0/Q_{pM0}) as well as other parameters are assumed to be constant.

The pretwisted helicoidal tube multivibrator.

FR-IIIqtu (Table 1-6). These tubes find an effective use as sensitive elements in the manometers (*see* Chap.3) for high pressure measurement. They can also be used as double-helicoids in kinematic systems. These helicoids have a smaller longitudinal rigidity $j_{\theta tu}$ than the solid helicoids with the same outer parameters that is clear from the following expression $j_{\theta tu} = 69.6\, (G\lambda\pi/ L\kappa_p)[\, a_{1(1)}b_{1(1)}(1+ \nu_{jf}b^2_{1(1)}\, \kappa_p^2/\lambda)$
$- a_{1(2)}b_{1(2)}(1+ \nu_{jf}b^2_{1(2)}\, \kappa_p^2/\lambda)]$, (5-80)
where $a_{1(1)}; b_{1(1)}$ and $a_{1(2)}; b_{1(2)}$ are the larger and smaller semi-axes of an outer surface with subscript 1 and an inner surface with subscript 2, $\kappa_{p1} = \kappa_{p2} = \kappa_p = 2\pi / S_o$ if S_o is the helix pitch, and $\lambda_1 \approx \lambda_2 \approx \lambda = (b_{1(i)}/a_{1(i)})^2$. The pretwisted double-helicoid tube has a significantly smaller transmission ratio than the thin double-helicoid from a pretwisted thin strip.

The spatial shroud system multivibrator.

FR-IIIqca (Table 1-6). The most convenient use for the quasi-helicoidal multivibrator is the aforementioned second version of the spatial shroud system with the two crossed 2 (Fig. 5-10d) connected to base 4 and one 2'straight middle wire which is located in the magnetic field (*N/S*) and connected to the movable spindle 3. Wires 2 and 2' are clamped to rotatable output element 1.
Natural frequency. The natural frequency of the wire 2' vibration may be calculated as for a regular round string with the formula

$(n_f/ 2L)\, \sqrt{Q_s/\rho_\gamma A}$, (5-81)

where L is the length of wire 2', Q_s is the force of its tension, $A = \pi\, d_w^2/ 4$ is its cross-section area, ρ_γ is the density of the wire's material, and n_f is a harmonic's number. $|Q_s| = [0.15G\, d_w^4\theta X_0 /(L\, L_a L_b \sin\theta)] + 0.6\, EL_a\, d_w^4\theta X_0/ L^3 L_b$, (5-82)
or if $\theta \ll 1$ and $\sin\theta \approx \theta$, then
$|Q_s| = [0.15\, G\, d_w^4\, X_0 /(L\, L_a L_b)] + 0.6\, EL_a\, d_w^4\theta X_0/ L^3 L_b$. (5-83)
The derivative of (83) in respect to the untwist angle yields the expression for driving force gradient in this mechanical transformation system

$j_\theta = dQ_s / d\theta = 0.6\, EL_a\, d_w^{\,4}\, X_0 / L^3 L_b$, (5-84)

where X_0 is the initial value of the distance x_1 (Fig. 5-10d), other notations as in Fig. 4-19g.

5-4. The Largest Attainable Fundamental Frequencies

General approach.

Old estimations. The natural frequency f_0 for a beam can be represented by the expression $f_0 \approx (1/2\pi)\sqrt{Q_F/\delta_s m_{eq}}$, (5-85)

where Q_F is the force applied to the beam, δ_s is the sag, m_{eq} is the point-load mass equivalent to the elastic element's mass of unit length. The formulae presented in Table 5-3 for different elastic elements were obtained by substituting the expressions of δ_s and m_{eq} into (85).

Table 5-3. Formulae for the largest attainable natural frequency

Structure's number, name, and load	Structure and type of loading*	Natural frequency f_0	Product $f_0 \times \delta_{smin}$*
1.Prismatic cantilever, concentrated force	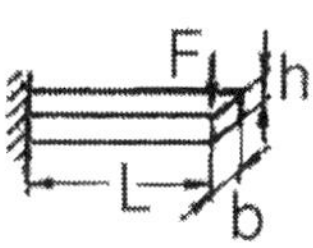	$(h/1.97\pi L^2)\sqrt{E/\rho_\gamma}$ **	$\sigma_{max} / (3\pi\sqrt{E\rho_\gamma})$
2.Prismatic cantilever, distributed load	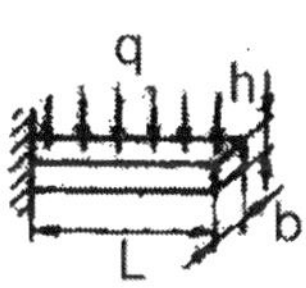	$(h/1.2\pi L^2)\sqrt{E/\rho_\gamma}$ *	$\sigma_{max} / (4\pi\sqrt{E\rho_\gamma})$
3.Doubly clamped beam, distributed load	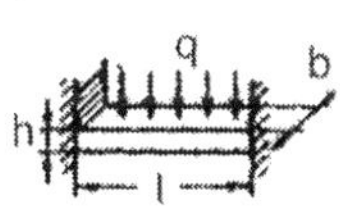	$(h/0.3\pi L^2)\sqrt{E/\rho_\gamma}$ ***	$\sigma_{max}/ (3.2\pi\sqrt{E\rho_\gamma})$
4.Round plate, distributed load	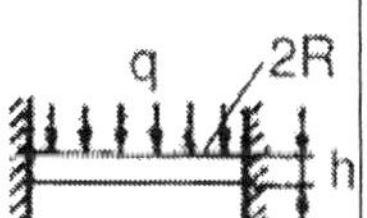	$q = gh\rho_\gamma$ $(h/2R^2)\sqrt{E/\rho_\gamma} \times \gamma_f$, $\gamma_f = 1$, or ****	$\sigma_{max} / (7.8\sqrt{E\rho_\gamma})$

*After I.N. Dolitskii *et al.*, 1972.

Notes to Table 5-3: ** This formula is based on the frequency equation $\cos \alpha L \cosh \alpha L$=-1 with the first root αL=1.875. ***The formula is based on the frequency equation $\cos\alpha L\cosh\alpha L$=1 with the first root αL=4.730 (*see* Y.G.Panovko, 1968); if the beam has one fixed and one simply supported end as in the contact mode of AFM (*see* Chap. 8), its frequency equation is $\tan\alpha L = \tanh\alpha L$ with the first root αL_n =3.927, and the natural frequency of the simply supported beam increases with addition of axial tensile force Q_{ax} and decreases with the axial compression force - Q_{ax} in accordance with (S.P.Timoshenko *et al.*, 1964) the following expression : $f_0 = \frac{hn_f^2}{0.702\pi}\sqrt{E/\rho_\gamma}\sqrt{1+\frac{Q_{ax}}{n_f^2 EJ\pi^2}}$, where we accept $J = bh^3/12$ and n_f is the mode number . ****Influence of surrounding air or liquid (Weaver *et al.*, 1990) with density $\rho_{\gamma 1}$ reflects in the factor $\gamma_f = 1/\sqrt{1+\beta_f}$, where $\beta_f = 0.6689\frac{\rho_{\gamma 1}}{\rho_\gamma}R/h$.

Stress σ_{max} is the largest acceptable stress for the material under working conditions: σ_e, σ_f, or σ_{en} (*see* Table 1-14).

The last column in Table 5-3 shows the relationship between natural frequency f_0, minimal sag δ_{smin}, and maximum stress σ_{max} without any elastic element's geometrical dimensions figuring in these formulae. It is important to note that in the above reasoning a complete utilization of the elastic element's strength characteristics was assumed. The similar procedure was used for the round plane diaphragms subjected to a distributed load (pressure) whose formulae are also shown in Table 5-3's row 4. Analysis of the formulae in Table 5-3 shows that the product of the natural frequency f_0 of the elastic element on its minimal sag δ_{min} does not depend on the element's size, but is completely determined by the selected material features. Moreover, this product is structurally the same for all studied elastic systems. The computation's results of f_0 for elastic elements made of the different materials (steel, beryllium bronze, titanium, and aluminum alloys), in the case when minimal sag δ_{smin} =10 μm, show that the natural frequencies are in the range from 50 to 113 kHz. However, we should recall that the value of the minimal sag was selected in this case as a typical for the pressure electromechanical transducers of different types known in 1970s. It should be noted that this value is a conditional one and is changed in new structures, which have much higher fundamental frequencies. The largest natural frequency for the flat strings can be defined from the expression $f_n = 0.5 n_f L^{-1}\sqrt{\{\sigma_e\}/\rho_\gamma}$, (5-86)

where n_f is the harmonic's number and $\{\sigma_e\}$ is the permissible yield stress for the string's material. Then at the minimum value of the string's length L=1 mm(this number is conditional with respect to the mesoscopic structure's options), the largest yield stress value $\{\sigma_e\} \approx 1800$ MPa of the elastic materials in the bulk state, and the density of the material $\rho_\gamma \approx 8\cdot 10^{-6}$ kg/mm^3, we obtain $f_{n=1} \leq 230$ kHz.

Revisited estimations. Today these options have large progress in the micro and nanotechnology field. The nanowire-based very-high-frequency electromechanical resonator is already fabricated with natural (fundamental) frequency of 105.3 MHz (A. Hussain *et. al.*, 2003). The resonator is built of suspended platinum nanowire and has a diameter of d_w=2R_d=43 nm, length of L=1.3 μm with a suspended mass of 40 fg and a volume of 1.9×10^{-15} cm^3. This microresonator is among the smallest *NEMS* structures whose motion has been detected directly. Advanced sensing applications require both high responsivity and ultrahigh frequency operation. Among the most challenging of these problems are attainment of subattonewton, high frequency force sensing for magnetic resonance force microscopy, and the study of mechanical motion in the quantum regime. Let us recall from Table 1-12 that a platinum microstrip has yield stress σ_e = 1570 MPa, Young's modulus E=169 GPa, and density ρ_γ=21090 kg/m^3. If we assume that the microresonator is a flat string, its maximum fundamental frequency estimated by (86) is equal to:

$f_0 = \frac{1}{2\times1.3\times10^{-6}}\sqrt{\frac{1570\times10^6\times9.8}{210900}}$ = 103.9 MHz.. This number is within 1.3% close

to the measured by A.Husain *et al.*(2003) nanowire resonator's natural frequency. An inapt modeling the response of this microresonator as that of an unstrained doubly clamped beam with formula borrowed from S.Timoshenko *et al.*(1974) $f_0 = \frac{22.4}{2\pi}\frac{R_d}{2L^2}\sqrt{\frac{E}{\rho_\gamma}}$, the creators of the resonator (A.Husain *et al.*, 2003) got a calculated fundamental frequency of 64 MHz, which is much lower than the measured response. They realized, however, that this is likely "due to the differential thermal contraction of the beam and the substrate, which should cause the beam to be under tension." Therefore, the right modeling of this resonator's response is a string version with formula (86) or with formula for beam under tension (*see* note ** to Table 5-3). This example shows how important the correct theoretical model for structural analysis and synthesis as well. Indeed, the fundamental frequency for nanoresonators can be even higher because a whisker of many materials has much higher mechanical properties than in a bulk state. Magnetomotive detection, in which an ac current drives a string in transverse magnetic field, has been employed to successfully actuate and read *NEMS* resonators at frequencies up to and exceeding 1 GHz. The measured quality factor Q of the microresonator is approximately 8500-8800 in vacuum and decreases slightly with increasing magnetic field from 1 to 8 T. Previous studies have shown that quality factors measured for *NEMS* generally decrease with increasing surface area-volume ratio, apparently indicating that surface processes contribute strongly to energy dissipation. D.W. Carr *et al.*(1999), for example, measured the quality factors of single-crystal silicon beams and found that Q-factor decreases from 3000 for beams with a surface-volume ratio 0.02 nm^{-1} to a value 1000 for devices with the ratio of 0.06 nm^{-1}. The response of the platinum nanowire resonator as the frequency is swept both upward and downward confirms the expected hysteretic nature in the nonlinear regime. In the simplest model for the phenomenon, the critical oscillation amplitude x_c above which instability occurs is dependent only upon the geometry, Q-factor, and Poisson's ratio ν_p of the beam, and is given by as $x_c = \frac{h_{ge}\sqrt{2}}{\sqrt{0.528Q(1-\nu_p^2)}}$ (*see* L.D. Landau and E.M. Lifshitz , 1969),where for round beam h_{ge}= $d_w\sqrt{3/4}$ and d_w is the diameter of the beam; for the rectangular cross-section beam h_{ge} =h and h is the thickness of the beam at equal normal stress. The critical amplitude for the platinum microresonator evaluated on this basis is in the order of nanometers that pose a significant challenge for detecting the motion of nanowire and nanotube devices.

Microcantilevers. They are of different shapes: (i) rectangular (Fig. 5-11a); (ii) triangular (Fig. 5-11b), (iii) V-shaped with the triangular (Fig. 5-11c) and trapezoid (Fig. 5-11d) recesses, double-tapping and wedge type.

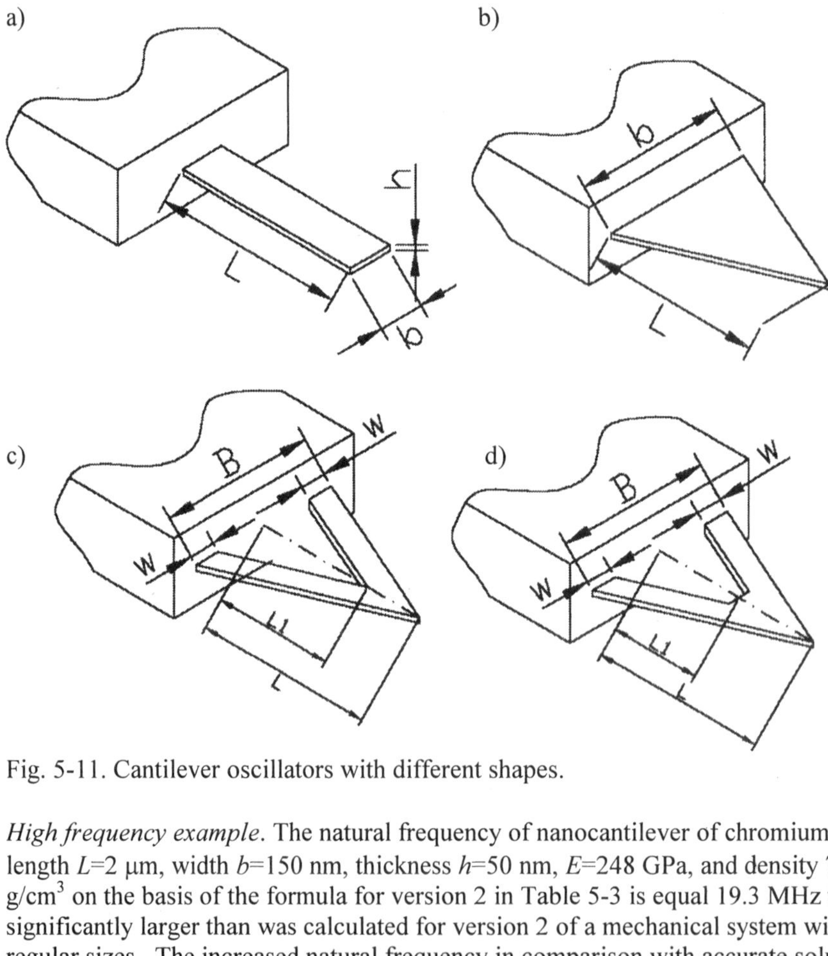

Fig. 5-11. Cantilever oscillators with different shapes.

High frequency example. The natural frequency of nanocantilever of chromium with length L=2 μm, width b=150 nm, thickness h=50 nm, E=248 GPa, and density 7.19 g/cm^3 on the basis of the formula for version 2 in Table 5-3 is equal 19.3 MHz that is significantly larger than was calculated for version 2 of a mechanical system with regular sizes. The increased natural frequency in comparison with accurate solution for the rectangular cantilever has a sharp-end wedge cantilever whose natural frequency one can estimate on the basis (after W. Weaver *et al.*, 1990) of the following formula $f_0 = \frac{h_r/2}{0.65\pi\, L^2}\sqrt{E/\rho_\gamma}$, where h_r is the height of the wedge in the fixed end.

Triangular cantilevers. Let us estimate the natural frequency of a solid triangular cantilever. This can be done by Rayleigh's method with application of a force on the movable end of the cantilever in which case the flexure curve is represented by the following formula

$X(x)=\frac{QL^3}{6EJ}\left(\frac{x^3}{L^3}-\frac{3x}{L}+2\right)$, where L is the length of the cantilever (Fig. 11b), x is the longitudinal coordinate with the origin in the movable end, and the flexural moment of sectional inertia equals $J= b(x)h^3/12$, where $b(x)=bx/L$.

In accordance with Raleigh's method, the square of natural frequency of the triangular cantilever is equal to:

$$f_0^2=\frac{1}{(2\pi)^2}\frac{Q^2L^3/3EJ}{\int_0^L \rho_\gamma b\frac{x}{L}h(QL^3/6EJ)^2\left(\frac{x^3}{L^3}-\frac{3x}{L}+2\right)^2 dx}=\frac{68.571EJ}{(2\pi)^2\rho_\gamma bhL^4}$$

because the integral $\int_0^L I_f=\int_0^L\frac{x}{L}\left(\frac{x^3}{L^3}-\frac{3x}{L}+2\right)^2 dx=0.175L$. If one assumes that the radius of gyration $r_g^2= J/bh = h^2/12$ of the triangular cantilever cross-sectional area is constant, we obtain the following expression for the fundamental resonant frequency of the solid triangular cantilever

$$f_0\approx\frac{h}{0.837\pi L^2}\sqrt{E/\rho_\gamma}\approx 0.38\frac{h}{L^2}\sqrt{E/\rho_\gamma}. \tag{5-87a}$$

Similar expression may be found in W. Weaver *et al.* (1990), where the following formula is presented for the cantilever beam with variable cross section form

$$f_1=\frac{\alpha_j r_g}{2\pi L^2}\sqrt{E/\rho_\gamma}. \tag{5-87b}$$

Here r_g is the radius of gyration of the built in section, α_j is a constant depending on the shape of the beam and on mode of vibration.

If the variation of the cross-sectional area and of the moment of inertia along the x axis can be expressed in the forms $A=A_0(1-c_j\frac{x}{L})$ & $J=J_0(1-c_j\frac{x}{L})$, which is typical for the triangular cantilever at c_j=1 with x being measured from the built-in end, then the quantity α_j=7.16. Substituting this quantity and expression for the constant r_g into the above shown formula, we obtain the natural frequency of the triangular cantilever, which equals

$$f_1=\frac{h}{0.967\pi L^2}\sqrt{E/\rho_\gamma}\approx 0.33\frac{h}{L^2}\sqrt{E/\rho_\gamma}. \tag{5-87c}$$

Expression (87a) differs within 15 % from presumably more accurate (87c). This difference may be caused by not accurate enough approximation of elastic curve used by us above in the Rayleigh's method. The numerical coefficients on right-hand side of these expressions are close to the coefficient 0.31 which has been chosen for curve fitting the plot to the results obtained for the triangular cantilever by FEA with software *ANSYS*, where the experimental results, the analytical plot, and the *ANSYS* plot are matched within 10%. If the cantilever has a tapering trapezoid form, then the corresponding quantity c_j and the natural frequency of the cantilever are smaller than for the triangular one with the sharp free end. An interesting reader can find the fundamental frequency coefficients for spring hinged cantilever and single- and double-tapered beam in publication by E. B. Magrab (1979).

V-shaped cantilevers. These cantilevers have attainable fundamental frequency, which is an intermediate between that for rectangular and triangular cantilevers. We can find many publications devoted to this subject because V-shaped cantilevers are widely used in AFM. The simplest theoretical model describes V-shaped cantilever by two parallel rectangular cantilevers. Some authors showed good agreement between this model and experimental data. But this model does not reflect specifics of V-shaped cantilevers' geometry. Other publications are devoted to more complicated and presumably more accurate models: (i) "correction factor" with numerical integration (*see* H.-J. Butt *et al.*, 1993) and (ii) "variational" representation (*see* G.Y. Chen *et al.*, 1994). Let us recall that a proper calculation of the cantilever's fundamental frequency is essentially important for the synthesis of a precision system in AFM. It should be noted, however, that not all of the more complicated models demonstrate notable increase in accuracy (*see* Sect. 8-4), and most of them are difficult to use for the digestible structural synthesis. Therefore, we intend here to find a simple model for V-shaped cantilever, which reflects specifics of its geometry and provides satisfactory accuracy. This model may be built by Raleigh's method on the similar basis as we used it for a solid triangular cantilever because the radius of gyration remains constant for all cross sections in the accepted by us model (Fig, 5-11c,d). The main parameters of V-shaped cantilever that have influence on its fundamental frequency are thickness h, overall length L, recess length L_1, Young's modulus E, and density ρ_γ of material. The following formula reflects all of these parameters

$$f_0^2 = \frac{1}{(2\pi)^2}\,\frac{F\dfrac{FL^3}{3EJ}}{\rho_\gamma Bh\displaystyle\int_0^{L-L_1} I_f + \int_{L-L_1}^{L} \rho_\gamma B\left(1-\frac{L_1}{L}\right)h\left(\frac{FL^3}{3EJ}\right)^2\left(\frac{x^3}{L^3}-\frac{3x}{L}+2\right)^2 dx}\text{, where}$$

$$\int_0^{L-L_1} I_f = \int_0^{L-L_1} \frac{x}{L}\left(\frac{x^3}{L^3}-\frac{3x}{L}+2\right)^2 dx = 0.175(L-L_1) \text{ and } \int_{L-L_1}^{L}\left(\frac{x^3}{L^3}-\frac{3x}{L}+2\right)^2 dx = 0.943L_1 .$$

Hence, after substituting and cancellations, we finally obtain a quasi-universal and simple enough formula for the fundamental frequency of V-shaped cantilever as follows

$$f_0 = \frac{h}{2\pi\, L^2}\sqrt{\frac{E}{\rho_\gamma}\,\frac{1}{(0.175+0.768L_1/L)}} \,. \tag{5-87d}$$

One can use this quasi-unified formula for a solid triangular cantilever, if to assume that L_1=0, and for the rectangular cantilever as well if to accept that L_1=L. The calculation results on the basis of (87d) are close within few percents with the experimental data and more complicated theoretical methods presented by other authors (*see* Sect. 8-4).

Recent studies show that nanomechanical resonators can achieve fundamental resonance frequencies exceeding 1 GHz with quality factor Q in the range $10^3 \le Q \le 10^5$ in vacuum. However, the recent experimental measurement of their Q-factor in air revealed very low single numbers. It is suggested that the Q-factor of the nanocantilever system can be improved by operating nanocantilever at a high-resonance frequency, or by the active Q-control feedback method (*see* Chap. 8).

Plates. Vibrations of plate studies are based on the solutions of the differential equation similar to $N^{\underline{o}}$ 16 in Table 1-10.

Circular plates. The fundamental frequency of free vibrations for circular plates clamped at the boundary may be determined (after S.P. Timoshenko *et al.*, 1964) on the basis of the following formula

$f_0 = \frac{K_{mn}}{R^2}\sqrt{\frac{D}{\rho_\gamma h}}$,where K_{mn} is the constant whose value depends on the number of circular nodal lines m and diametric nodal lines n at the plate's vibration (Table 5-4, upper numbers).

Table 5-4. Nodal lines and constants K_{mn} for circular platcs and W_{mn} for membranes (absolutely flexible diaphragms).

m \ n	0	1	2	3
0	K=1.625 W=0.3826	3.377 0.6099	5.545 0.8173	8.123 1.0153
1	6.331 0.8785	9.680 1.1167	13.461 1.3396	17.671 1.5534
2	14.149 1.3773	19.111 1.6191	24.480 1.8494	30.287 2.0717
3	25.175 1.8768	31.678 2.1204	38.628 2.3549	46.023 2.5821

After W. Weaver *et al.*, 1990; J.W.S. Rayleigh ,1945; and R. Szilard, 1974.
Boundary circle is not included in m values.

In the case of elliptic plate clamped at the boundary with semiaxes $a>b$, we can use the same formula with $R=a$ and the constant K_{mn} value (after Szilard, 1974) which depends on the ratio b/a as follows: K_{mn}= 1.625 (b/a=1); 2.100(b/a=0.8); 3.215(b/a=0.6); 6.462(b/a=0.4); 23.873(b/a=0.2).

Thin plates (*membranes*). The fundamental frequency of free vibrations for the thin circular membranes (absolutely flexible diaphragms) clamped at the periphery may be determined (after J.W.S. Rayleigh , 1945; Szilard, 1974; Weaver *et al.*, 1990) on the basis of the following formula

$f_0 = \frac{W_{mn}}{R}\sqrt{\frac{Q_s}{\rho_\gamma h}}$, where W_{mn} is the constant whose value depends on the number of circular nodal lines m and diametric nodal lines n at the membrane vibration (Table 5-4, lower numbers), $Q_s = \sigma_n h$ is the uniform tension per unit length of the membrane's boundary and σ_n is the normal stress.

If the middle surface of a circular plate is stretched, its natural flexural frequency increases (S.P. Timoshenko *et al.*, 1964) in the proportion to the right member of the following formula

$f_0 = \frac{h}{2.1R^2}\sqrt{\frac{E}{\rho_\gamma(1-\nu_p^2)}}\sqrt{1+1.464\frac{U_1^2}{h^2}}$, where U_1 is the plate's deflection under applied uniform pressure p. In the case when $U_1 = ½h$, $U_1 = 0.89\frac{pR^4}{64D}$ .If we assume that plate's deformed elastic surface in this case is close to a spherical form, then the corresponding stretch of its middle surface is equal to:

$U_{st} \approx 2(\sin^{-1}2U_1/R)\frac{R^2+U_1^2}{2U_1} - 2R$. Therefore, at $2U_1/R \approx 0.15$ for example, stretch of the plate's middle surface is approximately $U_{st} \approx 0.0033R$ because the inverse sine function of X in this range can be expanded (*see* Table 1-9) by the minimal polynomial $1.011X$. Hence, the superposition method allows us to assume that the application of this initial stretch U_{st} per unit length of the circular plate middle surface boundary will increase its natural frequency in correspondence to $U_1=0.075R$. In general, the absolutely flexible diaphragm (membrane) with the different shapes that are fixed only at the boundary have natural frequency, which depends (W.Weaver *et.al.*, 1990) on applied to it uniform tension Q_{us} per unit length of the boundary, as $f_0 = \alpha_f\sqrt{\frac{Q_{us}}{\rho_\gamma hA}}$,where A is the membrane area, factor

$\alpha_f = 1.202/\sqrt{\pi} = 0.6782$ for the circular membrane, $\alpha_f = \sqrt{2}/2 = 0.7071$ for the square membrane, and $\alpha_f = 0.9129$ for the rectangular membrane with the ratio 3/1of its boundary sides length. The allowable largest tension Q_{us} is limited by the elastic stress of material which mean that $Q_{us}/h < \{\sigma_e\}$.

Rectangular plates. Exact solutions for the free vibration of rectangular plates are more complicated. Their behavior depends on the ratio of side dimensions, methods of clamping, uniformity of thickness, and number of modes in free vibrations.

Table 5-5. Nodal lines (dashed) and constants Z_{HV} corresponding to different modes and clamping options for free vibration of a square plate.

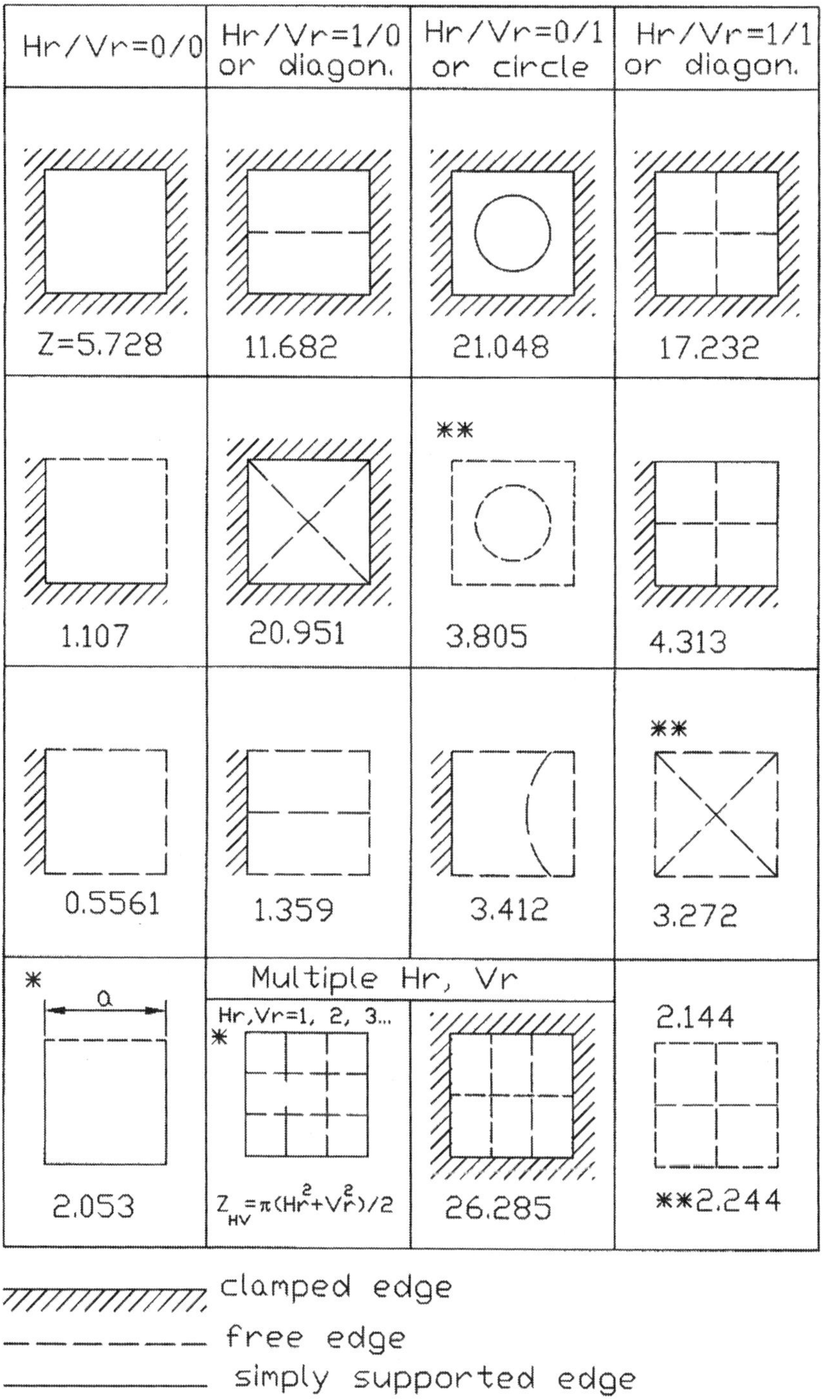

Note: Table 5-5 is after D. Young, 1950; V.V. Bolotin, V.N. Moskalenko, 1968, except * which is after R. Szilard, 1974, and ** which is at ν_p=0.225 after W. Weaver *et al.*., 1990.

However, there are several approximate methods for the solution of the differential equation $N^{\underline{o}}$ 16 (Table 1-10) in this case as well. The square plates free vibration may be determined (V.V. Bolotin, V.N. Moskalenko, 1968) on the basis of the following formula

$f_0=\frac{Z_{HV}}{a^2}\sqrt{\frac{D}{\rho_\gamma h}}$,where a is the width of plate's side and constants Z_{HV} depend on the clamping of edges methods and number of possible "horizontal" *Hr* and "vertical" *Vr* modes of vibration (*see* examples in Table 5-5).

In the case of rectangular plate at simply supported sides with dimensions $a \times b$ (*see* Szilard, 1974), the free vibrations frequencies are as follows

$f_{HV}=\frac{1}{2}\pi\,[(Hr/a)^2+(Vr/b)^2]\sqrt{\frac{D}{\rho_\gamma h}}$, where node number Hr=1,2,3... corresponds to side a and number Vr=1,2,3... corresponds to side b.

Frames.

Table 5-6. Boundary of instability for spring parallelogram carriage vibration

Schematic	Lower boundary of instability amplitude s_{1a}	Example
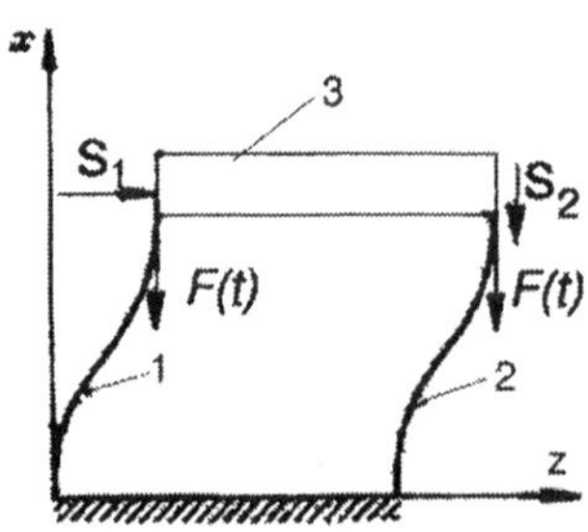	$1.18L\sqrt{\frac{F_t}{F^*+F_0}}$	$s_{1a}\approx 0.236L$ at F_t=0.04 F^* and F_0=0

$F^*=\frac{2(EJ}{L^2}$ is Euler's critical force (*see* Table 3-3) , J is the sectional moment of inertia for the leaf springs, and L is their length. After V.V Bolotin , 1956.

V.V Bolotin (1956) investigated the dynamic stability of statically indeterminate frames, including the frame similar to spring parallelogram (ELKS of type TG-I, Table 4-3). He showed that the differential equation for the vibration of crossbar-carriage 3 in the leaf springs 1 and 2's axes direction s_2 (depicted in Table 5-6) might be presented in the form $\ddot{s}_1+\omega^2(1-\frac{N_Q}{N_*})s_1=0$, where N_Q denotes the longitudinal force in the springs, $\omega^2 = \dfrac{24EJ}{(M_m + \frac{13}{35} m_{\gamma L} L)L^3}$ is the angular natural frequency of the frame vibration, $N* = \dfrac{10EJ}{L^2}$ is the critical Euler's force in the spring, M_m denotes the mass of the crossbar-carriage, and $m_{\gamma L} L$ denotes the mass of the spring. To determine the dynamic longitudinal force, the vertical vibrations s_2 of the crossbar are considered. In this case, $s_2 = ½ \int_0^L (\partial s_1 / \partial x)^2 dx = \dfrac{3s_1^2}{5L}$ (*see* Table 4-3).

If the longitudinal dynamic forces in the springs are $N_Q = F_0 + F_t \cos \Omega t - M_m \ddot{s}_2$, then the amplitude of the crossbar-carriage vibration is equal to $s_{1a} = \aleph^{-1/2} \sqrt{\dfrac{F_t}{F* - F_0}}$ and $\aleph = \dfrac{18}{25L^2} \dfrac{M_m}{M_m + \frac{13}{35} m_{\gamma L} L} \approx \dfrac{18}{25L^2}$, from which follows the linear approximation for the amplitudes on the lower boundary of the principal region of instability shown in Table 5-6.

The helicoidal pretwisted string. This string has a lower longitudinal rigidity (stiffness) than the flat string with the same dimensions, shape of the cross section, and of the same material. Therefore, the maximum attainable natural frequencies for the helicoidal strings are lower as well. This is easy to show, for example, on the basis of the helicoidal multivibrator. If the maximum attainable natural frequency for the flat string equals $f_{nmax} = (n_f/2L)\sqrt{\{\sigma_e\}/\rho_\gamma}$, then for the helicoidal multivibrator we have the largest attainable natural frequency, which is equal to:

$$f_{nmax} \approx = (n_f/2L)\sqrt{0.36 j_\theta \theta_{\lim} / \rho_\gamma A}\ . \tag{5-88}$$

Here $\theta_{\lim}$ is the limit of untwist angle for the used pretwisted strip in the helicoidal multivibrator; j_θ is the pretwisted strip stiffness at the longitudinal extension (in cN per angular degree), and A is the area of the strip's cross section. Substituting the expressions for $\theta_{\lim}$ and j_θ from Chap. 4 into (88) yields the calculation formula for the attainable largest natural frequency in the helicoidal multivibrator such as

$$f_{n\max} \approx (n_f/2L)\sqrt{\frac{\{\sigma_e\}}{\rho_\gamma}\frac{25.6(1+\nu_{jf}\chi_{cr})}{8\pi(1+0.38\chi_{cr})}}\ . \tag{5-89}$$

Hence, for the used pretwisted strips in the helicoidal multivibrator with $\chi_{cr}=8$ and $\nu_{jf}=0.18$, we have $f_{n\max} \approx 0.78(n_f/2L)\sqrt{\{\sigma_e\}/\rho_\gamma}$ which corresponds to the lowering of the string's upper limit frequency by 22%. In deriving the relationship (89), we neglected the nonlinearity in the helicoidal multiplicator's characteristics and a very small possible change of its density at the initial elastic-plastic twist. But the above shown lowering of the limit for attainable natural frequency in the helicoidal multivibrator is not important because with the increase of its pretension, as mentioned above, the sensitivity of the frequency transformation for the longitudinal displacement decreases. This shows that using the range of frequencies, which are close to the maximum attainable one for string, is pointless in the case of the helicoidal multivibrator.

Flexural wave propagation. Recent studies of nanostructures' dynamics, such as dispersion of phase velocity v_{pc} in carbon nanotubes at high frequency vibration, are based on the different beam models: the traditional Euler-Bernoulli and Timoshenko elastic beams as well as their nonlocal elastic versions. All four models are effective at wavelengths that are larger than 6.3×10^{-9} m. Predictions of flexural waves dispersion at very high frequencies (in THz region) with wavelengths less than 3×10^{-9} m more effective on the basis of the Timoshenko nonlocal elastic beam vibration (*see* L. Wang and H. Hu, 2005). The Timoshenko nonlocal elastic beam takes not only the rotary inertia (Rayleigh correction) and shear deformation, but also the second-order gradient of strain which characterizes the microstructure of the system under consideration. These considerations are important for high frequencies when a vibrating beam is subdivided by nodal cross sections into comparatively short portions. For example, the influence of shear and rotation inertia on the simply supported short rectangular beam's frequency of oscillations may be represented (*see* Ya.G. Panovko, 1968) by the following formula: $f_{kl}^{*}=f_1\frac{k_w\lambda_{\rho L}}{\pi}\sqrt{\frac{1}{2}B_{kL}\pm\sqrt{B_{kL}^2-4\beta_{GE}^2}}$,

where $f_1=\frac{\pi^2}{L^2}\sqrt{\frac{EJ}{m}}$ is the lowest resonance frequency calculated without account for the shear and rotation inertia, m is the mass per unit length of the beam,

$B_{kL}=1+\beta_{GE}^2+\frac{\beta_{GE}^2\lambda_{\rho L}^2}{\pi^2k_w^2}$; β_{GE} is the factor depending on the cross section geometry (β_{GE}=1.2 for rectangular cross section); $\lambda_{\rho L}=L/\rho_{gr}$ is the beam's flexibility, ρ_{gr} is the radius of cross section gyration, and k_w is the number of half-waves in the beam at oscillations. Let us recall that the shear deformation in static contributes under uniform load only few percents to the comparatively short prismatic cantilever free end total deflection which depends on the square of the ratio of depth of the beam to its length L (*see* J. T. Oden, E. A. Ripperger , 1981).

Practical applications of the problems discussed in Part A of the book are shown in Chap. 8 of Part B.

B.

Application of Precision Elastic Systems

The most important application of the structural synthesis in precision elasticity is related to precision machinery, measurement means, and instrumentation where problems of precision, accuracy, and uncertainty limitation are the main priority. Let us recall that uncertainty and error propagation analysis is developed in a theory for tests, measurement, and calibration (Figliola, R.S., 2000; Dieck, R. H., 1997), in basic ISO documents such as the *Guide to the Expression of Uncertainty in Measurement* (*GUM*), and in ANSI /ASME standards.

Another important application of the structural synthesis in precision elasticity is related to micro-, nano-, opto-electromechanical systems (MEMS, NEMS, MOEMS) where problems of precision are connected to the goals of miniaturization. In this part of the book we discuss quasi-static and dynamic problems related to the limitation of uncertainty in elastic structures caused by their intrinsic features and the influence quantities of different physical nature: scalar, vector, bias and random influence fields. We also analyze the comprehensive requirements of the structure's synthesis under standard reference and operating conditions. Unknown operating conditions are assumed to be predictable with statistical regularity.

6.

Uncertainties and Conditions

Precision starts with the limitation of uncertainty. Any precision structure has a number of elastic and rigid elements, each of which may contribute to the system's combined uncertainty. Combined uncertainty usually increases with the increase of the element's number in the system, but some elements can work as compensators, which decreases the combined uncertainty. Moreover, the system can have a permanent structure or a changeable one. The latter is typical for universal machines and measuring systems that include the changeable object of treatment or measurement, which has different informative (to be measured) and uninformative features than the calibration standard used for the adjustment and calibration.

6-1. Uncertainties Propagation, Negligibility Criterion

Standard, combined, and expanded uncertainty.

Uncertainty is an estimate of the range of a possible or probable error or errors. There are two types of uncertainty: bias and statistical. One can verify the bias uncertainty (systematic or type B in ISO classification) by comparison or calibration, and predict their possible limits on the basis of experience and calculation (concomitant methods). The latter can be used in the structural analysis or synthesis process with experimental verification. The typical source of the bias uncertainty in elastic systems is their physical or geometrical nonlinearity, systematic deflections of functional features (pitch, yaw, roll angles, non-rectilinear motion in translational guides, etc.), as well as deflections in elastic (E, G) and geometrical properties. The precision uncertainties (random or type A in ISO classification) are statistically defined by repeated functional tests of the system under certain reference or operating conditions. They are supposed to be experimental. Nevertheless, in some cases they can be predicted by probability methods for the system under structural synthesis as well. The ISO *GUM* specifies methods for evaluation of standard uncertainties in type A and B cases, combined standard uncertainty from A and B cases, and for determining expanded uncertainty with a level of confidence for intervals with a coverage factor. This factor and the level of confidence depend on the type of random variable distribution that may be known or unknown. The central limit theorem is stated that the distribution of a combined large number of independent random variables at constant set of conditions is normal (Gaussian). The level of confidence is connected with the degree of freedom for random variables in Student's distribution, or the t-distribution.

In general, the t-distribution can be used for the variable y with the variance $u_c^2(y)$, which is a sum of two or more estimated variance components u_i^2 with degree of freedom ν_i if we use the effective degree of freedom ν_{eff} obtained from the Welch-Satterthwaite formula $\frac{u_c^4}{\nu_{eff}} = \sum_{i=1}^{N} \frac{u_i^4}{\nu_i}$, where $u_c^2(y) = \sum_{i=1}^{N} u_i^2(y)$. The expanded uncertainty $U_p = k_{up}\, u_c(y) = t_{up}(\nu_{eff})\, u_c(y)$ provides an interval $Y = y \pm U_p$ which has an approximate level of confidence p, k_{up}, $t_{up}(\nu_{eff})$ are the coverage factors of interval (latter at Student's distribution).

Variables distribution. The above described mathematical apparatus is developed in mathematical statistics very well, where the standard deviation and its variance are considered the main and stable values for random quantities characterization. However, for the uncertainties of type B, the standard uncertainty or its variance can be estimated only on the basis that the variable x's possible distribution is assumed because for this variable we usually know only the limits $a_\pm$. For this case in the *GUM*, it is conditionally recommended to use the rectangular, triangular, or normal distribution. As a result, the standard uncertainty corresponding to type B of mean value $\mu_a = (a_+ + a_-)/2$ may be equal to $u(\mu_a) = a/\sqrt{3}$, $u(\mu_a) = a/\sqrt{6}$, or $u(\mu_a) = a/3$, respectively. There are estimations for digital indication and hysteresis with $u(\mu_a) = 0.29a$. One can recall that for Cauchy's distribution a standard deviation or standard uncertainty cannot be estimated at all. In addition, we should know that for the structural synthesis of precision machinery, instruments, miniature systems, and measurement techniques one is concerned with the quality of their work and the problems that the uncertainty or errors can cause in their functions. As will be shown below, the most important character in this respect is the expanded uncertainty U_p whose estimation requires knowledge of the type of distribution and the level of appropriate confidence. The latter value in mathematical statistics is not a certain one and can be chosen in an arbitrary way. We will show how this value can and should be limited in precision applications.

A recommendation found in the literature on measurement uncertainty, and often used in practice to obtain an expanded uncertainty that is intended to provide an interval with a 95 percent level of confidence, may be expressed as

$$U_{95} = t_{95}(\nu_{eff})[\tilde{s}^{\,2} + u^2]^{1/2}, \tag{6-1}$$

where the statistically evaluated uncertainty from repeated observations is

$$\tilde{s}^{\,2} = \Sigma c_i^2 s_i^2;\; c_i \equiv \partial f/\partial x_i, \text{ and } u^2 = \Sigma u_j^2(y) = \Sigma c_j^2(a_j^2/3) \tag{6-2}$$

accounts for all other components of uncertainty with the exactly known upper $+a_j$ and lower $-a_j$ bounds of X_j relative to its best estimate. Expression (1) is effective only for distributions, which are normal or Student's type. Similar recommendations may be found for special cases with some other types of distribution, but the confidence level of 95% or another value of this level is still arbitrary or based on common practice with certain risk factors. The known Chebyshev's inequality $Pr\,[y \pm ku(y)] \leq 1\text{-}k^{-2}$ for any probability Pr distribution with statistical coverage k and Gauss inequality for symmetric-unimodal distributions $Pr\,[y \pm ku(y)] \leq 1\text{-}(4/9)k^{-2}$ do not provide accurate enough information on the possible risk factor in the certain applications.

Risk factor. For example, estimation of the influence of the expanded uncertainty U_{mes} on the results of the dimension's sorting in the set of manufactured parts is required to establish a connection between the probability for erroneously accepted rejects m_χ (Fig. 6-1a), the falsely rejected good (acceptable) parts n_χ (Fig. 6-1b), and the probable value δC (Fig. 6-1c) of exceeding tolerance δ_p for the erroneously accepted rejects. The compound variance of this distribution is equal to the sum of manufacturing uncertainty's variance σ^2_{ma} and measurement uncertainty variance σ^2_{mes}. It is assumed that the compound distribution and its constituents are normal; the expanded uncertainty in the manufacturing process may be expressed with confidence level factor k_{ma} as $U_{ma} = k_{ma}\sigma_{ma}$; and the expanded uncertainty of measurement is usually equal to $U_{mes} = 2\,\sigma_{mes}$ which corresponds to a confidence level of 95%. The ratio of expanded uncertainty in the measurement to the parts' dimension tolerance δ_p is called the coefficient of measurement accuracy $A_{mes} = U_{mes}/\delta_p$. If the distribution of the part set's dimensions is normal, then the probability that the actual dimension is in the tolerance boundaries may be expressed through the probability integral $2\,\Phi(z) = \frac{1}{\sqrt{2\pi}}\int_{-z}^{z} e^{-u^2/2}du$, where $z = \delta_p / 2\sigma_{ma}$.

The probability that the rejects are in the zone equal to the normal distributed expanded uncertainty of measurement is equal to: $2[\Phi(z') - \Phi(z)]$, where

$z' = (\delta_p / 2\sigma_{ma}) + |U_{mes}| / \sigma_{ma}$. Part of these rejects can be erroneously accepted as good ones if the measurement uncertainty at the moment of their checking occurs with an absolute value that is larger than their output from the tolerance and has the opposite sign. Charts $m_\chi(\delta_p / \sigma_{ma})$; $n_\chi(\delta_p / \sigma_{ma})$; $\delta C / \delta_p (\delta_p / \sigma_{ma})$ shown in Fig. 6-1a,b,c have been calculated by numerical integration (GOST 8.051, N.N. Markov, 1973).

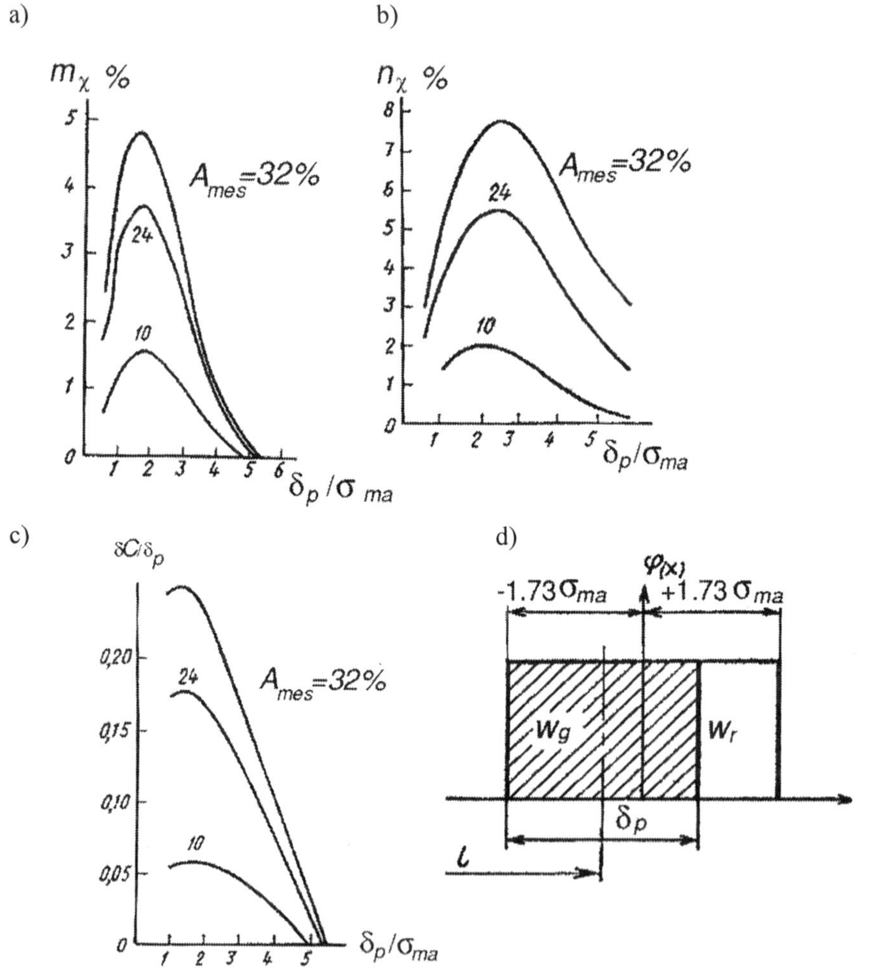

Fig. 6-1. Dependence of sorting quality on an uncertainty of measurements

Let us note that all the curves with a different measurement accuracy coefficients A_{mes} have points that correspond to the maximum value of m_χ, n_χ, and $\delta C / \delta_p$. In addition, $(\delta C_{\max} / \delta_p) < A_{mes}$. Hence, we can approximately from the top estimate the quality of the sorting process even if the precision parameters of the manufacturing process are unknown. However, the maximum values of $m_{\chi\,\max}$ and $n_{\chi\,\max}$ will be larger if the expanded uncertainty distribution in measurement corresponds to the rectangular type (equal probability, uniform). Thus, the expanded uncertainty and its type of distribution have a direct influence on losses to the manufacturer (n_χ) and possible losses in the area of application (m_χ).

Productivity factor. Another example of the expanded uncertainty's influence in measurement may be presented for the parts of an individual or small set manufacturing where the difference is in the necessary number of manufacturing passes that has influence on the manufacturing rate. Let us assume that the distribution of the parts' dimensions in the manufacturing process corresponds to the rectangular (equal probability) one. It is evident that the manufacturing process should provide one limit of the distribution field ($+ 1.73\ \sigma_{ma}$ or $- 1.73\ \sigma_{ma}$) that does not exceed the boundary of the specified tolerance, which corresponds to the non-fixable reject (Fig. 6-1d). Clearly the probability of obtaining a good part after the first manufacturing pass is equal to w_g (hatched area), and the probability of obtaining a reject after the first manufacturing pass is $w_r = 1 - w_g$. If the distribution type of the part's dimensions after the following manufacturing passes does not change, then the volume of rejects or the probability of obtaining a reject after *k*-th manufacturing pass may be calculated by the formula $w_{rk} = (w_r)^k$, where *k* is the number of executed manufacturing passes, and $w_r = 1 - (\delta_p / 3.46\ \sigma_{ma})$. The number *k* of manufacturing passes necessary to obtain the specified value w_{rk} with the known value of w_r is equal to: $k = \ln w_{rk} / \ln w_r$. The above discussed relationships do not yet reflect the measurement uncertainties. If the probability of rejecting the good (acceptable) part as a result of measurement uncertainty is equal to χ_n, then the probability of finding a reject after the first manufacturing pass will be $w_{rm} = w_r + \chi_n$, and the necessary number of manufacturing passes equals $k' = \ln w_{rk} / \ln (w_r + \chi_n) > k$.

If the probability in acceptance of rejects is forbidden by the customer, then $\chi_m = 0$ and the manufacturing tolerances should be installed ($\delta_{ma} = \delta_p - U_{mes}$) which may cause an additional problem for the manufacturer. Hence, the discussed examples and many more known practical situations show the importance of the expanded uncertainties' knowledge and certain limitation in the precision applications. However, the level of their permissible confidence cannot be effectively obtained from the probability m_χ of the erroneously rejected parts' acceptance and additional good parts' rejection n_χ, which nobody mandatory specifies, nor through the number of manufacturing passes and manufacturing tolerances because these probabilities are controversial and may be the subject of complicated discussions. In addition, the actual type of variables distribution is very often unknown, which makes this problem as a whole more complex and speculative. Moreover, the direct square summation of the limits of type-*B* uncertainties, the expanded uncertainties, or errors is appropriate only if all variables have or suppose to have normal or fraction-linear distributions.

Information criterion.

Foundation of the criterion. The author of this book has found (*see* GOST 8-050-73, Y.M. Tseytlin, 1974, Y.M. Tseytlin *et al.*, 1976) an independent and objective solution to the problem of expanded uncertainty limitation for precision applications and structural synthesis problems. This solution is based on the following postulate (or axiom). Precision machines and measurement means are used to achieve and define the specified actual values of the certain objective parameters. The precision machines' function is based on the internal measurements (internal information).

For these purposes their uncertainties should be small enough for the specified goals. However, it has to be established which small uncertainties may be considered negligible.

Let us recall that the measurement information I_q is equal to the difference between the entropy $H_э(x)$ of the precision variable parameter to be executed, defined, or measured and the conditional entropy $H_э(\Delta)$ of the random measurement uncertainty Δ distribution such that $I_q = H_э(x) - H_э(\Delta)$,

where $H_э(x) = -\int_{-\infty}^{+\infty} \varphi(x)\ln\varphi(x)\,dx$ is the entropy or measure of variable's uncertainty. It is natural requirement (Y.M. Tseytlin, 1974) that the measurement information should be positive, or $H_э(x) \geq H_э(\Delta)$, (6-3)
because otherwise we will have misinformation that is not an option for precision systems. In latter illogical case, the larger uncertainty will occur after measurement, monitoring, or sorting rather than before these operations are performed. The goal of measurement and sorting, on the contrary, is to obtain more accurate and definite knowledge about the object or set of objects under evaluation.
The entropy of expanded uncertainty distribution represents the so-called entropy error $\Delta_э$ (P. V. Novitskii, 1968) expressed through the exponential function

$\Delta_э = A_э e^{H_э(\Delta)} = K_э\sigma$, where $A_э = ½$ is a normalizing factor for the symmetrical distributions and $A_э=1$ for asymmetrical ones; $K_э$ is an entropy coefficient, and σ is a standard uncertainty of the distribution. Hence, in accordance with (3), positive measuring information is obtained if the following relationships are satisfied

$$A_э e^{H_э(x)} \geq A_э e^{H_э(\Delta)}. \quad (6\text{-}3a)$$

Now we assume that if we divide the left-hand and right-hand sides of (3a) by the probable scattering distribution interval or corresponding tolerance-value δ_p for the object-value x of measurand, we obtain a limit for the permissible relative expanded uncertainty in measurement which is equal to:

$$\eta_U = \frac{U_p}{\delta_p} \leq \frac{A_э}{\delta_p} e^{H_э(x)}. \quad (6\text{-}3b)$$

Table 6-1. Calculated limits for the criterion of uncertainty's negligibility

Distribution of x	$K_э$	δ_p/σ	η_U	$\eta_v = \eta_U \times v$	$A_э$
Normal (N)	2.07	6.0	0.35=v	0.12	0.5
Uniform (U)	1.73	3.46	0.5	0.17	0.5
Maxwell's(M)	1.38	5.50	0.25	0.09	1.0
*Triangular (Tr)	2.0	4.90	0.41	0.14	0.5

Note:*Simpson's distribution; $K_э$ is the entropy coefficient, η_v is the negligibility of uncertainty component; hyperbolic(loguniform) distribution at wide range of value (X_1-X_2) may have an uncertainty limit criterion which is equal to:

$$\eta U = \frac{\exp[\ln(\sqrt{X_1 X_2}\ \ln X_2/X_1)}{X_2 - X_1}.$$

Thus calculated, the upper limits for the permissible relative expanded uncertainty for different distributions are shown in Table 6-1, where the confidence level for the scattering field δ_p is assumed to be 0.997. This limit we called the information criterion of measurement uncertainty negligibility.
The values of entropy are known for many distribution types or may be calculated on the basis of its mathematical definition (*see* Y.M. Tseytlin,1981b).
It is interesting to note that for Cauchy's distribution, which does not have the standard deviation, criterion η_U has a definite value that is equal to $\eta_U|\ 0,\ 0\ | \approx 0.09$ if the confidence level for the distribution equals 0.997.
Comparison of information criterion η_U values with the other known recommendations for the limit of permissible errors (uncertainties) shows a good correspondence especially in the precision linear measurement and metrology. It should be noted that the permissible errors are previously established on the partial consideration of the corresponding metrological practice: $\eta_U \le 0.35$ was used in industrial applications in USSR, $\eta_U = 0.1$ is a typical recommended limit in Germany and the US. The latter permissible limit of uncertainty is practically acceptable for all known possible probability distributions. Nevertheless, its implementation in all cases of precision measurements and precision structures may require certain methodical efforts and even expensive special solutions. Therefore, one can find more suitable value for η_U on the basis of (3b) limitation especially if the corresponding distribution and tolerance are known or specified. Values of the component's negligible criterion η_v for different distributions are also shown in Table 6-1.
Surface of the criterion. The value of the criterion for negligible uncertainty depends on the type of distribution of variables to be measured, and its diagram for the whole set of probable distributions can be presented in the form of a surface (Fig. 6-2a) whose equation corresponds to:

$$\eta_U(K_э;\ \sqrt{\psi_2+3} = \varepsilon_k^{-1}), \tag{6-4}$$

where ψ_2 is the corresponding excess (kurtosis, flatness) of the distribution under consideration and ε_k may be called a contra-excess.
Fig. 6-2a shows that the largest value of criterion η_U corresponds to the uniform U (rectangular) distribution. The continuous probability distributions have decreasing criterion value η_U when excess ψ_2 increases, especially if this excess is positive. In latter case, criterion value η_U is negatively correlated with the positive excess. The value of η_U is approximately constant and equal to 0.35- 0.33 (from 1/3 to 0.35) for the continuous probability distributions with negative excess when $\varepsilon_k = 0.58$ to 0.9. Let us recall that contra-excess ε_k=0.58 corresponds to the normal (Gauss) distribution.
Dimensional sorting precision criterion. It was shown above that the dimensional sorting of objects into the "good" (acceptable) in tolerance (G_κ) and "rejects" (B_r) is one of the problems in precision machine manufacturing that may even be applicable to a wider field of objects and signals qualification. This procedure accompanied by some percent m_χ of erroneously accepted rejects and some percent n_χ of falsely rejected good elements. Values m_χ and n_χ in metrology and mathematical statistics are called the errors of the first and second kind. We can estimate the quantity of the measurement information I_{qs} obtained in the sorting procedure by comparison of an

uncertainty in the two systems: before sorting (*a priori*) and after sorting (*a posteriori*). This may be expressed by the following formula

$$I_{qs} = H_э(G_к) + H_э(B_r) - [G_к H_э(G_{к1}/G_к) + B_r H_э(B_{r1}/B_r) + G_к H_э(n_χ/G_к) + B_r H_э(m_χ/B_r)], \quad (6\text{-}5)$$

where $H_э(G_к)$ and $H_э(B_r)$ are the entropy of the good (acceptable) elements and rejects, respectively, in their set before sorting; $H_э(G_{к1}/G_к)$, $H_э(B_{r1}/B_r)$, $H_э(n_χ/G_к)$, $H_э(m_χ/B_r)$ are the conditional entropy of good elements $G_{к1}= G_к - n_χ$, rejects $B_{r1}= B_r - m_χ$, falsely rejected $n_χ$, and erroneously accepted $m_χ$ elements after sorting. Calculated values of I_q for the normal distribution 1 at η_U =0.32, Raleigh's distribution 2 at η_U= 0.24, and uniform distribution 3 at η_U=0.50 are shown in Fig. 6-2b. These charts prove that I_q>0 for normal distribution at $\delta_p < \gamma_p \sigma_{ma}$ =3.5σ_{ma} which corresponds to $G_к$=0.92; B_r=0.08; I_q>0 for uniform distribution at $\delta_p < 3.0\ \sigma_{ma}$ that corresponds to $G_к$=0.87; B_r=0.13 ; and I_q>0 for Raleigh's distribution in the whole scattering interval of monitoring values. Formula (5) does not have any empirical stipulations and reflects all *a priori* and *a posteriori* information for the measuring means and monitored objects. It is evident that $I_{qs} = H_э(G_к) + H_э(B_r)$ if $m_χ$=0 and $n_χ$=0.

a)

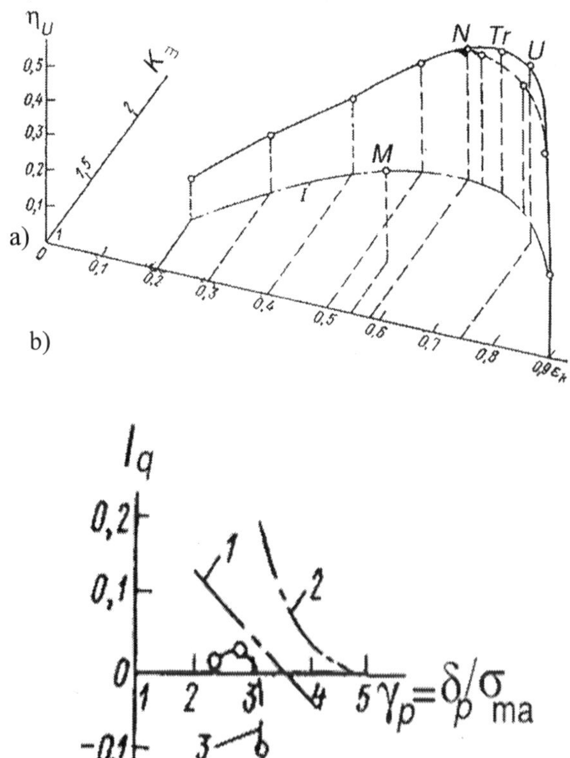

Fig. 6-2. The information criterion of negligible uncertainty.

The confidence of information estimations. Some authors have discussed the legitimacy of the information estimations in the measurement techniques and statistics (*see* P.V. Novitskii, 1978). Let us study this problem in more detail on the basis of the fundamental mathematical / statistical formula of maximum likelihood (Wilks S. S., 1963), which requires the largest value for the expression

$L_\theta = P^{m_1}_1(\theta) P^{m_2}_2(\theta) \ldots P^{m_r}_r(\theta)$ with the fixed number of variable θ tests $n = \sum_1^r m_r$ which is equivalent to the requirement of the largest value for the function

$$\frac{1}{n}\ln L_\theta = \frac{m_1}{n}\ln P_1(\theta) + \frac{m_2}{n}\ln P_2(\theta) + \ldots \frac{m_r}{n}\ln P_r(\theta) = \sum_i^r \frac{m_r}{n}\ln P_r(\theta), \tag{6-6}$$

where $P_i(\theta)$ is the probability density of the samples with dimension θ_i and number m_i in the finite set. The right-hand side of this expression corresponds to the entropy because in the limit $m_r / n = P_r(\theta)$, and the largest value of the function (6) corresponds to the entropy with the opposite sign:

$$\frac{1}{n}\ln L_\theta = \sum_1^r P_r(\theta)\ln P_r(\theta) = -H_э \tag{6-7}$$

Therefore, the absolute value of the entropy should be the maximum. This approach corresponds to the Bayesian theory of measurement uncertainty. However, the largest absolute value in the negative field corresponds to the minimum of the function that provides a less chaotic and more precise outcome.

In addition, in accordance with the criterion of the likelihood ratio λ_H (S. Wilks, 1963) for large statistical samples $n \to \infty$, we have

$$\lambda_H = \frac{L_\theta(x,\theta_0)}{L_\theta(x,\theta_n)} \quad \text{and} \quad \ln L_\theta(x, \theta_0) - \ln L_\theta(x,\theta_n) = \ln\lambda_H.$$

Therefore,

$$\frac{1}{n}\ln L_\theta(x,\theta_0) - \frac{1}{n}L_\theta(x,\theta_n) = -H_э(x,\theta_0) + H_э(x,\theta_n) = \lim_{n\to\infty} \frac{1}{n}\ln\lambda_H = 0 \tag{6-8}$$

Hence, expression (8) is a condition of the asymptotic power for the criterion of likelihood. This condition coincides with requirement (3), which was used by the author for estimation of the criterion of uncertainty negligibility.

Quantitative estimation. It follows from (7) that $L_n(\Theta) = e^{-H_э n}$. Therefore, we have for the uniform probability (rectangular) distribution of random variables the following relationship

$$e^{H_э} = [L_n(\Theta)]^{-1/n} = 1/P(x) = \delta_p, \tag{6-9}$$

and the criterion $\eta_U = \Delta_э / \delta_p = 0.5\, e^{H_э} / e^{H_э} = 0.5$.

One can define the criterion of uncertainty negligibility for other distributions by the same method on the basis of (8) and (9).

Fundamental research on the connection between the information theory and the extension of the maximum likelihood principle was also provided by Akaike Hirotogu (*see* E. Parzen *et al.*, 1998), V.Y. Rosenberg (1978), L. Brillouin (1962), and others but for different applications.

6-2. Uncertainties and Environment

Stability of Test Conditions.

A theory of tests and precision measurements as well as theory of probability are based on the requirement for definite observation conditions (*see* N. Wiener,1961, H. Cramer, 1971). Nevertheless the absolutely invariable or identical conditions of statistical experiments cannot practically be conducted and we usually have some approximation of these conditions. The probability distribution of the measuring system's output can be changed under influence quantities with a wide range of values. However, the elastic system should keep its precision properties.

Distribution's Transformation.

Distribution homogeneity. If the measurement results have a distribution with normal (Fig. 6-3a) density f($\bar{y}$, σ_0), standard uncertainty σ_0, and mathematical expectation $\bar{y}$ when they are made in the presence of influence quantities (*IQ*) with certain mean values, then the distribution of the measurement results may become Laplace (sharp top) one (Fig. 6-3b) if the *IQ* act over a sufficiently wide range of values and consequently tend to maximize the entropy of such a distribution.

c)

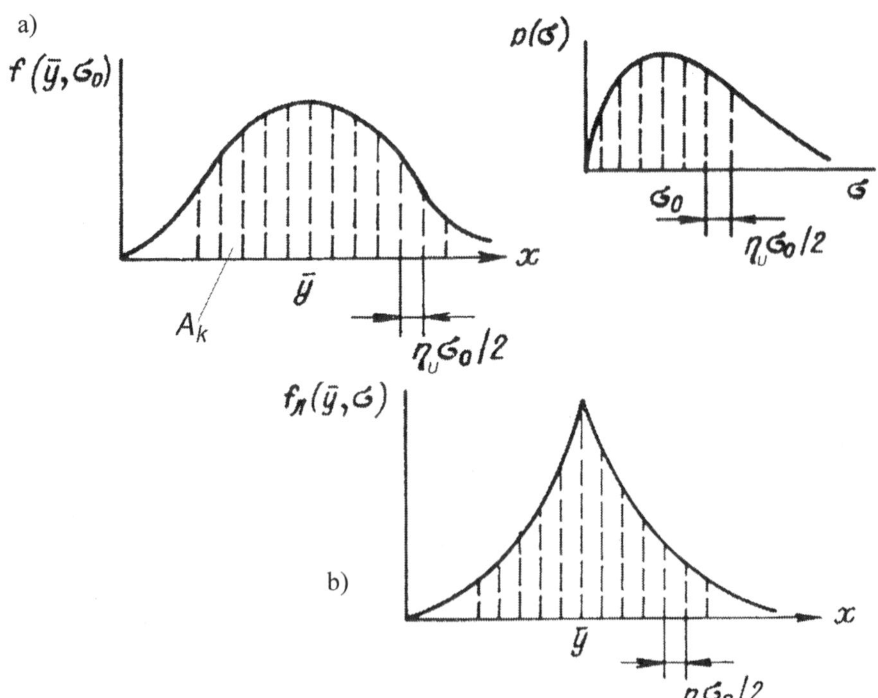

Fig. 6-3. Probability distribution of measurement results.

In this case, the sample standard uncertainty shows a variation relative to the mean value σ_0 and has a density of distribution (Fig. 6-3c) as follows

$$p(\sigma) = 2\sigma \exp(-\sigma^2 / \lambda_\sigma^2)/\lambda_\sigma^2, \quad \text{where } \lambda_\sigma = 2\ \sigma_0 / \sqrt{n}\ . \tag{6-10}$$

This distribution in the range $0 \le \sigma \le \infty$ has a following form

$F(\sigma) = 1 - \exp(-\sigma^2/\lambda_\sigma^2)$ and for $\sigma = \sigma_0$ its value is $F(\sigma_0) = 0.544$.

Cochran criterion. Let us show (Y.M. Tseytlin, 1985) on the plot (Fig. 6-3c) the distribution's density (10) a set of $2m$ bands that are symmetrical relative to the mean value σ_0 and of width equal to $(\eta_U/2)\sigma_0$, where η_U is the criterion of negligible uncertainty (see Sect. 6-1). We consider each such band a certain statistical group with number k and variance $\tilde{S}_k^2 = (1+k\eta_U/2)^2\sigma_0^2$, and then we try to estimate the statistical homogeneity of the set of groups with Cochran's criterion, which is equal to: $Gc_k = \dfrac{\tilde{S}_{k\max}^2}{\sum_{k=-m}^{m} \tilde{S}_k^2} = \dfrac{(1+m\eta_U)^2}{\sum_{k=-m}^{m}(1+k\eta_U/2)^2}$. Statistical homogeneity will occur if $Gc_k < G_{\max}$ where $G_{\max}(2m, n\text{-}1)_q$ is the tabulated value (Cochran, 1963) of the statistic for $2m$ groups, volume n = constant of a sample in each group, and the level of significance of q. It is easy to show that for $\eta_U = 0.35$ and $2m = 4$, we have $Gc_{k=4} = 0.42 < G_{\max}(4,16)_{q=0.05} = 0.44$; for $2m=6$, we have $Gc_{k=6} = 0.34 < G_{\max}(6,16)_{q=0.01} = 0.35$; and for $2m=8$, we have $Gc_{k=8} = 0.29 > G_{\max}(8,16)_{q=0.05} = 0.25$. Hence, the permissible changes of standard uncertainty under the action of influence quantities are within

$\pm\, 0.5\sigma_0$ in accordance with Cochran's criterion at $n=16$ in each group, and the whole volume of the sample is equal to: $n_w = 96 / \sum_{k=-m}^{m} A_k(\tilde{S}_k)$. This estimation is an approximate one because Cochran's criterion is used to estimate the homogeneity of variances in groups with normal distribution of data.

Bartlett's criterion. In Bartlett's test for estimation of the statistical homogeneity, it is necessary to define the effective volume n_k of the samples in each group .The reader can see notes about Bartlett and Cochran tests in measurement with probability distribution different from normal in Y.M. Tseytlin (1981b) and in *Measurement Techniques*, 2004, vol.**47**, No 10,p.960-968 publication by B.Yu. Lemeshko and E.P. Mirkin.

Chi-square criterion. Finally, we perform the χ^2 test for goodness-of-fit by grouping data with normal distribution and the assumed Laplace distribution in bands with width $\sigma_0\eta_U/2$ (Fig. 6-3a,b).

Dynamic influence. Dynamic deformation of an input signal distribution in the linear inertial system's output follows from Malakhov's (1978) equation

$\psi_{2out} = \psi_{2in}(t_{in} / t_{out})$, where ψ_{2in}, ψ_{2out}, t_{in}, and t_{out} are the excess on the input and output, time of the input signal, and the time of response on the exit of the system, respectively. *Denormalization* will occur if $t_{in} > t_{out}$. One can see the same results from differential equations with changeable matrix components when the Fokker-Plank equation is applied. It can be shown that even under dynamic influence the constraint with the information criterion η_U turns out to be sufficient. For example, if the dynamics of a precision system is described by a first-order differential equation, $\dot{x} + [h_\xi + \xi(t)]x = F(t)$, then Mises' omega-square (ω^2)criterion can be estimated by transforming this equation into the corresponding Fokker-Planck equation $F(x) = \mu_u^{\Theta} e^{-\mu_u/x}/[\Gamma(\Theta)x^{1+\Theta}]$.

If $\Theta = 1 \Rightarrow \omega^2 = \Gamma^{-3}(1)[\frac{\mu_{u2}}{\mu_{u2}+2\mu_{u1}} - \frac{2\mu_{u2}}{\mu_{u1}+2\mu_{u2}} + 1/3]$, where $\Gamma(\Theta)$ is the gamma-function; $\mu_u = F/D_\xi$, $\Theta = h_\xi / D_\xi$,and $D_\xi \delta(\tau) = \overline{\xi\xi_\tau}$ is a delta-correlated variable. The plot of ω^2 criterion is presented in Fig. 6-4a for $\mu_{u2} = 1$; $\mu_{u1} = 1$- 3, from which it is seen that criterion $\omega^2 = 0$ for $\mu_{u1} = \mu_{u2}$, and the constraint $\mu_{u2} / \mu_{u1} \leq \eta_U$ corresponds to good homogeneity at $n\omega^2(\Theta) > Z_q \ \exists \ q_s = 5$, $\Theta = 4$ with a representative enough random sample of size $n = 46$. Operator $\exists$ (the existential quantifier) shows that the condition is correct at least for presented data with the level of significance q_s and the volume of samples n.

a) b)

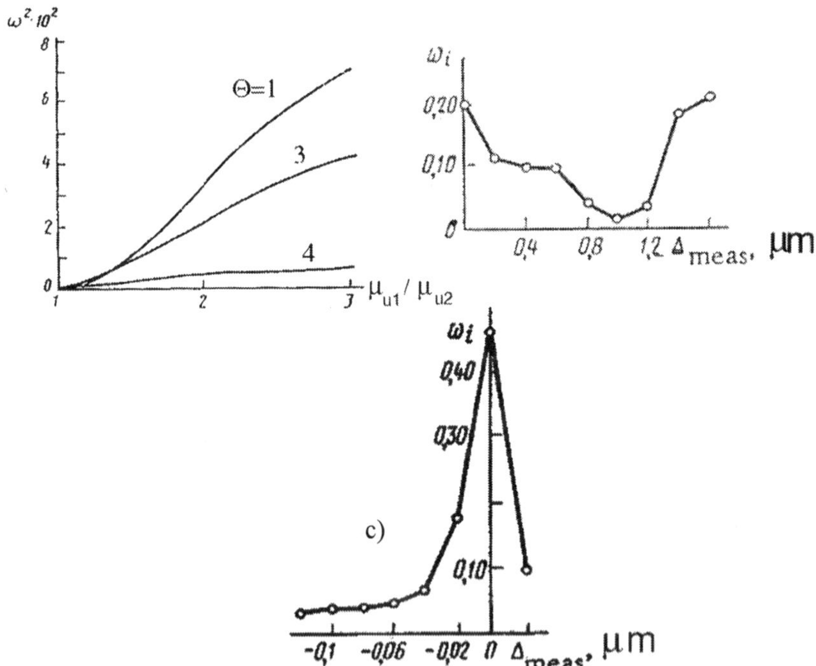

Fig. 6-4. Change of distribution under the influence of dynamic variable.

Robustness. It is often recommended to process the measurement results with robust methods. However, it can be shown that even when using the statistical processing "median" method (J.E. Walsh, 1949; F.R. Hampel *et al*., 1986), some restraint of influence factors under the information criterion of uncertainty negligibility is also useful. In this case, we consider normal truncated distribution, which is obstructed by asymmetric distribution of the triangular type with amplitude of 25% < η_U (*see* Table 6-2). Hence, the information criterion η_U of negligible uncertainty is necessary (providing stability of a probability distribution) as well as sufficient (small increase of uncertainty under influence factors).

Table 6-2. Limited disturbances at structural optimization

Method of optimization	Type and limit of uncertainty	Reference conditions with η_U limit
Gradient	Relative, deterministic, $\leq$ 100%	Random $10 \leq \eta_U \leq 50\%$
Newton	Absolute in ill-condition system is not acceptable	Well-condition number in the range $1 \leq K_{cTI} \leq 1+ \eta_U$; $\eta_U \leq 0.5$
Maximum likelihood	Change of distribution's type is not acceptable	Influence distribution close to normal at $\eta_v \leq 0.12$
Least squares	Change in numerical parameters of distribution	$\sigma(x, y) \leq 1.06\sigma(x)$
Robust "median"	Disturbance less 0.5	Disturbance $\leq$ 0.25

The test. The following distributions' changes were found with measurement of the gage blocks of length L=100 mm on the fixed table by spring measuring head with dial graduation 0.2 μm installed on the stand with the round steel column of diameter 65 mm:

1. The reading of the instrument Δ_{meas} with a single resetting to the calibration standard and frequency ω_i has a distribution (Fig. 6-4b) which is similar to beta distribution Be (ν_1, ν_2) with "anti-modal" function of its density

$\omega(x) = \dfrac{\Gamma(\nu_1+\nu_2)}{\Gamma(\nu_1)\Gamma(\nu_2)} x^{\nu_1 -1}(1-x)^{\nu_2 -1}$ at $\nu_1 < 1$, $\nu_2 < 1$ and standard uncertainty

u_1 = 0.63 μm , when the surrounding temperature changes sharply and quasi-periodically in the range 18-22 °C . The readings Δ_{meas} have a quasi-normal distribution with standard uncertainty u_2 =0.075 μm when the temperature value is quasi-stable. Ratio u_2^2/u_1^2 =70 does not satisfy Fisher's criterion, which requires $P\{ F_{lo} < u_1^2/u_2^2 < F_{up} \} = 1\text{-}q_s$, where u_1^2 *and* u_2^2 are the larger and smaller variance, respectively; F_{up} and F_{lo} are the upper and lower limit values of Fisher's F-distribution (L.N. Bol'shev, N.V. Smirnov, 1983) with a degree of freedom (n-1)=181 and a level of significance $q_s/2$ =0.05.

2. The readings' distribution becomes close to normal (Fig. 6-4c) with a standard uncertainty u_2= 0.05μm when regular periodical resetting after each series of 10 measurements is accomplished with a standard that has the same coefficient of thermal expansion, shape, and dimensions as the gage blocks to be measured. The variance ratio in this case is $u_2^2/u_1^2 \rightarrow 1$. Let us note that the periods of resetting to the measurement standard are not specified for non-automatic measurements. Therefore, it is possible to obtain different uncertainties of measurement by the same measuring means under the "same operating conditions", especially with the intense changes of IQ over time even within the specified limits of their values. We can now discuss this problem in more detail in connection with the systems' and algorithms' optimization.

Signal optimization. This optimization (Table 6-2) is provided on the basis of the objective (criterion) function's (F_{oc}) minimization with the estimation of adequate argument values x_* (*see* B.T. Poliak, 1987, Iu. V. Linnik, 1961).

Robust methods. If the type of distribution for the measurement results' probability is unknown, robust methods should be used. The robust (stable) estimators for the regression problem may be as follows

$$x^*_m = \arg\min F_{com}(x);\ F_{com} = \sum_{i=1}^{m} F[y_i-(u^i,x)]\text{ ,where }\ F(z) = -\ln p^*(z);\ p^* = \arg\min J(p);$$

and $J(p) = \int \frac{p'(z)^2}{p(z)} dz$ is the Fisher's information. In other words, it is necessary to find a satisfactory distribution $p^* \in P$ that has the smallest Fisher's information, and then use the maximum likelihood estimator that corresponds to it. If P is a class of all non-degenerative distributions and nothing is known about it, the method of the least absolute values is recommended, in which $F_{\text{com}}(x) = \sum_{i=1}^{m} |y_i-(u^i,x)|$. If P is a class of approximately normal distributions, then

$$F_{\text{com}} = \sum_{i=1}^{m} F[y_i-(u^i,x)];\ \text{ and } F(z) = \begin{cases} z^2/2 & |z| \le d_C \\ |z|d_C - d_C^2/2 & |z| > d_C, \end{cases},$$

where d_c depends on the level of disturbance of the basic distribution. If the distribution is close to normal, it is recommended to use the intermediate method, which lies between the method of least squares and the least absolute values. Robust absolute stability of continuous systems is a very important factor in structural synthesis (*see* Ya.Z. Tsypkin, B.T. Polyak , 1992).

Recurrent estimators. Recurrent filters and correlators are useful for the treatment of the sequential input data in real time.

The statistical risk for the problem of reference conditions identification is in the errors of the first m_χ and second n_χ kind with minimization of the difference $[B_r - (m_\chi + n_\chi)]$. The information criterion of negligible uncertainty $\eta_U \approx 1/3$ is a common solution for all possible statistical distributions (*see* Fig. 6-2a) with contra-excess in the range $\varepsilon_k = 0.58\text{-}0.9$. Therefore, this solution is supposedly the most uniformly powerful one. Let us recall that Gauss distribution has contra-excess $\varepsilon_k = 0.577$; Simpson's distribution corresponds to $\varepsilon_k = 0.645$; uniform distribution has $\varepsilon_k = 0.745$.

Amplitude and frequency modulation.

Summation of the signal u_s and a disturbance u_d corresponds to the output reading of a structure as $u = u_s / \cos \varphi_{sd}$, where φ_{sd} is the phase angle between the signal and the mix of signal and a disturbance, $\tan \varphi_{sd} = u_s^{-1} u_d \sin(\omega_d - \omega)t$.

Amplitude modulation. In the case of the amplitude modulation, the difference between the reading and the signal, at a small value of angle φ_{sd} and signal frequency $\omega = 0$, is equal to:

$$\Delta u_s = u - u_s = u_s(1/\cos\varphi_{sd} - 1) = u_s\,\varphi_{sd}^2/2 \approx \frac{u_d^2 \sin^2 \omega_d t}{2u_s}.$$

If the disturbance phase angle $\omega_d t$ has the uniform (rectangular) distribution in the range 0 to π, the mathematical expectation of the amplitude distortion equals

$$\left\langle \frac{u - u_s}{u_d} \right\rangle = \frac{\Delta u_s}{u_d} = \frac{u_d}{2\pi\, u_s} \int_0^{2\pi} \sin^2 \omega_d t\, dt = ½\, u_d / u_s.$$

Frequency modulation. In the case of the frequency modulation, the relative uncertainty of the signal is equal to:

$$\delta\omega = \varphi_{sd} / \omega = \frac{u_d \sin(\omega_d - \omega)t}{u_s \omega}.$$

The mathematical expectation of the disturbance of a frequency signal with a uniform distribution of signal's frequency equals

$$\langle \delta\omega \rangle = \frac{u_d}{2\pi u_s} \int_0^{2\pi} \frac{\sin(\omega_d - \omega)t}{u_s \omega} = \frac{4}{3\pi} \frac{u_d}{u_s \omega}.$$

The system with the frequency signal has a low sensitivity to the changes of the disturbance's amplitude when $\omega >> 1$. This shows the effectiveness of the frequency modulation. It is also important that frequency primary standard with precision 10^{-16} in metrology is much more accurate than the length or other primary standards.

Changeable connections. All of the above described methods of optimization are general for the synthesis of precision systems, in which information and signals are transformed and transmitted. More specific problems for the structural synthesis in precision elasticity are linked to the multiply-connected and changeable mutual position of the system's elements. For example, the structure for linear measurement at calibration is singly linked with the standard (e.g., gage blocks), which has specified informative (length, parallelism) and uninformative (form, electrical and thermal properties) features. The same structure at the working measurement conditions is linked to the object to be measured (measurand) with approximately the same length as for the standard, but whose uninformative parameters may be quite different and distributed in the wide range of their values. In the linear measurements, the object may have different forms, materials, and positions relative to the instrument structure. These are important specifics in the precision procedures under operating conditions. Hence, the system of instrument structure-measurand may have a random operator in relation to the action of the "same" measurement conditions. This situation is similar to the circulation of the vortex free influence field on the multiply-connected contours. In this case the potential of the vortex free field is a multi-valued function because the measurement system or another precision structure that consists of the changeable elements is approximated by multiply-connected regions often with changeable connectivity. Each contour has a corresponding circulation of the *IQ*'s flow. Let us recall that the circulation is a line integral around any closed path (*L*) of the force field vector's projection $A_{\varphi\tau}$= (grad $V_{\kappa\varphi})_\tau$ on the tangent to the line (*L*), which follows in its positive sense

$$C_r = \oint_{(L)} A_{\varphi\tau} dL = \oint_{(L)} grad V_{\kappa\varphi}\, d\,\boldsymbol{r},$$

where $V_{\kappa\varphi}$ is the field acting on the surface with the influence inside the volume that has density which is equal to grad $V_{\kappa\varphi}$, and $\boldsymbol{r}$ is a radius-vector. The system under standard calibration by gage blocks or another standard device may be represented as a two-connected region 1 (Fig. 6-5) which has one independent circulation C_r in the vortex free field.

The same system for measuring different objects 2, which differ by uninformative features, has a changeable degree of connections. As a result, the formula for the circulation at calibration may be presented as follows

$$\oint\limits_{(L)} grad\, V_{K\varphi}\, d\mathbf{r} = \oint\limits_{(L)} dV_{K\varphi} = C_r\,. \qquad (6\text{-}11a)$$

And for the working measurement with a multiply-connected region, we have

$$\oint\limits_{(L+L_1+...L_n)} gradV_{K\varphi}\, d\mathbf{r} = C_{r1} + C_{r2} + \ldots C_{rn}, \qquad (6\text{-}11b)$$

where C_r; C_{r1} ; C_{r2}; …;C_{rn} are the independent circulations on contours L, L_1, L_2…,L_n. If the contours L, L_1, L_2…L_n may be continuously transformed one into another, the circulations become equal

$C_r = C_{r1} = C_{r2} = \ldots C_{rn}$, i.e., $\oint gradV_{K\varphi} \bullet d\mathbf{r} = n\, C_r$. (6-11c)

Hence, the difference between the influence uncertainties for the measurement (working) Δ_c^m and for the calibration (adjustment) Δ_c^c under the same physical conditions corresponds to: $\Delta_c^m - \Delta_c^c \Leftarrow C_r\,(n\text{-}1)$. (6-11d)

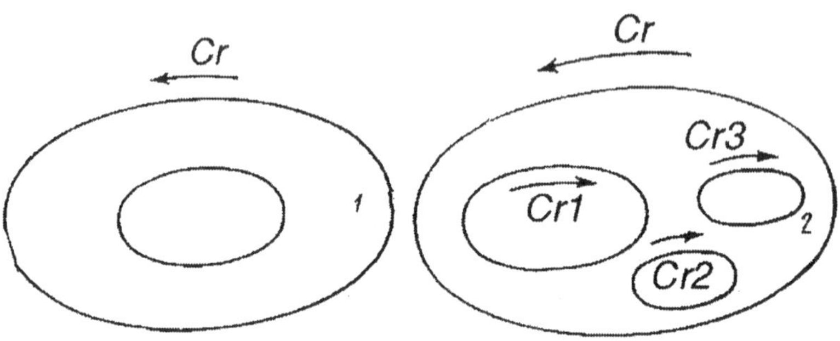

Fig. 6-5. Circulation of influence vector.

A similar effect can be shown in a structure with different element positions and some auxiliary elements that are not in the measuring (direct functional) chain. *Experimental data.* There are charts in Fig. 6-6 that show the change in readings Δ_{cm} of spring measuring head (*see* Chap.8) with dial graduation 0.1μm in the measurement of the gage blocks of length $L = 20$ mm under the same jump-like change in the surrounding temperature (R. Noch, H. Huhn, 1959) within 2 °C (from 20 to 22 °C). These curves show the behavior in time t of non-thermocompensated I and thermocompensated II and III heads (Fig. 6-6a), the change of their behavior at installation of plastic (Fig. 6-6b) and steel (Fig. 6-6c) additional tubes on the stem, for lower IV-IX and upper IV'-IX' clamping of their base stem in the small stand (Fig. 6-6d) with the diameter of column 49.5 mm and the large stand (Fig. 6-6e) with the diameter of column 88 mm. The difference in the reaction of the spring measuring head (curves I…III, Fig 6-6a) is caused by their design specifics and change in the reaction (Fig. 6-6b,c) is caused by installation of special bushings on the stem with the thickness of their wall equal to 10 mm. This is simulation of the temperature influence function change by the additional elements or recesses in the objects to be measured. The curves of changes in the heads' readings depend on the type of stand and the clamping position. The lower position of the spring head without thermal compensation IV (Fig. 6-6d) has a compensation influence, while the better results occur for the heads with thermal compensation II and III (Fig.6-6b,c, d) in their upper position in the stand's bracket (curves VI'-IX', Fig. 6-6 d, e). The lower position of these heads has not a compensating influence (curves VI-IX).

Analogous tests were done by the author (Y.M. Tseytlin, 1981b), which showed significant difference in temperature influence function at the measurement of the steel and quartz gage blocks.

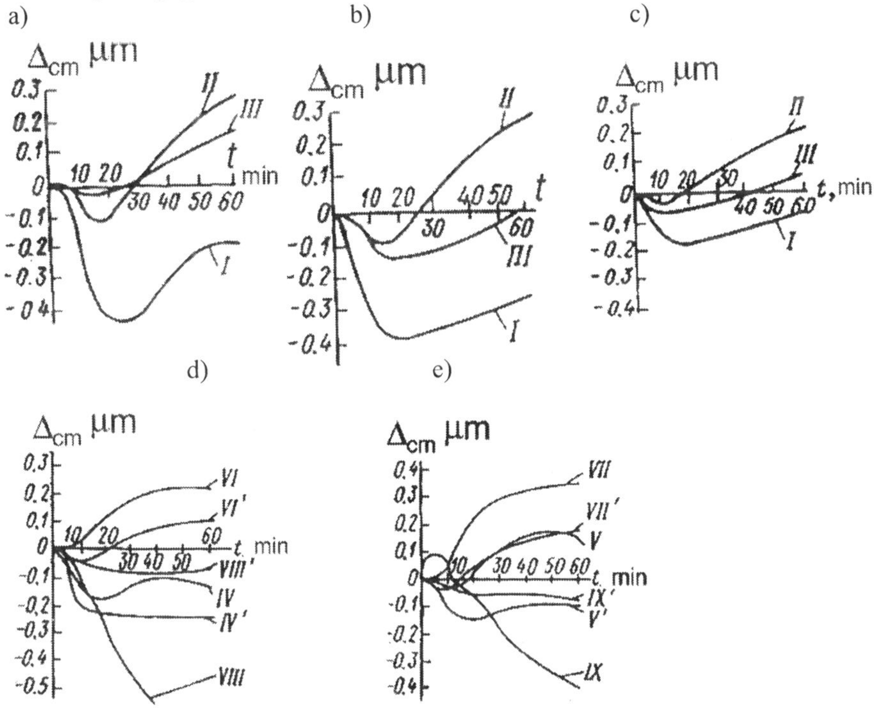

Fig. 6-6. Change in measuring heads' readings under change of temperature. (After R. Noch, H. Huhn).

The breaks of the thermal conductive chain in a measurement system with the optical spring measuring head (*see* Chap. 8)at the measuring spindle retraction cause the temperature jumps of 0.05 to 0.15 °C because the average coefficient of the thermal conductivity for a steel body is equal to $k_{T.con}$=0.24 W/(cm·K) and for air $k_{T.con}$=0.0002 W/(cm·K). The quartz gage blocks have $k_{T.con}$= 0.008 W/(cm·K) that causes less temperature jumps at the measuring spindle retraction. Note that for solid bodies we should rather consider in this analysis the coefficients of the temperature conductivity $a_T = k_{T.con}/(c_T\rho_\gamma)$, where c_T is the specific heat of material.

All this experimental data prove the necessity of a systemic approach to the influence problem at structural synthesis in precision elasticity. The influence of the surrounding temperature and other *IQ* on precision elastic systems is shown in more detail in Chap. 7.

6-3. Reliability of Structural Links

The precision elastic system's function mathematical description may be presented in the form of a matrix equation (Y. M. Tseytlin, 1984d) as follows

$$\boldsymbol{C} = A\,\boldsymbol{X}, \qquad (6\text{-}12)$$

where $\boldsymbol{C}$ is the vector of input (to be measured, measurand) values; $A(a_{ij})$ is the system's transform matrix of the n-th dimension; $\boldsymbol{X}$ is the vector of informative parameters in the system's output (results of measurements).

Uncertainty in Matrix equation.

Table 6-3. Formulae of uncertainty estimators in the matrix equation solution

$N^{\underline{o}}$	Type of uncertainty	Relationship	Notes
1	Deterministic	$\Delta X_k = \sum_{i=1}^{n}\sum_{j=1}^{n} X_j b_{ki}\varepsilon_{ij} + \sum_{i=1}^{n} b_{ki}\varepsilon_i$	$\Delta a_{ij} = \varepsilon_{ij}$ $\Delta C_i = \varepsilon_i$
2	Random uncorrelated with normal distribution	$\mathring{\Delta X}_k = [\sum_{i=1}^{n}\sum_{j=1}^{n} (X_j b_{ki}\mathring{\varepsilon}_{ij})^2 + \sum_{i=1}^{n} (b_{ki}\mathring{\varepsilon}_i)^2]^{1/2}$	$\mathring{\Delta A} = \varepsilon_{ij}$ $\mathring{\Delta C} = \varepsilon_i$
3	Same as in 2 with rectangle distribution	$X_k \in [X_{kH} - \sqrt{3}\,\mathring{\Delta X}_k;\ X_{kH} + \sqrt{3}\,\mathring{\Delta X}_k]$	-
4	Multiple n_i, n_{ij} with σ_i, σ_{ij}	$\sigma_{xk} = [\sum_{i=1}^{n}\sum_{j=1}^{n} (X_j b_{ki})^2 \frac{\sigma_{ij}^2}{n_{ij}} + \sum_{i=1}^{n} b_{ki}^2\sigma_i^2/n_i]^{1/2}$	$\mathring{\Delta C} = \varepsilon_i(\sigma_i)$ $\mathring{\Delta A} = \varepsilon_{ij}(\sigma_{ij})$
5	Random uncorrelated	$\frac{RMS\,\Delta X}{RMS\,X} = K_{cT1}\frac{RMS\,\Delta A}{RMS\,A}$	$\Delta C = 0$

The formulae for clause 1-4 are after I.B. Kuperman, 1971, expression for clause 5 is after A. M. Turing, 1948.

We can estimate the uncertainties of solution for (12) by the formulae (J.B.Kuperman, 1971 , A. M.Turing, 1948) that are shown in Table 6-3, where values b_{ki} are the components of the reciprocal matrix $B(b_{ki}) = A^{-1}(a_{ij})$. If transform matrix $A(a_{ij})$ is expressed in the homogeneous coordinates, then the matrix $A(a_{ij})$ members uncertainty $\varepsilon_{4j} \equiv 0$. This allows us to provide a summation in the first member for the right-hand side of expressions 1-4 (Table 6-3) with j which is only equal to j=3, and in the second member with i only to i =3. If $\Delta C = \varepsilon_i = 0$, it is convenient to use link *N*o5 (Table 6-3) between the root mean square of the solution's uncertainty ($RMS\Delta X$) and the root mean square of transform matrix [A]'s uncertainty ($RMS\Delta A$) through Turing's first N- condition number K_{cT1}, which may show ill-conditioning of the matrix if number K_{cT1} is much larger than one.

Condition numbers. The theory of matrices has a first and second condition Turing's numbers, Todd's condition numbers, and Faddeev's (1959)condition number (Table 6-4). Use of any of these condition numbers K_c yields close results. These numbers in algebra have a limitation only from the bottom $K_c \geq 1$. Nevertheless, in approximate estimations for problems of a certain physical nature, the upper well-condition limit (elimination of ill-conditioning) may also be established.

Table 6-4. Condition numbers (cond) of matrix

Name and notation	Expression*
First Turing's number $K_{cT1} = N_T(A)$	$N_T(A)=$ $\frac{1}{n}N(A)N(A^{-1})=\frac{1}{n}[A]_I\,[A^{-1}]_I;\ [A]_I-(\sum_{\mu=1}^{n}\sum_{\nu=1}^{n}a_{\mu\nu}^2)^{1/2}$ $[A]_I=\|A\|$
Second Turing's number $K_{cT2} = M_T(A)$	$n\,M(A)\,M(A^{-1}) = \frac{1}{n}[A]_2[A^{-1}]_2;\ [A]_2=n\max_{\mu,\nu}\|a_{\mu\nu}\|$
Todd's number K_{c3}	$\max\|\lambda_i\| / \min\|\lambda_i\|$
Tood's P-condition number $K_{c4}(A)$	$\max\lambda_\mu / \min\lambda_\mu;\quad \lambda_\mu = Tr\,(A) = \sum_{i=1}^{n}a_{ii}$
Faddeev's H_F-condition number $K_{c5}(A) = H_F$	$\|A\|\,\|A^{-1}\| = \sqrt{\lambda_1/\lambda_2}$

Note: λ_1, λ_2 are the eigenvalues of matrices $[A]$, $[A^{-1}]$, respectively; $\|A\|$ and $Tr\,(A)$ are the norm and the trace(spur) of n x n matrix $A=[a_{\mu\nu}]$. * After A.M. Turing, 1948; D.K. Faddeev, 1959.

Probability and Statistical content. The probability content of the H_F- condition number (Table 6-4) corresponds to the ratio of the large and small ellipsoid's axes which we can find in the scattering results of equation (12)'s solution if the uncertainty of matrix $[A]$ members have normal distribution. It can be shown that for diagonal matrices $H_F = K_{cT2} / n$. In addition, let us recall that RMS corresponds to the second moment a_2 of probability distribution, i.e.,

$RMS = \sqrt{a_2}$, and the second moment can be expressed through variance $\tilde{S}^2$ and the

first moment a_1 of probability distribution as follows $a_2 = a_1^2 + \tilde{S}^2$. (6-13)

Matrices with uncorrelated members. A. M. Turing (1948) expressed the relationship between the root mean square of the solution's uncertainty ($RMS\Delta X$) and the root mean square of transform matrix $[A]$'s uncertainty ($RMS\Delta A$) through N-condition number as

$$\frac{RMS\Delta X}{RMSX} = \frac{1}{n}N(A)N(A^{-1})\frac{RMS\Delta A}{RMSA}, \tag{6-14}$$

Matrices with correlated members. We have found (Y.M. Tseytlin, *et al.*, 1989), however, that expression (14) is applicable for matrix [A] with uncorrelated members. If we have matrix with correlated members, then it is better to use the following relation

$$\frac{< dX_k\, dX_P >}{< \sum_{jq} X_j X_q >} = K_{cT1}^2(A)\frac{< da_{ij}\, da_{pq} >}{(1/n)\, N(B_2)}. \tag{6-15}$$

If the members of matrix $B_{(2)}$=A⊗ A are uncorrelated, then Cov $(da_{ij}\, da_{pq}) = 0$ and matrix B_2 converts into matrix A. Thus, we made in (15) the continuation of Turing's formula for the case of correlated members of the transformation matrix.

Convergence. Let us discuss this problem on the basis of the matrix equation

$\boldsymbol{b} = A\,\boldsymbol{X}.$ (6-16)

If for the square matrix we denote $\partial X_k / \partial a_{ij} = -\alpha_{ki} X_j$, (6-17)

then $dX_k = -\Sigma\alpha_{ki} X_j\, (da_{ij})$ and $ME(dX_k) = -\Sigma\alpha_{ki} X_j\, ME\,(da_{ij})$. (6-18)

It is clear that in the case $ME\,(da_{ij}) = 0$ we have $ME(dX_k) = a_1 = 0$. (6-19)

This is also correct for the diagonal matrix because

diag X = diag (a^{-1}) diag (b); (6-20)

ME diag ($X + dX$) = ME diag (a +da)$^{-1}$ diag (b) =

$$\frac{diag(b)}{ME\ diag(a)} - \frac{diag(b)\ ME\ diag(da)}{[ME\ diag(a)]^2}; \quad (6\text{-}21)$$

and ME diag $dX = \dfrac{diag(b)\ ME\ diag(da)}{[ME\ diag(a)]^2}$. (6-22)

Therefore, if ME diag(da) = 0, then ME diag(dX) = 0. (6-23)

Statistical extrapolation. Extrapolation of the calculated data for experimental sample's standard uncertainty $\tilde{S}$ to the population (infinite sample) distribution may be provided if we find the confidence boundaries for corresponding variance σ^2 or standard uncertainty σ on the basis of the estimation $\tilde{S}$, which is calculated from the limited sample of matrix n members. These boundaries for σ^2 can be found on the basis of accepted confidence probability P from the following equation

$$P\{\tilde{S}^2 f_{\aleph}/\chi^2_{P2} < \sigma^2 < \tilde{S}^2 f_{\aleph}/\chi^2_{P1}\} = P_2 - P_1, \quad (6\text{-}24)$$

where $f_{\aleph} = n$ is the degree of freedom; P_1 and P_2 are the confidence probabilities; χ^2_{P1} *and* χ^2_{P2} are the table values of χ^2 (chi-square) criterion at the corresponding values of P and $f_{\aleph}$. If we assume P_1 = 2.5%, P_2 = 97.5%, and $f_{\aleph}$ = 3 for the matrix of the third order, then $\chi^2_{2.5(3)} = 0.0720$; $\chi^2_{97.5(3)} = 3.1160$; and $P\{0.32\ S^2 \le \sigma^2 \le 13.9\ S^2\}$ = 95% (L.N. Bol'shev, N.V. Smirnov, 1983). The latter expression means that the predicted theoretical value of population standard deviation σ with a confidence probability of 95% is in the range 0.57 $S \le \sigma \le$ 3.74 S. Such a wide range obtained through the estimation of standard uncertainty is not representative enough. In the case of the third order quasi-diagonal matrix, $f_{\aleph} = 9$; $\chi^2_{2.5(9)} f^{-1}_{\aleph} = 0.3000$ & $\chi^2_{97.5(9)} f^{-1}_{\aleph} = 2.1137$. Hence, the value of population standard uncertainty σ is in the range 0.69 $S \le \sigma \le$ 1.82 S, and the use of the quasi-diagonal matrix yields more reliable results in the statistical relationship.

The estimation of the largest deviation ΔX_{max} from the mean value has technical applications. Let us recall that for normal distribution $\Delta X_{max} = 3\sigma$ with a confidence probability of 0.997. Therefore, we finally obtain $\Delta X_{max} = 3 \times 1.82\ S = 5.5\ \tilde{S}$ with an upper limit $\sigma = 1.82\ \tilde{S}$.

The family of calculated charts $\eta_{\chi}(P_2/P_1) = \dfrac{\chi^2_{P2} f^{-1}_{\aleph}}{\chi^2_{P1} f^{-1}_{\aleph}}$ and the charts for upper values $\eta_u(P_2) = \chi^2_{P2} f^{-1}_{\aleph}$ are shown in Fig. 6-7, where $f_{\aleph}$ is the limited degree of freedom.

Analysis of these charts reveals the following principles: matrices with larger rank $n = f_{\aleph}$ yield more representative values of $\tilde{S}$; and confidence probability $P = P_2 - P_1$ should be as low as possible but enough with respect to certainty. Here $P_1 < P_2$ and

χ^2_{P1} & χ^2_{P2} are the table values of χ^2 (chi-square) criterion at the corresponding values of $P_{i=1,2}$ and $f_{\aleph}$.

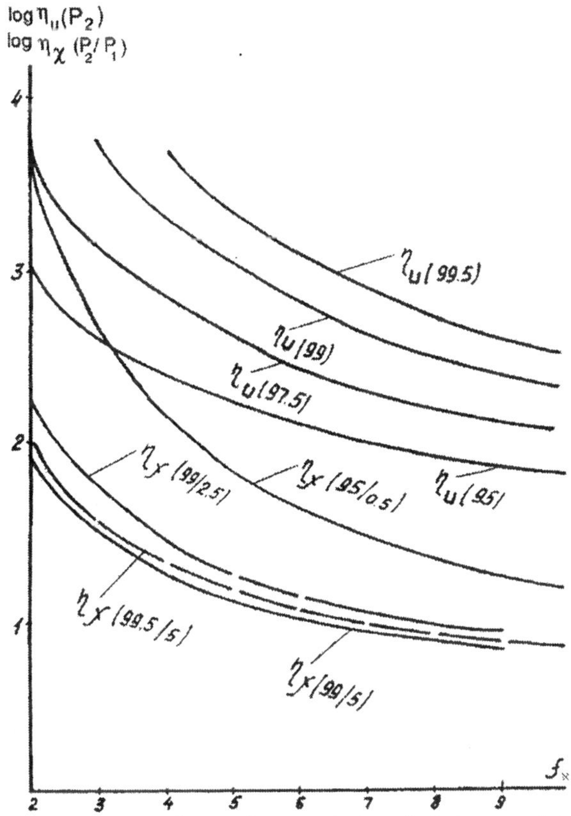

Fig. 6-7. Expansion to the population (infinite sample) of standard uncertainty calculated from the limited samples.

Influence uncertainty and structures' adaptation.

We differentiate three types of systems in their relationship to the *IQ*: 1) non-adaptive, which do not purposely use the current information about influence quantities and their values; 2) adaptive, which effectively use such information to correct their accuracy and precision features; and 3) invariant systems, which are practically not sensitive to changes of ambient *IQ* and their values.

Non-adaptive systems. These systems have certain phenomenological relationships between theirs schematic (Δ_{sch}), technological (Δ_{tec}), adjustment (Δ_{adj}), calibration (Δ_{cal}), permissible intrinsicΔ_{pitr}, and small influence δ_{inf} errors in reference conditions

$$\Delta_{pitr} \leq (\Delta_{sch} + \Delta_{tec} + \Delta_{adj} + \Delta_{cal}) \,/\, (1-\nu) \text{ if } \delta_{inf} \leq \nu\Delta_{pitr}\,, \tag{6-25}$$

where $\nu = \eta_U$ is the criterion of the negligibility of the error's (uncertainty's) component. This approach corresponds to the worst case of the estimation for the expansion operator Ξ in the limited set of random variables $\Omega_{\aleph}$ and the unspecified measure of probability, when Ξ coincides (Rosenberg V. Y., 1978) with the upper

bound of the set (i.e., $R_\omega = \sup \rho(\omega)$ where $\sup \rho(\omega)$ is the largest value of corresponding uncertainty R_ω). A similar approach allows us to find the limitation for the instrument's permissible schematic error (uncertainty) Δ_{psch} on the basis of specified tolerance δ_P for the measurand
$\Delta_{psch} \le \eta_U (1- \nu)^2 \delta_P$ if we assume that all other components of (25) are negligible in accordance with the information criterion. Calibration of non-adaptive systems should be provided by the standards whose influence functions are close to one of the measurand or the system's functional output.

Upper limit. The principles of the precision measurement system's synthesis may be linked with the upper limit of matrix N-condition numbers. Let us show that the upper limit of Turing's first condition number can be found on the basis of the criterion of negligible uncertainty. Turing's first number of matrix condition K_{cT1} characterizes, in accordance with case 5 in Table 6-2, the ratio, which in the numerator and denominator has estimators of the influence uncertainty in the measurand X and the transformation $A[a_{ij}]$system:

$$K_{cT1} = \frac{RMS\Delta X \cdot RMS\ A}{RMSX \cdot RMS\Delta A} = \frac{\tilde{S}_\sigma(X) \cdot RMS\ A}{RMSX \cdot \tilde{S}_\sigma(A)} \approx \frac{|\Delta^{(1)}_{\inf}|}{|\Delta^{(2)}_{\inf}|}, \qquad (6\text{-}26a)$$

where $\Delta^{(1)}_{\inf}$ and $\Delta^{(2)}_{\inf}$ are the two estimations of the influence uncertainty whose difference in the reference conditions should be negligible with respect to the information criterion; $\tilde{S}_\sigma(X)$ and $\tilde{S}_\sigma(A)$ are the standard.deviations of random variables X and $A[a_{ij}]$, and RMS is the root mean square value . The components of matrix $[A]$ represent the transient and transfer responses, transformation functions, and coefficients of influence in the structure under consideration.

Hence, in accordance with the information criterion of negligible uncertainty, the upper limit of Turing's first condition number for the matrix $A[a_{ij}]$ can be found from the following inequality,

$$K_{cT1} \approx \frac{|\Delta^{(1)}_{\inf}|}{|\Delta^{(2)}_{\inf}|} \le 1 + \eta_U\ . \qquad (6\text{-}26b)$$

For practical applications, this upper limit can be doubled and considered equal to 3.

S. G. Mikhlin (1988, 1991) showed that the functional estimation for the upper limit of the matrix's condition number could be performed using the characteristic polynomial. S.G. Mikhlin, however, suggests that this polynomial estimates may be rather coarse. He recommends apply the Sturm theorem for their improvement by separating roots of the polynomial.

Adaptive precision systems. These systems have an extended range of reference conditions for the *IQ* because they are automatically self-adjustable and can decrease the influence errors Δ_{inf} by certain automatic corrections (see Y.M.Tseytlin, 1981b). Let us remember that it is important to have a measurement adaptive system whose reaction is stable and adequate to the reaction of the measurand to the same changes of *IQ*. It is useful to have standard elements with known influence functions in the dimensional chain of measuring instrument or precision machine. Differential measurements are preferred for compensation of base displacement by gravitational field and other *IQ*.

Invariant systems. This precision systems are effective if the measurand is insensitive ($\Delta_{inf} \equiv 0$) to the changes of *IQ*, e.g., the measurand is of material with approximately zero temperature expansion coefficient; its quantity is invariant to the changes of temperature as well as force, mass, etc.

Structural synthesis and *Π-theorem.*

For synthesis of a precision system, one should specify the range of structure's function (e.g., measurement) R_m, the permissible intrinsic uncertainty $u_{p\cdot itr}$, or the permissible intrinsic expanded uncertainty ($U_{p.itr}$), and *n* significant *IQs*, whose dimensionalities are different from the dimensionality of the measurand. In this case, we can find, in accordance with *Π*-theorem (L.I. Sedov, 1993) that $k = (n+2) - (n+1)$ corresponds to one non-dimensional combination of specified data which includes the ratio $U_{p.itr} / R_m \leq \eta_0 = \sqrt{\eta_U^2 - \eta_V^2}$. If, in addition, we specify the permissible uncertainty $U_{p\cdot inf}$ or error from the action of the *IQ*'s combination, then it is possible to have $k = (n+3) - (n+1)$ which corresponds to 2 non-dimensional combinations in specifications, such that $U_{p\cdot inf} / R_m \leq \eta_v$ and $U_{p.itr} / R_m \leq \eta_0$. In this case, the non-dimensional system of requirements is a complete one.

Let us recall that *Π*-theorem states that any physical relationship between dimensional quantities may be presented as the relationship between non-dimensional quantities. Link between $n+1$ dimensional quantities $a, a_1, \ldots a_n$, *k* of which are specified with the independent units, may be represented in the form of relationship between $(n+1-k)$ dimensionless quantities which represent the combination of $(n+1)$ dimensional ones. If we specify *m* permissible influence functions ψ_i and the range of the corresponding *IQ*, then it may already be that $k = (n+3) - (n+1-m)$ corresponds to $m+2$ necessary dimensionless combinations in the complete system of requirements. In this case, one will add *m* weight coefficients of influence to the requirements for negligible intrinsic uncertainty (error) and its components. If the errors ΔA of the transformation's matrix $[A]$ are caused by the uniform influence quantity *V*, then one can find the permissible ΔV_{rr} limit for this *IQ* with the criterion of negligible component η_{vi} of the permissible measurement uncertainty. However, we should provide an extrapolation of the *RMS* estimator from the limited sample to the standard uncertainty of the population. If the measurement system's transformation matrix is an ill-condition, then $\delta L_{mrd} \neq 0$, even with system's adaptive options, and the problem of synthesizing such a precision system may be mathematically incorrect.

The *filter and correlator.* The filter and correlator in the precision system allow us expand the limits of the reference conditions' range because in this case we have the option to increase the ratio between noise *Ns* and signal *Sn* in the input without a loss of precision and accuracy: $(Ns / Sn)_1 = (Ns / Sn)_2 / \rho(\tau)$, where $\rho(\tau) \leq 1$ is the coefficient of correlation, and indices 1 and 2 correspond to the input and output, respectively. Ratio Ns / Sn should be less than criterion η_U if the system does not have a filter or correlator. The latter is helpful but it increases the time of gathering data. The narrow-band filter is effective under stationary stochastic disturbances and does not require increased time for measurement (acquisition data). The recurrent estimator, Kalman's filter, and similar methods (*see* averaging filters in G.A. Korn *et al.*, 1968) are used for the treatment of data (sequentially following in real time) when it is not necessary to remember the collected data, which otherwise should repeatedly be used for calculations. The defects of the recurrent estimators are revealed by rejecting rough (abnormal) data. The danger exists that one false or rough observation can deviate an *a posteriori* estimator so far that the next classification and identification with the object to be measured become impossible.

Reference limits for the moduli of elasticity.

Precision systems should be elastic under the action of influence quantities in the reference and operating conditions of their application because these influences should be reversible. The general relationships of the theory of elastic medium deformation usually correspond to its natural unstressed state. The definition of the term "natural" state in this case is an idealized abstraction because in any real initial state the elastic medium condition is a result of a superposition of residual internal and influencing external stresses from gravitation, vibration, temperature, clamping, etc. However, the actual values of moduli of elasticity are linked with the definite state of the medium. Therefore, we need to find a range of the dispersion for the modulus of elasticity values under certain conditions, which may be considered "real initial" or "reference".

General relations. Relationships between forces and displacements or between stresses σ_{ij} and strains ε_{ij} in the linear elastic system are linked through the potential energy Π_p as follows: $\sigma_{ij} = \partial \Pi_p / \partial\varepsilon_{ij}$. In accordance with Hook's law, we can express these links for the linear elastic medium of a deformable solid body by diagonal matrices as

$$diag(\overset{0}{\sigma}_{ii},\overset{1}{\sigma}_{ii},\overset{-1}{\sigma}_{ij}) = diag(\overset{0}{E},\overset{1}{E},\overset{-1}{E})diag(\varepsilon_\Sigma, \varepsilon_{ii}, \varepsilon_{ij}) = diag(\overset{s}{E})diag(\overset{s}{\varepsilon}), \tag{6-27}$$

where $(\overset{0}{\sigma}_{ii}+\overset{1}{\sigma}_{ii})$ are the components of normal stresses; $\overset{-1}{\sigma}_{ij}$ are the tangential stresses, $\overset{s}{E}$ are the generalized moduli of elasticity with $\overset{0}{E} = \nu_p E/[(1+\nu_p)(1-2\nu_p)]$; $\overset{1}{E} = E/(1+\nu_p)$; $\overset{-1}{E} = G$, E is the modulus of normal elasticity (Young's modulus); G is the shear modulus, ν_p is the Poisson's ratio; ε_{ii} are the relative elongations; ε_{ij} are the shifts of angles with deformation, $\varepsilon_\Sigma = \sum_{i=1}^{3}\varepsilon_{ii}$; "*diag*" denotes the diagonal matrices with null members except those positioned on the diagonal, *and i, j* = 1, 2, 3 are the notations for the respective Cartesian coordinates. Let us recall that (27) is correct for the initial natural unstressed state of elastic medium. If the initial state of the elastic medium differs from the natural one by the transformation of similarity with scale factor K_a, then the generalized moduli of elasticity $\overset{0}{E},\overset{1}{E},\overset{-1}{E}$ can change and accept values as follows

$$\overset{0}{E}{}^{*} = \tilde{f}_0(K_a)\overset{0}{E};\ \overset{1}{E}{}^{*} = \tilde{f}_1(K_a)\overset{1}{E};\ \overset{-1}{E}{}^{*} = \tilde{f}_{-1}(K_a)\overset{-1}{E}. \tag{6-28}$$

Then the components of stress $\overset{s}{\sigma}$ and strain $\overset{s}{\varepsilon}$ are connected by the following

equation $diag(\overset{s}{\sigma}^{*}) = diag(\overset{s}{E}^{*})\, diag(\overset{s}{\varepsilon}^{*})$ .Their reciprocal matrices are linked in a similar way, $diag(\overset{s}{\sigma}^{*})^{-1} = diag(\overset{s}{E})^{-1} diag(\overset{s}{\varepsilon})^{-1}$. (6-29)
By multiplying the left-hand and right-hand sides of (27) and (29), we yield the dimensionless matrices of relative change in stress and strain

$$diag(\overset{s}{\sigma})diag(\overset{s}{\sigma}^{*})^{-1} = diag(\overset{s}{E})\, diag(\overset{s}{E}^{*})^{-1}\, diag(\overset{s}{\varepsilon})\, diag(\overset{s}{\varepsilon}^{*})^{-1} = diag(\delta \overset{s}{E})\, diag(\overset{s}{\varepsilon})\, diag(\overset{s}{\varepsilon}^{*})^{-1}.$$

If we assume (Lurie, 1970) that $\tilde{f}_0\,(K_a) = K_a^3;\ \tilde{f}_1\,(K_a) = 1/K_a^2;\ \tilde{f}_{-1}\,(K_a) = K_a$,

and denote the change in the relative deformation and shift as $\Delta\overset{s}{\varepsilon} / \overset{s}{\varepsilon}$, then we finally have the sought expression

$$diag(\overset{s}{\sigma})\, diag(\overset{s}{\sigma}^{*})^{-1} = diag(K_a^3, 1/K_a^2, K_a)\, diag(1 - \Delta\overset{s}{\varepsilon} / \overset{s}{\varepsilon}). \qquad (6\text{-}30)$$

On the basis of the estimator for the first Turing's condition number K_{cT1}, the *RMS* for relative error $\Delta\overset{s}{\varepsilon} / \overset{s}{\varepsilon}$ is equal to

$$RMS\left(\frac{\Delta\overset{s}{\varepsilon}}{\overset{s}{\varepsilon}}\right) = 2\Delta K_a \frac{1}{n}\sqrt{\sum_s (\delta\overset{s}{E})^2}\sqrt{\sum_s (\delta\overset{s}{E}^{-1})^2} \qquad (6\text{-}31)$$

if the ratio of *RM.S* for errors $\Delta\delta\overset{s}{E}$ to the *RMS* of coefficients $\delta\overset{s}{E}$ equals

$\dfrac{RMS\Delta\delta\overset{s}{E}}{RMS\delta\overset{s}{E}} = \dfrac{RMS(1-\delta\overset{s}{E})}{RMS\delta\overset{s}{E}} \approx 2\Delta K_a$, where $\Delta K_a = 1 - K_a$, $n = 3$ is a rank of the matrix,

and $\delta\overset{s}{E}^{-1}$ are the members of reciprocal matrix diag $(\delta\overset{s}{E})^{-1}$. If the *RMS* in the reference conditions with extrapolation for the population of the largest deflection should be less than $\eta_v / 6$, then we find from (30), (31), and the value of K_a close to but less than one that $\Delta K_a \leq \eta_v / 12$. (6-32)
Substituting into (32) value $\eta_v = 0.12$ for the normal distribution of uncertainty, we will find the sought numerical estimator for permissible transformation scale K_a of the similarity in the initial state of the elastic medium under reference conditions, $K_a = 1 - \Delta K_a \approx 0.99$. Therefore, the boundaries of the permissible changes in the elastic moduli under the reference conditions should correspondingly be less than or equal to $|\Delta \overset{0}{E}| \leq 0.03\overset{0}{E};\ |\Delta \overset{1}{E}| \leq 0.02\overset{1}{E};\ |\Delta \overset{-1}{E}| \leq 0.01\overset{-1}{E}$.

The generalized effective moduli of elasticity $\overset{0}{E}_e, \overset{1}{E}_e, \overset{-1}{E}_e$ are the functions of the medium material's properties and its stress-strain state, as well as depend on the action of *IQ*, including temperature. Particularly, effective Young's modulus *E* and shear modulus *G* have temperature expansion functions which can be expressed by the following polynomial of the third degree, $\alpha_{TE\aleph}\,\Delta t_T^{\aleph}$, where $\aleph = 1, 2, 3$. It is enough to use the first member of the polynomial for the approximate estimation of the boundary of reference temperature, which corresponds to the following limitation $\Delta t_{Tref} < 100\eta_v / 12\alpha_{TE1}$. In the case of normal distribution, we have a requirement $\Delta t_{Tref} < (0.12 \times 100) / 12\alpha_{TE1}$,where α_{TE1} is the change in the elastic modulus in percent per 1°C. If the distributions for the measurand and permissible uncertainty are normal and, for example, α_{TE1} =0.2, then Δt_{Tref}< 5°C that is usual requirement for common purpose laboratory.

Isothermal and adiabatic moduli. The elastic moduli can be isothermal under the constant temperature of the elastic medium and adiabatic. The latter with the sufficient in practice accuracy differ (L.D.Landau and E.M. Lifshitz, 1970) from the isothermal E and ν_p by the following additions

ΔE_{ad}=$E^2 T_T \alpha_{TL}^2 / 9c_p$ and $\Delta \nu_{p.ad}$= $(1 + \nu_p) E T_T \alpha_{TL}^2 / 9c_p$,where T_T is the temperature, K; c_p is the thermal capacity of the material, J/(kg·K); and α_{TL} is the thermal expansion coefficient of the body. Let us recall that the adiabatic and isothermal rigidity (shear) moduli are the same.

Linear elastic system. Relationships between displacements and forces in a liner elastic system may be expressed through the following matrix equation (*see* Part A):

$$\begin{bmatrix} S_1 \\ Q_1 \end{bmatrix} = \begin{bmatrix} a_{11} & a_{12} \\ a_{21} & a_{22} \end{bmatrix} \begin{bmatrix} S_2 \\ Q_2 \end{bmatrix}, \tag{6-33}$$

where S_1, Q_1, S_2, and Q_2 are displacements and forces in the input (index 1) and output (index 2) of the system. Equation (33) changes with the change of environment around the system that we denote with an asterisk:

$$\begin{bmatrix} S_1 \\ Q_1 \end{bmatrix}^* = \begin{bmatrix} a_{11} & a_{12} \\ a_{21} & a_{22} \end{bmatrix}^* \begin{bmatrix} S_2 \\ Q_2 \end{bmatrix}^*. \tag{6-34}$$

Let us multiply from the left the left-hand and right-hand parts of expression (34) by matrix $\begin{bmatrix} S_1^{-1} & 0 \\ 0 & Q_1^{-1} \end{bmatrix}$, which is reciprocal to matrix $\begin{bmatrix} S_1 \\ Q_1 \end{bmatrix}$. We also assume that

$\overset{*}{S_1} S_1^{-1} = 1 + \delta S_1; \overset{*}{Q_1} Q_1^{-1} = 1 + \delta Q_1$, where δS_1 and δQ_1 are the relative changes of displacements and forces in the input. We accept similar notations for the output (index 2). As a result we obtain the sought matrix non-dimensional equation

$$\begin{bmatrix} 1+\delta S_2 & 0 \\ 0 & 1+\delta Q_2 \end{bmatrix} = \begin{bmatrix} A_{11} & A_{12} \\ A_{21} & A_{22} \end{bmatrix} \begin{bmatrix} 1+\delta S_1 & 0 \\ 0 & 1+\delta Q_1 \end{bmatrix}, \tag{6-35}$$

where

$$A_{11} = \frac{S_2^{-1} \overset{*}{a}_{22}}{\overset{*}{a}_{11} S_1^{-1} \overset{*}{a}_{22} - \overset{*}{a}_{21} S_1^{-1} \overset{*}{a}_{12}}, \quad A_{12} = \frac{(-1)^3 S_2^{-1} \overset{*}{a}_{12}}{\overset{*}{a}_{11} \overset{*}{a}_{22} Q_1^{-1} - \overset{*}{a}_{21} Q_1^{-1} \overset{*}{a}_{12}}, \tag{6-36}$$

$$A_{21} = \frac{(-1)^3 \overset{*}{a}_{21} Q_2^{-1}}{\overset{*}{a}_{11} S_1^{-1} \overset{*}{a}_{22} - \overset{*}{a}_{21} S_1^{-1} \overset{*}{a}_{12}}, \quad A_{22} = \frac{\overset{*}{a}_{11} Q_2^{-1}}{\overset{*}{a}_{11} \overset{*}{a}_{22} Q_1^{-1} - \overset{*}{a}_{21} Q_1^{-1} \overset{*}{a}_{12}}. \tag{6-37}$$

Let us recall from (26) and case 5 in Table 6-2 that the *RMS* and standard deviation σ in a matrix equation similar to (33) have the following relationships

$$\frac{S_\sigma(\delta S_2, \delta Q_2)}{RMS(1+\delta S_2, 1+\delta Q_2)} = K_{cT1}(A_{ij}) \frac{S_\sigma(\delta A_{ij})}{RMS(A_{ij})}. \tag{6-38}$$

Estimation of $K_{cT1}(A_{ij})$ requires a matrix reciprocal to A_{ij} which is equal to:

a)

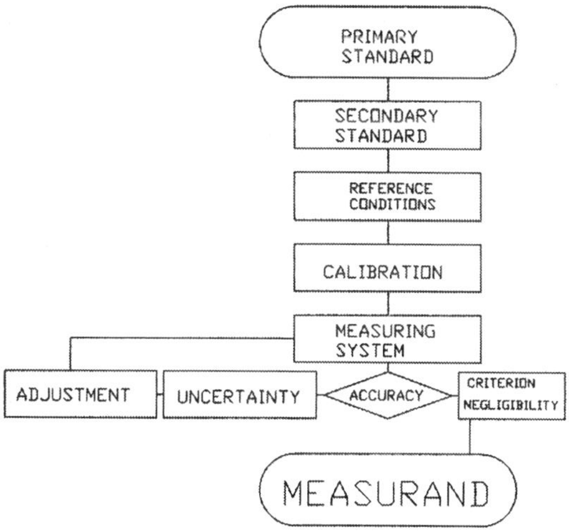

b)

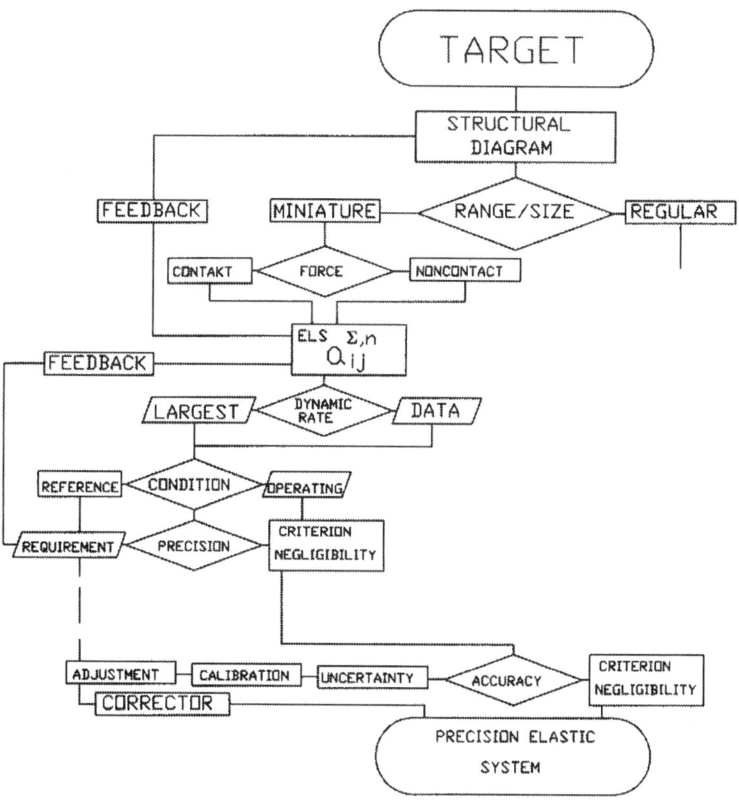

Fig. 6-8. Examples of the flowcharts for calibration and structural synthesis.

$$A^{-1}_{ij} = \begin{bmatrix} a^*_{11}S^{-1}_1 & a^*_{12}S^{-1}_1 \\ a^*_{21}Q^{-1}_1 & a^*_{22}Q^{-1}_1 \end{bmatrix} \begin{bmatrix} S_2 & 0 \\ 0 & Q_2 \end{bmatrix} = \begin{bmatrix} a^*_{11}S^{-1}_1 S_2 & a^*_{12}S^{-1}_1 Q_2 \\ a^*_{21}Q^{-1}_1 S_2 & a^*_{22}Q^{-1}_1 Q_2 \end{bmatrix}. \tag{6-39}$$

Turing's first condition number can then be limited by the information criterion of negligible uncertainty η_U as follows:

$$K_{cTI}(A_{ij}) = \frac{1}{n}\|A_{ij}\|\,\|A^{-1}_{ij}\| \leq (1+\eta_U)\,. \tag{6-40}$$

Requirement (40) is an important one for the structural synthesis of a linear precision elastic system. In addition, the standard uncertainty $S_\sigma[\delta S_2, \delta Q_2]$ should also be limited by the criterion η_v of negligible partial uncertainty as follows

$$\{S_\sigma[\delta S_2, \delta Q_2] \,/\, RMS.\ [1+ \delta S_2,\ 1 + \delta Q_2]\} \leq \eta_v. \tag{6-41}$$

Hence, $$\frac{S_\sigma(\delta A_{ij})}{RMS\ (A_{ij})} \leq \frac{\eta_v}{1+\eta_v}\,. \tag{6-42}$$

The physical aspects of *IQ*'s action on the elastic structures are discussed in the following chapters. We should note that the approach discussed in this chapter is an addition to well known *GUM* recommendations for uncertainty estimation, which shows the importance, goal, and possible methods for the legitimate limitation of the expanded uncertainty and its constituents at the structural synthesis in precision elasticity. See also detailed analysis of the GUM problems in R. Kacker and A. Jones (2003) publication.

Flowcharts.

Let us recall that a flowchart is a pictorial representation of all the steps of a process, including path of calibration and measurement traceability (Fig. 6-8a), as well as steps of the structural synthesis (Fig. 6-8b) and a structure of computer program. Short descriptions of the steps in Fig. 6-8b are shown in Table 8-8. All steps in flowchart not shown in Fig. 6-8b for the system with regular size are the same as for the miniature one. However, resulting matrix $a^{\Sigma,n}_{ij}$, the applied structural elements, and their assembly methods are totally different.

7.

The Main Influence Quantities

This chapter is dedicated to the main *IQ*, which are important for the structural synthesis in precision elasticity; boundaries of influence factors at reference and operating conditions in instrumentation, machine building, and nature. *IQ* may have different physical properties and mathematical representations. Most of them are three-dimensional dynamic heterogeneous fields with random constituents and specific features. Their influence occurs within the working space whose definition is discussed below.

7-1. The Working Space and its Monitoring

Space. The working space is a part of space surrounding a precision elastic structure and the measurand (if the system is a measuring one). The action of *IQ* beyond this space on the system's functions and results of measurement (when applicable) should be negligible. The working space of a precision elastic system may be isolated by thermal protection covers, vacuum (pressure) seals, magnetic shields, electrical filters, light protecting covers, vibro-isolating supports, etc. The minimal sizes of the working space (*see* Y.M. Tseytlin, 1981b) can be defined on the basis of the transient functions' $h_i(\xi_V)$ topological diagram or "topogram"(detail *see* in the next section) and the partial influence functions $\chi_i(\xi_V)$ obtained by the local influence ξ_V action on the quasi-uniform area segments of the precision structure.
Monitoring. The main problem with monitoring the real non-stationary, non-homogeneous influence field with random constituents is the choosing a necessary number and correctly positioning the sensors for certain influence quantities to acquire information complete enough for estimation of the field's state. All influence quantities may conditionally be split into quantities with post (after)-action and without post-action. Influence of post-action in the latter case continues for short time in comparison with the system's response dynamic constants. So this post-action can be neglected. Temperature influence usually has some significant post-action.
Positioning of sensors. V. V. Bolotin formulated principles of the measurement sensors' effective distribution in the random influence field for monitoring. In accordance with those principles, the minimum number of sensors is equal to the number of members in the series, which are necessary for the field under monitoring accurate enough approximation. In the one-dimensional system with length l and orthogonal base functions $\varphi_\alpha(x)$ on the segment $(0, l)$, the number of sensors is

defined from the following inequality, $n \geq l / \chi_{\lambda f}$; and for the monitoring of surface, the minimum number of sensors is $n = (l_1 / \chi_{\lambda f1}) + l_2 / \chi_{\lambda f2}$, where l_1, l_2 are the length of the surface coordinate directions; $\chi_{\lambda f}$, $\chi_{\lambda f1}$, $\chi_{\lambda f2}$ are the scale of non-homogeneity in the monitoring field which is defined through the intervals of the *IQ*'s autocorrelation function. If the influence field is homogeneous and ergodic along the coordinates, then it is enough to position the sensors on the distances, which are equal to several intervals of the field's autocorrelation function. The measure of covariance for the field's values in two points is the correlation coefficient $\tilde{\rho}_{12} = K_{12}/(K_{11}K_{22})^{1/2}$, where covariations K_{ij} form the correlation matrix

$$K_k = \begin{bmatrix} K_{11} & K_{12} \\ K_{12} & K_{22} \end{bmatrix}. \qquad \text{(a)}$$

The best distribution of monitoring points in the random field can be defined through minimizing the condition number of the correlation matrix. However, the condition of matrix K_k of the rank n as a whole does not yet guarantee an absence of the correlation between monitoring parameters in some points. To exclude this option of partial correlations, V. V. Bolotin has involved generalized condition numbers, which are equal to the largest values in the set of condition numbers k_c for this correlation matrix and all its principal submatrices:

$$\min \zeta(K_k) = \max \{k_c(K_k), k_c(K_1) \ldots k_c(K_{1,2..(n-2)})\}. \qquad \text{(b)}$$

The whole number of submatrices for the matrix of the n-th rank is equal to $(2^n - n - 2)$. If n=2, the generalized criterion coincides with the requirement for the best condition of the matrix (a). If n >2, the generalized criterion (b) also includes the requirement of the minimum mutual correlation between data in all monitoring points and the requirement of minimization of multiple correlation. Let us recall that the well condition of the matrices-factors yields the well condition of their product.

Discretizing. The problem of the discretizing the monitoring information arises when the resolution of the measuring system and its "traffic" capacity is not good enough in the cases with a very large volume of acquisition data. In these cases, we can use the Nyquist–Kotelnikov-Shannon sampling theorem (*see* G.A.Korn, T.M. Korn, 1968), which states that any function $x(t)$ with a bandwith $\omega = 2\pi B_f$ of the frequency B_f limited from the top is uniquely determined for all t by samples $x(t_k)$ spaced $1/(2B_f)$ t-units apart. Its mathematical expression is as follows

$$x(t) = \sum_{k=-\infty}^{\infty} x(t_k) \frac{\sin 2\pi B_f(t - t_k)}{2\pi B_f(t - t_k)},$$ where $t_k = (k / 2B_f)$ and $k = 0, \pm 1, \pm 2, \ldots$

The generalized sampling theorem (with Walsh square waves) allows us to find the frequency $B_f = f_c$ if the influence quantities are described through the continuous functions, which satisfy Dirichlet's conditions.

Statistical weights. Table 7-1 shows the approximate statistical weights of the *IQ* most common for precision measuring systems, as defined by the author on the basis of expert estimations for precision measurements. Similar expert estimations for other precision structures may be developed on the basis of the specific experience in their analysis, synthesis, production, and practical application. Special tests can be provided for the typical influence functions and reference conditions evaluation.

Table 7-1. Statistical weight* of influence quantities

Influence quantity	Typical precision system	Approximate weight of influence
Temperature	All types without thermal compensators, silicon cantilevers	$W_T = 2\text{-}3$
Forced vibration	All types without vibration isolators	$W_{va} = 1\text{-}2$
External forces: gravitational, Casimir, others	All resilient types, MEMS for Casimir normal and lateral force	$W_G = 1/2\text{-}2$
Orientation	With resolution less than 0.2 μm	$W_\psi = 1/3 \text{-}1$
Surrounding air pressure	Pneumatic, interference	$W_P = 1/3\text{-}1/2$
Humidity of air	All types, especially interference and non-hermetic electronic	$W_\varphi = 1/3\text{-}1/2$
Air density	Vibrators	$W_{ad} = 1\text{-}2$
Dust distribution	Precision with flexible elements	$W_{dd} = 1/3\text{-}1$
Speed of air	Interference, flexible elements	$W_{sa} = 1/3\text{-}1/2$
Strength of magnetic field	With ferromagnetic elements, capacitors	$W_M = 1/3 \text{-}1$
Strength of electrostatic field	With dielectric elements, microcantilevers	$W_E = 1/3 \text{-} 1$
Noise	With microphone-like elements	$W_{snd} = 1/3 - 1/2$
Lightning	Extremely flexible, photoelectric	$W_{lv} = 1/3 \text{-} 1$
Power supply parameters	Automatic without stabilizers	$W_{ep} = 1\text{-} 2$

* Y.M. Tseytlin, 1997.

Table 7-2 shows the internationally accepted nominal values of typical *IQ* for the reference conditions in precision measurements.

Table 7-2. Nominal values of influence quantities

No	Influence quantity	*Nominal reference value
1	Temperature for all types of measurement	20 °C (293.15 K)
2	Pressure of the ambient air for measurement of pressure and motion parameters	100 kPa
3	Pressure of the ambient air for the linear, angular, spectroscopic measurements	101,325.24 Pa
4	Relative humidity of the air for linear, angular measurements, measurement of mass, and in spectroscopy	58% (1332. 22 Pa) at 20 °C
5	Relative humidity of the air for the measurement of temperature, force, hardness, parameters of motion	65%
6	Air density	1.2 kg/m^3
7	Acceleration of gravity	9.8 m/s^2
8	Magnetic induction (strength of magnetic field) and strength of electrical field for the measurement of motion parameters, magnetic and electrical quantities	0
9	Magnetic induction (strength of magnetic field) and strength of electrical field for all types of measurement except mentioned in Clause 8	Corresponds to the characteristics of the Earth field in the certain geographic region

* *see* Comptes Rendus, 1961.

7-2. Temperature

Conditioned space. We may approximately represent the random constituents of the thermal field in conditioned space by the multi-dimensional function as follows:

$$\theta_{Tk} = A_{k1} \sin(\omega t + \gamma_l) + A_{k2}[1 - e^{-q(t-t_{l2})}] + A_{k3} U_e(t - t_{l3}) + A_{k4}\delta(t - t_{l4}),$$

where $\gamma_l = \omega t_{l1}$; $t_{l\upsilon}$ is the lag time for heating; γ_l is a phase angle; U_e and δ are the unit-step and impulse functions (*see* Table 1-11); $A_{k\upsilon}$ are independent random variables which characterize the amplitude of random changes in temperature from different sources: $\upsilon = 1$ is the convection change of the air temperature with the working conditioner; $\upsilon = 2$ is the thermal radiation from the body of an operator with wavelength $\lambda_q \approx 9$ μm or from other local inertial heaters; $\upsilon=3$ corresponds to a on/off switch of the low-inertial heaters; and $\upsilon=4$ corresponds to the work of an impulse low-inertial lighter. Variable A_{k1} in the thermostatic spaces has normal distribution in the range from ±1 to ±0.5 °C. Conditioners with high precision controllers of temperature provide the distribution of A_{k1} within the limits ± 0.05 °C. Variable $A_{k2} \leq 0.2$ °C. Index $k = 1, 2, 3$ reflects the non-homogeneity of an influence thermal field in the space.

Non-conditioned space. In a space without thermal conditioning, the ranges of variables $A_{k\upsilon}$'s distribution may increase, especially for A_{k1}, A_{k2} whose values approach tens of degrees Celsius. In addition, the multi-dimensional approximating function is different:

$$\theta_{Tk} = A_{k\upsilon} \sum_{1}^{n} \sin(\omega_{\upsilon} t + \gamma_{l\upsilon}) + A_{k(n+1)}[1 - e^{-q(t-t_{l2})}] + A_{k(n+2)} U_e(t - t_{l3}) + A_{k(n+3)} \delta(t - t_{l4}).$$

The first member of this expression reflects the approximation of polyharmonic periodic changes in temperature components with frequency ω_υ and phase angle $\gamma_{l\nu}$ including daily, seasonal, and annual changes. The distributions, variances, and mathematical expectations of variables $A_{k\upsilon}$ with $\upsilon = 1$to $(n+3)$ are also different (see Y.M. Tseytlin, 1981b).

Temperature-specified parameters.

The necessary and sufficient parameters for the characterization of thermal field and temperature deformations of elastic structure can be defined from the Fourier equation for the heat transmission

$$\partial u_T / \partial t = a_T \frac{\partial^2 u_{Ti}}{\partial x_i^2}, \qquad (7\text{-}1)$$

where $a_T = K_h / (C_p \rho_\gamma)$ is the coefficient of temperature conductivity (thermal diffusivity) which is equal to ratio of thermal conductivity K_h (*see* Table 7-6) and heat capacity $C_p \rho_\gamma$. It follows from (1) that for the influence temperature field the speed of its changes in time $\partial u_T / \partial t$ and the functions of its distribution in space $\partial^2 u_{Ti} / \partial x_i^2$ should be specified. Let us recall that instead of function $u_T = f(t)$ describing temperature changes in time, it is enough to specify the mean value and the amplitude of temperature changes in the specified time duration. Hence, the representative estimators in time are equal to sup $f(t) = \delta t_{Tm} + \Delta t_T$ and $\langle f(t) \rangle = \Delta t_T$, which correspond to the amplitude and mean deflection of temperature from its nominal value. Instead of the temperature parameters being represented with a continuous distribution in space $\partial^2 u_{Tm} / \partial x_i^2$, we can use their difference representation Δt_{Td}. The preference in values of specified ranges for reference temperature parameters are usually as follows: $\Delta t_T > \Delta t_{Td} > \delta t_{Tm}$. These preferences show the intent to decrease space and dynamic changes of temperature by constraining their non-homogeneity, amplitude, and speed. If the specified calculation time is not known *a priori,* it is assumed to be equal to one hour or another appropriate time interval.

Statistical thermodynamic fluctuations.

Deflections at equilibrium. Macroscopic body's physical models should have the smallest size that is larger than microscopic size L_{mic} (see Sect. 1-3). In an equilibrium state, systems of macroscopic bodies and external physical medium have parameters, which are approximately equal to their mean values. However, random small deflections from mean values take place as a result of thermodynamic fluctuations even at the macroscopic equilibrium. These thermodynamic fluctuations have a standard deviation, which may be equal to (Y. M. Tseytlin, 1972):

$$\sigma_{xa} = (K_B T_T X_a^2 / 2R_{v\,min})^{1/2}, \tag{7-2}$$

where $R_{v\ min}$ is the minimal work necessary for reversible small changes of the thermodynamic quantity X_a, K_B is Boltzmann's constant, and T_T is an absolute temperature. For a straight linear beam the tensile stiffness equals EA/L, and the minimal work is equal to: $R_{v\,min} = EA\ X_a^2 / 2L$, (7-3)

where E is Young's modulus of the beam's material, A is the area of the body's cross section, and L is the beam's length. Substituting (3) into (2) yields

$$\sigma_{xa} = (K_B\ T_T L / EA)^{1/2}, \tag{7-4}$$

and subsequently the limit of these fluctuations with normal distribution is equal to: $\Delta_{\varepsilon M} = 3\sigma_{xa}$. From this formula, it is evident that increasing length and decreasing the cross-section area of the body yield increasing statistical fluctuation of the beam's length. Fluctuations at room temperature can practically be larger than 10^{-10} for steel body, and would also be higher for materials with smaller elastic modulus. The expanded uncertainty U_{th} of thermal fluctuations at Gauss distribution with confidence level 0.95 are equal to $U_{th}=2\sigma_{xa}$. Therefore, we have factor 4 in the numerator under square root after substitution the expression for σ_{xa} from (4) as follows:

$$U_{th} = (4K_B T_T X_a^2 / 2R_{v\,min})^{1/2}. \tag{7-5}$$

Microcantilever tip thermal oscillation. The differential equation of the forced motion for the mechanical harmonic oscillator with viscose damping and the Langevin's stochastic thermal force F_{th} (t)=$j_k U_{th}(t)$ may be presented (*see* E. Majorana et al., 1997) as $\ddot{x} + \frac{\omega_0}{Q_{fac}}\dot{x} + \omega_0^2 x = j_k U_{th}(t)/m$, where m is the mass of the tip on the cantilever-oscillator. Let us recall that for a mechanical harmonic oscillator with viscose damping $X_a^2 / 2R_{min} = B_\omega / j_k \omega_0 Q_{fac}$, (7-5a)

where Q_{fac} is the Q-factor, j_k is the stiffness of the cantilever, B_ω is the bandwidth of measurement (*see* T.R. Albrecht *et al.*,1991), and ω_0 is the resonant frequency of the oscillator. Substituting (5a) into (5), we obtain an expression for the limit of expanded uncertainty in thermal fluctuations:

$$U_{th} = (4K_B T_T\ B_\omega / Q_{fac} j_k \omega_0)^{1/2}. \tag{7-5b}$$

If the thermal fluctuations correspond to harmonic excitement with the stationary amplitude U_{th} and an angular frequency ω, then the amplitude of the oscillator fluctuations in the bandwidth $B_\omega = (\omega - \omega_0)$ is equal to:

$$U_{th,\omega}=U_{th}\frac{\omega_0^2}{\sqrt{(\omega_0^2-\omega^2)^2+(\omega\omega_0)^2/Q_{fac}^2}}. \tag{7-5c}$$

Substituting (5b) into (5c), we obtain the equation for the mechanical oscillator thermal fluctuations in the bandwidth B_ω as follows:

$$U_{th,\omega}=(4K_BT_TB_\omega/Q_{fac}k\omega_0)^{1/2}\frac{\omega_0^2}{\sqrt{(\omega_0^2-\omega^2)^2+(\omega\omega_0)^2/Q_{fac}^2}}, \tag{7-5d}$$

where $\frac{\omega_0^2}{\sqrt{(\omega_0^2-\omega^2)^2+(\omega\omega_0)^2/Q_{fac}^2}}=\mu_{\omega/\omega_0}$ is the transfer function (or the dynamic response factor). Expression (5d) is similar to that presented by N.V. Lavrik *et al.* (2004) for the resulting thermal noise density spectrum.

Normal temperature revision. The nominal of reference temperature is commonly equal to t_{Tn} = 20 °C (Comptes Rendus des Séances ,1961) or in the absolute temperature representation T_{Tn} =293.15 K ≈ 293 K. It can be shown that the increase of temperature $t_{Tn} + \Delta t_T$ in precision elastic structures can lead to an increase of the thermodynamic fluctuations of their elements' dimensions and increase of their standard uncertainty $\Delta\sigma_{\Delta T}$, such that

$\Delta\sigma_{\Delta T}=\sqrt{\frac{\Delta t_T}{T_{Tn}}}\sigma_{xa}(T_{Tn})$,where $\sigma_{xa}(T_{Tn})$ is a standard uncertainty in the system element's thermodynamic dimensional fluctuations under established normal temperature T_{Tn}. For example, the expanded uncertainty of thermodynamic fluctuations is increased by 40% with an increase of temperature by 5 K because

$$3\,\Delta\sigma_{\Delta T}=3\sigma_{xa}(T_{Tn})\sqrt{5/293}\approx 0.40\,\sigma_{xa}(T_{Tn}).$$

The expected limits of statistical fluctuations for different elastic mechanisms are shown in Table 7-3. These fluctuations correspond to normal statistical distribution and in agreement with the known equipartition theorem, which states that the object's fluctuation variance is proportional to Boltsman's constant, absolute temperature (in K), and is inversely proportional to the object stiffness.

Table 7-3. Limits of statistical fluctuations for ELKS

Type of ELKS	Fluctuation limits $\Delta_{\varepsilon M}$	Example of estimation at normal temperature 293 K
PM-II	$\Delta_{\varepsilon M}=\pm 3\sqrt{\frac{k_BTL}{2Hb_0^{-2}}}$	b_0= 0.59 mm; L =12 mm; h=0.16 mm; b=3 mm; H=200 MPa · mm^4 $\Delta_{\varepsilon M}=\pm 6.2\times 10^{-10}$ mm
PM-IV	$\Delta_{\varepsilon M}=\pm 3\sqrt{k_BTL/4Q_s}$	L=14 mm; Q_s =3 cN $\Delta_{\varepsilon M}\approx\pm 6.5\times10^{-8}$ mm
SM-VI	$\Delta_{\varepsilon M}=\pm 3\sqrt{\frac{2k_BTL(\lambda+v_{if}\,b_1^2k_p^2)}{EA(\lambda+0.18\,b_1^2k_p^2)}}$	SM-VI 0.004×0.06×π× 20 E=112.8 GPa $\Delta_{\varepsilon M}\approx\pm$ 0.8×10^{-8} mm
SM-VII	$\Delta_{\varepsilon M}=\pm 3\sqrt{\frac{k_BT\pi D_r^3L}{0.36S_{av}\,bh^3E}}$	D_r =0.2 mm; S_{av} =1 mm; b=0.5 mm; h =0.01 mm; E=200 GPa; L = 76 mm $\Delta_{\varepsilon M}=\pm 4.4\times 10^{-8}$ mm

Note: S_{av} is an average pitch of the helix.

Medium with shape memory. Most elements of precision elastic systems proportionally change their dimensions with the change of temperature. The hysteretic damping of this process in elastic systems is usually small if the change of temperature is not very large. However, there are some metal alloys, which have a so-called shape memory (e.g., titanium – nickel or nitinol (*Ti-Ni*), *Ti-Pd*, *Cu-Al*, *Au-Cu*, *Fe-Mn* alloys). In these cases, small changes of an ambient temperature relative to the specific level may cause significant certain changes in the shape of elastic element, which is made of the material with the shape memory. This process is not an accurate enough for precision systems and more applicable to servo-motion systems, grippers, and medical injection applications.

Methods of temperature deformations' calculation.

The heat transfer in the temperature influence field can be performed by thermal conductivity, convection, and radiation. This corresponds to three specified types of boundary conditions:
(i) temperature distribution in any point of the body's surface at any moment of time;
(ii) density of heat flow through the body's surface at any moment of time;
(iii) a certain temperature of the surrounding medium and the convection type of heat transfer.

The problem at calculation of temperature distribution in the solid body and its temperature deformation is a complicated one if the temperature field is dynamic and not uniform (not homogeneous). The most universal contemporary methods for these calculations are based on the computerized finite element analysis and certain software programs (e.g., ANSYS, etc.).

Similarity. The analytical calculation methods for thermoelastic problems are based on the use of the so-called similarity criteria (dimensionless numbers): Fourier's thermal homochronism , Bio , Nusselt's , Prandtl's , Predvoditelev's, Reynolds', and Grashof's ,etc (*see* V.M. Marchenko, 1965, for more detail).

Distribution. The temperature distribution with the presence of internal heat sources is described by the following Fourier thermal diffusion equation,

$$\frac{\partial \theta_T}{\partial t} = a_T \nabla^2 \theta_T + \frac{W_{H.S}}{c_T \rho_\gamma}, \tag{7-6}$$

and the thermoelastic potential is described by the Laplace equation,

$$\nabla^2 \psi_T = \frac{1+\nu_p}{1-\nu_p} \alpha_{TL} \Delta\theta_T, \tag{7-7}$$

where $W_{H..S}$ is a specific power of the internal sources of heat and θ_T represents the temperature field. Mean temperature of the body is calculated with an integral over inside the considered volume of the body.

Precise solution. The precise analytical solutions of the thermoelastic problems have been obtained only for semi-bounded bodies of so-called classical or basic forms: infinite plate, infinite cylinder, and sphere (see A. V. Lykov, 1968). A semi-bounded body is one in which a heat exchange is carried out only through one surface separating it from the medium.

Approximation. Several approximate methods have been developed for the estimation of temperature distribution in bodies with complicated forms including: (i) a method based on the property of "heat flow stability," (ii) a method of basic bodies, (iii) a finite element analysis. Methods (i) and (ii) have an uncertainty of usually more than 50%. The finite element method provides more accurate calculations with an uncertainty of less than 35%. However, the accuracy of this method depends on the proper simulation of boundary conditions for the heat transfer and material features (isotropic, anisotropic, orthotropic). Verification of the calculation results may be performed by experimental-calculation methods.

Topogram method.

This method of *topographical diagram* (*topogram*), developed by G.G. Berdzenishvili, Y.M. Tseytlin *et al.* (1974), based on the representation of a real precision system as a set of finite elements on which partial transient temperature functions θ_{ei} are sequentially defined by the application of local surface jumps of temperature Δt_{Ti} and registration in time of the corresponding ΔL_{Ti} responses in the system's functional output. The phenomena on the surface and inside the volume limited by it are mathematically interrelated through the gradient or flux of an *IQ*. If we find a consecutive *topogram* of partial transient functions, then the temperature influence correction can be evaluated on the basis of a convolution and Duhamel's integral:

$$\Delta L_T = \sum_{i=1}^{n} \Delta L_{Ti}(t)\,, \tag{7-8a}$$

where

$$\Delta L_{Ti}(t) = \int_0^t \dot{\theta}_{ei}(t-\tau)\Delta t_{Ti}(\tau)d\tau\,. \tag{7-8b}$$

Special contact and non-contact local dynamic heaters have been developed for the realization of this method. The former can create a simulation of heating with the boundary conditions of the first and third types, i.e., heating a surface directly or via an air pocket. The latter heater may simulate a radiation heat transfer, or the boundary conditions of the second type.

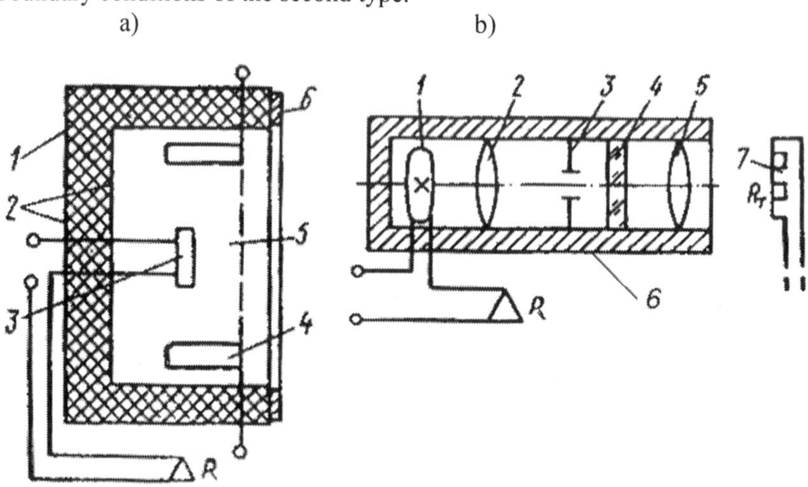

Fig. 7-1. Local dynamic heaters

Contact heater. The superposed heater with air pocket 5 (Fig. 7-1a) has thermometers 4 for monitoring the temperature in the pocket where connected with controllable by relay *R* resistor 3's power supply provides a jump in temperature in the near-surface air layer by approximately 5 °C. This eliminates the influence of a small fluctuation of the environment temperature from which the body of the heater is isolated with foam sponge 6, plastic case box 1, and thermoinsulating metalized fabric 2. In the first moment of sharp change in temperature, the heat impact occurs on the local segment of the system which is adjacent to the heater. However, the time period of this impact is small and close to that time of elastic wave front passing by. Let us recall that the quantity of heat translated through the wall with thickness h_w is equal in quasi-static process to $Q_{stat} = \frac{2\lambda_T}{h_w}(t_{T1}-t_{Ta})A_{out}\tau$, and the quantity of heat exerted by the heater with power W_R equals $Q_h = W_R\,\tau$. Hence, increase of temperature in the heater's pocket in the stabilized regime is equal to:

$$(t_{T1}-t_{Ta}) = \frac{W_R h_w}{\lambda_T A_{out}},$$

where $(t_{T1}$-$t_{Ta})$ is the finite increment of temperature in the pocket of the heater, A_{out} is the area of the output window in the heater's pocket, λ_T is the heat conductivity in the structure under investigation which has wall with thickness h_w. At regular heating regime, the cooling process of any body has exponential character in time.
Non-contact heater. Conditions of distant heat transfer by radiation can be simulated with an optical heater (Fig. 7-1b) consisting of a light source 1connected with controllable relay *R*, condenser 2, diaphragm 3 regulating the size and shape of the light's spot, frosted glass screen 4, and objective 5. A thermocouple or thermo-resistor can be used for the temperature monitoring by thermometer 7 (R_T). The optical heater is enclosed in a heat-protective casing 6 to eliminate heat dispersion into the ambient space. The heater may be installed at a considerable distance from the object being tested. This non-contact heater is very useful for the investigation of elastic compliant elements. A specified jump in heat is provided with source of rays and thermometer R_T, whose signal transfers to relay *R*, which controls the intensity of the source of rays. It is convenient in this case to use halogen lamp as a source of light. Now the heaters with laser and electron beam are also available. Time for the realization of the heat jump should be less than 10% of the time constant for the object under investigation.
Calculations. The method of partial *topogram* is effective for the objects with a complicated shape. Experimental data can be tabulated and treated with convolution algorithms. The advantage of using this method is the possibility of accounting for real temperature expansion coefficients in a structure, real types of strains, and boundary conditions. Tests showed that an uncertainty of this method in its application to precision linear measuring systems is less than 35%. However, the heating of an *i*-th local element by Δt_{Ti} (when we obtain the partial transient functions) is followed by the heating of other elements of the structure and even near-surface air layers with $\Delta t_{Tj}\,(t)$ which is less than Δt_{Ti}, i.e.,
$\Delta t_{Tj}\,(t)\,/\,\Delta t_{Ti} = a_{ij}\,(t) \le 1$. Hence, when heating one element, it is necessary, simultaneously with the registration of the partial transient function for temperature displacement, to register the heat portrait of the other structure's elements, which allows us to build matrix $[a_{ij}]$ for induced temperature on the conjugate elements of the surface. Then for the calculation of temperature influence corrections, we should substitute the calculated temperature changes $\Delta t_{Tic}(t)$ instead of the real once Δt_{Tj} in the convolution formula as follows:

$$\Delta t_{Tic}(t) = \int_0^t [\Delta t_{Tj}(t-\tau)]\frac{d}{d\tau}[a_{ij}(\tau)]d\tau\,. \tag{7-8c}$$

In this case, $a_{ii} = 1$; $a_{ij} < 1$ if $i \neq j$; and for remote elements of the surface $a_{ij} \approx 0$. The heat portrait may be registered by infrared imager, high-resolution scanning thermal probe for small objects, and heat-reactive dyes for a rough approximation. For example, the typical partial transient functions (temperature responses) in the structure with a spring measuring head of mikrokator type (*see* Chap. 8) on the stand are shown in Table 7-4.

Table 7-4. Typical partial transient temperature responses

Response	Chart and mathematical expression	Features
Aperiodic 1st-order with lag	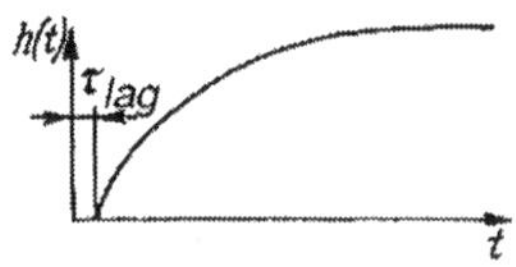 $\theta_{eT} = h(t) = k_\tau \left[1 - \exp(-\frac{t - \tau_{lag}}{T_{con}})\right] = k_\tau f(t, \tau_{lag}, T_{con})$	Close to exponential at $t > \tau_{lag}$; equals 0 at $t \leq \tau_{lag}$
Aperiodic 2nd-order with lag	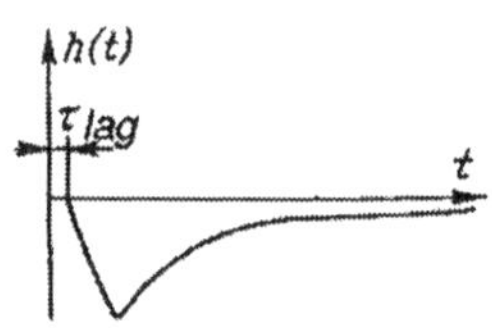 $h(t) = k_{\tau 1} f(t, \tau_{1lag}, T_{1con}) - k_{\tau 2} f(t, \tau_{2lag}, T_{2con})$	*V*-function after $t = \tau_{lag}$; equals 0 at $t \leq \tau_{lag}$
Aperiodic monotonic 2nd-order with lag	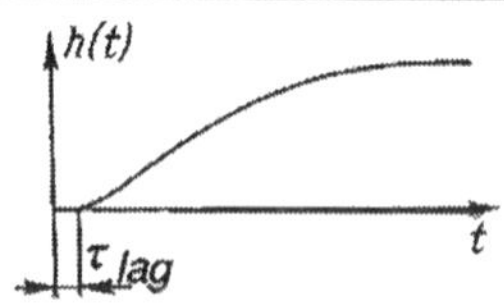 $h(t) = k_\tau \left[1 - \frac{T_{1con}}{T_{1con} - T_{2con}} \exp(-\frac{t - \tau_{lag}}{T_{1con}})\right] + k_\tau \left[\frac{T_{2con}}{T_{1con} - T_{2con}} \exp(-\frac{t - \tau_{lag}}{T_{2con}})\right]$	*S*-function
Zero	--	Equals 0 at any t

Note: T_{con} is a time constant; k_τ is a steady-response coefficient; τ_{lag} is a time lag (delay).

Minimization of the temperature influence.

Area, volume, mass. The transient temperature function for elastic displacements has an exponential character, $\theta_{Te} = h(t) = K_\tau [1 - e^{-(t - \tau_{lag})/T_\tau}]$,

where $T_\tau = K^* V_\Omega / (A \times \ln 10)$ is a time constant; K_τ is a transition coefficient; K^* is a coefficient that depends on the conditions of cooling (heating) the body; V_Ω (m) is the volume (mass) of the body, A is the area of the body's surface, and τ_{lag} is the lag time for the temperature transient function in the precision elastic structure.

The change of the transient function value during the precision system's working time t_m should be restricted by the criterion of negligibility η_U that corresponds to the following expression:

$1\text{-}\exp[-(t_m - \tau_{lag})\, A\times \ln 10 \,/\, K^* V_\Omega] \le \eta_U$. By subtracting 1 from both sides of this expression we find that $\exp[-(t_m - \tau_{lag})\, A\times\ln 10 \,/\, K^* V_\Omega] \ge (1 - \eta_U)$.

And after logarithming both sides of the latter expression, we have the relationship

$$(t_m - \tau_{lag}) \le \frac{K^* V_\Omega[-\ln(1-\eta_U)]}{A\times \ln 10}\,. \tag{7-9}$$

It is clear from (9) that increasing the working time, within which the transient function's change is a small one compared to the criterion η_U of negligibility, is possible by increasing the ratio between volume V_Ω and the area A of an elastic body. In this case, we suppose that K^* = constant. The volume of a solid body is proportional to its mass, which corresponds to the positive effect on the thermal stabilization from the increase of the body's mass. Therefore, we should use in the structural synthesis the optimal forms of elastic bodies with the largest possible relations V_Ω / A. It is known that the largest ratio V_Ω / A is in the spherical body, a relative extremum is provided by a cylinder. The structure as a whole should have the smallest possible summary of topogram (7-8a,b,c) in its working chain where

$$\Delta L_T = \sum_{i=1}^{n} \Delta L_{Ti}(t) \to 0.$$

Thermal correctors. Effective thermal correctors are based on the bimetallic elements (*see* Table 7-5), which yield approximate temperature adaptiveness to the elastic system. Formulae shown in this table provide good results for elastic elements with regular sizes in mechanical engineering as well as for microcantilevers (*see* P.G. Datscos *et al.*, 2004). One can see that influence of temperature on the precision elastic system is a complicated one.

Table 7-5. Bimetallic thermal correctors

Name, schematic	Formulae*	Notes
1. Straight beam 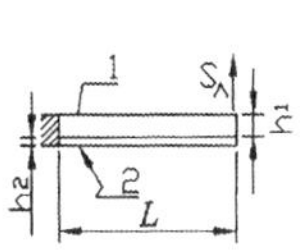	$s_\lambda = \dfrac{3L^2(h_1+h_2)^{-1}(\alpha_{TL2}-\alpha_{TL1})\Delta t_T}{3+(1+m_h^{-1}n_E^{-1})(m_h^3 n_E+1)/(1+m_h^2)}$	1-chrom-hickel Steel or Brass 2-invar b/L< 1/3 , $L\aleph \le 0.1$; $*\delta\Delta\aleph \le 10\%$; $S_\lambda = L^2\aleph/2$, for larger argument ($L\aleph$) see Table 1-9
$S_\lambda = \aleph^{-1}$ $\times\tan(L\aleph/2)\sin(L\aleph)$ $m_h = h_1/h_2$ $n_E = E_1/E_2$	$S_{\lambda\max} = \dfrac{3L^2}{4}(\alpha_{TL1}-\alpha_{TL2})\dfrac{\Delta t_T}{h}$	$*\ h_1/h_2 = \sqrt{E_2/E_1}$; $h = h_1+h_2$; * $\sigma_1 = E_1\dfrac{\Delta t_T}{h}(\alpha_{TL1}-\alpha_{TL2})(\frac{3}{2}y - h_1)$; $\sigma_2 = E_2\dfrac{\Delta t_T}{h}(\alpha_{TL1}-\alpha_{TL2})(\frac{3}{2}y + h_2)$ $L\aleph \le 0.1$.
	$S_\lambda = \dfrac{5}{4}(\alpha_{TL1}-\alpha_{TL2})\dfrac{h_1+h_2}{h_2^2 K_\lambda}\dfrac{L^3 W_p}{\lambda_{T1}h_1+\lambda_{T2}h_2}$ $K_\lambda = 4+6m_h+4(m_h)^2+n_E(m_h)^3+m_h^{-1}n_E^{-1}$	After J.R. Barnes *et al.*(1994) W_p is absorbed power with the thermal flux along the cantilever, λ_{T1} and λ_{T2} are the thermal conductivity of layers 1 and 2, respectively.

Table 7-5 Cont'd.

Name, schematic	Formulae*	Notes
	$<\delta S^{2}_{\lambda T}>^{1/2} = \frac{T\sqrt{4k_B B_\omega}\,\mu_\omega/\omega_0}{\omega_0^2 C_T^{1/2}\sqrt{(1/\tau_{tT})+\omega^2\tau_{tT}}}$	After P.W. Kruse, 1997; N.V. Larikov *et al.*, 2004. μ_ω/ω_0 is dynamic response factor; $<\delta S^{2}_{\lambda T}>^{1/2}$ is RMS response from temperature thermodynamic fluctuations; τ_{tT} is thermal response time, C_T is the heat capacity of the system.
2. *U*-beam 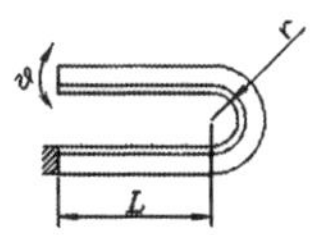	$\vartheta = \frac{3}{2}(\alpha_{TL1}-\alpha_{TL2})\frac{\Delta t_T}{h}L_w$	$L_w=2L+\pi r$
3. Equivalent moment M_t 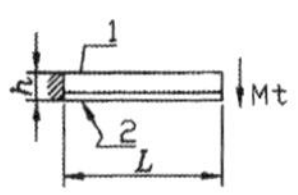	$M_t=\frac{3}{2}(\alpha_{TL1}-\alpha_{TL2})\frac{\Delta t_T}{h}EJ$	* $E=\frac{4E_1E_2}{(\sqrt{E_1}+\sqrt{E_2})^2}$ $J=\hat{J}_1+\hat{J}_2$; $\hat{J}_2=J_2+(\frac{h_1+h_2}{2}-h_n)^2 bh_2$; $\hat{J}_1=J_1+(\frac{h_1}{2}-h_n)^2 bh_1$; $h_n=\frac{E_1h_1+E_2(2h_2+h_2^2/h_1)}{2(E_2h_2+E_1h_1)}-h_1$
4. Intermediate support 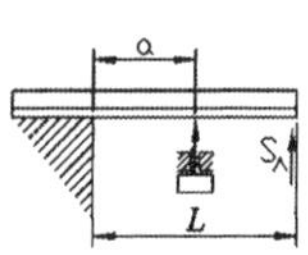	$S_\lambda=\frac{3}{4}(\alpha_{TL1}-\alpha_{TL2})\frac{\Delta t_T}{h}L^2\times$ $(1-\frac{3}{2}\frac{a}{L}+\frac{a^2}{2L^2})$	

Note: $\aleph$ is the curvature of a heated bimetal beam, b is the width of bimetal beam, Δt_T is beam's temperature change, σ_1 and σ_2 are stresses in layer 1 and 2, respectively, at distance y from the joint bonding layer; h_n is the distance of the neutral layer from the free boundary of layer 1; relative curvature deflection $\delta\Delta\aleph$ in more accurate solution with account for Poisson ratio from the values by the formula for $\aleph$ showed in this Table I increases with the increase of ratio b/L; practically α_{TL1}-$\alpha_{TL2}\approx$ from 10 to 24 μm/m·K^{-1}. Multistage microcoolers are discussed in R. Yang et al., 2004. * After S. Timoshenko, 1925; L.E. Andreeva, 1962.

Material properties. Thermal properties of elastic materials are shown in Table 7-6. It should be stressed that all data of these properties are temperature sensitive. The most data presented in the table are related to "room" temperature range. Let us also recall remarks about attention to the units of measurement when you use reference sources.

Table 7-6. Thermal properties of elastic materials

Material, profile	Designation	Temperature limits, °C		Specific heat, C_p J·kg^{-1} K^{-1}	Thermal conductivity, K_h, W m^{-1} K^{-1}	α°_{TE} / α°_{TL}
		Melting	Application			
Micro-strips	Br4Sn-3Zn **, (550)*	1020	*/3	385	390	1.0 /1.0
	Br Be2*** (320)*	915	*/2	420	118	0.67/ 0.99
	PtAg20****	1200	<200	165	143	0.56/ 0.93
Steel	4340(538)* 4340(204)* 4340(427)*	1420	*/2	460	24-33	0.77/ 0.77
Stainless steel strips	420 (315)*	1370	*/2	460	25	0.77 /0.7
Steel wire					32.9	
Invar	Fe-Ni36	1427	< (315)*	515	11	1.1/0.05^^
Copper Alloys, Hard	70Cu-30Zn 90Cu-10Zn -	870	...	375	120-389	1.07-1.28 /1.17
Brass-strip	12Zn-Cu	1050	...	375-393	116	.../1.21
Aluminum alloys	7075-T6 6061-T6	476	450	962 596	167 (237 for *Al*)	.../ 1.48
Titanium alloy	*Ti-6Al-4V* (700)*	1500	*/3	502-795	7.2-16.9	.../0.6
Nitinol	*Ti-55Ni* ^*TEP*=9-13	1310	70-130	470-620	10-18	.../0.65
Carbon clathrates whisker, nanotubes	C ^*TEP* = 22	3652	<750	~600	14-22$ 86-200 >2000	.../0.6 /0.04$
Boron carbide	B_4C	2450	<600	950	27-90	.../ 0.36
Silicon single crystal	Si (110)	1415	<1100	707-799	125-156	0.22/ 0.17
Fused quartz silica	SiO_2	1710	<1000	703-825	1.4-1.5 4.6	-0.22/ 0.032
Silicon carbide	SiC	2700	<600	715-975	42-150	.../0.3
Silicon nitride	Si_3N_4, Ceramic at room t_T sintered	1800-1900	<1000	700	19-50	.../0.18to/ 0.21
Poly-silicon	Si	1415	<1050	707.6	30-140	... /0.18

Table 7-6 Cont'd.

Material, profile	Designation	Temperature limits, °C		Specific heat, C_p J kg^{-1} K^{-1}	Thermal conductivity, K_h, W m^{-1} K^{-1}	α°_{TE} / α°_{TL}
		Melting	Application			
Tungsten	W	3410	...	132.7	108-178	...
			...			/0.29
	WC	2777			84-120	/0.32

*Tempering temperature; ** Thermal electromotive force (*TEMF*) with copper $TEMF = 2.0$ μV/K, resistance temperature coefficient $\alpha_{T\rho}$=9.5×10^{-4} K^{-1}, elasticity temperature coefficient α_E =- 4.5×10^{-4} K^{-1}; temperature coefficient of linear expansion α_L = 15.5 × 10^{-6} K^{-1}; *** *TEMF*= 1.0 μV/K, $\alpha_{T\rho}$=15.5×10^{-4} K^{-1} **** *TEMF*= 8.0 μV/K, $\alpha_{T\rho}$=10.5×10^{-4} K^{-1}; ^TEP is the thermoelectric power in μV/K, ^^ . α_{TL}°=0.003 for Super-Invar; $for Thornel fiber ∅ 11 μm . Zerodur has . α_{TL}°=0.008. Superscript °indicates relative values; sign ... shows that this parameter is not applicable or the author does not have reliable information and recommends to verify data in handbook by J.M. Shackelford and W. Alexander, 2001; Smithells, 2004; publications by D. Munz *et al.*, 1999; M.J. Madou, 2002; V.N. Popov, 2004; and B. Bhushan ed., 2004.

7-3. Reference Air

Pressure and humidity, Dalton's law.

If P_x, P_y, and P_z represent the respective partial pressures of gases *X*, *Y*, and *Z* in mixture, Dalton's law states that gas pressure $P_m = P_x + P_y + P_z$. Air must be considered a mixture of gases: oxygen, nitrogen, carbon dioxide, and water vapor (neglecting the minor constituents). All discussions of humidity can then be reduced to discussions of water vapor pressure, and all definitions encountered in humidity can be expressed in terms of vapor pressure. Let us recall that humidity may be characterized by dew point, frost point, and relative humidity. By employing relative humidity φ_H, temperature t_T , atmospheric pressure p_{aT}, or psychometric measurement data on wet t'_T and dry t_T bulb temperature, it is possible , for example, to determine the elasticity φ_{Ha} of the water vapor in the air (*see* Fig. 7-2a) as follows: $\Delta\varphi_{Ha}(p_{aT}) \approx 0.5\,(p_{aT}-1000)\,10^{-2}$; $p_{AT} \in [80, 106]$, kPa; $|t_T - t'_T| \approx 5$ °C.

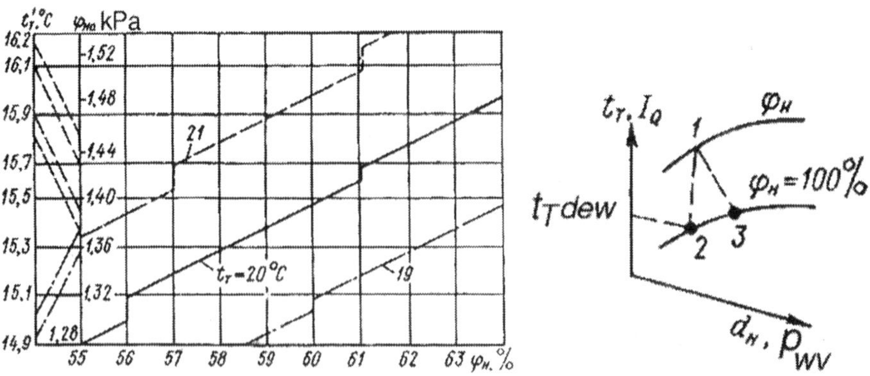

a) b)

Fig. 7-2. Diagrams of air state

To estimate the correspondence of φ_H and φ_{Ha}, the temperatures of the wet-bulb t'_T and the dry-bulb t_T, lines of the same type on the diagram are used (solid, dashed, dash-dot). Temperature and humidity environments for precision dimensional measurement are specified in American National Standard ANSI B89.6.2-1973 (later reaffirmed) and in many international documents.

$I_Q - d_H$ chart of air state. The relationships between the physical parameters of the air (temperature t_T, relative humidity φ_H, enthalpy I_Q, moisture content d_H, and partial pressure $\varphi_{Ha}=p_{wv}$ of water vapor) are expressed by I_Q - d_H diagram (Fig. 7-2b). Let us recall that enthalpy (heat content) is a function of the state of a thermodynamic system, which is equal to the sum of its internal energy and the product of pressure by volume. If the pressure is held constant, the enthalpy is a measure of the total heat in the system. Enthalpy I_Q of a perfect gaseous mixture is equal to the enthalpies of all gases entering into its composition. For example, the enthalpy of moist air is equal to:$I_Q = I_c + I_p$, where $I_c = c_p t_T$ is the enthalpy of dry air, kJ/ kg; I_p =(2500 + 1.88 $t_{T.p}$) d_H is the enthalpy of d_H kg of water vapor in kJ /kg; 2500 kJ / kg is the water evaporation heat at 0 °C, 1.88 kJ /(kg · K) is the heat capacity of water vapor, and $t_{T.p}$ is the temperature of water vapor in °C. Moist air is regarded as a homogeneous mixture with the same temperature. We have then an expression in kJ/kg for wet air

$I_Q = c_p t_T + (2500 + 1.88\ t_T)d_H$, where $c_p = 1$ is the heat capacity of air under constant pressure, in kJ/(kg · K). The I_Q-d_H diagrams are constructed for a particular barometric pressure in the oblique coordinates. On the margins of the I_Q-d_H diagram there is an angular scale that reflects the heat-humidity ratio, such that $\varepsilon_{ID} = \frac{\Delta I_Q}{\Delta d_H} = 240\frac{I_{Q2}-I_{Q1}}{d_{H2}-d_{H1}}$, where I_{Q1}, d_{H1} and I_{Q2}, d_{H2} are the enthalpy and moisture content of air in the initial 1 and final 2 states, respectively. By employing the I_Q-d_H diagram and using any two parameters of air, one may find all of its other parameters. The dew point temperature t_{Tdew} of air corresponds to intersection 2 of curve φ_H = 100% (Fig.7-2b) with a vertical line drawn from point 1 corresponding to the state of the air. The wet-bulb temperature corresponds to point 3, which coincides with the intersection of curve φ_H = 100% and the straight line running parallel to the enthalpy lines in point 1, which corresponds to the air parameters. The least practical uncertainty in the measurement of the temperature of ambient air has a random component within ±0.05 K, i.e., $1.7\times10^{-4}\ T_{Tn}$ and a systematic component within ±0.01 K, i.e., $3.4\times10^{-5}\ T_{Tn}$, where the normal absolute temperature is T_{Tn} = 293.15 K.

Density of air. The following estimation formula for air density $\rho_{\gamma a}$ is derived from the equation for the state of a real gaseous mixture

$$\rho_{\gamma a} = p_{at.a} M_a R_{g.c}^{-1} T_T^{-1} k_z^{-1} [1+(\varepsilon_{v.v}-1)\varphi_e / p_{at.a}],$$

where $p_{at.a}$ is the summary air pressure; $\varphi_e = \varphi_{Ha}$ is the partial pressure of the water vapor; M_a is the molelar mass of the dry air which is equal to:

$M_a = M_{a0.4} + 12.011[(V_{CO_2}/V) - 0.0004]$, and for the content of $CO_2 = V_{CO_2}/V$, which is equal to 0.04% in the normal reference air, $M_{a0.4} = 28.9635\times10^{-3}$ kgmol^{-1}; $R_{g.c} = 8.31451 \pm 8.4\times10^{-6}$ J·mol^{-1} K^{-1} is the gas constant; T_T is the temperature (in K) of air and water vapor; $\varepsilon_{v.v}$ is the ratio of the molecular mass of water vapor and dry air which is equal to 0.6220 if the content of CO_2 equals 0.04 %, and $k_z \approx 0.99960$ is an air compression coefficient at temperature T_T = 293.15 K, $p_{at.a}$=101325 Pa (~760 mm of the mercury column) , and $\varphi_H = 50\%$. Let us recall that the value of k_z is less dependable on humidity at lesser temperatures. For calculation purposes in practice, the following expression is employed

$$\rho_{\gamma a} = 0.12027\, p_{at.a} M_a T_T^{-1} k_z^{-1} (1 - 0.3780\, c_{pt}\, \varphi_e / p_{at.a}),$$

where $c_{pt}\varphi_e/p_{at.a}=d_H$ is the mole fraction of water vapor and $c_{pt} \approx 1.004$ is a correction factor. The density of the normal air therefore is equal to $\rho_{\gamma a} = 1.199$ kg/m^3 with air temperature $T_T = 293.15$ K (20 °C), pressure $p_{at.a} = 101325$ Pa, $\varphi_H = 58\%$, which corresponds to the partial pressure of water vapor $\varphi_e = 1333.22$ Pa, and the content of carbon dioxide (CO_2) of 0.04%. The random component of the relative uncertainty in the estimation-experimental appraisal of air density equals $\pm\, 3\times10^{-5}$ with the bias being within $\pm4\times10^{-4}$. The CO_2 content in the laboratories and shops is variable, and a change of 0.01% in the CO_2 content produces a relative change of 4×10^{-5} in the air density, which calls for checking this quantity when high-precision measurements of mass are involved. At the same time, the sum of the CO_2 and oxygen concentrations is assumed to be sufficiently constant and equal to 0.20979. For practical applications in mass measurement and air damping estimation in precision mechanical oscillators, the conditional density of normal air $\rho_{\gamma a} = 1.2$ kg /m^{3}can be used. For more details on the determination of the air density with International Committee equation (1981/91) *see* R.S. Davis, *Metrologia* (1992), **29**, 67-70. Air viscosity, its density, and Reynolds number are important for the calibration of AFM microcantilevers' flexural and torsional spring constants as well.

Refractive index. The refractive index n_λ is one of the most important parameters employed in optical interference measurement which enables a transition from the wavelengths in the vacuum λ_{vac} to their values in the ambient air $\lambda_{air} = \lambda_{vac} / n_\lambda$. The standard dry air has a specified temperature of 15 °C with the other parameters corresponding to the normal air but with $\varphi_H = 0$. To determine the index n_λ of refraction in the standard air, which is free of CO_2, the following Edlen's dispersion formula (B. Edlén, 1953, 1966) satisfying metrological requirements over the entire optical range had been adopted with the necessary correction (*see* K.P.Birch and M.J. Downs, *Metrologia*, 1994, **31**, 315-316) for the change of temperature scale in 1990,

$(n_{\lambda s}-1)\times10^8 = A_E + B_E(130-\tilde{\nu}^2)^{-1} + C_E(38.9-\tilde{\nu}^2)^{-1}$, where $\tilde{\nu} = 1/\lambda_{vac}$ is the vacuum wave number in µm^{-1}, $A_E = 8342.54$; $B_E = 2406147$; $C_E = 15998$.

The index $n_{\lambda s1}$ of standard air refraction, as corrected for the content of CO_2, is equal to: $(n_{\lambda s1}-1) = \left(1 + 0.0003\dfrac{n_{CO_2} - n_{\lambda s}}{n_{\lambda s}-1}\right)(n_{\lambda s}-1)$.

An uncertainty in the approximate expression for ($n_{\lambda s1}$-1) does not exceed 5×10^{-8} for the wavelength of 500 nm or more. The update of the latter Edlén formulae for the refractive index of the normal air (20°C, CO_2-0.04%) in accordance with CIPM-1981 definition for air density is given by R. Muijlwijk (*see Metrologia*, 1988, **25**, 189). To determine refractive index $n_{\lambda,\, p,t,\varphi}$ in air under any values of p, t_T, and φ_H, we employ corrected Edlén formulae such that

$$(n_{\lambda,p,t,\varphi}-1)=(n_{\lambda s1}-1)A_{\lambda}C_{\lambda}^{-1}(1+t_{T}\alpha_{T})^{-1}-\varphi_{Ha}B_{\lambda},$$

where $A_{\lambda}=1+10^{-8}(0.601-0.00972t)p$; $B_{\lambda}=[3.7345-0.0401\,\tilde{\nu}^{2}]\times10^{-10}$

C_{λ}–96095.43, $\alpha_T = 0.003661$. Pressure p is here in Pa and temperature in °C. Experimental data show that the behavior of quantities ($n_{\lambda,p,t,\varphi}$-1) is the same as that for air densities, i.e., such quantities follow Boyle-Mariott's and Gay-Lusaak's laws, and the index of air refraction minus one is obtained by adding together quantities (n_{λ}-1) for the individual components. As defined by standards, the reference air under normal atmospheric conditions is comprised of 20.9476 % oxygen, 78.087 % nitrogen, 0.917to 0.934 % argon, and 0.04 % carbon dioxide. The standard air contains no humidity data. The effect of CO_2 content on the change Δn_{CO_2} of refractive index can be determined from the following relationship

$\Delta n_{CO_2}\times10^{6}=(0.0145+0.00011\tilde{\nu}^{2})(k_{CO_2}-3)$, where $\Delta n_{CO_2}=n_{CO_2}-n_{\lambda s}$; $k_{CO_2}\times10^{-4}$ is the content of CO_2 in the air.

Refractometer. The most accurate values for the index of air refraction are provided by direct measurements taken with the help of a refractometer.

Reference conditions for length's measurement. Non-contact interference measurements of length are the most accurate in this field, but they are very sensitive to the conditions of their execution if the difference between the interference rays' paths is more than 1 mm (see Y.M. Tseytlin, 1981b). The wavelength of a primary reference radiation of krypton 86 in normal air is λ_1 =605.61574 nm with the refractive index $n_{\lambda 1}$ =1.00027159; the wavelength of secondary reference radiation is λ_2 =546.07819 nm with the refractive index $n_{\lambda 2}$= 1.00027259, and the wavelength of the laser stabilized with the iodine cell is λ_{las}= 0.632819505 μm with the refractive index n_{λ}= 1.00027128.

It is also significant for interferometric instruments and other elastic precision structures the limitation of the speed of air motion (can cause refraction, deflection, vibration), its dust content, strength of electromagnetic fields, spatial orientation, etc.

Dust content and settling.

Dust may influence the behavior of elastic systems by adding some mass to a thin elastic element and increasing damping in small functional gaps. Soil erosion, wear of asphalt pavement, aero plankton, microorganisms, and all production processes are the sources of dust in the air.

Particles and aerosols. The air contains particles and the vapors of petroleum oil and other organic liquids, which may be referred to as aerosols. Dust particles may arbitrarily vary in shape and chemical content. Air within production areas contains numerous fine fibers, including metallic fibers. In modern industrial cities, dust is comprised of mineral (70%) and organic (30%) substances. The fractional composition of dust in air is shown in Fig. 7-3a. A steep descent of the curve with an increase in the particle size, a_{ch}, is due to their rapid settling. Calculations based on Stock's formula for the free fall of a body in perfectly still air yield the dust with dimensions a_{ch} ≤ 1 μm settling velocity equal to $v_{ach} \leq 3.5\times10^{-5}$ m / s. The particles of at least 10 μm in diameter settle down rapidly enough with the velocity v_{ach} > 0.003 m/s. The concentration of the dust particles in air increases as their size decreases (Fig.7-3b).

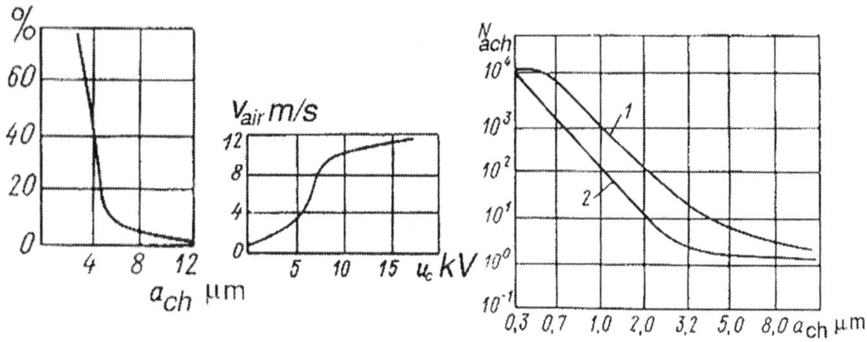

a) b) c)

Fig. 7-3. Features of dust particles in air.

With an increase of dust particle concentration N_{ach}, however, the probability of their mutual collision and coagulation under the action of attracting forces (Van der Waals and electric type) also increases. These forces are dependent on the size of particles. The binding energy of two ball-shaped particles is estimated as 10^{-17} J. Such binding energy substantially exceeds that of atoms in chemical compounds. Consequently, when in contact, the particles tenaciously adhere to one another. According to aerosol mechanics, the coagulation velocity of dust particles is proportional to their concentration. Therefore, with an arbitrarily large concentration of dust at the instant of its formation, even in a short length of time, the concentration may not exceed 10^6 particles in 1 liter of air. The coagulation of dust particles is speeded up in moist air since the absorbed film of moisture increases the binding forces of particles in collision. Therefore, concentration of particles at rain (curve 2 in Fig. 7-3c) is less than in dry air (curve 1).

The unlike charges of particles also encourage coagulation. The limits for splitting of dust particle cover the range down to molecular size. Particles of 0.5 to 1.0 nm in size were detected with the help of an electronic microscope.

Cyclone, moisture, adhesion. Another mechanism of airborne dust settling is due to inertia of dust particles (cyclone effect) when eddy (cyclonic) currents of air pass in the neighborhood of objects. In this case, the dust particles are thrown onto the objects and adhere to them. Under substantial humidity, the particle-to-surface binding forces abruptly increase. The surface adhesion of the particles of 1 to 2 μm in diameter, which have settled on the object, features such tenacity that the particles cannot be blown off by the compressed air jet at a velocity of 200 m/s.

Charge. Charge accumulation by particles is due to a variety of causes: the sorption of ions from air, friction of particles against objects, photoelectric effect, thermal electron emission, etc. The charge value is a function of the airflow velocity v_{air} (Fig. 7-3b) and gas composition. As it is seen from Fig. 7-3b, the charge of dust particles u, kV, can reach several hundreds volts even when the velocities are comparatively low. The polarity of dust particles can be both positive and negative. Attraction of the charged particles to objects is caused by either the induced forces of "mirror reflection" or the charges on the object's surface. The term "dust-repelling material", which may be illustrated by referring to films, paints, and coatings, is only taken to signify minimum dust-retaining properties, i.e., the materials poorly absorbing water vapor and discharging dust particles in collision.

7-4. Forced and Natural Vibrations, Impact Action

Sources and parameters of forced vibration

Sources. Vibration sources are various. Effects of the wind, sea waves, fluctuations of the ground due to the traffic, technology equipment in operation (presses, hammers, compressors, fans, blowers, machine tools) cause oscillation of buildings, structures, and their elements.

Self-excited oscillations. Unlike vibrations due to runout and eccentricity, self-excited oscillations may be of inharmonious type. The initial cause of self-excited oscillations is a nonlinear nature of the relationships between the cutting forcc and the cutting speed as well as due to the presence of dry (Coulomb) friction in the system of machine tool-fixture-part (*MTFP*). For example, the part has surface under treatment with radius r.

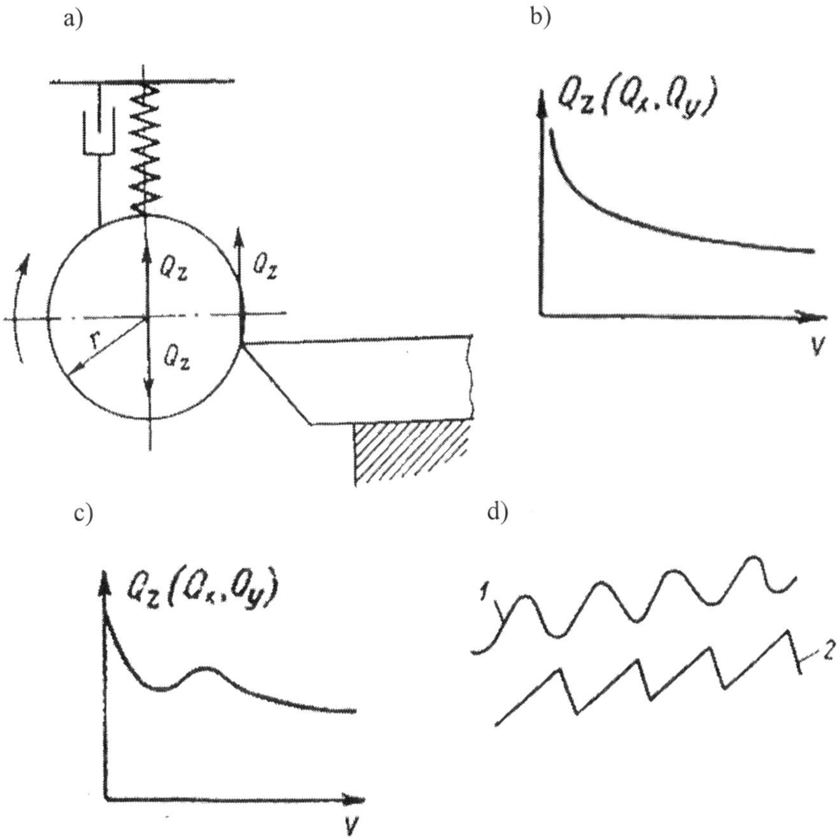

Fig. 7-4. Self-excited vibrations caused by cutting force

The differential equation of vibrations for *MTFP* in turning system (Fig. 7-4a) with one degree of freedom may be presented as

$$M_m \ddot{Z} + C_h \dot{Z} + j_{s.p} Z = Q_z(\dot{Z}), \tag{7-10a}$$

where M_m is the datum mass of *MTFP*, C_h is the motion resistance factor (friction); $j_{s.p}$ is the stiffness of *MTFP*, and $Q_z(\dot{Z})$ is a nonlinear characteristic of the cutting force along *Z*-direction. Depending on the work material, the cutting force Q_z (Q_x, Q_y) characteristic may be either 1) a decreasing monotonic one with dropping relationship against the cutting speed v (Fig. 7-4b), or 2) with two dropping segments along the relationship curve against the cutting speed v (Fig. 7-4c). The approximation of such characteristic may be executed using the expansion with polynomial of the third degree. As a result of this approximation, the differential equation (10a) may be represented in the following nonlinear form

$$\ddot{Z} - C_\theta \dot{Z} + \beta \dot{Z}^3 + \omega^2 Z = 0, \tag{7-10b}$$

where $C_\theta = H_{pz}/M_m$; $\beta = H_{pz}/3M_m v_{ap}^2$; $\omega^2 = j_{s.p}/M_m$; $H_{pz} = (Q - Q_0)/[(\dot{Z} - \dot{Z}^3)/3v_{ap}^2]$.

Here $Q = Q_z(\dot{Z}) + C_h(\dot{Z})$ is the summarized cutting and resistance force, $Q_0 = Q_z(0) + C_h(0)$ is the summarized cutting and resistance force in conditions without vibration, v_{ap} is a half of nonlinear segment's extent on the curve describing relationship between cutting force and cutting speed v. Equations analogous to (10b) may be derived for the systems having torsional and bending vibrations as well. The solution of differential equation (10b) with small parameter $\varepsilon_{pz} = H_{pz}\,\omega/j_{s.p} < 1$ may be delivered (L. S. Murashkin, S. L. Murashkin, 1977) as

$$Z(t) = (2V_{ap}/\omega)[\cos v_\omega t - \frac{1}{8}(\sin v_\omega t - \frac{1}{3}\sin 3v_\omega t)\varepsilon_{pz} + \frac{5}{64}(\cos v_\omega t + \frac{2}{5}\cos 3v_\omega t - \frac{1}{15}\cos 5v_\omega t)\varepsilon_{pz}^2], \tag{7-11}$$

where the frequency of vibrations $v_\omega = \omega(1 - \frac{1}{16}\varepsilon_{pz}^2)$ is different from the natural angular frequency ω of *MTFP*. The oscillation form may be close to the sinusoidal 1 (harmonic vibrations when $\varepsilon_{pz} \le 1$) and to saw-tooth type 2 (Fig. 7-4d), i.e., with inharmonious character when $\varepsilon_{pz} > 1$.

Oscillation parameters consideration. Recognition of harmonics with considerable amplitudes is carried out by means of Parseval-Lyapunov's equality:

$$s_{s.p} \approx \sqrt{\sum_{n=1}^{p} \frac{A_n^2}{2}}, \tag{7-12}$$

where $s_{s.p}$ is a standard deviation from mean line of the oscillation process realization curve, $\sum_{n=1}^{p} \frac{A_n^2}{2}$ is a half-sum of the squares of considerable p-harmonic amplitudes.

As (11) shows, the self-excited process of oscillations in the *MTFP* system with relative large stiffness has first, third, and fifth harmonics with respective amplitudes:

$A_1 = (2V_{ap}/\omega)\sqrt{(1 + \frac{5}{64}\varepsilon_{pz}^2)^2 + \frac{1}{64}\varepsilon_{pz}^2}$; $A_3 = \frac{2V_{ap}}{\omega}\varepsilon_{pz}\sqrt{\frac{1}{16}\varepsilon_{pz}^2 + \frac{1}{9}}$; $A_5 = \frac{1}{96}(V_{ap}/\omega)\,\varepsilon_{pz}^2$. As for a general periodic process with a random phase, the covariance function is equal to $R_{xx}(\tau) = C_0^2 + \frac{1}{2}\sum_{n=1}^{\infty}(a_n^2+b_n^2)\cos k\nu\tau$, (7-13)

and the variance of oscillation process equals

$$s_{s.p}^2 = D_{s.p} = R_{xx}(0) - C_0^2 = \frac{1}{2}\sum_{n=1}^{\infty}(a_n^2+b_n^2) = \frac{1}{2}\sum_{n=1}^{\infty} A_n^2 , \qquad (7\text{-}14)$$

then Parseval-Lyapunov's condition for the solution of (11) may be satisfied even without account of the fifth harmonic if $\varepsilon_{pz} \leq 1$. The fifth harmonic becomes considerable when $\varepsilon_{pz} > 5$.

a) b)

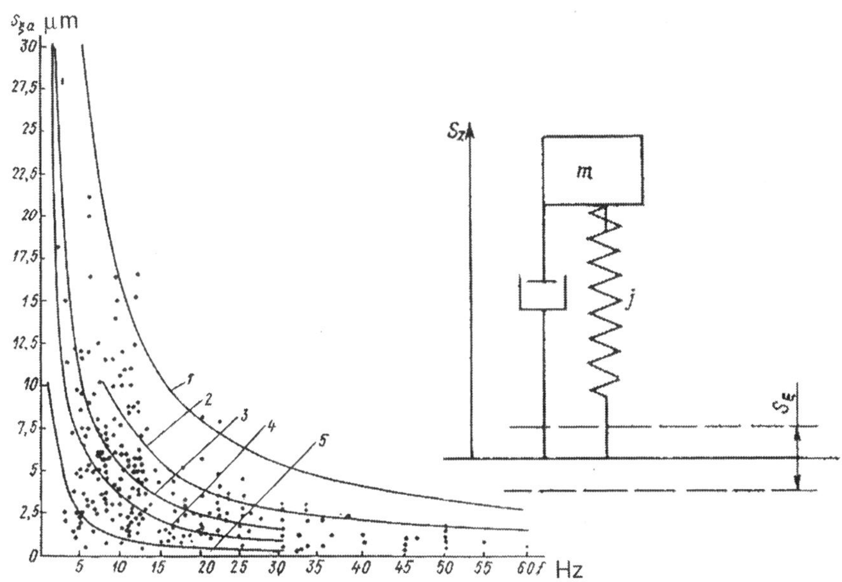

Fig. 7-5. Amplitude of forced vibrations distribution and single-mass model of oscillations

Speed amplitude. Experimental diagrams of forced vibrations acting on the measuring instruments in different Russian industries revealed their speed quasi-constant amplitude because the curves in frequency f-displacement amplitude $s_{\xi a}$ coordinates for these vibrations are close to the hyperbolic shape.

Fig. 7-5a shows distribution of forced vibration amplitudes $s_{\xi a}$ acting on the linear measuring instruments (curve 1) with amplitude of speed V_{oa}=1 mm/s, on the production shop floor (curve 2) with V_{oa}=0.5 mm/s, and curves 3-5 represent the limits for the permissible amplitude of vibration speed at reference conditions in GOST 8.050 with V_{oa}=0.315, 0.189, and 0.63 mm/s, respectively. The amplitude function $s_{\xi a}(f)$ decreases monotonically in the range of frequencies from f=2 to 60 Hz. The hyperbolic relationships between the amplitudes and the frequencies in the forced vibrations are typical in industry if there are no special vibro-isolation systems. In these cases, we can choose the amplitude of speed as the norm for the forced vibrations. Such method of specification was chosen by the author in Russian interstate metrological standards GOST8.050, GOST8.395 for the conditions of precision measurements and traceability of calibration. Russian Central Research Institute of Building Construction (TSNIISK, Moscow) later (1977) used the same method of the constant amplitude of forced vibrations from 0.0315 to more than 10 mm/s with frequency in the range from 1 to 100 Hz in the specifications for calculation of the dynamic loads at the construction of buildings with a wide range of installed gages, machine tools, and technological equipment. The hyperbolic character in the relationship between amplitude and frequencies is found in experimental studies on the vibrations of different sand-clay grounds during ramming of piles as well.

The limits of forced vibrations. We can show that the constant amplitude of speed in the forced vibrations is a natural phenomenon in industrial applications. Let us examine an elementary dynamic single-mass model with concentrated mass m, stiffness j, and viscose damper (Fig. 7-5b). The differential equation of this mass motion under action of the support forced vibration $\ddot{S}_z+2\upsilon_c\dot{S}_z+\omega_0^2 S_z=2\upsilon_c\dot{S}_\xi+\omega_0^2 S_\xi$ has a solution with the amplitude that is equal to: $S_{za}=S_{\xi a}\sqrt{\dfrac{1+(\omega^2/\omega_0^2)\varepsilon_\upsilon^2}{(1-\omega^2/\omega_0^2)^2+(\omega^2/\omega_0^2)\varepsilon_\upsilon^2}}$,

where $\varepsilon_\upsilon=2\upsilon_c/\omega_0$; $\upsilon_c=C_h/2m$ is the damping coefficient, C_h is the coefficient of resistance; $\omega_0=\sqrt{j/m}$ is the natural angular frequency of the system's oscillations at the absence of resistance, and j is the stiffness of an elastic link. By balancing of technological equipment, one is achieving the restriction of vibrations under the specified level of the acceleration's amplitude, which is an analog for the force action of the equipment on the system of its installation, i.e., this is securing the condition $S_{\xi a}\omega^2\le A_\omega$ and in the limit $S_{\xi a}\omega^2=A_\omega=$ constant. Because of the very large mass m of the installation and transferring vibration systems to the buildings, structures, and the precision arrangements, such systems can transfer only resonance frequencies. Hence, when $\omega=\omega_0$ and $\varepsilon_\upsilon<1$, we have the following relationship

$S_{za}=S_{\xi a}\dfrac{\omega}{2\upsilon_c}$. If $\upsilon_c=$ constant, the product of the amplitude S_{za} and the frequency

ω_0, which is equal to the amplitude of the forced vibrations' speed acting on the precision elastic structure, can be expressed as follows

$$S_{za}\omega_0=\frac{S_{\xi a}\omega_0^2}{2\upsilon_c}=\frac{A_\omega}{2\upsilon_c}=V_{0a}=\text{constant.}$$

We consider that the υ_c is constant as a result of using practically the same materials and typical structures for ceilings, floors, and other elements of the buildings and constructions. Thus, the constancy of the forced vibration's amplitude is not a random empirical case. The examined model also reveals the boundaries of the typical asymmetric range for the normal forced vibrations.

The vibro-sensitivity.

Estimations and *tests.* This estimation may be produced analytically and experimentally. We use the analytical method during the first steps of synthesis and design of structures, and also in the cases when experimental means are impossible or ineffective. The experimental tests may be executed by vibrating installations

(Fig. 7-6a) based on a mechanical (I) with eccentric 1 and unbalanced 2 exciters, electrical (II), hydraulic (III) principles and attachable superposed vibrators (Fig. 7-6b) of eccentric (I), eccentric-pendulum (II), electrodynamics (III), and piezoelectric (IV) types.

a)

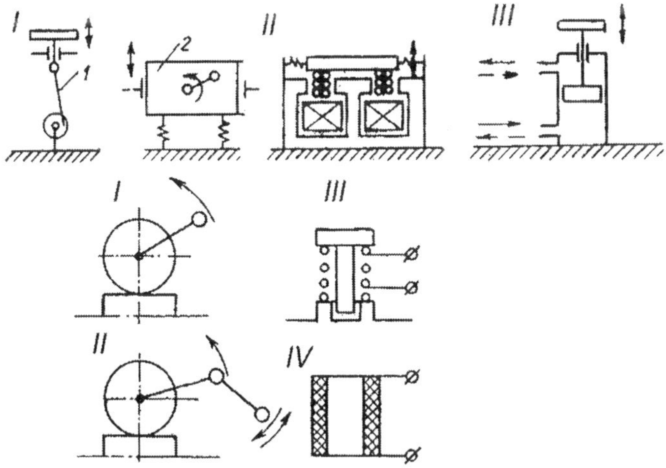

b)

Fig. 7-6. Vibration influence testing equipment

Oscillating mechanical systems. Mechanical oscillating systems are divided into linear and nonlinear, without damping and with damping. We describe the linear oscillations of a mechanical system by the system of differential equations, such as

$$m_i \ddot{x}_i + C_i \dot{x}_i + j x_i = Q_0 \sin \omega_i t \, . \tag{7-15}$$

Similar differential equations for a nonlinear system may be expressed in the following form $m_i \ddot{x}_i + C_i \dot{x}_i + j_i(x_i) x_i = Q_i(t) \sin \omega_i(t)$, (7-15a)

where $j_i(x_i)$ is a stiffness which depends on displacements.

Frequencies and amplitudes of natural oscillations for a system without damping may be estimated with the frequency-response equation expressed in the determinant form

$$\begin{vmatrix} m_1\omega^2 - \rho_{11} & -\rho_{12} & -\rho_{1n} \\ -\rho_{21} & m_2\omega^2 - \rho_{22} & -\rho_{2n} \\ -\rho_{n1} & -\rho_{n2} & m_n\omega^2 - \rho_{nn} \end{vmatrix} = 0 \, ,$$

where ρ_{ik} is the unit response of the system to the forced displacement of an impulse type. A linear system has single-valued stable states. A nonlinear system may have unstable states, which depend on the direction of change in the forced vibrations' frequency. Nonlinear models, due to complexity of their analysis, are used only in the cases when the system cannot even approximately be considered as a linear one.

Electrical analog. The mechanical system with one degree of freedom and viscose damper corresponds to the elementary oscillating contour, which consists of inductance coil L_{ind}, capacitor C_j, and resistor R_{res} (Fig. 7-7a) where current I is conducted. If the variable is equivalent to charge g_I on the capacitor and derivative $\dot{g}_I = I$, then the differential equation of ringing oscillations in the contour is as follows $L_{ind}\ddot{g}_I + R_{res}\dot{g}_I + \frac{1}{C_J}g_I = 0$. This equation is analogous to the equation of a mechanical system's oscillations.

a) b)

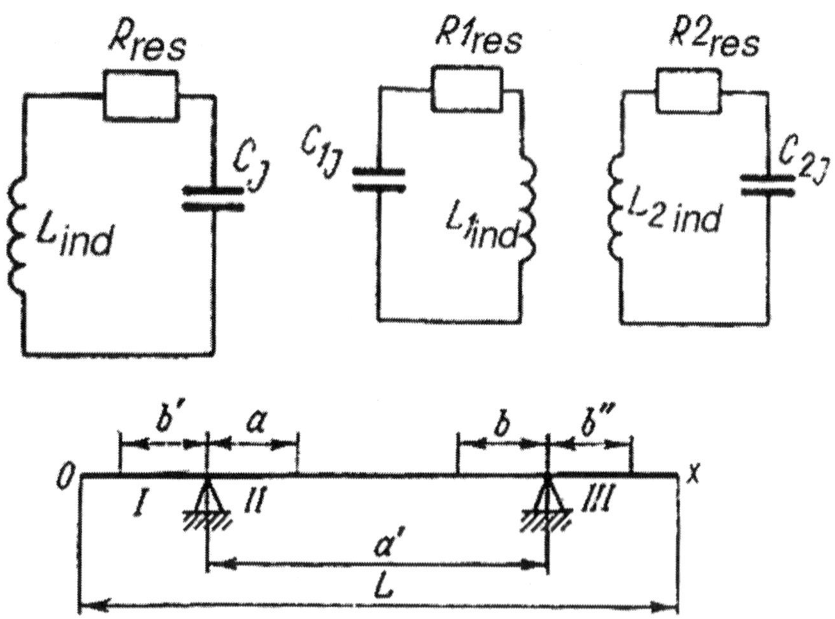

c)

Fig. 7-7. Similarities in electrical and mechanical oscillators.

The electrical analog of a mechanical system with two degrees of freedom may be represented by the system which consists of two (Fig. 7-7b) mutually linked contours R_{1res}, C_{1J}, L_{1ind} and R_{2res}, C_{2J}, L_{2ind}.

Approximate calculation methods. Many elastic structures consist of beams or similar elastic elements. The typical approximate methods for analysis of natural oscillations in beams with variable stiffness and mass are as follows: substitution of distributed parameters by concentrated ones, using formulae of Rayleigh, Grammel, Prager, and Dunkerley (Table 7-7), a successive approximation method, Ritz and Galerkin methods, and S. A. Bernshtein's (Table 7-8) estimators (*see Ya. G.* Panovko, 1968).

Table 7-7. Rayleigh, Grammel, Prager, and Donkerly methods for estimating the square of lower angular natural frequency p_1^2

Vibr.	Formulae			
	Rayleigh's	Grammel's, expected more accurate result	Prager's	Dunkerley's, expected lower value
Long.	$\frac{\int_0^L EA(X')^2dx}{\int_0^L mX^2dx+\sum m_iX_i^2}$	$\frac{\int_0^L mX^2dx+\sum m_iX_i^2}{\int_0^L N^2dx/EJ}$	$\int_0^L \frac{(N')^2dx}{m} : \int_0^L \frac{N^2dx}{EA}$	$\left[\int_0^L m(x)\delta(x,x)dx + \sum m_i\delta(x_i,x_i)\right]^{-1}$
Bend	$\frac{\int_0^L EJ(X'')^2dx}{\int mX^2dx+\sum m_iX_i^2}$	$\frac{\int_0^L mX^2dx+\sum m_iX_i^2}{\int_0^L M^2dx/EJ}$	$\int_0^L \frac{(M'')^2dx}{m} : \int_0^L \frac{M^2dx}{EJ}$	
Tors.	$\frac{\int_0^L GJ_p(X')^2dx}{\int_0^L IX^2dx+\sum I_iX_i^2}$	$\frac{\int_0^L IX^2dx+\sum I_iX_i^2}{\int_0^L M_{Tr}^2dx/GJ_p}$	$\int_0^L \frac{(M'_{Tr})^2dx}{I} : \int_0^L \frac{M_{Tr}^2dx}{GJ_p}$	$\left[\int_0^L I(x)\delta(x,x)dx + \sum I_i\delta(x_i,x_i)\right]^{-1}$

After Ya.G. Panovko, 1968. Note: Long.=longitudinal, Tors.= torsional; Bend =lateral bending.

Table 7-8. Sequantial, Ritz, Galerkin, and Bernshtein methods for estimating the square of lower angular natural frequency p_1^2

Vibr.	Formulae			
	Sequential approximations	Ritz*	Galerkin's**	Bernshtein's ***
Bend	$q^{(0)}(x)=m(x)X^{(0)}[p^{(0)}]^2$ $\frac{\int_0^L q^{(0)}X^{(1)}dx}{\int_0^L m(x)[X^{(1)}(x)]^2dx}$	$T_{ik}=\int_0^L mX_iX_kdx$ $U_{ik}=\int_0^L EJX''_iX''_kdx$	$W_{ik}=\int_0^L (EJX''_i)''X_kdx$	$\frac{1}{\sqrt{B_2}}<p_1^2<$ $\frac{2}{B_1+\sqrt{2B_2-B_1^2}}$ $B_1=\int_0^L m(x)\delta(x,x)dx$ $+\sum_i m_i\delta(x_ix_i)$ $p_2^2>\frac{2}{B_1-\sqrt{2B_2-B_1^2}}$
Long.	Superscript index shows approximation order	$V_{ik}=\int_0^L \frac{N_iN_k}{EA}$	$W_{ik}=-\int_0^L (EAX'_i)'X_kdx$	
Tors.	-	-	$W_{ik}=-\int_0^L (GJ_pX'_k)'X_kdx$	

After Ya.G. Panovko, 1968. Note: p_2^2 corresponds to the second natural circular frequency.

*Use with frequency-response equation

$$\begin{vmatrix} T_{11}p^2-U_{11}, T_{12}p^2-U_{12},\ldots,T_{1n}p^2-U_{1n} \\ \ldots\ldots\ldots\ldots \\ T_{n1}p^2-U_{n1}, T_{n2}p^2-U_{n2},\ldots,T_{nn}p^2-U_{nn} \end{vmatrix}=0\,; \qquad (a)$$

and

$$\begin{vmatrix} T_{11}-p^2V_{11},\ T_{12}-p^2V_{12},\ldots,T_{1n}-p^2V_{1n} \\ \ldots\ldots\ldots\ldots \\ T_{n1}-p^2V_{n1},\ T_{n2}-p^2V_{n2},\ldots,\ T_{nn}-p^2V_{nn} \end{vmatrix}=0\,. \qquad (b)$$

** Use with the frequency-response equation such as (a), but substitute U_{ii} by W_{ii}, respectively.

*** $B_2=\int_0^L\int_0^L m(x)m(s)\delta(x,s)dxds+\sum_i\sum_k m_im_k\delta(x_i,x_k)$.

Notations in Tables 7-7 and 7-8 are as follows: $m(x)$ is the intensity of the distributed mass; m_i are the concentrated masses; $I(x)$ is the mass moment of inertia for unit length of the beam relative to its axis; I_i are the mass moments of inertia for clamped discs; $X(x)$, X', X'' are the assumed approximate functions describing the beam displacement at vibration and its first and second derivatives along the beam's length, respectively; $\delta(x,x)$ is the function of influence for the displacement at longitudinal and bending deformation or the angle of section rotation at torsional deformation under action of the unit force or unit torque in the same section, respectively; M is the acting bending moment in the lateral vibrations which can be found from the differential equation $M''= m(x)X(x)$ or $M=EJX''$; $N=EAX'$ and N' are the internal longitudinal force due to longitudinal vibrations and its derivative; $M_{Tr}= GJ_pX'$ and M_{Tr}' are the internal torque due to torsional vibrations and its derivative, and J_p is the polar moment of inertia for the beam's cross section. Examples of these methods effective application are shown in this chapter, in Chaps 5 and 8.
Application of Dunkerley's method. Let us discuss, for example, the application of Dunkerley's formula for the estimation of the length standard's natural frequency. This standard (Fig. 7-7c) is a platinum-iridium or steel beam horizontally positioned on the two supports in Airy's points. We assume that standard's beam consists of three segments I-III which functions of influence under small deformations may be presented by the following expressions

$$\delta^I(x,x)=\frac{(a'+b')b'^2}{3EJ};\quad \delta^{II}(x,x)=\frac{a^2b^2}{3EJ(a+b)};\quad \delta^{III}(x,x)=\frac{(a'+b'')b''^2}{3EJ},$$

where $b' = (0.21L-x)$; $a'=0.58L$; $a= (x-0.21L)$; $a' \geq 0$; $b' \geq 0$; $b \geq 0$; $b=0.79L-x$; $b''=x-0.79L$.

We split integral $\int_0^L m(x)\delta(x,x)dx$ into three integrals, respectively, such as

$$\int_0^L m(x)\delta(x,x)dx=\int_0^{0.21L} m(x)\delta^I(x,x)dx+\int_{0.21L}^{0.79L} m(x)\delta^{II}(x,x)dx+\int_{0.79L}^{L} m(x)\delta^{III}(x,x)dx.$$

After substituting expressions for a, b, a', b', b'', $m(x) = A\rho_{\gamma M}/g$ and finding solutions for corresponding integrals, we derive the following approximate relationships for the natural frequency f_0 of the length standard:
(i) if the beam installed on the narrow side of cross section with size h_M, then

$$f_0\approx\frac{3b_M}{4\pi L^2}\sqrt{\frac{Eg}{3\rho_{\gamma M}}};\tag{7-16}$$

(ii) if the beam installed on the wide side of cross section, then

$$f_0\approx\frac{3h_M}{4\pi L^2}\sqrt{\frac{Eg}{3\rho_{\gamma M}}},\tag{7-17}$$

where $\rho_{\gamma M}$ is the density of the beam's material, b_M is the beam's width.

Methods for the synthesis of vibro-isolation means.

Quality of the transmission of a working force or motion may be defined by the following functional $I_s=\int_0^{\infty}[x(t)-u(t)]^2dt$, (7-18)

where $x(t)$ is the actual displacement, and $u(t)$ is the specified one. Thus, the problem of synthesizing the optimal vibro-isolating system may be formulated as the reproduction of the useful signal u under the present of vibration disturbances. In this case, the frequency-response function $H(j\omega)$ of the dynamic system, which consists of the structure and damping, should minimize the integral estimator (18).

Dynamic vibro-damper. In the two-mass system with certain relations between masses m_1, m_2 (Fig. 7-8a) and stiffness j_1, j_2 of springs, one can obtain a so-called anti-resonance with which the point of application of the periodic force remains fixed (motionless).

a)

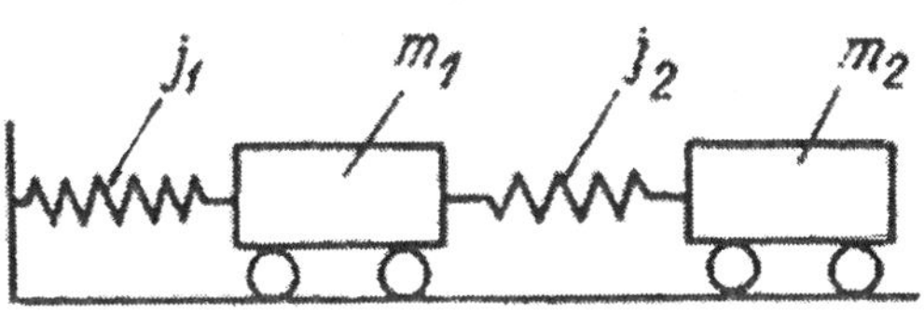

b) I. II. III. IV.

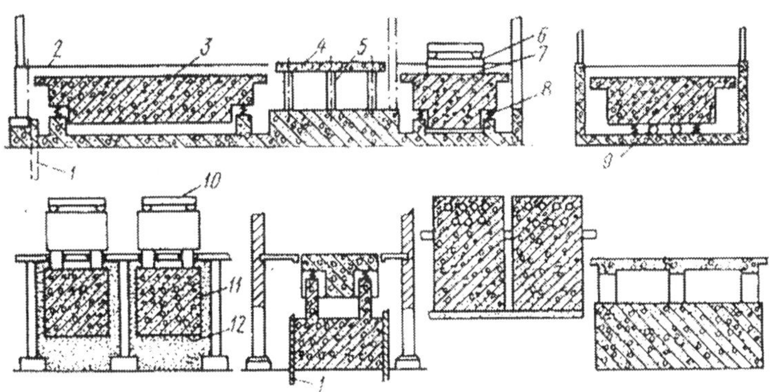

V. VI. VII. VIII.

Fig. 7-8. Mechanical damping systems.

The equations of mass m_1 and m_2's motion may be expressed in the following form

$$-j_1y_1+j_2(y_2-y_1)+Q_0\sin\omega t=m_1\ddot{y}_1; \tag{7-19a}$$

$$-j_2(y_2-y_1)=m_2\ddot{y}_2, \tag{7-19b}$$

where $y_1=y_1(t)$, $y_2=y_2(t)$ are the displacements of corresponding masses in time. The system of simultaneous equations (19a,b) is satisfied by the solution

$$y_1=a_1\sin\omega t;\ y_2=a_2\sin\omega t, \tag{7-20}$$

which shows that the oscillations have the same frequency as that of the exciting force with amplitude Q_0. This partial solution describes the stationary part of the process. Substituting (20) into (19a,b) yields two equations with two unknown amplitudes $-a_1m_1\omega^2+j_1a_1-j_2(a_2-a_1)=Q_0$; $-a_2m_2\omega^2+j_2(a_2-a_1)=0$.

Hence,

$$a_1 = \frac{Q_0(j_2 - m_2\omega^2)}{(j_1 + j_2 - m_1\omega^2)(j_2 - m_2\omega^2) - j_2^2}; \quad a_2 = \frac{Q_0 j_2}{(j_1 + j_2 - m_1\omega^2)(j_2 - m_2\omega^2) - j_2^2}. \tag{7-21}$$

The resonance of this system occurs when the denominators in (21) are equal to zero and amplitudes a_1 and a_2 tend to infinity. In this case, the resonant frequencies are equal to the natural frequencies of the system, and their number is always equal to the system's degree of freedom. At anti-resonance, the numerator of the first expression in (21) is equal to zero. It is necessary in this case to have the forcing frequency $\omega = \sqrt{j_2/m_2}$. With this frequency value, we have from formulae (21a,b) that a_1=0 and $a_2 = -Q_0/j_2$. Hence, the first mass remains immovable despite that periodical exciting force $Q = Q_0 \sin \omega t$ is applied to it. The possibility of anti-resonance is used practically in the dynamic damper of oscillations. The additional mass on the elastic link should be added to the structure for this reason. Certain adjustment of the vibro-damper provides cancellation of vibrations in the main structure while additional mass m_2 vibrates intensively. This damper works properly if the forcing frequency is constant. Otherwise, the ratio of stiffness j_2 and mass m_2 should have feedback adjustment.

Passive vibro-isolators. For special precision manufacturing needs and precision measurement centers, the whole room or the entire building may be vibro-isolated. Fig. 7-8b depicts foundations with: I-spring damper, II- round columns, III-additional plate, IV- rubber-spring dampers, V- heavy block in the sand, VI-concrete columns, VII with the ventilation channels for heating, and VIII-the forced concrete columns, where 1is a metallic rabbet, 2 is the hanging floor, 3 is a concrete block, 4 is the concrete plate, 5 is a metallic or wood column, 6 is a damping insert, 7 is the plate of cast iron, 8 are the spring dampers, 9 are the rubber dampers, 10 is the precision structure, 11is a solid concrete block, 12 is the sand foundation. Low vibration noise of the local foundation may be reduced by pneumatic system, with the damped magnetic and air springs. Such solution is used even for the improvement of stability of known Besocke-type scanners in scanning tunneling microscopes. Layered structures (gaskets) from flexible graphite, neoprene rubber, zinc-aluminum alloy (ZA-27) are also useful for vibration damping. The details of these solutions are beyond the scope of this text.

The reader may find some recommendations in Y.M. Tseytlin (1981b) publication and many other special sources. Active vibro-protective systems with automatic control and general problems of vibration in machines with elastic transmission mechanisms are discussed in M. Z. Kolovskii (1976), M.Z. Kolovsky *et al.* (2000) publications.

Impact. The action of impact on the precision elastic structure is a complicated one. This action, which does not cause the damage of a structure, may be estimated by equalizing the potential energy of the impact and the energy of the caused elastic deflection. For example, one can estimate the largest deflection s_δ of an elastic element with flexural rigidity j_s at the impact by some free falling mass m_δ from height h_δ above this element on the basis of the following equation

$m_\delta g(h_\delta + s_\delta) = \frac{1}{2} j_s s_\delta^2$. Therefore, $s_\delta = \frac{m_\delta g}{j_s} + \sqrt{\left(\frac{m_\delta g}{j_s}\right)^2 - \frac{2 m_\delta g h_\delta}{j_s}}$.

Acoustic noise.

Noise influence. It was shown that uncertainty level of the metrological large range scanning probe microscope is approximately 3 nm and is principally due to the vibrational noise originating from the building and other sources of acoustic noise. A noise reduction requires the use of improved acoustic isolation.

7-5. Electric and Magnetic Influence Fields

Electrostatic field.

Source, action, and measurement. Electrostatic field is caused by friction of non-conducting elements, sources of electrical discharge in non-conducting media, ambient air ionization, etc. The action of this field on the precision elastic system is produced by force, which occurs between the electrostatic field and electric charges accumulated on the dielectric elements. The influence force of an electric field with strength of 1 V/m per the 1 C charge equals 1 N. This force is used, for example, in electrostatic chucks for clamping the semiconductor wafers in microfabrication.

Magnetic influence field.

Sources and *action.* The sources of magnetic field are (i) closed electric currents, and (ii) magnetized bodies. In their action, the latter are equivalent to a collection of elementary magnets or the elementary circuits of current called dipoles. The mechanical action with the attractive force Q_{mag} of the magnetic field is determined by the magnetic inductance B_{mag} as $Q_{mag} = 4.06\ B_{mag}^2 A_{mag}$, or by a magnetic flux Φ_δ in the gap as Q_{mag}=4.06 Φ_δ^2 / A_{mag}, where A_{mag} is the area through which the magnetic flux Φ_δ flows. In the engineering industry, typical sources of significant magnetic fields are magnetic plates, and other magnetic grippers: cubes, blocks, stand's bases, faceplates, chucks, jigs, etc. Magnetic plates are also a source of heat, which is equal to $Q_{hT} = 0.24\, I_E^2 R_r t$, where I_E is electrical current and R_r is the chain resistance. Fig. 7-9 shows the relationship of magnetic inductance B_{mag} for some brands of steel placed in the magnetic field with strength H_{mag}.

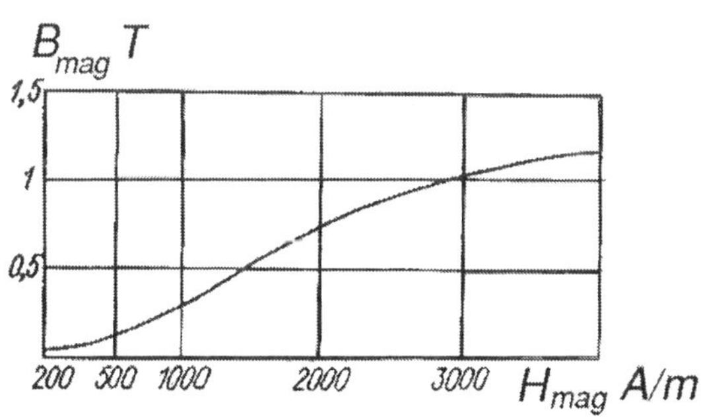

Fig. 7-9. Typical relationship between strength of magnetic field and its inductance in steel.

Magnetostriction. High saturation strain has such materials (C.H. Joshi, Energen, Inc., Compact Magnetostrictive actuators and linear motors, presented at Actuator 2000 conference, Bremen, Germany, June, 2000) as *Ni* (-50×10^{-6}, Curie temperature 630 K), *Fe* (-14×10^{-6}, 1040 K), $SmFe_2$ (-2340×10^{-6}, 690 K), $TbFe_2$ (2630×10^{-6}, 700K), and $Tb_{0.6}Dy_{0.4}$ (6300×10^{-6} at 75 K). Here *Sm* is samarium, *Tb* is terbium, and *Dy* is dysprosium. All of these chemical elements belong to the lanthanum group. In recent time the giant magnetostrictive film material such as Terefenol-D is developed.

Electromagnetic field.

Protection against LF and HF. Relative permeability μ_{mag} for the tape material from 0.02to 2.5 mm thick of iron-nickel alloys ranges from 10000 to 35000. In practice, complete magnetic saturation for such alloys occurs with the field strength of 600 to 800 A/m. Ferromagnetic screens are only used for protection against low-frequency (*LF*) interference carrying fields since at high frequencies (*HF*) the actual cross section of ferromagnetic walls becomes negligible as a result of surface skin effect. For the purpose of protection against *HF* interference carrying fields with at least 1 kHz frequency, the conducting shields of copper, aluminum, and other nonmagnetic materials are used.

The strength of electric field $\boldsymbol{H}_E$, the magnetic component $\boldsymbol{H}_{mag}$, the magnetic inductance $\boldsymbol{B}_{mag}$, the electric bias $\boldsymbol{D}_E$, the density of electric current j_l, and the density of electric charges σ_{pl} are related by Maxwell equations for the electromagnetic field. This field is a vortical.

7-6. Radiation and Gravitational Influence Fields

The radiation pressure of light.

Relationship. Let us recall from the classical and quantum physics that the pressure of light p_{lt} can be expressed by the following formula

$$p_{lt}=\langle w\rangle\,(1+R_{ref})\cos^2 i_{\varphi}, \qquad (7\text{-}22)$$

where $\langle w\rangle = 2\pi\hbar\, n_0\, \nu_{lgt}$ is the mean value of the spatial density of incidence light, R_{ref} is the coefficient of light reflection, i_φ is the incident angle of light on the flat surface, n_0 is the concentration of photons in incidence light, $2\pi\hbar$ is Planck's constant, and $\nu_{lgt}=\dfrac{c n_{ref.i}}{\lambda_{vac}}$ is the frequency of light in a medium. In the latter expression, we have constant velocity of light c=299 792 458 m/s, a wavelength of light in vacuum λ_{vac}, and a refractive index $n_{ref.i}$ of a medium.

Experimental test. R. V. Jones in 1950's provided very careful research of light pressure in dispersive media with the help of a special precision elastic installation, which was based on the thin torsional suspension of a thin light mirror. He concluded that within the limits of his measurements' accuracy (standard uncertainty of order 0.05%), the pressure exerted by a steady beam of light on a mirror in an optically dense medium is directly proportional to the phase refractive index of the medium, and that it is independent of the group velocity ratio and the plane of polarization in oblique incidence, at least up to angles of 20°. We can see that the main conclusion of Jones' fine experiments about direct proportionality between the light's pressure and refractive index of the medium are correspondent to (22) after substituting into it certain expressions for spatial density <*w*> and light frequency ν_{lgt}, such that $p_{lt}=2\pi\hbar n_0 c n_{ref} / \lambda_{vac}$.

In a pioneering work, Braginsky and Manukin (1977) demonstrated that radiation pressure acting on the mirrors of a Fabry-Pérot cavity leads to a measurable coupling between the optical and mechanical properties of the resonator. In recent years submicron micromanipulators driven by light are developed and used.

Action of penetrating radiation.

Radioactive irradiation affects the electric resistance of metals employed in transducers depending on the structure of their crystal lattice and interatomic bond energy. When exposed to irradiation, the chemical content of copper, platinum-rhodium, and tungsten electrodes featured instability. Copper converts to nickel and zinc, rhodium converts to palladium, and tungsten converts to osmium. As a result, irradiation may produce changes in adjustment, graduation, and other technical and metrological characteristics of a precision structure.

Action of the gravity field.

Gravitational bending. Typical examples for installation and orientation of precision beams are represented by the methods of basing the gage blocks exceeding 100 mm in length and the toolmaker straight edges in the horizontal plane.

a) d)

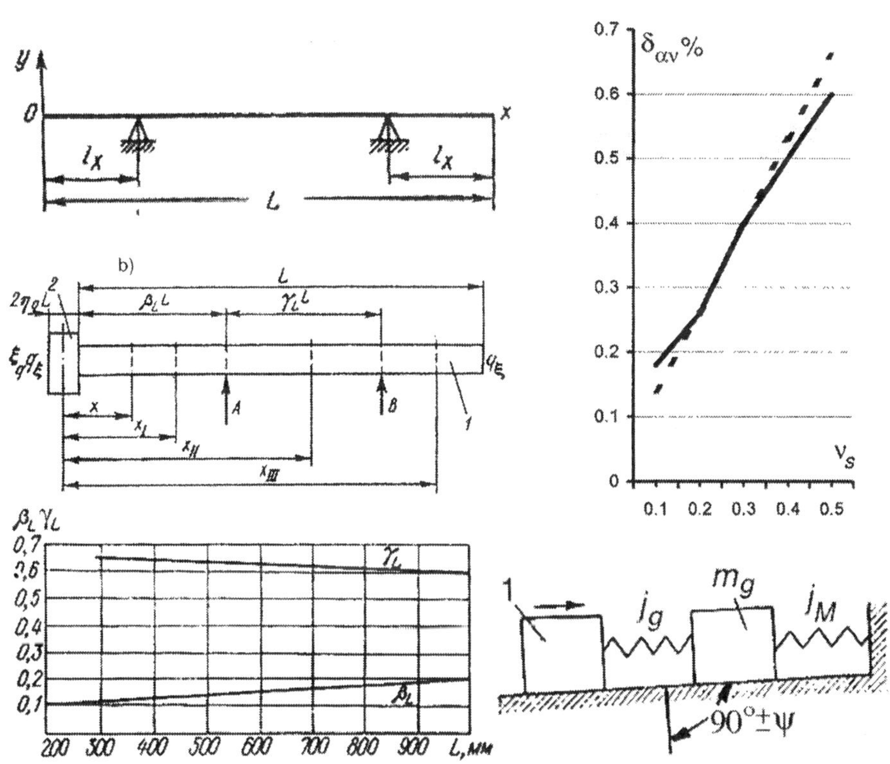

c) e)

Fig. 7-10. Position of supports and orientation in gravitational field influence.

For the gage blocks, where it is necessary to ensure that the edges are parallel, the installation supports (Fig. 7-10a) should be in the Ary's points at a distance of $l_x = 0.2113\,L$ from the gage blocks edges. This follows from the equation of the beam's bending deflection y on two supports by the distributed mass load, which we apply to two beam's segments:

(i) for the beam's overhang segment $x \in [0, l_x]$, we have the following expression

$$y_I = \frac{m_\gamma l_x^4}{24EJ}\left(3 - 4\frac{x}{l_x} + \frac{x^4}{l_x^4}\right) + \frac{m_\gamma}{24EJ}[(L-2l_x)^3 - 6l_x^2(L-2l_x)](x-l_x),$$

(ii) for the beam's segment between supports $x \in [l_x, (L-l_x)]$, the expression is as follows

$$y_{II} = \frac{m_\gamma (L-2l_x)^4}{24EJ}\left[\frac{x-l_x}{L} - 2\frac{(x-l_x)^3}{L^3} + \frac{(x-l_x)^4}{L^4}\right] + \frac{m_\gamma l_x^2}{4EJ}[(x-l_x)^2 - L(x-l_x)].$$

In this case, a slope angle of face sections is determined by the formula

$$\vartheta_{yx} = y'_{lx=0} = -\frac{4m_\gamma l_x^3}{24EJ} + \frac{m_\gamma (L-l_x)^3}{24EJ} - \frac{6m_\gamma l_x^2 (L-2l_x)}{24EJ}, \tag{7-23}$$

where m_γ is the mass per unit length. From (23), we can find that the angle $\vartheta_{yx} = 0$ if

$$l_{x\vartheta=0} = \frac{L}{2} - \sqrt{\left(\frac{L}{2}\right)^2 - \frac{L^2}{6}} = \frac{(\sqrt{3}-1)}{2\sqrt{3}}L = 0.2113\ L.$$

Plate addition. Sometimes, for example in the interferometric measurement, it is necessary to wring a parallel-sided glass plate 2 (40×40×11.5 mm) to the gage blocks 1. In so doing, the bent shape of the gage blocks and the optimum arrangement of its supports become slightly different (Fig.7-10b, c).

Rulers and straight edges. The length standards and rulers should have minimal change in the distance between extreme marks when they are positioned on two supports. In this case, the bending equation implies the requirement $l_x = 0.22031\ L \approx (2/9)L$ which corresponds to Bessel points. The change of distance between extreme marks with this condition is equal to: $\Delta L = 0.653\dfrac{L^5 \Im_m^2}{E^2 J^2}10^{-6}$, where $\Im_m$ is a uniformly distributed mass of the measure. For the least distortion of the straightness of the toolmaker straight edge, the bending equation yields the least ends deflection from a stright line condition with l_x=0.2232 L, and for the least deflection in the middle of the measure $l_x = 0.2386\ L \approx (6/25)\ L$.

Stretching. If the beam with length L is set upright from the horizontal position, it becomes stretched (shortened or lengthen) by the quantity $\delta L_\Psi = \dfrac{1}{2}\dfrac{\rho_\gamma L^2}{E}$, where ρ_γ is the density of material.

For the steel standard of the meter, we have $\delta L_\Psi = \dfrac{7.8 \cdot 100^2}{4 \times 10^9} = 0.2\ \mu\text{m}.$

The heavier platinum-iridium standard has a larger change in length with conversion from horizontal to the upright position, such as $\delta L_\Psi \approx 0.4\ \mu\text{m}$.

Strain by gravity. Table 7-9 shows the calculating relationships for the strains of ring and snap gage caused by the gravity force $\Im$ and the forces Q_p of reaction at the supports with the change of orientation ($\Delta\psi$) in the gravity field. Changes of the ring diameters are denoted by δ_x and δ_y along with axes x and y, respectively. The minus sign corresponds to a decrease in diameter. Let us recall that rings have been used in the precision transducers of force standards and displacement measuring instruments with transformation ratio δ_x / δ_y.

Table 7-9. Strains in ring and snap gages at different orientation in the gravitational field

Schematic	Calculating relationships
Ring 	$\delta_x = \frac{Q_p R^3}{\pi EJ}[\frac{\pi}{2}(1 - 4\sin\alpha + \sin^2\alpha) + 2(\alpha\sin\alpha + \cos\alpha)]$; $\delta_y = -\frac{Q_p R^3}{\pi EJ}[-2.4674 + \frac{\pi}{2}(\sin\alpha\cos\alpha + \alpha - 2\sin\alpha)$ $+ 2(\alpha\sin\alpha + \cos\alpha)]$; $Qp = ½\, \Im \cos\Delta\psi$
Linear snap 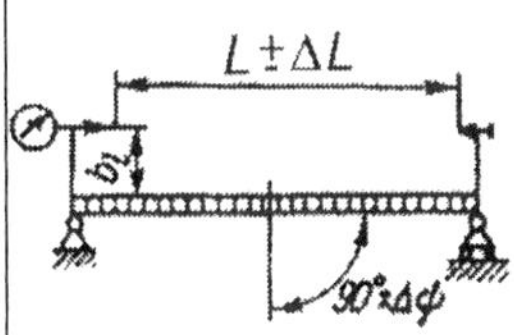	1. $\Delta\psi < 0.1$ $\Delta L = \frac{\Im L^2[1-(\Delta\psi)^2/2]}{12EJ} b_l$
	2. $\Im \sin\Delta\psi > 0$ $\Delta L = \frac{\Im\cos\Delta\psi}{12EJ(1+\nu_{sL})}$; $\nu_{sL} = \frac{L^2 \Im \sin\Delta\psi}{\pi^2 EJ}$
	3. $\Im \sin\Delta\psi < 0$ $\Delta L = \frac{\Im\cos\Delta\psi}{12EJ(1-\nu_{sL})} b_l$; $\nu_{sL} = \frac{L^2 \Im \sin\Delta\psi}{\pi^2 EJ}$

The change of structure's orientation in the gravity field can change the frequency of oscillations in elastic transducers and tips of measuring machines (*see* Chap.8).

Change in the distance between tips in the snap gage under changeable direction towards the gravity force reminds us the well known and important in metrology Abbe's error. We neglect the tips' length for simplification. However, it is only a certain geometric relationship. In case 1 of Table 7-9 for the linear snap, we consider a small deflection in orientation where the influence of the longitudinal force constituent $\Im \sin \Delta\psi$ on the rotation of beam's ends may be neglected. In case 2 at the beam larger clockwise rotation, we consider the appearance of a significant longitudinal tensile force $F_{Qs} = \Im \sin \Delta\psi > 0$. We derive presented in this case formula by differentiation of approximate expression (3-29) for beam deflection under simultaneous transverse $F_{TR} = \Im \cos \Delta\psi$ and longitudinal force $F_{Qs} = \Im \sin \Delta\psi$. We recall that derivative of the beam's ends deflection is equal to the tangent of the angle α_v of their rotation. Therefore, it follows that

$$y'_x(0;\,L)=\frac{d}{dx}\left(\frac{\nu TR}{1+\nu_S}\right)_{|0;L}=(1+\nu_S)^{-1}\frac{d}{dx}\nu_{TR}(0;L)\ =\tan\,\alpha_\nu(0;\mathrm{L})\approx\alpha_\nu,\tag{7-24}$$

where the derivative $\frac{d}{dx}\nu_{TR}(0;\,L)=\frac{\Im\cos\Delta\psi}{24EJ}$ as the angle of the beam ends rotation under distributed transverse force alone with intensity (F_{TR}/L) and $\nu_s=F_{Qs}/F_{eul}$ which does not depend on coordinate x. The critical longitudinal load F_{eul} depends on the type of the beam supports (*see* Table 3-3) and $F_{eul}=\pi^2EJ/L^2$ for the case of our study with two pivots in the ends.

Let us verify an uncertainty of derived (24) by comparing it with the solution for the similar beam under simultaneous distributed transverse and axial tensile longitudinal load, which we obtain through differentiation of the expression for the beam's deflection found (*see* I.A. Birger, 1968) with the method of initial parameters. The tangent of the beam's ends rotation angle in this case is equal to:

$$y'_x(0;L)=\frac{d}{dx}\left\{\frac{F_{TR}}{EJLk_S^4}\left[1-\frac{\cosh k_S(x-L/2)}{\cosh(k_SL/2)}-\frac{k_S^2Lx}{2}(1-x/L)\right]\right\}_{|0;L}=$$

$$\frac{\Im\cos\Delta\psi L^2}{EJk_{s0}^4}\left[\frac{k_{s0}\sinh k_{s0}/2}{\cosh(k_{s0}2)}-k_{s0}^2/2\right]=\tan\alpha_{ks}(0;L)\approx\alpha_{ks}\,,\tag{7-25}$$

where we consider for the comparison purposes that $k_{s0}=\pi\sqrt{\nu_s}$.

The calculated chart $\delta_{\alpha\nu}=100\times|\tan\,\alpha_\nu(0;\mathrm{L})-\tan\,\alpha_{ks}(0;L)|/\tan\,\alpha_{ks}$ with influence of tensile longitudinal force is shown in Fig. 7-10d (solid line). One can see from this chart that the uncertainty of (24) is less than 0.6 % at $\nu_s\leq 0.5$. We assume that this accuracy of calculation is sufficient enough in many cases for the structural synthesis purposes.

In case 3 at the beam counterclockwise rotation, we consider the appearance of a significant longitudinal compressive force $F_{Qs}=\Im\,\sin\,\Delta\psi<0$ and assume that $\nu_s<0$. Therefore, we obtain the following expression for the angle of beam's ends rotation in this case by substituting $-\nu_s$ into (24)

$$y'_x(0;\,L)=\frac{d}{dx}\left(\frac{\nu_{TR}}{1-\nu_S}\right)_{|0;L}=(1-\nu_s)^{-1}\frac{d}{dx}\nu_{TR}(0;\,L)\ =\tan\,\alpha_{\nu c}(0;\mathrm{L})\,.\tag{7-26}$$

Let us verify an uncertainty of derived (26) by comparing it with the solution for the similar beam under simultaneous distributed transverse and axial compressive longitudinal load, which we obtain through differentiation of the expression for the beam's deflection found (*see* I.A. Birger, 1968) with the method of initial parameters. The tangent of the beam's ends rotation angle in this case is equal to:

$$y'_x(0;L)=\frac{d}{dx}\left\{\frac{F_{TR}}{EJLk_S^4}\left[1-\frac{\cos k_S(x-L/2)}{\sin(k_Sl/2)}+\frac{k_S^2}{2}(1-x/L)\right]\right\}_{|0;L}=$$

$$\frac{\Im\cos\Delta\psi L^2}{EJk_{so}^4}\left[-\frac{k_{so}\sin(k_{s0}/2)}{\cos(k_{s0}/2)}+k_{s0}^2/2\right]=\tan\alpha_{kc}(0;L)\approx\alpha_{kc}\tag{7-27}$$

The calculated chart $\delta_{\alpha\nu}=100\times|\tan\,\alpha_{\nu c}(0;\mathrm{L})-\tan\,\alpha_{kc}(0;L)|/\tan\,\alpha_{kc}$ is shown in Fig. 7-10d (dashed line). One can see from this chart that the uncertainty of (26) with compression longitudinal force is less than 0.7 % at $\nu_s\leq 0.5$. We assume that this accuracy of calculation is sufficient enough in many cases for the structural synthesis purposes. Let us recall that our approximated tractable and digestible formulae (24) and (26) for angular deflection are based on the formula (3-29) for a beam displacements under transverse and longitudinal force that was initially presented by S. Timoshenko and which good precision is known. Here we verified that precision of estimations for the derivatives of this approximate expression is also satisfactory.

Friction in guides. The change of orientation of the structure with the sliding guides in the gravitational field can cause a change of friction forces ΔQ_{fr} as a result of a change in the support reaction Q_k.. Therefore,

$$\Delta Q_{fr}= \mu_{fr}\Delta Q_k= \mu_{fr}Q_{k0}\,(1\text{-}\cos\psi),$$

where ψ (Fig. 7-10e) is an angular deflection of orientation from the normal one and μ_{fr} is the coefficient of friction. Let us recall that at any change of friction the Q-factor of such system with mass m_g, stiffness of driver j_g and driven system j_M will be changed as well. This leads to intermittent stick-slip motion with different strokes on the sliding guides and probes with a dry and semi-dry friction [*see* Fig. 1-8a, expressions (1-51), (1-52)] especially at low speed.

8.

Precision Elastic Structures Development

This chapter is dedicated to the experience in the effective application of optimized elastic structures for precision measuring means, automation, miniature micro- and nano- systems, and the examples of effective analytical procedures in their structural synthesis.

8-1. Elastic Systems in Measuring Means: displacement, mass, and force sensitivity

This section reflects useful achievements in the field by many researchers and developers. Despite that some of the solutions are well known, we present them in more effective detail, which will help for the further development of precision elastic structures.

Spring measuring heads with pretwisted strip.

Mikrokators. The mikrokator amplifying mechanism was patented by H. Abramson (Johansson Gage Co.) in Sweden Patent 82522 (1931). This mechanism is based on the application of a thin pretwisted strip (elastic double-helicoid, ELKS of type SM-VI) which is similar to that introduced by J. Perry much earlier in 1890. However, the mikrokator mechanism's structure also has elastic-lever (ELKS PM-I*el*) mechanism 5 (Table 8-1, scheme S1), called the "spring knee", which provides motion transmission to the elastic helicoid 6 with relaxation of its tension after quick release of the retracted measuring spindle 2. As a result, this structure minimizes the possible asymmetry (damping factor $\psi_\Delta \leq 0.01$) of micro-plastic shifts in the elastic helicoid (*see* Chap.5) at a fast and slow spindle's release after retraction and practically eliminates the corresponding bias uncertainty. Fortunately, the mikrokator with the correctly designed elastic lever–pretwisted strip transmission (therefore, we can call it Abramson's mechanism) does not practically have a hysteretic damping asymmetry defect at the correct application, and this accurate instrument and its modifications with graduations in the range between 0.00001 and 0.01 mm have found very wide application in precision industry and laboratory measurements in the 1930-1980s. They are rarely but effectively in use at the present time as well. Although this is an impressive choice of amplifications in mechanical system up to $100{,}000^{\times}$, the mikrokators are not intended for measuring wide dimension's variations. The increase of measuring range in mikrokator was achieved by increase of dial's angle to 250°, with elastic compensator in the elastic-lever mechanism whose bending deformation decreases the amplification rate (SU Patent 167308, 1965), and with multi-revolution structure. The gaging force of mikrokators is regularly between 1.6 and 2.5 N. Other models of this type of instrument are available with adjustable gaging force, ranging from 0.3 to 11.5 N for general measurements, or from near zero to about 0.3 N for the gaging of soft materials or components with very thin cross sections. These low gaging forces are feasible because of the practically frictionless operation of the pretwisted strip and flexible guiding plates. The linearity of transmission can be adjusted by changing the pretwisted strip's initial tension. Most models have uncertainty within ±1 percent, and even for the models with the highest degree of amplification an uncertainty does not exceed interval ±1.5 percent of the measuring range. Because there are practically no wearing elements in the gage, the original accuracy can be maintained during a long period of continuous service and can be adjusted if necessary. This mechanism has been used not only for linear dimensions measurements but also for measuring parameters of surface finish, forces, pressure, and strains in extensometers and other measuring systems. Although the above discussed and shown in Table 8-1 data provide a wide range of potential use for the mikrokator-like instruments, these instruments have comparatively small graduation length on the dial (1 mm or less) that makes the operator's eyes tired at continuous work with pointer 8. The pointer has limiters on the dial. The possible impact force Q_{cim} between the pointer and the limiter stop at a fast retraction of the measuring spindle depends on the angular speed ω_{ap} of the arrow pointer in the beginning of the impact. From the equality of the kinematical energy of the moving arrow pointer and the potential energy of its bending at impact follows that $Q_{cim}=\sqrt{3EJJ_{dm}\,\omega_{ap}^2/R_{lc}^3}$, where J is the moment of inertia for the pointer's cross-section area, J_{dm} is the mass moment of inertia for the arrow pointer with its flag and balance bead in relation to the axis of rotation. The periodicity of these impacts and corresponding ringing frequency of the pointer have been found on the oscillogram where they equal 30 ms and 33 Hz, respectively.

Table 8-1. Archive of effective elastic mechanisms in the instruments

Instrument name, application, developer/ Manufacturer, Scheme S number	Dial graduation, a_g, pulse, sort group size (measuring range), mm	Main transmission links		i_θ, °/μm (i), thou-sand	$\{\Delta\theta\}$* $(\delta\theta)$*, μm
		Pos.	Type and parameters of ELKS (approx.)		
Mikrokator, LIZ** 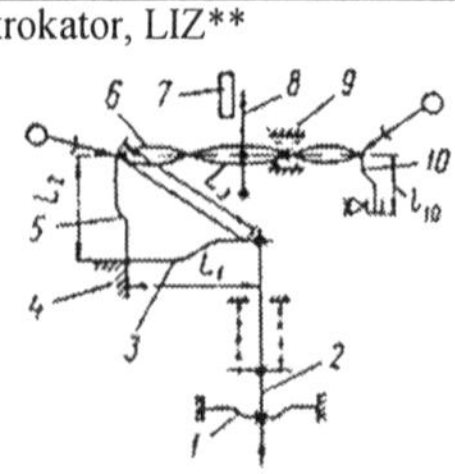 1, 3- guiding plates, 2- spindle , 5- "spring knee", 6- elastic helicoids,7-dial, 8- pointer, 9-damper, 10- cantilever, 4-base S1	0.0001 (±0.004)	6 5	SM-VI 0.004×0.08× π×18 PM-I L_1=9; L_2=11.8 L_3=16 mm	12 (7)	0.1 (0.03)
	0.0002 (±0.008)	6 5	SM-VI 0.006×0.1× π×18 PM-I L_1=7; L_2=11.8; L_3=17 mm	7.9 (4.5)	0.15 (0.06)
	0.001 (±0.03)	6 5	SMVI 0.008× 0.12× (2.5-4.1)×18 PM-I L_1=12; L_2=6	1.73 (.9)	0.5 (0.3)
	0.005 (± 0.15)	6 5	SM 0.016×0.17×π×18 L_1=10; L_2=6 PM-I type B (Fig. 4-9b)	0.35 (.18)	0. 25 (0. 17)

Table 8-1 Cont'd.

Instrument name, application, developer/ Manufacturer, Scheme S number	Dial graduation, a_g, pulse, sort group size (measuring range), mm	Main transmission links		i_θ, °/μm (i), thou-sand	$\{\Delta\theta\}^*$ $(\delta\theta)^*$, μm
		Pos.	Type and parameters of ELKS (approx.)		
CEJ Mikrokator type 510 (Sweden, Patent 153887), similar to S 1 with additional input lever transmission	0.0001-0.01($\pm a_g\times30$ or $\pm a_g\times100$) 0.00001-0.00005	6 5	SMVI 0.0025×0.06× 3.5×18 (approx.) or SM-VI 0.0045×0.12 PM-Iel with i up to10	(9 - 0.09) (70 - 14)	Min ± 0.01 (0.02)
Spring head LIZ** 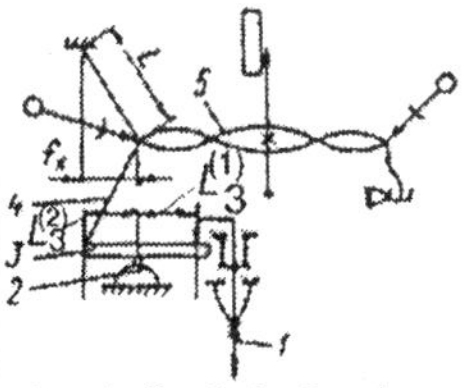 1-spindle, 2 -leaf spring hinge, 3-lever, 4-cable, 5-pretwisted strip S 2	0.00002 (±0.0008)	5 4 3	SM-VI 0.004×0.08× 2.8×18 PM-IV L=7; f_k=0.7; b_4=2; h_4=0.07 mm $L^{(1)}_3 : L^{(2)}_3 = 4:9$	60 (35)	0.02 (0.006)
	0.00005 (±0.002)	5 4 3	SM-VI 0.004×0.08× 2.8×18 PM-IV L=7; f_k=1.0; b_4=2; h_4=0.03 mm $L^{(1)}_3 : L^{(2)}_3 = 6:7$	24 (14)	0.05 (0.02)
Multi-revolution spring head on mikator base, LIZ** (Patent ***) 1-counter dual, 2-PM-Iel; 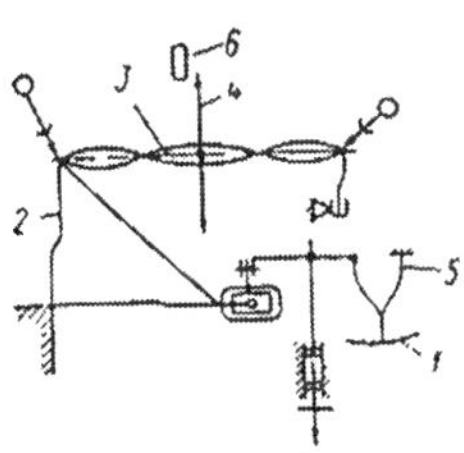 3-SM-VI, 4, 5-pointers S 3	0.0005 (0.2)	3 2 5	SM-VI 0.006× 0.1× π × 18 PM-I PM-II	(1.4)	1.5 (0.5)
Lever-spring head 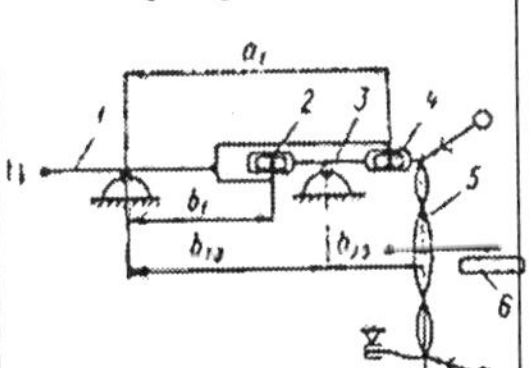 Minikator, LIZ** S 4	0.0005 (±0.015)	5 1 3	SM-VI 0.004×0.08× 4.2 × 7 PM-I b_1=15; a_1=30; PM-I b_{13}=20; b_{35}=12 $\frac{b_{13}}{b_1} - 1 = -\frac{b_{13}}{a_1} + 1.$	6.4	0.5

Note: Mikrokator mechanism (scheme S1) transmission ratio may be expressed by formula $i_\theta = i_{ELKSM1} \times i_{\theta 6} / [1 + l_{10}^3 j_{\theta 6} i_{\theta 6} / (3E_{10} J_{10})]$.

Table 8-1. Cont'd

Instrument name, application, developer/ manufacturer Scheme *S* number	Dial graduation, a_g, pulse, sort group size (measuring range), mm	Main transmission links Pos	Type and parameters of ELKS (approx.)	i_θ, °/μm (i), thou-sands	{Δθ}* (δθ)*, μm
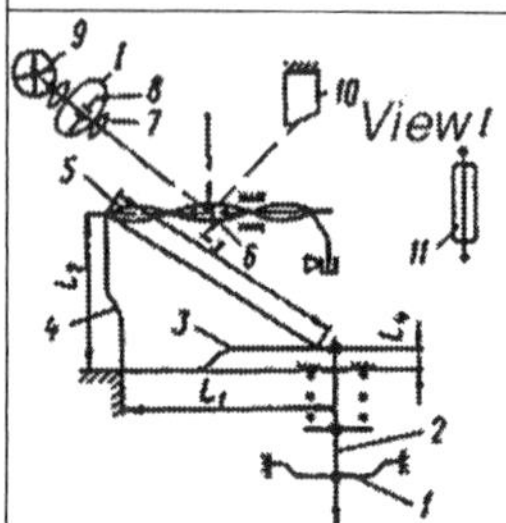 Spring-optical head; Optikator, LIZ** S 5	0.0001 (±0.012) 0.0002 (± 0.025) 0.0005 (±0.06)	5 3-4 6 4 4 1	SM-VI 0.008×0.12× (2.5-π) × 17 PM-I L_1=12; L_2=12 L_4=0 mm Mirror 1.5×1.5×0.06 mm L_1=12;L_2=6 L_1=12; L_2=6; L_4=3 TG-VIII-ar	7 (10) 3.5 (5.0) 1.8 (2.0)	0.06 (0.03) 0.16 (0.08) 0. 4 (0.2)
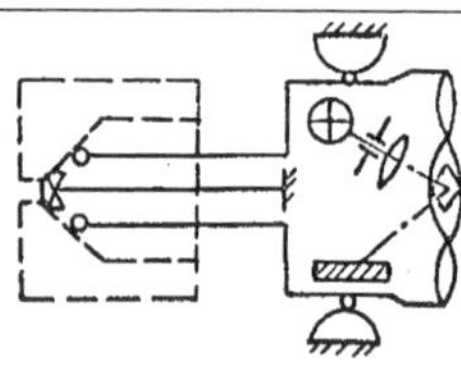 S 6 Differential system for conical-cylindrical surfaces; TSNITA****	0.0001	-	SM-VI 0.008×0.12× π × 18	-	-
Sorting photoelectric transducer; LIZ** 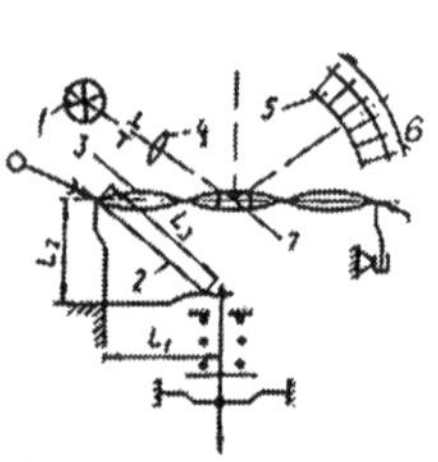 S 7	0.0005, 0.001, 0.002 Type 05PFS, 1PFS, 2PFS	2 3 7	PM-I L_1=12; L_2=12,6,3 L_3=17,13.5,12.5 SM-VI 0.008×0.12× π × 18 Mirror 5× 5×0.1	7; 3.5; 1.7 (6.0), (3.0), (1.5)	0.5 (0.2), 1(0.3), 2 (1)
	0.005 Type 5PFS	2 3 7	PM-I L_1=12; L_2=3 SM-VI 0.016×0.17× π × 18 Mirror 5×5×0.1	0.7 (0.6)	3.0 (1.5)
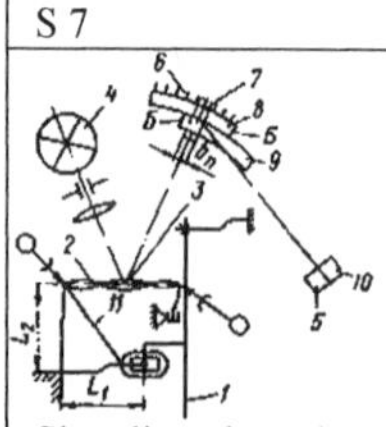 Signaling photoelectric,LIZ 1-spindel; 2-helicoid;3,4-optics;5, 10-PR; 6-dial; 7-index;8,9-sectors; 11-link S 8	0.001 Type DFP	11 2	MI L_1=8.5; L_2=8.5 MVI 0.008×0.12× π ×(18+5)	(2)	1.0

Table 8-1. Cont'd

Instrument name, application, developer/ manufacturer Scheme *S* number	Dial graduation, a_g, pulse, sort group size (measuring range), mm	Main transmission links		i_0, °/μm (i), thou-sands	{Δθ}* (δθ)*, μm
		Pos	Type and parameters of ELKS (approx.)		
Reversible photo-pulse transducer (Patent SU 168896) 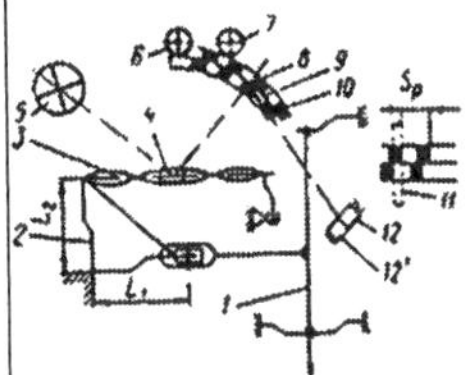 S 9	0.002; 0.001 (0.06) Type TP1	2 3	PM-I L_1=10;L_2=12 SM-VI 0.008×0.12× π × 18	(1.0)	0.5

* Permissible systematic {$\Delta\theta$} and random $\delta\theta$ uncertainty.
** Leningrad Instrumental Plant (Russia).
*** Patent SU 203926; USA 3,337,963; GB 1,104,460; France 1457986; Swiss 448543 (Y.M. Tseytlin, I.S. Amosov, A.A. Sokolov).
**** Institute for Fuel Apparatus, Russia, in 1970-1980s.

The latter number is close to the natural frequency of the pointer on the microkator's pretwisted strip (ELKS of type MVI with dimensions 0.008×0.12×π×20).
For the damping of the torsional vibrations, the pretwisted strip passes through the viscose damper 9 with a non-drying silicon oil of necessary viscosity (kinematic viscosity 12.45 m^2/s). Some models have the shellac bead on the pretwisted strip, which increases the damping quality and shake protection during instrument transportation. Mikrokators do not have electrical analog or a digital output, which makes measurement results more objective. Besides that, the inherent form and boundary dimensions with a stem tube diameter 28 mm of these instruments (base 4 on the schematic S1 in Table 8-1) could prove to be a limiting factor in applications where the gage must operate in a restricted area.

Mikators and minikators. Mikators (small size mikrokators) with a smaller stem tube diameter of 8 mm and minikators as the lever-spring test indicators (Scheme S4, Table 8-1) provide more convenient application in the restricted areas, but they also have small length of dial graduations and do not have any electrical analog or digital output.

Ultra-sensitive mikrokators. Scheme S2 LIZ with the cable mechanism 4 has a smaller gradient of measuring force (less than 18 cN). The analogous Johansson mikrokators of graduations 0.00002 and 0.00005 mm with pretwisted strip and two preceding levers had a smaller transmission ratio and a larger gradient of measuring force (up to 40 cN). The force compensator with sine spring (ELKS of type FC-V*b*) was used in some LIZ spring measuring heads with a very small measuring force.
Large range mikrokators. Modified LIZ mikrokators with graduation of 0.005 and 0.01 mm had kinematical elastic compensator (*see* Fig. 4-9b, type B) in the elastic lever 3-5, which allows for an expanded measuring range of 1.5 times in comparison

with the mikrokators of the regular structure. In mikrokators with a small measuring force, the force compensator (ELKS of type FC-I*pl*) is installed.

Multi-revolution spring measuring heads. Multi-revolution mikrokators and mikators are patented in Russia, US, France, and Switzerland (Y.M.Tseytlin *et al.*, 1966-1968). The structure of these instruments includes two-ended pointer 4 (Table 8-1, scheme S3) clamped to the crosspiece of the pretwisted strip (double-helicoid) which is linked to elastic mechanism 2. The ends of the pointer are coded and differ in length, shape, or in color to be matched to the certain segment on the dial 6 (concentric and colored). The multi-revolution springheads have special positional counter 5 for the evaluation of the pointer's half-revolutions on dial 1. It is also possible to use the second pointer on the pretwisted strip clamped close to strip's end (Poland patent 39875, 1956) as a half-revolution counter. As a result, the angle of rotation for this additional pointer is equal to θ_{count} which is much smaller than the angle $\theta > 360°$ of the crosspiece rotation with the main pointer, such that

$$\theta_{count}/\theta = (L - L_{count})/L,$$

where L_{count} is the distance from the middle of crosspiece to the position of the additional counter's pointer along the axis of pretwisted strip.

Minikators. These test indicators in accordance with scheme S4 have lever-link transmission 1-2-3-4 with the elastic rotational guide, which transmits motion to the shortened pretwisted strip 5 (L=7 mm). The measuring rod 1 of these instruments has an oscillating motion around the axis in both direction but the motion of the pointer's direction against the dial 6 remains the same. This solution is obtained with the known kinematical fork structure. Minikators are used for measuring dimensions and the deflections of surface geometry in restricted areas and in situ.

Spring-optical heads.

These systems have a light beam pointer, larger dial divisions, and larger measuring range than similar mechanical structures, but their size is larger as well. They can be divided into two structural groups: (i) with double-helicoid elastic transmission (ELKS of type SM-VI) and (ii) with double strip reed mechanism (ELKS of type PM-II). The first group includes regular optikators (*see* W. Eckerkunst, 1966) with the measuring range corresponding to ±125 graduations on the dial at graduation values from 0.0001 to 0.001 mm, and optikators with the extended measuring range and coded light pointers (Y.M. Tseytlin *et al.*, 1973). The second group (ii) includes instruments of Sheffield and MicroZIL visual gage types.

Optikators. Motion in optikators (scheme S5) from measuring spindle 2 attached to the translational guide (plate 1 with circular cutoffs and cantilever strip 3) transmits through the elastic-lever mechanism 4 (ELKS of type PM-I*el*) to pretwisted strip 5 (ELKS of type SM-VI) with the light mirror 6 (mass approximately 3-5 mg), which is clamped to the strip's crosspiece. The light ray from light source 7-9 falls on the mirror and reflects to dial 10 where the image of slot diaphragm 8 with the pointing index 11 appears. The transmission ratio i_θ of such a system with dial of 180° is twice larger than for the similar mechanical device. Besides that, the spring-optical system does not have parallax, which occurs in the arrow-pointer version. Optikator may have in the kinematic chain an elastic compensator similar to shown in Fig. 4-9b (type B) that allows the increase of the measuring range by 30%.

Differential systems. There has been developed the unified set of elastic mechanisms with pretwisted strip for measuring parameters of outer and inner cylindrical and conical surfaces where the pretwisted strip (ELKS of type SM-VI) is used differentially with inputs on its both ends (Scheme S6). This climinates the influence of the possible datum's displacement and eccentricity of rotating measurand's surface on a measurement uncertainty. Rigid measuring levers in these structures are hung on the circular elastic notch hinges (ELKS of type RG-IV) with a radius of 3.5 mm and the thickness of the crosspiece 0.15 mm. These measuring systems were widely used in industrial production of parts for monitoring plunger-ring pairs' dimensions and shape in fuel apparatuses with close tolerances in precision surface and permissible gaps within 0.1 μm. Possible improvement of these measuring systems is in their automation with the use, for example, of the helicoidal multivibrator principle for implementation of which it is only necessary to install a permanent or an electrical magnet on the case with its poles around the pretwisted strip segment. This allows us to connect these measuring devices to the computerized system for sorting measured parts on the dimensional groups and assembling the ring-plunger pairs with very close tolerances on the permissible gaps between mating parts (selective assembly).

Reed-optical systems. The spring-optical *MicroZIL* instrument is based on the reed mechanism (ELKS of type PM-II) with strips 8 and 9 (Fig. 8-1a). Strip 8 is attached to the instrument's case, and strip 9 is connected to measuring spindle 1 on the translational guide 2 (ELKS of type TG-I). Shield 5 with the straight edge is clamped to the reed mechanism output. Source of light 7 is positioned in the focal plane of condenser 6. Rays from the source become parallel after the condenser. Shield 5 in the parallel rays between condenser 6 and objective 4 produces the dark field on dial 3. The position of the straight edge of the shield image corresponds to the readout pointer on the dial. The transmission ratio of the instrument is equal to: $i_{opm}= i_m\, i_{op}$, where i_m =120 is the transmission ratio of the instrument's mechanical part, $i_{op}=b_f / f'$ =10 is the transmission ratio in the instrument's optical part, f' is the rear focus distance of objective 4, and b_f is the distance from the objective's focus to the dial. The graduation values are 0.001, 0.002, and 0.0025 mm, the graduation interval (length) is close to 3 mm, and the measuring force Q_m =0.5 N.

The *Sheffield visual gage*. This measuring gage utilizes the combined amplification potential of the mechanical linkage with the reed mechanism (ELKS of type PM-II) and a light ray lever (Fig. 8-1b). Functionally, the gage is not dependent on a power whose fluctuation could not affect the measuring accuracy, and it uses electricity only for the lamp, which produces the light ray. The increase of the transmission ratio here is achieved by increasing the length of the "optical lever"(dash-dot line). Additional reflecting prism 10 helps to prevent significant increase in the instrument's size.

MicroZIL and Sheffield visual gages can be used only on the vibroisolated bench and do not have digital or analog automatic output. There are known instruments in which the dial and pointer are clamped to two reed mechanisms, which have common driver but rotate in opposite directions. As a result, the transmission ratio is doubled.

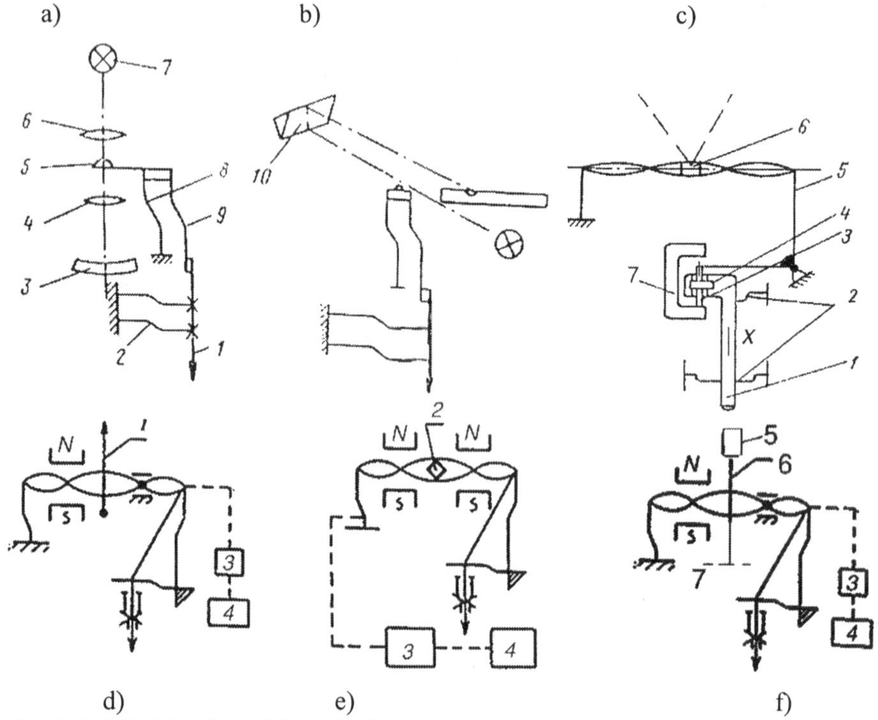

Fig. 8-1. Models of precision elastic measuring systems.

Photoelectric transducers.

Several types of photoelectric transducers with the elastic helicoidal mechanism were developed: (i) sorting multi-command with up to 50 dimensional groups; (ii) signaling with adjustable limits;(iii) functional with continuous amplitude output, (iv) photo-pulse with irreversible count; and (v) photo-pulse with reversible count (*see* Y. M. Tseytlin *et al.*, 1962b, 1968).

Photoelectric sorting- "quantized" transducers.

The position type multi-command transducers (patent SU 179954) have the set of up to 50 light sensing photoresistors 5 (scheme S7) under the dial 6, which can be switched by the light index. These transducers for mass production were built on the basis of the optikator's mechanism. The group intervals are as follows: 0.0005, 0.001, 0.002, and 0.005 mm. The transformation chain consists of the measuring spindle on the elastic translational guides with the perforated plates, link 2 (ELKS of type PM-I*el*), and link 3 (ELKS of type SM-VI) with rotatable mirror 7, which controls the switching of photoresistors. The source of light rays consists of lamp 1, diaphragm, and condenser 4. The direction of light rays from the source to the mirror is constant, but their direction from the mirror to the set of photoresistors, the position of the light spot on the set 5 and on the dial 6 depends on the angular position of the mirror, which is kinematically linked to the measuring spindle. Each of the photoresistors can have independent output to the secondary transducer and to the service net; they may be connected in grouping relay combination (Y.M. Tseytlin *et al.*, 1962b, 1968) on the link principle.

Photo-pulse transducer. These transducers with reversible count have measuring spindle 1 (scheme S 9) on the elastic guides of type TG-VIII, spring mechanisms 2 and 3 of type PM-I*el* and SM-VI with rotatable mirror 4. The direction of light rays between the source 5 and mirror 4 is constant, and it depends on the angular motion of mirror 4 in the space between this mirror and the annular raster grids 9 and 10 with radius R_{gr} and raster's pitch intervals S_p. Moving light spot 11 on these reflecting segment grids produces rays amplitude modulation with the shifted phase. Their phase shift depends on the mutual angular position γ_s of grids 9 and 10. Each of the modulated rays is focused on a certain light sensing element 12 and 12'. Changing the phase shift of the raster grids by pinions 6 and 7, we can change the value of pulse from $c_i = S_p/i_{om}$ to $c_i = 0.5\ S_p/i_{om}$, where i_{om} corresponds to the transmission ratio of the installed opto-mechanical system. A random reverse or parasitic vibration of the light spot on the grids does not cause additional uncertainty. However, in photo-pulse transducers with only one raster grid and irreversible pulse count (Patent SU 165315), a random reverse or parasitic vibration of the light spot on the grid will cause a large uncertainty in the measurement.

Table 8-2.Transient function $\theta_e[U(t)]$ and dynamic rate of spring-optical mechanism

Type of transducer	η_s	Approximate expression	Rate, Hz $\alpha_{T\theta}$ °/°C
05 PFS 1 PFS 2 PFS 1 DFI Case 1 of (4-77a)	1.8 0.9 0.45 0.9	$\theta_e[U(t)] = \eta_s s\ [\ 67\text{-}74 \exp(-100t) + 6.7 \exp(-1100t)]$ 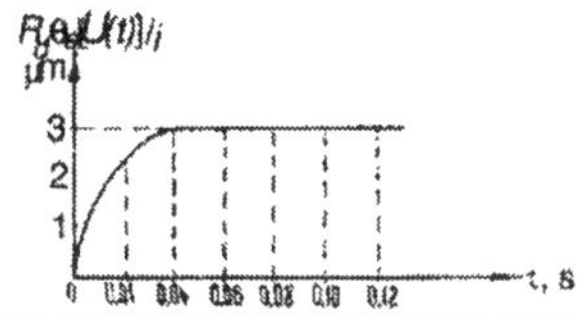	10/ 2.2
5 PFS Case 3 of (4-77a)	-	$\theta_e[U(t)] = s\left[28 - \frac{338\cdot 10^5}{z}\exp(-n_\theta t)\frac{z\cos zt - n_\theta \sin zt}{n_\theta^2 + z^2}\right];$ $n_\theta = \frac{\varsigma_\theta}{2J_{mr}} = \frac{1}{\tau}\ln\frac{A_{i+1}}{A_i};\ z = \sqrt{1.2\cdot 10^6 - n_\theta^2}$ 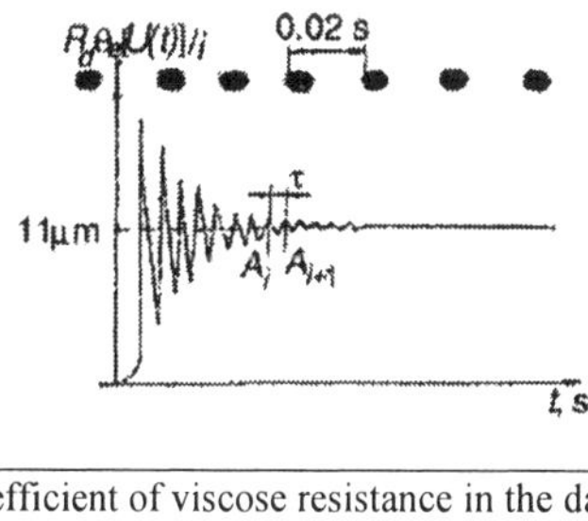	2/ 0.87

Note: $U(t)$ is a unit step; ζ_θ is a coefficient of viscose resistance in the damper; J_{mr} is the mass moment of inertia for rotatable mirror and shellac bead; A_i and A_{i+1} are sequential amplitudes of the transducer's index vibration on the dial; $\alpha_{T\theta}$ is the temperature coefficient for crosspiece of the pretwisted strip untwist in angular degree per 1 °C ; number in the transducer notation corresponds to the group and dial-value intervals in μm.

Photoelectric signaling transducers. These transducers with adjustable limits (Patent SU 136894) have reflecting segments 8,9 (Scheme S 8) similar to the photo-pulse transducer's reflecting segments but without a grid. Signaling occurs when index 7 crosses edges *Б*.

Dynamic rate. The approximate transient functions and dynamic working rate for photoelectric sorting transducer spring-optical mechanism with the parameters shown in Table 8-1 (scheme S7) are presented in Table 8-2.

These functions may be used for the pulse and signaling transducers with the similar mechanisms. R_d on the experimental charts denotes the radius of the dial (faces of photoresistors set's position), i is the transmission ratio of the spring-optical mechanism, A_i and A_{i+1} represent the consequent amplitudes of the optical index ringing oscillations, and τ is the oscillations period. Chart for the transducer 5PFS allows us to find parameter $n_\theta = \frac{1}{\tau}\ln\frac{A_{i+1}}{A_i}$. It shows also that the system of the mechanism has ringing frequency in the range 150 to 170 Hz during the stabilization period. Now one can calculate the motion of the transducers optical index by using the known transient functions, but in addition we need to know motion of the measuring spindle in time. This motion in the regular measurement is defined by the shape of its tip and a measurand surfaces.

Problem I. It is necessary to estimate a mechanism's lag time in the transducer 1PFS (Table 8-1, 8-2). We assume that the depth of the mechanism's spindle retraction is equal to 0.02 mm. The measuring voltage on the light sensing elements can be switched on only after the light index comes to a new position corresponding the dimension of a measurand. After cancellation of the spindle retraction, the switching light index should pass an angular distance, which is equal to: $\theta = \eta_s \cdot 0.02 \cdot 67 = 0.9 \cdot 0.02 \cdot 67 = 1.2$ rad, and the time of its motion in accordance with the transient function of Table 8-2 equals $t_M \approx 0.06$ s. This should be reflected in the cyclogram of the automatic measuring system. The process of the light index displacement is considered finished after an angle of the mirror rotation corresponds 0.98θ.

Problem II. Accurate wedge moves under the measuring tip with the constant speed V_{wed} in the direction that is perpendicular to the spindle of the transducer 1PFS (Table 8-1, 8-2). Therefore, we can represent the input motion of the spindle by the following expression $s_{in} = V_{wed}\, t \tan\psi_{wed}$, where ψ_{wed} is the slope angle of the wedge plane. It is necessary to estimate the difference Δs in the position of light pointer of transducer on the dial and set of photoresistors corresponding to the quasi-statical and dynamical measurement of the same wedge's section dimension. Substituting into the Duhamel's integral (4-74b) the spindle's motion presentation and the derivative of the transducer's transient function (Table 8-2), we obtain the following expression for the transducer's pointer output motion

$$\theta_{out}(t) = \int_0^t \eta_s [7400e^{-100(t-\tau)} + 7330e^{-1100(t-\tau)}]\, V_{wed}\, \tau \tan\psi_{wed} d\tau \approx$$

$$\eta_s V_{wed} \tan\psi_{wed} \times (67t - 0.74 + 0.74\, e^{-100t}).$$

The actual dimension of the wedge section, above which the measurement tip passes in the moment t, is equal to $V_{wed}\, t \tan\psi_{wed}$, and the position of the light pointer on the dial that corresponds to this dimension in the statical state is

$$\theta_{out.st} = 67\,\eta_s t\, V_{wed} \tan\psi_{wed}.$$

Let us note that number 67 represents the quasi-statical transmission ratio of the transducer's mechanism. As a result, the sought difference is equal to:

$\Delta\theta = \theta_{out}(t) - \theta_{out.st} \approx \eta_s t\, V_{wed}(0.74 - 0.74e^{-100t})\, tan\,\psi_{wed}$.

It is clear from this equation that dynamic uncertainty of the transducer will increase with an increase of the wedge's motion velocity or with increase of its slope angle ψ_{wed}. These conclusions were of course expected.

Problem III. It is necessary to estimate the requirement for the largest working frequency of the light-sensing element in the pulse transducer of type TPI. The function of the light index motion on the raster grid may be expressed as follows: $s_{gr} = R_{gr}\,\theta$. Hence, the largest frequency of light's modulation in this transducer is equal to the ratio of the largest speed of the light spot's displacement on the raster grid to the pitch of the grid. The largest speed equals the largest value of the transient function's derivative $\frac{\partial}{\partial t}\theta_e[U(t)]_{t_0}$. We assume that t_0=0.005 s corresponds to the automatic measurement practice. The top limits of the light sensing element's frequency f_{ls} have to be larger than the frequency of light modulation, such that

$$f_{ls} > \frac{R_{gr}\,\frac{\partial}{\partial t}\theta[U(t)]_{0.005}}{S_p}.$$

If R_{gr} = 75 mm, S_p= 5 mm, and s= 50 μm, then f_{ls} > 4000 Hz.

In the case of the spindle's free motion without contact with a measurand, this motion may be presented by the following expression

$u_{spindle} = \frac{F_s + F_m}{K_{gd}}[1 - \cos(t\sqrt{K_{gd}g/F_s})]$, where F_s is the weight of the spindle, F_m is the measurement effort, K_{gd} is the elastic constant of the helical spring connected to the measuring spindle and elastic guide of the spindle.

Difference-amplitude. An attempt to built precision "difference-amplitude" transducers with the measuring spindle 1(Fig. 8-1c) guided by elastic elements 2 (TG-VIII) and "floating" cylindrical pin 3 on it, which motion limited by frame 7 for transmission only the difference part of the measurand's variable value to the mechanism 5 (PM-I*el*) – 6 (SM-VI), was not successful as a result of the influence of an acceleration and dry (Coulomb) friction between the pin and keeping it in the nest strip 4 which is clamped to the spindle. The following differential equation corresponds to the pin with mass m_p probable motion relative to the spindle after impact, $m_p\frac{d^2x}{dt^2} = F_{fr} - F_{mec} - \delta(x - x_f)F_f$,where F_{fr} is the friction force between moving pin 3 and strip 4, d^2x/dt^2 represents pin's acceleration, F_f is the reaction of the frame in point x_f of a contact with the pin, $\delta(x-x_f)$ is the Dirac's delta-function, $F_{mec} = F_{me0} + j_m x$ where F_{me0} is the initial tension and j_m is the stiffness of the transmission mechanism.

It is clear from this equation that the displacement of the pin may be not smooth (intermittent, stick-slip with "jumps") and depends on its acceleration and the presumably dry friction in kinematic pair 3-4, which depends on the speed of motion and the time in immovable state. The latter two parameters are variables in the measurement process. Indeed, this situation is similar to one described in the example with the verification of measuring effort in the mechanism with an external friction [*see* equations (1-51), (1-52) in Chap.1].

In nonlinear system, an elastic transmission mechanism may "stick" to the velocity that is reached during the starting acceleration when the latter passes the zone of elastic resonance. This phenomenon is called Sommerfeld effect (*see* M.Z. Kolovsky *et al.*, 2000).

String helicoidal multivibrators.

The structure of the transducers with bifurcating helicoidal multivibrators (see Chap. 5) are: (i) mikrochator and michator based on mikrokator (Fig. 8-1d) and multi-revolution mikator mechanisms with pointer 1, respectively; (ii) optichator (Fig. 8-1e), which based on the optikator mechanism with mirror 2, (iii) mikrochator with diffraction analog output (Fig. 8-1f), and (iv) differential multivibrator (*see* Fig. 5-7). The transducers of type (i) have an arrow readout pointer 1, and the transducers of type (ii) have a light-spot pointer reflected from the rotatable mirror 2 with a hair-like mark in the amplitude output. Both the arrow pointer and the rotatable mirror are attached to the crosspiece of a pretwisted strip, which is electrically connected to self-excited oscillator 3 and frequency meter 4. The helicoidal multivibrator as a bifurcating transducer has specific influence functions in the amplitude and frequency outputs (*see* section 8-2). The amplitude output in the amplitude and frequency transducers with the pretwisted strip may be performed even without adding anything to the crosspiece by indicating its angle of rotation with the laser 5 (Fig. 8-1f) ray 6's diffraction picture 7. The mechanism in which the crosspiece has no additional mass is effective for a faster measuring rate than that with mirror.

Elastic Guides application.

Elastic rotational and translational guides found wide use in the measurement mechanisms of many precision instruments and automata. In these structures, ELKS of type RG-I-RG-IIa ,TG-I-TG-III, and TG-VIII, VIII*b* are typical. Fig. 8-2 depicts several examples of the translational guides: a) in automatic system for checking the diameters of bearing needle roles with possible oscillations after retraction (b, c); d) in pneumatic automatic measuring system for monitoring the height and parallelism of the bearing ring faces; e) and f) in the industrial monitoring devices, and g) in the pneumatic-photo-spring transducer.

Flexible leaf springs of ELKS TG-I in the mechanisms similar to presented in Fig. 8-2a are recommended to load with the stretching force in order preventing the loss of their stability. In addition, the lateral preliminary displacement of such springs should be the largest possible to increase their dynamic rate. The transient function of such system can be found with the input pulse from the gage blocks step. However, the step s_{st} should be limited by critical value $s_{st.cr}$ to prevent the transient speed which is larger than the critical $s_{st.cr}$ one:

$$s_{st} < s_{st.cr} = \sqrt{\frac{v_{cr}^2 + a_1^2 \Omega_\omega^2}{\Omega_\omega^2}} - a_1 = a_1 \left(\sqrt{\frac{1 + k_{jM}}{k_{jM}^2}} - 1 \right),$$

where a_1 is the spare loading of elastic elements; Ω_ω is the natural frequency of the system, and k_{jM} is the coefficient of elastic restoration of the measuring tip and gage blocks. The increase of measuring force from F_Q=1.5 N to 4 N by stretching spring 1 (this increases the value a_1) in the case of the measurement station for checking the diameters of bearing needle roles shown in Fig. 8-2a diminishes the duration and number of pulses of the measuring tip-knife's oscillations (Fig. 8-2b with F_Q=1.5 N, and Fig. 8-2c with F_Q= 4 N) after step input s_{st}. In Fig. 8-2b,c, plot 1 is the frequency 50 Hz indicator and 2 is the oscillogram of measuring tip-knife's motion.

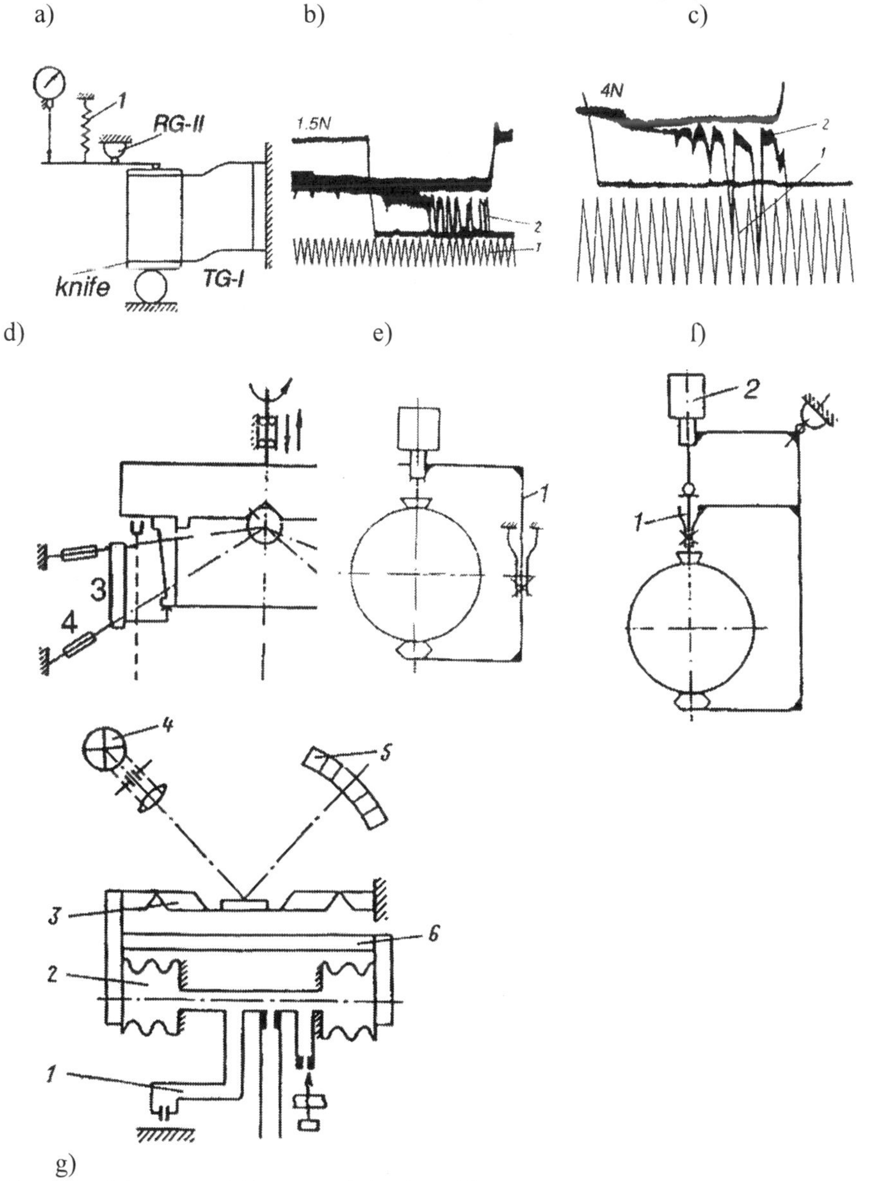

Fig 8-2. Schemes and features of elastic guides.

The translational motion of movable carriage 3 (*see* Fig. 8-2d) under some specified slope angle, i.e., its simultaneous motion in two directions, can be achieved with spring trapezoid 4 whose leaf springs in initial position are not parallel and are not under right angle to carriage 3 longitudinal axis. In the industrial monitoring snap gages, ELKS serve as the guides 1 (Fig. 8-2e) of the case, and as the guides for the intermediate transmission link 1 (Fig. 8-2f), which is necessary for eliminating a lateral force on the measuring spindle of transducer 2.

Pneumatic-photo-spring transducer. Schematic of such transducer is shown in Fig. 8-2g. Its transformation chain consists of differential pneumatic element 1, the bellows type spring sensitive elements 2, elastic transmission mechanism with ribbon coiled strip 3 (ELKS of type SM-VII), optical scheme 4, and set of light- sensitive elements 5. The pneumatic sensitive element changes air pressure in its chamber and in bellows 2 if the gap between the nozzle and the measurand's surface changes. This change in air pressure moves carriage 6 which is connected to bellows 2 (as in translational guide TG-VIII*b*, Table 1-2) and to the mechanism of coiled strip 3 with rotatable mirror on its crosspiece. The ray of light from source 4 is incident on the mirror and reflects to the set of light-sensing elements 5 that changes electrical current (voltage) in the servo-systems.

8-2. Specific Influence Functions

Elastic systems with different structure and principle of work have different main influence functions.

Joint mass of air.

Frequency influence. Let us approximately estimate this phenomenon using the equation for sound pressure (after Ph. M. Morse, 1948),

$P_{snd} = A_p \cos\varphi[J_1(k_{ps}a_p) + iN_1(k_{ps}a_p)]e^{-2\pi ift}$, where $k_{ps} = 2\pi f / c_{snd}$;

$A_p = c_{snd}^{-1} 2\pi^3 f^2 \rho_\gamma a_p^2 V_0$, $c_{snd} \approx 330$ m/s is the speed of sound in the air, V_0 is the amplitude of vibration, J_1 is the Bessel function, and $N_1(k_{ps}\, a_p) = -\dfrac{2}{\pi k_{ps} a_p} = -\dfrac{c_{snd}}{\pi^2 f a_p}$ is the Neumann function. In the case of helicoidal multivibrator (*see* Chap. 5) with $b_1 = 0.11$ mm and $f = 5000$ Hz,

$k_{ps}\, a_p = 2\pi \cdot 5000 b_1 / 330 = 0.0105$, and $p_{snd} = -i\omega\rho_\gamma a_p V_0 \cos\varphi e^{-i\omega t}$.

Here a_p is the distance to a point of the surface from the center of the vibrating symmetrical body, $\varphi = 2\pi x / S_0$, and S_0 is the pitch of elastic helicoid's pretwist. For a flat cross section with $h << b$, we approximately have $a_p = h / (2\cos\alpha)$, $\alpha \in [0, \tan^{-1}(b/h)]$. If $\Delta f / f = -½\, \Delta m / m$, then a possible change of multivibrator oscillation's frequency is equal to: $\Delta f / f = -\dfrac{1}{2\pi} \ln \sqrt{\dfrac{\sqrt{1+(b/h)^2} + b/h}{\sqrt{1+(b/h)^2} - b/h} \dfrac{\rho_{\gamma\, air}}{\rho_{\gamma\, str}}}$,

where $m = \rho_{\gamma str} A$, $\Delta m = \rho_{\gamma air} A$; $A = bh$ is the area of cross section, $\rho_{\gamma str}$ and $\rho_{\gamma air}$ are the density of the string material and ambient air, respectively. Substituting the typical value $b/h = 15$, we find $\Delta f/f = -1.08\, \rho_{\gamma air} / \rho_{\gamma str}$. For wider strips with $b/h = 25$ as in Johansson's mikrokators, we obtain a possible change in frequency as follows: $\Delta f/f = -1.25\, \rho_{\gamma air} / \rho_{\gamma str}$, which is approximately (*see* P.V. Novitskii *et al.*, 1970) two times more than for a round ($\Delta f/f = -0.5 \rho_{\gamma air} / \rho_{\gamma str}$) and two times less than for the flat strip-like strings ($\Delta f/f = -2\rho_{\gamma air} / \rho_{\gamma str}$). Despite the approximate type of estimators, they are good enough because the influence of air resistance is small as it is proportional to the small ratio of air and string material density. From the latter ratio, one can further estimate the influence of the content, temperature, pressure, and humidity of ambient air (*see* Chap.7) on the helicoidal multivibrator's and other string transducers' measurement uncertainty $\Delta f/f$. The real barometric coefficient for the flat bronze strings is equal to 4.95×10^{-6} 1/kPa (*see* P. V. Novitskii *et al.*, 1970). Hence, the transition from vacuum to atmosphere with normal pressure of 101.3 kPa will change the string frequency in proportion to 5×10^{-4}.

Transmission influence. Possible influence of air atmospheric pressure on the pretwisted strip's quasi-static transmission ratio is a result of changing the cross section's height and width under contraction. The change of the ambient air pressure practically has no influence on the amplitude transformation of the quasi-static strains in the helicoidal multivibrator unlike the possible change in its frequency output (*see* Y. M. Tseytlin, 1995).

Forced vibrations.

The experimental data (S.B. Tarasov, Y.M. Tseytlin, 1976) of the forced vibrations limited amplitude $\xi_a(f)$ for spring measuring heads with parameters shown in Table 8-1 are presented in Fig. 8-3a for instruments manufactured in Russia in 1980s (LIZ), where 1 represents lever-spring head (minikator) feature, 2 for mikator, 3 and 4 show the behavior of one- and multi-revolution indicators with dial graduation of 0.001 mm and rigid lever-gear transmission for comparison, 5 represents mikrokator, and 6 represents optikator behavior, respectively. ξ_{ac} chart corresponds to the amplitude of normal acting forced vibrations in the laboratory with frequency f. Let us look in detail of this problem.

Optikators (scheme S5, Table 8-1). Two oscillating models for this study may be found in the vibrating $\xi(t)$ optikator measuring mechanism: (i) the lateral oscillations of mirror on the stretched strip (Fig. 8-3b) and (ii) the torsional oscillations θ_φ of mirror around the axis of strip (Fig. 8-3c). Note that the lateral oscillations of the mirror (mass equals 5 mg) have a small influence on its angular position because the lateral stiffness of the stretched pretwisted strip is large. The torsional vibrations of the mirror may be presented by the following differential

equation $$J_g\ddot{\theta}_\varphi + C_\theta\dot{\theta}_\varphi + \frac{2j_{\tau\tau}}{L}\theta_\varphi = -m_g l_g \ddot{\xi}, \tag{8-1}$$

where J_g is the mass moment of rotatable mirror's inertia; C_θ is a damping coefficient; $j_{\tau\tau}=K_\theta$ is the torsional stiffness of the strip at torsion (*see* Chap.4); m_g and l_g is the mass and unbalance arm of the mirror, respectively. The solution of (1) yields the relationship between the amplitude ξ_a of the optikator case vibrations and the amplitude θ_a of its mirror angular oscillations

$$\theta_a = \frac{m_g l_g \omega^2}{J_g\sqrt{(\omega_0^2-\omega^2)^2+4\varepsilon_v^2\omega^2}}\xi_a, \tag{8-2}$$

where ω_0 is the natural frequency of the mechanism oscillations, ω is a frequency of forced vibrations, and $\varepsilon_v = C_\theta/2J_g$. The corresponding linear amplitude of the index-pointer vibration on the optikator's dial is equal to:

$$u_a = 2\,\theta_a R_d, \tag{8-3}$$

where R_d is the radius of the dial. From (2) and (3) one can find the relationship between amplitude ξ_a of the case vibrations and linear amplitude u_a of the index oscillations on the dial $\xi_a = \dfrac{u_a J_g \sqrt{(\omega_0^2 - \omega^2)^2 + 4\varepsilon_v^2 \omega^2}}{2 m_g l_g R_d \omega^2}$. It is evident that the vibrostability of the optikator increases with the decrease of the mirror's unbalance arm l_g.

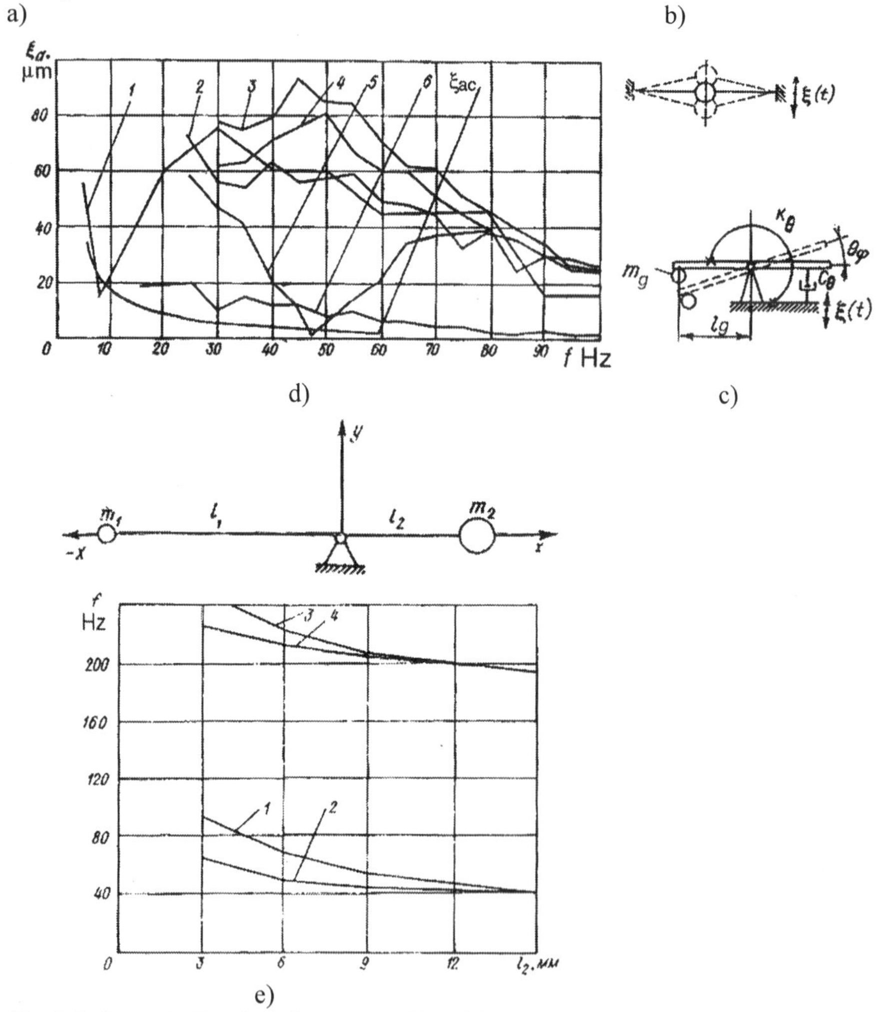

Fig. 8-3. Forced vibrations features and model schematics.

Minikators. Low rigidity of the clamping system is a typical feature for minikators (test indicators, scheme S4, Table 8-1). As a result of this, the head's case oscillates under the forced vibrations relative to the measurand's surface, which has contact with the measuring tip of the instrument. These vibrations transmit to the measuring mechanism and cause arrow-pointer oscillation, which may be expressed by the differential equation as follows

$$J_g\ddot{\theta}_\varphi + C_\theta\dot{\theta}_\varphi + \frac{2j_{\tau\tau}}{L}\theta_\varphi = \frac{j_{k\tau}}{L} i_{r.m} z(t)\,, \tag{8-3a}$$

where $j_{k\tau}$ is the torsional stiffness of the pretwisted strip under tension (*see* Chap. 4); $z(t) = \dfrac{\omega^2}{\omega_{01}^2 - \omega^2}\xi_a \cos(\omega t + \varphi_z)$ is the vibro-displacement of the springhead relative to the datum surface, ω_{01} is the angular natural frequency of the head's system vibration relative to the base and φ_z is a phase lag; J_g is the pointer's mass moment of inertia, L is the length of the pretwisted segment in the strip, and $i_{r.m}$ is the transmission ratio of the mechanism. The solution of (3a) yields the boundary of amplitude ξ_a under forced vibrations for the certain limit of the pointer linear displacement's permissible amplitude $\{u_{ap}\}$ relative to the dial, such that

$\xi_a \le \dfrac{\{u_{ap}\} L J_g(\omega_{01}^2 - \omega^2)\sqrt{(\omega_{02}^2 - \omega^2)^2 + 4\omega\varepsilon_V^2}}{2 j_{k\tau} i_{r.m} R_{pnt}\,\omega^2}$, where ω_{02} is the natural angular frequency of the arrow-pointer's oscillations around the axis of pretwisted strip, and R_{pnt} is the length of the arrow-pointer. In the normal reference measurement conditions, $\{u_{ap}\}$ is usually not more than one tenth of the dial graduation intervals.

Mikrokators. The investigation of mikrokators (Scheme S1, Table 8-1) with the different values of dial graduation in the range of 0.1 to 10 μm shows that experimental characteristics of their vibro-stability have a clear resonant feature as a result of the arrow-pointer (Fig. 8-3d) lateral oscillations. We (S.B.Tarasov, Y.M. Tseytlin, 1976) have found the relationship for the natural frequency of the pointer's lateral oscillations by solution of the oscillation form equations $y_1(x)$, $y_2(x)$ system which is presented as

$$y_1(x) = C_1 S(\alpha_w x) + C_2 T(\alpha_w x) + C_3 U(\alpha_w x) + C_4 V(\alpha_w x)\,;$$

$$y_2(x) = C_5 S(\alpha_w x) + C_6 T(\alpha_w x) + C_7 U(\alpha_w x) + C_8 V(\alpha_w x)\,,$$

where C_1, C_2, ...C_8 are the constants; $S(\alpha_w x)$; $T(\alpha_w x)$; $U(\alpha_w x)$; $V(\alpha_w x)$ are Krylov's beam functions (*see* Table 4-11), $\alpha_w = \sqrt[4]{\dfrac{m_\gamma \omega_0^2}{EJ}}$, m_γ is the mass of beam's length unit, ω_0 is the natural frequency of the oscillations, and EJ is the flexural rigidity of the pointer's cross section with diameter ~0.06 mm. The special interest for the structural synthesis in this case has relationship between the natural frequency and the position l_1 of pointer flag with mass m_1 and position l_2 of the balancing mass m_2 on the pointer's body. Figure 8-3e shows experimental 1, 3 and calculated 2,4 relationships between the pointer's second and third vibration modes and the length l_2 of the balancing part. The pointer's natural frequency increases with the decrease of l_2. The first fundamental frequency of the pointer on strip is 16 Hz in LIZ mikrokator. Note that the found second natural frequency for the pointer in this study is close to the above mentioned ringing frequency of its strikes with the graphite limiter after fast measuring spindle's retraction in mikrokator. The graphite stops are necessary to limit the rotation of the pointer and help to eliminate static electrical charges on the glass pointer and its possible adhesion and stiction to the stops and dial.

Microphone effect.

Noise and loud sound can cause an additional uncertainty in the elastic strip systems with electromechanical transducers. This action is analogous to the principle of the strip microphone's work, which sensitivity to the sound pressure (in 0.01 mV/kPa = mV/ bar≈ mV/atm) is equal to:

$$E_\theta = \frac{10^{-5}}{2c_{snd}} \frac{\omega B_{mag} A_b d_l}{Z + Z_k} \sqrt{R_n L A_{cs} \rho_\delta^{-1}} \cos\vartheta,$$

where $A_b = bL$ is the area of the thin strip's surface, B_{mag} is the magnetic induction in the gap between the strip and magnet pole pieces, L is the strip's length linked to magnetic field , b is the width of the thin strip, $A_{cs} = bh$ is the area of the strip's cross section, c_{snd} is the velocity of sound in ambient air, R_n is the electrical load-resistor, d_l is the shortest distance in air from one side of the strip to the other enveloping the magnet's pole pieces, ρ_δ is the specific electrical resistance of the strip, $Z_k = (10^{-9} B_{mag}^2 L A_b)/2\rho_\delta$; $Z = i\omega M_m(1 - \omega_0^2/\omega^2)$, $\omega_0 = (M_m/C_j)^{-1/2}$ is the natural frequency of the strip's oscillations; C_j is the strip's elastic flexural compliance, and ω is the frequency of the acting sound wave.

8-3. Pneumatic Systems and Ambient Air

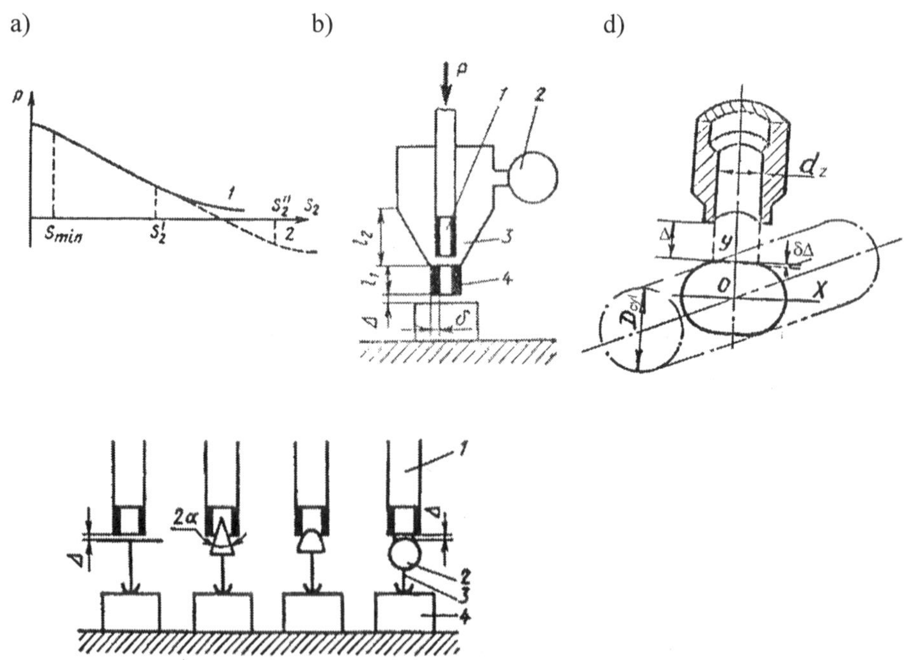

c) Fig. 8-4. Pneumatic measuring structures specifics.

The influence of a difference between inside and outside air pressure on the elastic diaphragms, bellows, and manometric (Bourdon) tubes is discussed in section 3-3. The function of pneumatic measuring systems is based on the principle of airflow through the hole or the gap between the face of the cylindrical nozzle and the shield or measurand's plane to the atmosphere. The readout systems react to the speed of airflow in rotameter type instruments or the change of pressure on the measuring nozzle input in manometric type instruments.

Double and ejector orifice. The double-orifice measuring structure (ELKS of type AM-I, AM-II, Table 1-6) has a smaller air escape clearance linear range S'_2 of pressure p (S_2) characteristic 1 (Fig. 8-4a) than linear range S''_2 for ejector structure of characteristic 2(dashed line). S_{min} denotes the minimal output clearance s in pneumatic measuring system. The ejector structure has input orifice 1with diameter d_1 (Fig. 8-4b) positioned close to the output nozzle orifice 4 with diameter d_2. The flow of air under constant pressure p from a source passes through the input orifice and sucks the air from chamber 3 where pressure is measured by manometer 2 and depends on the escape clearance Δ=S_2 against nozzle orifice 4 and the thickness of the nozzle wall δ because the air output surface in the flat gap changes from $\pi d_2\Delta$ to $\pi(d_2+2\delta)\Delta$. Usually $\delta \leq 0.5$ mm. The largest rectilinear segment of characteristic $p(S_2)$ at input pressure (1-2)$\cdot$ 10^2 kPa occurs when the length of the measuring output nozzle is equal to l_1= $(13.5\, d_1^2/d_2 - 1.5 d_2)$ and the length of a conical part of the chamber equals l_2=$(d_2$-$d_1)/0.63$.

Output gap shape. The shape of an object 4 at non-contact measurement or valve plunger–tip 3 with shield 2 (valve plunger) against the air output nozzle orifice 1 at the pneumatic measurement may be different (Fig. 8-4c): i) flat, ii) conical, iii) parabolic, iv) spherical or cylindrical.

The output area with a conical shield is equal to: $A_{cy} = \pi d_z\Delta + \pi\, d_z^{\,2}/(2\tan\alpha)$, where d_z is the diameter of the output nozzle. The decrease of angle α enlarges the measuring range, but the characteristic $p(\Delta)$ becomes more nonlinear. The output area with a spherical shield is equal to: $A_{sp} = 2\pi r[\Delta + R - (R^2 - r^2)^{1/2}]$, where R and r are radii of the sphere and the nozzle, respectively. The cylindrical shape of a shield may occur when the surface to be measured has this kind of shape (Y. M. Tseytlin, 1962a). The output gap for the air at the cylindrical measurand's surface (Fig. 8-4d) increases by the value

$\delta\Delta = (2000/\pi)\, k_d^{-1}[(\pi/2) - E(k_d, \pi/2)]$, and the actual escape clearance, in μm, is equivalent to $\Delta + \delta\Delta$, where $k_d = d_z / D_{cyl}$, D_{cyl} is a diameter of a cylindrical surface to be measured, d_z is the inner diameter of the orifice in mm, and $E(k_d, \pi/2)$ is the Legendre's complete elliptic integral of the second kind.

The output area for a parabolic shield is usually defined by experimental methods rather than by calculations.

Surface finish. For a random distribution of measurand's surface roughness, the variation in the flow of air through the gap between the face of the measuring nozzle and the measurand's surface corresponds to the increase in gap, in μm, which is equal to: $\Delta s_{sr} = (R_{zt}-R_{zr})/2$, where R_{zt} and R_{zr} are the heights of irregularities for tested and reference components, respectively. If we neglect the value of R_{zr}, which should be small in comparison with R_{zt}, we obtain the value for the possible surface roughness uncertainty in pneumatic measurement as $\Delta s_{sr} = R_{zt}/2$.

Distribution of pressure. The change of pressure along the orifice, at the same state of air before it with pressure p_n (Fig. 8-5a), may correspond to four different functions (A. Wiemer) depending on the counter-pressure p_{ac} of the ambient air.

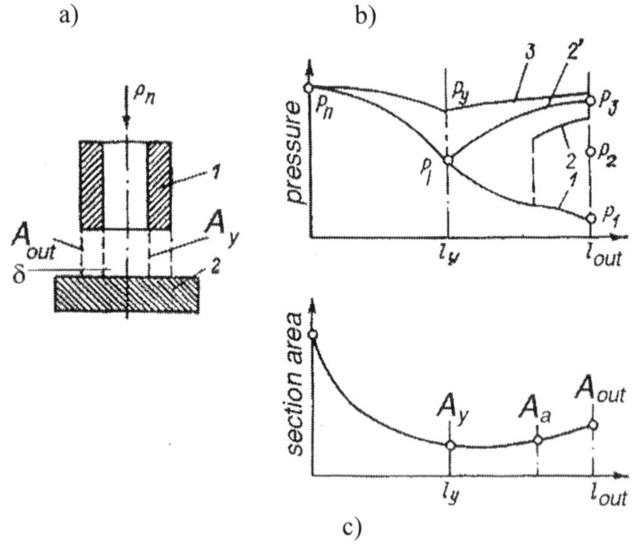

Fig. 8-5. Air escape areas and distribution of pressure (after A. Wiemer)

In this study, the output channel of air flow in the gap between the measuring nozzle 1's face and measurand surface 2 is modeled as the Laval's orifice where the narrowest area is equal to $\pi d_z\Delta = A_y$ and the output area equals $\pi(d_z+2\delta)\Delta = A_{out}$. It is assumed that the orifice consists of two parts: internal cylindrical part with constant diameter d_z and external continuously expanding part in the gap between a measurand or shield surface and the orifice face with outer diameter $d_z + 2\delta$. The ratio of the final pressure value p_l on the nozzle orifice outer face to the initial pressure p_n depends on the counter-pressure p_{ac} and ratio between the areas of the orifice's narrowest section A_y and A_{out} on the final output section. In this case, we consider $0<p_1<p_2<p_3<p_n$ (Fig. 8-5b). If value of counter-pressure p_{ac} is lower than p_2, then the process of pressure change along the orifice internal length (Fig. 8-5 b,c) and the gap length $l_{out}-l_y = \delta$ corresponds to smooth curve 1. Pressure in the narrowest section A_y of the orifice at its internal part end falls down to $p_l = 0.53\ p_n$, and it falls to $p_1 < p_l$ in the outer nozzle's face. If the counter-pressure $p_{ac} < p_1$, then a sudden expansion of the air occurs on the output of the nozzle. If $p_{ac} > p_1$, then a sudden condensation (the straight jump of condensation) occurs. In this case, the sonic velocity of airflow in the output of the narrowest orifice section l_y transforms to the supersonic one. If the counter-pressure is within the limits p_2 and p_3, then the straight jump of air condensation occurs inside the gap part of air flow in section A_a between the narrowest $l_y(A_y)$and output $l_{out}(A_{out})$ sections. The position of section A_a depends on the outer air pressure p_{ac}. The jump from supersonic to subsonic velocity occurs in this section on curve 2. If $p_{ac}=p_2$, then the jump from p_1 to p_2 occurs in the output orifice section. If $p_{ac} = p_3$, then the jump occurs in the narrowest section l_y of the orifice (curve 2'). Finally the pressure in the narrowest section equals p_y, which depends on the value of p_{ac} when the counter-pressure $p_{ac}>p_3$. The sonic velocity does not occur in this case (curve 3). Discussed change of pressure along the orifice can cause a difference in the pneumatic system internal friction. The degree of inside air cleaning (usually not less than 99.95%) is very important for the accurate pneumatic measurement. The zone of air rarefaction may occur in the gap between the nozzle and the shield (screen, measurand) that causes the suction inside it of ambient air. If this ambient air has airborne dust, then the dust can settle in the measurement gap and cause an increased measurement uncertainty.

Aerostatic bearings are used in many precision systems as rotary and linear guides. Here is an example of when these bearings with very small friction are used in connection with the elastic sling.

The *sling method* was invented by F. W. Witzke in 1972 (Patent USA 3,648,377) for out of the roundness centerless measurement on surfaces with large diameters, large sizes and developed by I.S. Amosov, O.A. Ivanov, and V.I. Anukhin (Russia, St. Petersburg Polytechnic University, Paten SU 839183) in the late 1970s with the application of several spotted air pressure bearings under the sling.

a) b)

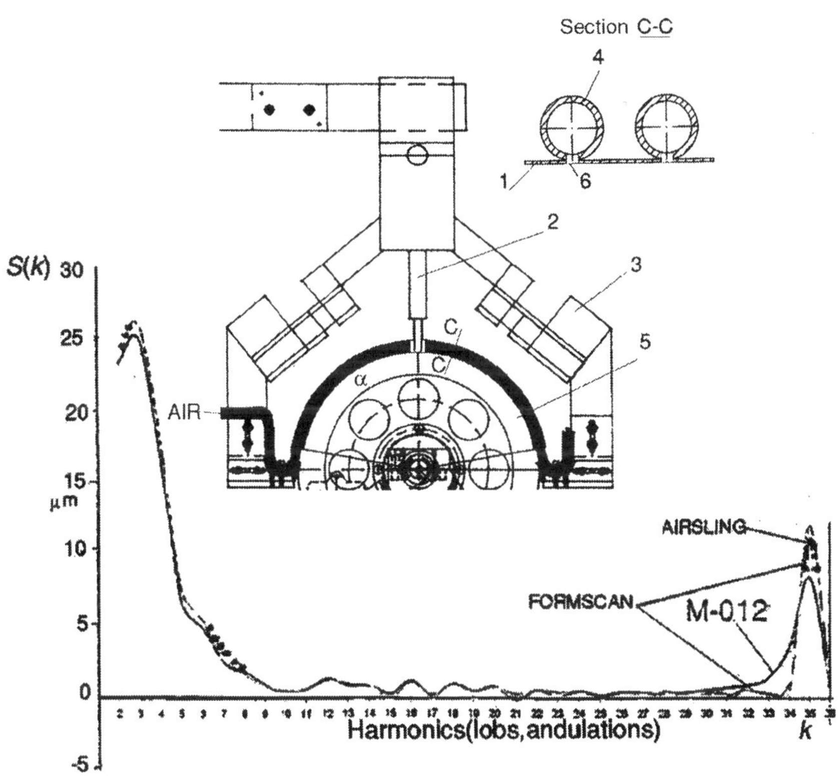

c)

Fig. 8-6. Sling centerless roundness measurement structure and its traceabilty.

Thin sling 1 clamped to hanged bracket 3 envelopes the surface profile 5 within the wrap angle α (Fig. 8-6a,b) and changes its sag in bracket corresponding to the position of the curvature center. Measuring probe 2 installed on the same bracket. The sling method is based on the representation of the out-of -roundness profile deflections $S(x)$ by the following formula

$$S(x)=\sum R_k C_{Fck}\cos(k\varphi+\varphi_k),\quad C_{Fck}=(1+\frac{\sin k\alpha/2}{k\sin\alpha/2}),\quad k\geq 2,$$

where k is the number of lobs (harmonics) on the surface and amplitude R_k's correction coefficients C_{Fck} depend on the wrap angle α. Hence the sling with a certain wrap angle has theoretically known correction coefficients for each harmonic whose values are significant only up to undulation (harmonic) number 14 (*see* Table 8-3 for wrap angle α=140°). Values of these correction coefficients for higher than #17 harmonics are within [1+0.05, 1-0.05] range.

Table 8-3. Air sling amplitude correction coefficients at wrap angle α=140°

Lobs, harmonic #	2	3	4	5	6
C_{Fck}	1.342	0.823	0.738	0.963	1.154

The sling should be thin enough (its thickness is approximately 0.1 mm) to properly envelop the surface with the highest undulations. The appropriate materials for the sling are steel and invar. The advantages of this method with one probe are the low cost and the small theoretical uncertainty for high harmonics. The disadvantage in the original Witzke's version was connected with large friction and fast sling's wear. The latest version of the sling with the inherent compensated air support (inherent air sling) was developed by Y. M. Tseytlin in 1990s for research purposes. Adjustable air pressure in this version is applied into two flexible plastic Tygon tubes 4 (Fig. 8-6b), which are cemented to the flexible sling 1. The air flows through the set of small holes 6 made in the sling and the tubes simultaneously. As a result, the uniform air bearing (calculation methods *see* in A.H. Slocum, 1992) takes place between the sling and surface 5 (Fig. 8-6a) to be monitored in order to eliminate dry friction and sling wear at the rotation of the object with surface 5. This system is compatible with the regular precision center FORMSCAN's amplifier and computer software of the Federal Products Company (USA), and does not require special interface. The experimental verification showed that the air sling measuring system does not have "leak or loss of harmonics (undulations)" in the whole investigated range from 2 to 36 undulations (*see* Fig. 8-6c), unlike the multi-contact centerless method with floating skids (equivalent to three probe in the three stages system) and the multi-probe method (*see* details in D. J. Whitehouse, 1994). The latter methods have significant uncertainty especially on high numbers of undulations. Chart in Fig. 8-6c represents the measurement results $S(k)$ of roundness deflections on the disk cylindrical surface with 12 inch (305 mm) diameter, where solid line belongs to the multiprobe method M-012, dashed line represents the airsling method, and black dots show some segments of FORMSCAN results which have slight deflections within 10% from airsling measurement results. In other segments FORMSCAN and airsling results practically coincide unlike multiprobe results, which have significant deflection up to 30% from FORMSCAN results in the region of harmonics #5-8 and especially # 29-36. Let us recall that FORMSCAN has precision axis for rotation of the disk under consideration (primary measurement method, but only for objects with limited size) unlike other two methods with the "centerless" measurement, or with so-called virtual center of rotation which is applicable to objects with large sizes.

8-4. Micro- and Nano-scale Systems

Typical elastic elements used in contemporary micro and nanoscale science and technology are presented in Table 8-4.

Table 8-4.Typical elastic micro- and nano-structures

Typical elastic elements	Specific features	Application & recommendations
Microcantilevers	Rectangular V-shaped Round, microfiber^ Composite	AFM, material silicon, SiC; Si_3N_4, Typical sizes: ~125 x 35 x 4 μm Cross-crystal with arms * Quartz, Q-factor 25 in air, 2000 in vacuum Coated with gold, platinum Sensing *in situ*, ellipsometry
Tuning fork	Quartz crystall	*AFM* scanner
Nanocantilevers	Rectangular	Material silicon, *CdO*, PbO_2; sizes in nm
Nanowires, nanotubes	Round	*Pt*, carbon *SWNT*,*MWNT*, suspensions
Flexure notch hinges	Corner-filleted Circular Elliptic	More motion, less accuracy More accuracy, less motion Intermediate
Arrays of microcantilevers	Rectangular Triangular	Linear transducers, tactile sensors

^ X. Chen *et al.*, 2004, * Z. Peng *et al.*, 2005.

Atomic force microscopy cantilevers.

Common features. Cantilever based scanning probe microscopy (atomic force microscopy-*AFM*) techniques are among the most powerful approaches for imaging, manipulating, and measuring nanoscaled properties and phenomena. Let us recall that a very sensitive tunneling microscopy can work only with conducting and semiconducting materials mostly under ultrahigh vacuum, whereas the AFM can work with any kind of substances in air, gas, and liquid. This microscopy generates surface, molecules, and atom images by measuring small forces between sample surface and the probe. Common forces detected using these techniques are van der Waals, Casimir, electrostatic forces, surface stress, etc. The effects of electrostatic field and Casimir force on the cantilever vibration and damping are thoroughly studied by many researchers.

Microcantilevers. Conventional microcantilevers are fabricated from isotropic and crystalline quartz, silicon, silicon carbide, and silicon nitride (Si_3N_4) using electron-beam or optical lithography and have typical length, width, and thickness of ~125, ~35, and ~4 μm, respectively.

Cantilevers with integrated (the tip height is a few orders of magnitude larger than the radius of its curvature) and colloidal (tip height is in the same order as the radius of curvature) probes should be calibrated for direct and the lateral force action. The latter is important because the coefficient of friction between the polysilicon measurand and a silicon probe can be in the range of 0.19 to 0.35. This friction can cause a torsional displacement of the probe, stick-slip motion, and corresponding additional measuring uncertainties.

The cantilever's tips are made from tungsten, diamond, silicon, cobalt, samarium, cobalt-samarium permanent magnets, and iridium for various purposes. New tips in scanning microscopy systems are based on the growing carbon nanotubes and implanted diamonds. These high aspect ratio tips are nearly ideal for atomic force microscopy imaging because of their size and durability. The atomic force microscope is routinely used to measure a number of interfacial phenomena including surface forces, friction, and adhesion (pull-off force of ~0.4 nN for hydrophobic tip at $RH \geq 30\%$, and ~ 4 nN for hydrophilic one). Of fundamental importance to these quantitative measurements is an accurate knowledge of the cantilever spring constant. Furthermore, by controlling the interaction geometry with attaching a particle to the end of the cantilever, it is possible to compare the measured interaction force with theory. The spring constant is dependent upon the cantilever dimensions, shape, and material properties. It is clear that the static AFM requires very soft cantilevers. The deflection of the cantilevers is most commonly measured by optical means through laser heterodyne Doppler, Michelson, or Fabry-Perot interferometer (threshold noise up to femtometer per $\sqrt{Hz}$) with spot size up from 3μm or by bouncing light ray off the cantilever. The detection in the latter system is provided by segmented or linear position sensing detectors (PSD), its calibration may be performed with the precision calibrated step sample or by the inverse calibrated piezolever. The deflection of the piezoresistive cantilevers is usually measured by making them part of a Wheatstone bridge. Let us recall that silicon piezoresistive transducers are nonlinear that require to use confluent statistical analysis at their calibration. If the cantilever in a variable-load *AFM* is tilted towards the specimen sample, its tip may have significant parasitic displacement along the cantilever axis direction that can be estimated on the basis of EKS and ELKS calculation methods (see Chaps 1, 3, 4).

Cantilevers with lesser cross-section sizes provide better sensitivity to piconewton (pN=10^{-12} N) forces. Cantilevers of AFM have been uniformly coated with gold and platinum thin films. These films are nanostructured with thickness between 18 and 73 nm.

Microcantilevers consisting of a silicon or silicon nitride substrate (relatively passive side) coated with a layer of special film (high affinity to the targeted analyte) deflect due to heat transfer or chemical sorption. They have been used, for example, to distinguish between oligonucleotide (DNA) molecules of different base sequences, to detect individual biomolecule, to measure *pH* changes, and to measure the surface stress associated with molecular adsorption or absorption. In many of these applications, the deflection of the cantilever is driven by the stress buildup in the layer of the metal film. Microcantilever sensors used for recognition of chemical or biological environment components work similar to AFM cantilevers in quasi-static and dynamic mode. In the dynamics they work in the resonant mode [*see* expressions (5-87d) and (5)]. The main difference follows from the assumed ability of some chemicals (for example *NaCl* solution) to change the cantilever's bending spring constant k_{fl} even at small oscillation amplitude. G.Y.Chen *et al.* (1995) showed the relationship between the changes of bending spring constant Δk_{fl} and the change of the surface stress on the top $\Delta\sigma_{top}$ and bottom $\Delta\sigma_{bottom}$ of the microcantilever as follows: $\Delta k_{fl} = \frac{\pi^2}{4n^*}(\Delta\sigma_{top} + \Delta\sigma_{bottom})$, where n^* is the effective cantilever's mass factor which depends on the cantilever geometry (*see* below). In quasi static mode, their work based on the cantilever's deformation which is detected by modulation of optical, electrical or magnetic field, modulation of mechanical stress, i.e. different surface stresses [*see* expression (a) relationship between the surface stress and strain]. The cantilever deformation can be caused by thermal effects as well (*see* Table 7-5).

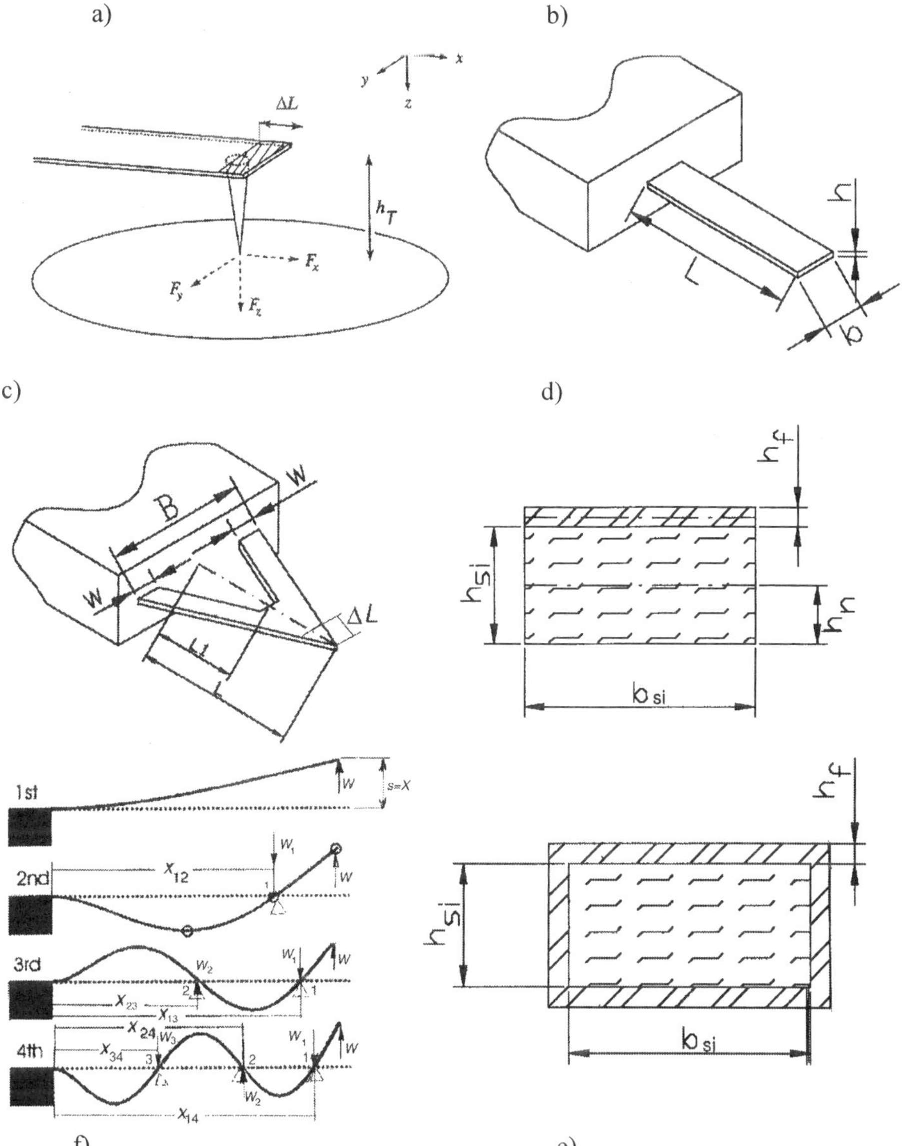

Fig. 8-7. Features of microcantilevers with different shapc and sructure

Measuring the cantilever deflection and knowing its geometrical parameters, the difference in surface stress per unit length $\Delta\sigma_{top/bot} = \sigma_{top} - \sigma_{bottom}$ can be determined using Stoney's equation (*see* G.G. Stoney, 1909), $\Delta\sigma_{top/bot} = \dfrac{Eh^2}{3(1-\nu_p)L^2} s_d$, (a) where s_d is the deflection of the free-end of the cantilever. Applicability of the classical curvature-stress relation for thin film was discussed with positive conclusions by F.J. von Preissig, 1989.

Preferred shape. It has been recommended to use V-shaped micro-cantilevers in *AFM* application, as they were assumed less susceptible to lateral forces than rectangular ones. However, in recent study, J.E. Sader (2003) has found that *V*-shaped cantilevers are generally more prone to the effects of lateral forces than rectangular AFM cantilevers. It is an important and interesting conclusion, which rather should be related to the moments from the lateral forces because the lateral spring constants k_x and k_y in Sader's formulae depend on the height h_T of probe with tip (Fig. 8-7a) . Later, however, J. E. Sader and C.P. Green (2004) on the basis of regular formulae for bending of rectangular beam (*see* Chap. 3) and wedge (S.P. Timoshenko, 1955) have found that the comparison of in-plane deformation of rectangular and V-shaped cantilever plates reveal the superiority of V-shaped cantilever which has a higher spring constant in this direction. Let us recall that unlike the wedge, V-shaped cantilever has some cut off (Fig. 8-7c) that requires solving a special differential equation of the cantilever bending as follows

$$y''(x) = \frac{12M}{EhB^3}\left\{(1-\frac{x}{L})^3 - (1-\frac{x}{L_1})^3(1-\frac{2w}{B})^3[1-U(x-L_1)]\right\}^{-1}.$$

The unit-step function $U(x\text{-}L_1) = 0$ at $x<L_1$ and it is equal one at $x \geq L_1$. If the cut off is assumed absent, $w=B/2$ and the second member in braces becomes equal to zero because $2w/B=1$. Only in this case, wedge model will be an exact one. We should also note that in-plane bending of the thin rectangular strip can cause its buckling (Table 3-3).

Another aspect of the problem that should be considered is the part of the cantilever, which remains in the space beyond probe (shaded area in Fig. 8-7a). This part is assumed a rigid one, but this part has a certain surface and mass, which will be many times larger for the rectangular cantilever than for V-shaped cantilever at the same distance ΔL from the axis of tip to the end of the cantilever. This may cause a larger damping in the ambient air, especially in a dynamic AFM. Note that ΔL cannot be equal to zero in any cantilever's shape due to the finite diameter of the probe base. It may conditionally be considered zero for the position of clamped spheres for calibration purposes.

Stick-slip effect. Possible stick-slip motion in AFM is similar to that in the system with balances (*see* Fig 1-8a), which we discussed in Chap. 1 (I.S. Amosov, Y.M. Tseytlin, 1969). It follows from equation (1-52) that possible slip jumps δS_k in the system with moving mass under the action of friction and elastic drive force is equal to: $\delta S_k = 2\ \delta Q_{fr} / j_{dr}$, where δQ_{fr} is the difference between the static and kinematic friction and j_{dr} is the spring constant of the drive. If accept that at low speed v_{dr} of the drive the stick time is enough to cause significant quasi-static friction F_s in comparison with much smaller kinetic friction, then we can assume that $\delta Q_{fr} \approx F_s$ and $j_{dr}\ \delta S_k = 2\ F_s$. One can see that after multiplying both sides of this equation by δS_k,, its left side will represent the potential energy in the drive's spring accumulated during the stick stop. Then on the basis of the Hamilton's principle with equating potential and kinetic energy, we obtain the relationship between velocity v_{dr} of drive motion and the probable slip jump of the driven probe with mass m_g as follows:

$\frac{1}{2} j_{dr}(\delta S_k)^2 = F_s \delta S_k = \frac{1}{2} m_g v_{dr}^2$. Therefore, the critical drive speed for stick-slip motion equals $v_{dr.c} = C_F \sqrt{F_s \delta S_k / m_g}$, where constant $C_F \approx 1$ is the factor of physical nature of sliding friction which is beyond the scope of this book (*see* B.N.J. Persson, 2000; B. Luan *et al.*, 2004).

Nanocantilevers. Nanowire, carbon nanotube, single crystal, and nanobelt technology allow us to manufacture cantilevers, which are ~35-1800 times smaller than conventional microcantilevers with above mentioned micrometer sizes.

AFM dynamics and parameters.

Techniques. Let us recall that AFM microcantilevers may work in the quasi-static regime and in dynamic oscillation regime. The latter one is more sensitive. Four techniques for atomic-resolution force microscopy in vacuum are in use today: (i) the classic frequency-modulation technique with large amplitudes (30 to 50 nm) and soft cantilevers (with spring constant in tenth of N/m) in which both parasitic attractive and effective repulsive action forces are possible, (ii) frequency modulation with small amplitude (less than 1 nm) for sensing only repulsive forces and stiff cantilevers (spring constant in the order of 10 N/m to several kN/m), (iii) the off-resonance technique, and (iv) amplitude-modulation method. To avoid cycle distortion at small amplitudes, the product between spring constant and the amplitude, $k_s A_c$, has to be larger than 100 nN that provide a sufficiently strong withdrawing force. The mode may be noncontact and tapping with fundamental and higher eigenmodes. The Kelvin probe microscopy method was developed to compensate different contact potentials of tip-sample materials at noncontact step height measurement.

Parameters. Let us recall that the more advanced frequency modulated "non-contact" AFM has five important operating parameters (for more detail *see* F.J. Giessibl, 2003): 1) the spring constant of the cantilever k_s, 2) the eigenfrequency (natural frequency) of the cantilever f_0 (tens kHz to GHz), 3) the quality factor value of the cantilever Q_c (up to tens of thousands), 4) the oscillation amplitude A_c, 5) the frequency shift of the cantilever oscillations

$\Delta f = -\frac{f_0}{k_s A_c^2} < F_{ts} A_c \cos(2\pi f_0 t) >$ from 4 to 100 Hz. The first three parameters (1-3) are determined by the type of cantilever that is used, while the last two parameters can be adjusted. In the fifth parameter, F_{ts} is the force acting between the cantilever's tip and the monitoring sample, and t is the time of signal averaging. The averaging is denoted by the angular brackets in the above shown formula. The long-range attraction forces between the tip of cantilever and the monitored sample have to be canceled by immersing the tip and sample in liquid or partly pulling off at the cantilever after its jump-to-contact occurred.

Let us recall that the cantilever itself dissipates energy (internal dissipation):

$\Delta E_{cL} = 2\pi E_c / Q_c$, where $E_c = k_s A_c^2 / 2$. When the tip oscillates close to the sample (the tip-sample contact stiffness k_{ts}), the oscillation amplitude changes according to the expression $A_c = \frac{A_{c0}}{1 + k_{ts}/k_s}$, additional damping occurs, and driving signal should be increased to keep the oscillation amplitude equal A_{c0}.

Resonance modes. Higher order resonance modes are expected to demonstrate higher sensitivity to the change in experimental conditions. They also enhance the sensitivity of atomic force microscopes operating in dynamical mode. Higher modes may also improve determination of force constant of a microcantilever with less uncertainty. The resonance frequency of each normal mode of vibration for an undamped rectangular cantilever with one end free can be expressed by

$f_{n}=\frac{h}{6.928\pi}\frac{(\alpha L)_n^2}{L^2}\sqrt{\frac{E}{\rho_\gamma}}$, where factor $(\alpha L)_n$=1.8751, 4.6941, 7.8548, 10.996 for n_f=1, 2, 3, 4, respectively, and for higher values of n_f, $(\alpha L)_n = (2n_f-1)\pi/2$. In addition $f_n/f_1=[(\alpha L)_n/(\alpha L_1)]^2=\sqrt{Q_{fac(n)}/Q_{fac(1)}}$. The higher modes of the cantilever lateral vibration have more nodal points with zero displacements (Fig. 8-7f). The position of these points (x_{ij}) can be found from the equation of oscillation form $X(x)$ amplitudes which may be presented through Krylov's functions (*see* Table 4-11) as follows:

$$X(x)=4[T(\alpha L)\,U(\alpha x)-S(\alpha L)V(\alpha x)]. \quad (8\text{-}4)$$

The nodal point positions correspond to $X(x)$=0 and the expression

$$U(\alpha x)[\,T(\alpha L)/\,S(\alpha L)]-V(\alpha x)=0. \quad (8\text{-}4a)$$

The numerical solution of the latter for the second mode of the clamped cantilever vibration with $(\alpha L)_2$=4.6941 and $T(\alpha L)_2/\,S(\alpha L)_2$= 0.98187 yields the nodal point position coordinate at αx_{12}=3.678 or x_{12}=0.7836L from the clamped end. Ratio $T(\alpha L)_1/\,S(\alpha L)_1$=1.36221 for the first mode, 1.00078 and 0.999966 for the third and fourth mode respectively, whose nodal points positions are αx_{13}=6.8144, x_{13}=0.8675L; αx_{23}=3.9553, x_{23}=0.5036L(for the third mode) and αx_{14}=9.95444, x_{14}=0.90528L ; αx_{24}=7.0823, x_{24}=0.64408L; αx_{34}=3.9401, x_{34}=0.35832L (for the fourth mode).

The higher roots of frequency equation cos (αL) cosh (αL) =1 for clamped on both sides (doubly clamped) beam are equal to 7.859, 10.906, 14.137, and $\pi(2n_f+1)/2$ at n_f= 2, 3, 4, and more, respectively. The nodal point positions correspond to the expression

$$S(\alpha x)[T(\alpha L)/U(\alpha L)]-V(\alpha x)=0. \quad (8\text{-}4b)$$

The higher roots of the frequency equation tan(αL)=tanh(αL) for the beam clamped on one side and simply supported on the other are equal to 7.069, 10.217, 13.352, and π (4 n_f+1)/4 at n_f= 2, 3, 4, and more, respectively. It is very important in these applications to isolate and discard the static component of a detector signal to enhance the resolution and sensitivity of an instrument. Higher modes n_f>1 frequencies for a tapering cantilever can be found on the basis of the following formula (after Ph. M. Morse, 1948) $f_n \approx f_{n\,unif}\left[1+\frac{2c_j}{\pi^2(n_f+1/2)^2}\right]$,

where index n=n_f, c_j is the tapering factor, and f_{nunif} is the corresponding frequency for a similar prismatic uniform cantilever (*see* Sect. 5-4).

V-shaped cantilever model.

Frequency. V-shaped cantilever (Fig. 8-7c) with a tip (*see* Fig. 8-7a) has fundamental resonant frequency, which is often expressed by the formula

$f_0 = \frac{0.162\,h}{L^2}\sqrt{E/\rho_\gamma}$ and the following its representation $f_0 = \frac{1}{2\pi}\sqrt{k_{ef}/m_{ef}}$,

where $k_{ef} = (EJ/3L^3)\text{-}F'$ is the effective bending spring constant (P.A. Arutyunov *et al.*, 2002), F' is the force gradient normal to the surface, $m_{ef}=m_c+ n^*m_{cn}$ is the effective mass, m_c is the probe mass, m_{cn} is the cantilever mass with mass factor n^*. The resonance frequency is altered on variation in a force pattern F' because the force gradient affects the effective rigidity. If for example there is an attraction, $F'>0$, the resonant frequency is reduced. The resonant frequency is increased at $F'<0$. One can thus record any change in f_0 to measure the force gradient in the interaction between the probe and the surface. Let us note that the accepted in the above mentioned reference value of mass factor n*=0.24 is the same as for rectangular cantilever because many authors assume that V-shaped cantilever frequency response is similar to the response of the corresponding two parallel positioned rectangular cantilevers (*see* H.-J. Butt *et al.*,1993). However, it was found in experimental studies and on the basis of a developed "variational" method (G.Y. Chen *et al.*, 1994) that for commercial V-shaped cantilevers n* ranges from 0.137 to 0.2. We will investigate this problem in more detail. We assume that the "variational" method as well as the "correction factor" method (H.-J. Butt *et al.*, 1993) are effective but not digestible and require a numerical integration. Therefore we prefer more tractable approach. Let us recall that in section 5-4 we derived simple formula (5-87d) for calculation of fundamental frequency and digestible synthesis of V-shaped cantilevers. Table 8-5 presents the comparison of the results of fundamental frequency calculation by the different methods and experimental data for silicon V-shaped cantilevers.

As we can see, the largest differences between the experimental data and calculation results by formula (5-87d) are not exceeding 9.4%. There is no influence of the width parameters (B, w) on the fundamental frequency of V-shaped cantilever. This of course follows from the accepted model with the constant cross-sectional radius of gyration and clearly manifested in formula (5-87d).

Table 8-5. Fundamental frequency of silicon V-shaped cantilevers

Parameter	Data			
L, μm	190	200	190	200
L_1, μm	87.5	115	87.5	115
B, μm	165	164	165	165
w, μm	33.3	20	33.5	20
h, μm	0.6	0.6	0.3	0.3
*K, N /m	0.094	0.05	0.012	0.006
*f_0 (exp.), kHz	31.50	25.80	17.46	13.10
*f_0 (var), kHz	34.00	25.80	17.30	12.90
f_0 (5-87d), kHz	31.65	26.38	15.82	13.20
δf_0(5-87d, exp), %	0.5	2.3	-9.4	0.8

All dimensional notations in the Table correspond to Fig. 5-11c,d; * data for these parameters, silicon modulus of elasticity E=179 GPa, its density ρ_γ= 2330 kg/m^3, experimental values f_0 (exp.) , and the calculation results of the "variational" method f_0 (var) are the same as presented by G.Y. Chen *et al.*, 1994; f_0 (5-87d) represents the calculation results by formula (5-87d), and δf_0(5-87d, exp.)= 100 [f_0 (5-87d) - f_0 (exp)]/ f_0 (exp).

Composite cantilever. Formula (5-87d) can be used for the calculation of the composite cantilevers as well, but in this case one should estimate the composite's modulus of elasticity and the average density (*see* D. Lange *et al.*, 2002) of the composite. For example, if V-shaped cantilever of silicon nitride with thickness h_{si}=0.54 μm, Young's modulus E_{si}=150 GPa, and density $\rho_{\gamma si}$ = 3100 kg/m^3 is coated with gold film (E_{gf}=80 GPa, $\rho_{\gamma gf}$= 19300 kg/m^3, and thikness h_{gf}=0.04 μm), then its composite elastic modulus E_c can be estimated from the following expression

$E_c=E_{si}+E_{gf}J_{gf}/J_{si}$, where J_{gf} and J_{si} are the moments of inertia with respect to the common axis of the cantilever's cross section, E_{gf} and E_{si} are the apparent Young's moduli of gold film and silicon nitride, respectively.

Table 8-6. Fundamental frequency of silicon nitride coated with gold film V-shaped composite cantilevers

Parameter	Data		
L, μm	202	195	119
L_1, μm	116	145	61
B, μm	205	198	121
W, μm	40.2	21.2	24.1
$h(Si_3N_4)$, μm	0.54	0.54	0.54
h_f (gold film)	0.04	0.04	0.04
*K, N/m	0.058	0.034	0.17
* f_0 (exp), kHz	18.25	14.88	50.28
*f_0 (var), kHz	18.35	16.26	51.93
$f_0(5)$, kHz	17.1	16.65	51.19
$\delta f_0(5, \text{exp})$, %	-6.3	11.8	1.8
$\delta f_0(5,\text{var})$, %	-6.8	2.4	-1.4

All dimensional notations in the Table correspond to Fig. 8-7; * data for these parameters, silicon nitride (Si_3N_4) modulus of elasticity E_{si}=150 MPa, its density $\rho_{\gamma si}$= 3100 kg/m^3, gold film modulus of elasticity E_{gf}=80GPa, its density $\rho_{\gamma gf}$=19000 kg/m^3, experimental values f_0 (exp.) and the calculation results of the "variational" method f_0 (var) are the same as presented by G.Y. Chen *et al.*, 1994; f_0 (5) represents the calculation results by formula (5) which is the version of formula (5-87d) for composite cantilever, $\delta f_0(5, \text{exp})= 100[|f_0(5) - f_0(\text{exp})]/f_0(\text{exp})$; and $\delta f0(5,\text{var})= 100[f_0(5)-f_0(\text{var})]/f_0(\text{var})$.

The position h_n of the neutral layer (Fig. 8-7d) for the double-layer composite beam may be estimated as the "center of elasticity" similar to the centroid, or center of mass, such that

$$h_n=\frac{E_{si}h_{si}\frac{h_{si}}{2}+E_{gf}h_{gf}(h_{si}+\frac{h_{gf}}{2})}{E_{si}h_{si}+E_{gf}h_{gf}}=\frac{E_{si}h_{si}^2+E_{gf}(2h_{gf}h_{si}+h_{gf}^2)}{2(E_{si}h_{si}+E_{gf}h_{gf})},$$ and therefore

$$h_n/h_{si}=\frac{E_{si}h_{si}+E_{gf}(2h_{gf}+h_{gf}^2/h_{si})}{2(E_{si}h_{si}+E_{gf}h_{gf})}.$$ After substituting the numerical data for all

members of this equation, one can find that h_n=0.52 $h_{si} \approx \frac{1}{2}h_{si}$. We will obtain a similar result if to suppose that the thin nanostructured gold film has slightly different Young's modulus which is equal to 69.1 GPa (M.C. Salvadori *et al.*, 2003). Let us also recall that the moment of inertia with respect to different parallel axes differs by product of cross-sectional area and square of the distance between respective axes. Hence, we have the following relationships with respect to the neutral axis in the middle of the silicon layer $J_{sf}=\frac{bh_{si}^3}{12}$;

$$J_{gf}=\frac{bh_{gf}^3}{12}+bh_{gf}\left(\frac{h_{si}+h_{gf}}{2}\right)^2=\frac{b}{12}(4h_{gf}^3+3h_{gf}h_{si}^2+6h_{si}h_{gf}^2)$$

and $J_{gf}/J_{sf}=3\frac{h_{gf}}{h_{si}}[1+2\frac{h_{gf}}{h_{si}}+\frac{4}{3}\frac{h_{gf}^2}{h_{si}^2}]$. Therefore, after substituting all of the constituents' data, one finds that $E_c\approx170.5$ GPa for the silicon nitride cantilever coated with the gold film. The average density $\rho_{\gamma c}$ of this composite cantilever is equal to: $\rho_{\gamma c}=\frac{h_{si}\times\rho_{\gamma si}+h_f\times\rho_{\gamma gf}}{h_{si}+h_f}=4196.6$ kg/m^3.

Table 8-6 shows calculated fundamental frequency for composite V-shaped cantilevers of silicon nitride coated with gold film by the following formula, which is an extended version of (5-87d) for the composite cantilever

$$f_{oc}=\frac{h_{si}}{2\pi L^2}\sqrt{\frac{E_c}{\rho_{\gamma c}}\frac{1}{0.175+0.768L_1/L}}\,. \qquad (8\text{-}5)$$

If the composite rectangular cantilever is coated uniformly on all its sides (Fig. 8-7e), then the position of its neutral axis remains in the center of the cross section and the composite Young's modulus can be estimated by the following expression

$E_c=E_{si}+2E_f[\frac{J_f}{J_{si}}+\frac{(h_{si}+2h_f)^3h_f}{b_{si}h_{si}^3}]$, where h_{si} and h_f are the thickness of the cantilever and the coating film, respectively; $E_{si}=162$ GPa for silicon cantilever, $E_{gf}=69.1\pm2.6$ GPa for a gold film, and $E_{pf}=139.7\pm2.7$ GPa for a platinum film; in addition,

$J_f=\frac{b_{si}h_f^3}{12}+b_{si}h_f\left(\frac{h_{si}+h_f}{2}\right)^2$ is the moment of inertia for the film cross section on one wide side of the cantilever, $J_f/J_{si}=h_f[h_f^2+3(h_{si}+h_f)^2]/h_{si}^3$, and b_{si} is the width of the cantilever without coating. The average density of this composite cantilever can be estimated as follows, $\rho_{\gamma c}=\rho_{\gamma si}+2\rho_{\gamma f}\frac{h_f}{h_{si}b_{si}}(b_{si}+h_{si}+2h_f)$, where $\rho_{\gamma si}=2.33$ g/cm^{3}and $\rho_{\gamma f}=\rho_{\gamma gf}=19.32$ g/cm^3 for gold film and $\rho_{\gamma f}=\rho_{\gamma pf}=21.44$ g/ cm^3 for platinum film are the density of the cantilever and the film materials, respectively. Comparison of the resonance frequency calculations with formula (5) and experimental data (M.C. Salvadori *et al.*, 2003) for a cantilever coated with gold or platinum f_c and uncoated f_u are shown in Table 8-7 and Fig. 8-8 a, b, where the experimental data are given by circles and the solid curve shows the theoretical results as a function of the gold (Fig. 8-8a) and platinum (Fig. 8-8b) film thickness h_f, respectively. The silicon cantilever dimensions in the case of gold coating are presented in the note to Table 8-7,and for the platinum coating they are as follows: L=138.8 μm, h_{si}=4.41μm, and b_{si}= 41.23 μm. The bias between the theoretical calculations by formula (5) and the experimental data is not more than 0.6 % in all cases presented in Table 8-7 and in Fig. 8-8a,b.

Table 8-7. Uncoated (f_u) and coated (f_c) resonance frequency of the cantilever

*Si coated with Au, uncoated f_u(exp)=308,745; f_u(5)=308.519 kHz			
h_f nm	** f_c (exp) kHz	f_c (5) kHz	$100\|f_c\text{ (exp)}-f_c\text{ (5)}\|/f_c\text{ (exp)}$, %
18.78± 0.4	300.849	298.4	0.53
28.55± 0.25	294.537	293.61	0.13
39.68± 0.32	288.612	288.5	0.04
54.19± 0.51	283.135	282.32	0.29
61.62± 0.40	277.756	279.28	0.55

*The silicon cantilever's for gold coating dimensions are L=136.9 μm, h_{si} =4.23 μm, b_{si}=39.38 μm; h_f is the thickness of gold coating.
** Experimental data for f_c(exp) resonance frequency are from M.C. Salvadori *et al*., 2003.

a)

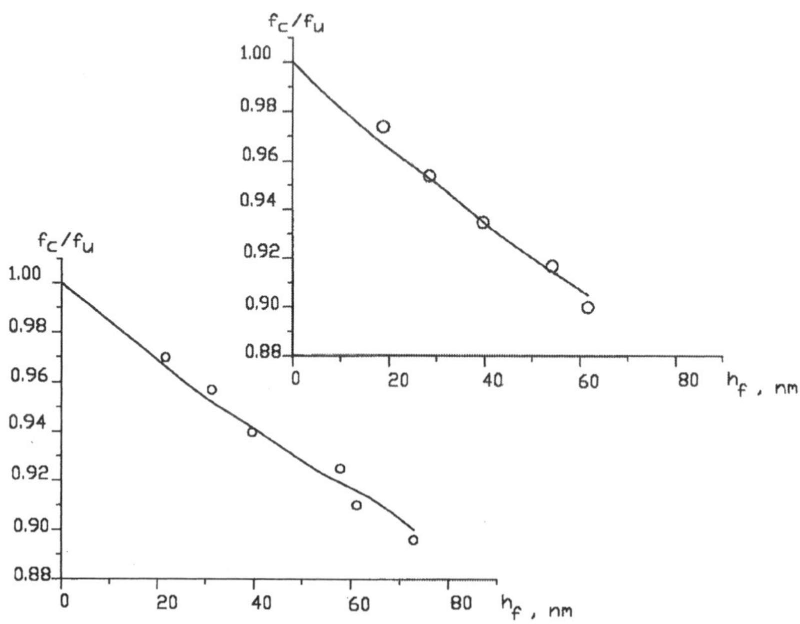

b)

Fig. 8-8. Theoretical and experimental charts of coated and uncoated cantilevers' frequencies.

The spring constants.

Bending (*flexural*). Let us now estimate the spring bending constant of a V-shaped cantilever, which has triangular part on the free end and two quasi-rectangular beams connecting the triangular part to the fixed base. If we apply a concentrated force F_c to the free end of the cantilever, each of its part develops a flexural deflection, which we can estimate separately and then summarize. One can find the deflection y_{tr} of the triangular part by integration of the following flexural differential equation,

$y''_{tr} = \dfrac{12\,F_cL(l-x/L)}{EB(1-L_1/L)(1-x/L)h^3} = \dfrac{12F_cL}{EBh^3(1-L_1/L)}$, where x is the V-shaped cantilever's longitudinal axis and B is its overall width (*see* Fig. 8-7c and Fig. 5-11c,d). The length of the V-shaped cantilever's triangular part equals $L_{tr} = L\text{-}L_1$. Therefore,

$y_{tr}=\dfrac{6L^3F_c(1-L_1/L)^2}{EBh^3}$. Hence, the spring constant of a solid triangular cantilever

equals $K_{tr}=\dfrac{EBh^3}{6L^3}$, because, in this case $L_1= 0$ (*see* D. Sarid, 1997). The deflection of the quasi-rectangular part of the cantilever with its length L_1 consists of (i) the deflection $y_{rec}(F_c)$ under action of force F_c,
(ii) the deflection $y_{rec}(M_{Fc})$ induced by the corresponding bending moment $M_{Fc} = F_c(L-L_1)$, and (iii) the angular deflection $\vartheta(F_c, M_{Fc})$ of this part, which causes an additional motion $y_{tr\vartheta} = \vartheta\, L_{tr}$ of the triangular part's end as a solid body. Let us recall that $y_{rec}(F_c) = \dfrac{2F_cL_1^3}{Ewh^3}$, $y_{rec}(M_{Fc})= \dfrac{3F_c(L-L_1)L_1^2}{Ewh^3}$, and

$y_{tr\vartheta}= \dfrac{6(L-L_1)^2L_1+3L_1^2(L-L_1)}{Ewh^3}F_c$. Finally, we find the spring constant of the V-shaped cantilever with a trapezoidal or rectangular recess $K_{vsh} = F_c / [y_{tr}+ y_{rec}(F_c)+ y_{rec}(M_{Fc}) + y_{tr\vartheta}]$ as a result of all shown deflection constituents by the following expression,

$$K_{c\,vsh} = \frac{Ewh^3}{L^3}\left[\frac{6(1-\varsigma_L)^2}{B/w} + 2\varsigma_L(\varsigma_L{}^2-3\varsigma_L+3\right]^{-1}. \qquad (8\text{-}5a)$$

Here we denote the ratio $L_l/L = \zeta_L$ for the more convenient presentation of this and other related formulae. If we assume that for the rectangular cantilever without a recess $\zeta_L=1$ and $w=b/2$=B/2, then its spring constant calculated with expression (5a) equals the common value for the rectangular cantilevers, $K_{crec} = EBh^3/\,4L^3$. Expression (5a) works for the triangular cantilever without a recess as well if we assume that in this case $w=b/2= B/2$ and ζ_L=0. The spring constant of the trapezoidal part of the V-shaped cantilever with a recess may be similarly estimated on the basis of the following approximate formula,

$$K_{c\,trap}=\frac{EBh^3m_B^3}{6L^3[1-(1-m_B)^2](1-m_B\varsigma_L)^2}, \qquad (8\text{-}5b)$$

where $m_B = 1\text{-}\ B_1/B$, $B_1/B \ll 1$, and B_1 is the width of the trapezoid's free smaller end. If we assume that this cantilever is a triangular one and does not have a recess, then B_1=0, ζ_L =0, m_B =1, and the formula converges to the expression for the spring constant of a common triangular cantilever K_{tr} shown above. One can use the same spring constant formula for the triangular cantilever with the probe shifted by δ_L from its sharp end. In this case, $m_B = 1\text{-}\ \delta_L/\,L$. A more accurate estimation of the spring constant for a trapezoidal cantilever or a triangular cantilever with a probe shifted by δ_L can be found on the basis of the following differential equation for the cantilever's free end or the shifted probe displacement under applied force F_c,

$y''=\dfrac{12F_c(L-x-\delta_L)}{EBh^3(1-x/L)}$, with the boundary-value conditions on the cantilever's fixed end y'=0 and y=0 at x=0. The solution of this equation leads to the following expression for the spring constant of the above mentioned cantilevers,

$$K_{c\,tr\delta}=\frac{F_c}{y}=\frac{EBh^3}{6L^3(m_L^2L^2)}\{1+\frac{2(1-m_LL)}{m_L^2L^2}[m_LL+(1-m_LL)\ln(1-m_LL)]\}^{-1}, \tag{8-5c}$$

where $m_L L$=1-δ_L/L, and for a V-shaped cantilever with a recess, the width B should be multiplied by (1-ζ_L). Expression (5c) is presumably more accurate than (5b) but it is more unwieldy. This expression at δ_L=0 converges to the common formula for the spring constant of a triangular cantilever. Similar unified formula for the spring constant of a trapezoidal cantilever without a recess has been derived by D. Sarid (1997) who shows that his formula is applicable to a V-shaped cantilever without a recess, where B_1=0, and for rectangular cantilever, where B_1=B. Unfortunately, Sarid's formula is also an unwieldy one and is difficult to digest. However, if we assume that 1-m_LL=δ_L/L<<1, then at the approximation with the series expansion ln (1-m_LL) ≈-m_LL, the member of (5c) in braces significantly simplifies and becomes equal to 1+2 (1-m_LL). Therefore, after substitution of the latter expression into (5c), we have a more accurate equation than (5b), and a digestible approximation for the bending spring constant of a trapezoidal or triangular cantilever with the probe shifted by δ_L = ΔL. This expression is as follows $K_{c\,tr\delta}\approx\frac{EBh^3}{6L^3(m_LL)^2[1+2\,(1-m_LL)]}$.

Torsional. The torsional spring *constant* $K_{c.tors}$ for a thin rectangular cantilever may be found from the well-known expression $K_{c.tors}$= $GLbh^3/3$.

Calibration of AFM rectangular microcantilevers' torsional spring constant may be provided on the basis of the following (*see* A.E.H. Love, 1944) formula

$K_{c.tors}=3^{-1}\pi^{-2}\rho_\gamma bh^3L\,f_{t,vac}^2$, where $f_{t,vac}$ is the fundamental frequency of the cantilever torsional vibration in vacuum. This formula, however, is of limited use due to the difficulty in measuring both the cantilever mass ($\rho_\gamma bhL$) and resonant frequency in vacuum.

Formulae for torsional spring constant of V-shaped cantilever are presented by J.M.Neumeister and W.A.Ducker (1994) and also by J.E. Sader *et al.*(2003) in the form, which shows the complexity of this problem. The approximate solution was found by combining of torsional spring constant for the two inclined quasi-rectangular cantilevers K_{tors}^{q-rec} (parallel beam approximation), which undergo simultaneous bending and twisting under applied torque on the axis of symmetry close to the vertex of V-shaped cantilever with a small shift ΔL (Fig. 8-7a, c), and the torsional spring constant of the triangular part of V-shaped cantilever K_{tors}^{trian}. The resulting torsional spring constant K_{tors}^{V} for V-shaped cantilever is equal to:

$$K_{tors}^{V}=\left(\frac{1}{K_{tots}^{q-rec}}+\frac{1}{K_{tors}^{trian}}\right)^{-1}. \tag{8-6a}$$

One can find the torsional spring constant for the rectangular part of the "parallel beam approximation" by applying torque M_{tq} in the section of V-shaped cantilever with x=L_1, i.e., on the end of the recess. This torque produces bending forces Q_i and twisting moments M_i in each of the inclined by angle $\tan^{-1}(B/2L)$ strips with width w and distance between their axes $\left(B\frac{L-L_1}{L}-w\right)$.

Therefore, $Q_i=\dfrac{M_{tq}-2M_i/\cos\tan^{-1}(B/2L)}{(B\frac{L-L_1}{L}-w)}$ and $K^{q-rec}_{tors}=\dfrac{Ewh^3(B\frac{L-L_1}{L}-w)\Re(B)}{8L_1^3}$,

where $\Re(B)=(B\frac{L-L_1}{L}-w)+\dfrac{16L_1^2\cos\tan^{-1}(B/2L)}{6(1+\nu_p)(B\frac{L-L_1}{L}-w)}$. (8-6b)

The torsional spring constant for the triangular part of V-shaped cantilever is equal to: $K^{trian}_{tors}=\dfrac{Eh^3(B/2L)}{3(1+\nu_p)}(\log\dfrac{w}{\Delta LB/2L})^{-1}$ (8-6c)

at $L_1 = L(B\text{-}2w)/B$ (triangular recess) because torsional section rigidity I_{tors} of the thin bar with constant thickness h and varying width

$b(x_1)=(B\frac{L-L_1}{L})(1-\frac{x-L_1}{L-L_1})$ approximately equals $I_{tors}\approx\dfrac{Gb(x_1)h^3}{3}$, where $x_1=x\text{-}L_1$.

If we recall that modulus $G=E/2(1+\nu_p)$, the angle of rotation for this triangular bar under torque M_{tr} shifted from its vertex by ΔL is

$\theta_{tr}=\int_0^{L-L_1-\Delta L}\dfrac{6(1+\nu_p)M_{tr}dx_1}{Eh^3b(x_1)}$. The latter integral yields logarithm as a result of the integrand with the variable x_1 on the first power in the denominator.

The approach to the problem discussed here as well as in referenced sources is indeed a rough approximation for very small strains. However, more accurate calculations will not provide significant progress because of approximate knowledge for the cantilever's real tiny geometric and anisotropic elastic parameters. Meantime, expressions (6b,c) correctly show the main functional relationships for V-shaped cantilever's constituents that are useful in its structural synthesis for precision elastic systems, e.g., large relative influence of the cantilever thickness, effect of Poisson's ratio, etc. The necessary accuracy in this case can practically be achieved only on the basis of a traceable calibration and adjustment, including the microcantilever laser bending for high-precision curvature change.

Effective mass factor. Let us recall that the resonance frequency f_0 of the cantilever may be presented through its bending spring constant $K_{c\,t}$ and its effective mass as follows, $f_0=\dfrac{1}{2\pi}\sqrt{\dfrac{K_{ct}}{n^*m_c}}$, where m_c is the cantilever's mass and n^* is a mass factor which depends on the cantilever's shape. The mass of the V-shaped cantilever may be presented as $m_{cn} = Lh\,\rho_\gamma[\frac{1}{2}B(1\text{-}\zeta_L)^2+2w\zeta_L]$ and the mass factor n* can be found from the equation $n^* = K_{ct}/[m_{cn}(f_0/2\pi)^2]$. Substituting the expressions for spring constant, mass of cantilever, and its fundamental frequency found through (5-87d), we obtain, after simple cancellations, the following quasi-unified formula for the V-shaped cantilever's mass factor, which is equal to:

$n^* = \dfrac{2(0.175+0.768\varsigma_L)}{[4\varsigma_L+(1-\varsigma_L)^2 B/w][\dfrac{6(1-\varsigma_L)^2}{B/w}+2\varsigma_L(\varsigma_L^2-3\varsigma_L+3)]}$. This expression for the rectangular cantilever with ζ_L=1 yields a commonly used value, n^*=0.24.
If $\zeta_L < 1$ for the triangular cantilever with a trapezoidal recess, then the mass factor n^* becomes significantly smaller than 0.24.
Sensitivity to additional mass. Let us now estimate the shift of the resonant frequency for the system of a cantilever with an additional mass Δm clamped to its free end (probe, specimen, etc.). In this case, we have to represent the shift $\Delta f_0 = f_0(m_c) - f_0(m_c + \Delta m)$ of the resonance frequency $f_0(m_c)$ in the system with the cantilever, which has mass m_c , and the resonance frequency $f_0(m_c + \Delta m)$ with the additional mass Δm attached to the cantilever. It is convenient first express the difference of these frequencies' squares through the bending spring constant K_{ct} of the cantilever and the corresponding masses as follows,
$f_0^2(n^* m_c) - f_0^2(n^* m_c+\Delta m) = (\frac{1}{2\pi})^2(\frac{K_{ct}}{n^* m_c} - \frac{K_{ct}}{n^* m_c+\Delta m})$. If $(\Delta m/m_c) << 1$, the following approximation is helpful,
$f_0^2(n^* m_c) - f_0^2(n^* m_c+\Delta m) = \Delta f_0\,[f_0(n^* m_c)+f_0(n^* m_c+\Delta m)] \approx 2f_0\Delta f_0\,(n^* m_c)$.
Therefore, we can find the simplified expression for the shift of frequency Δf_0 in the form, $2\pi\,\Delta f_0 = \frac{K_{ct}}{4\pi f_0}\frac{\Delta m}{n^* m_c(n^* m_c+\Delta m)}$ or $\Delta f_0 / \Delta m \approx 2.08 f_0 / m_c$ for the rectangular cantilever. A similar expression for a V-shaped cantilever has a larger numerical factor before the ratio f_0/m_c as a result of lesser mass factor n^*. If we assume that, for a V-shaped cantilever, n^* is in the range from 0.135 to 0.2 (*see* G.Y. Chen *et al.*, 1994), then this cantilever has a larger frequency shift $\Delta f_0/\Delta m \approx (3.73-2.5) f_0 / m_c$, and therefore, may be considered more sensitive to change of an additional attached mass. Effective mass factor for a doubly clamped prismatic beam is equal to 0.735.

The higher resonance frequency modes, the higher the sensitivity of cantilever detector to the additional mass on its moving end. However, its sensitivity to the additional mass at the corresponding nodal points is close to zero. We can show this on the basis of known formula for the resonant frequency $f_n = \frac{1}{2\pi}\sqrt{k_{neff}/m_{neff}}$ in n mode by evaluation of corresponding effective spring constant, k_{neff}, effective mass, m_{neff}=n^*_nm, and effective mass factor n^*_n. Let us assume that vibrating at higher mode cantilever can be modeled as an indeterminate beam to the degree which is equal to the number of nodal points (virtual removable supports). Then the virtual redundant forces W_j (see Fig. 8-7f) in the nodal points can be estimated with the force superposition method by matrix equation such as $[d_{ij}][W_j]=[d_{iw}]$, where d_{ij}, d_{iW} are the flexibility coefficients, W is the quasi-end force at the coordinate $L^*=Lq$ with $q<1$, the first index i=1to (n-1) corresponds to the number of nodal point, and the second index j=1to (n-1), w corresponds to the acting virtual force. The solution of this matrix equation in respect to redundant forces may be obtained with the inverse matrix $[d_{ij}]^{-1}$ as follows: $[W_j] = -[d_{ij}]^{-1}[d_{iw}]W$, where $d_{ii}=\frac{(x^{**}_{in})^3}{q^3}$; $d_{ij}=d_{ji}$; $d_{ij}=\frac{(x^{**}_{jn})^3}{2q^3}(3x^{**}_{in}-x^{**}_{jn})$ at $i<j$; $d_{iw}=\frac{(x^{**}_{in})^2}{2q}(3-\frac{x^{**}_{in}}{q^2})$.

Note that all coordinates marked with two asterisks are equal to corresponding nodal points' coordinates [derived from (4a)] divided by the length of cantilever L that makes them dimensionless. Now one can estimate the deflection at the cantilever's quasi-end offset section caused by superposition of forces W, W_1, W_2, W_j action which is equal to: $s=s_{qend}(W)-\sum_j^{(n-1)} s_{qend}(W_j)=A_{qn}\frac{L^3}{3EJ}$,

where $s_{Wqend}=\frac{WL^3q^3}{3EJ}$; $s_{W_jqend}=\frac{W_j(x^{**}_{jn})^2L^3}{6EJq}(3-x^{**}_{jn}/q^2)$. Substituting these expressions, we obtain the total deflection in the offset moving end section and the effective statical spring constant for the cantilever in the n-th mode of resonant oscillations as $K_{neff}=\frac{3EJ}{A_{qn}L^3}$. This allows us to estimate the ratio between the effective spring constant K_{neff} at n resonant mode and the same parameter $K_{1eff}=3EJ/(L^3)$ for the first resonant mode of prismatic cantilever that equals $K_{neff}/K_{1eff}=A_q^{-1}$. Therefore the ratio between effective mass factor for the n-th mode and the first mode of oscillations is $n^*_n/n^*_1=\frac{f_1^2}{f_n^2}K_{neff}/K_{1eff}=\frac{(\alpha L)_1^4}{(\alpha L)_n^4}A_{qn}^{-1}$.

Let us recall that $n^*_1\approx0.24$ for prismatic rectangular cantilever. Therefore, the effective mass factor for n-th mode of oscillations of the same cantilever is equal to:

$n^*_n=0.24\frac{(\alpha L)_1^4}{(\alpha L)_n^4}A_{qn}^{-1}$. Hence the ratio between sensitivity to additional mass Δm at the n-th mode of resonant oscillations and the first resonant oscillation is as follows: $\frac{\Delta f_n(\Delta m)}{\Delta f_1(\Delta m)}=\frac{f_1(K_{neff}/K_{1eff})n^*_1(n^*_1m_c+\Delta m)}{f_n n^*_n(n^*_n m_c+\Delta m)}\approx\frac{f_n n^*_1}{f_1 n^*_n}$, where m_c is the mass of the cantilever. As a result, the sensitivity to additional mass at the higher mode equals

$\Delta f_n/\Delta m=\frac{f_n n^*_1}{f_1 n^*_n}\Delta f_1/\Delta m=\frac{(\alpha L)_n^2 n^*_1}{(\alpha L)_1^2 n^*_n}(\Delta f_1/\Delta m)$. It is clear that at the nodal points $K_{neff}=\infty$, $n^*_n=\infty$, and $\Delta f_n/\Delta m=0$. See Y.M. Tseytlin (2005) for more detail. See also M.B. Bauza et al., (2005), *Rev. Sci. Instrum.*,095112 with application of vibrating shank virtual probe.

Analysis indicates that nanomechanical resonators offer immense potential for mass sensing ultimately with the resolution at the level of individual molecules. A mass responsivity of approximately 5 fg/ Hz has been observed for micrometer sized cantilever when operating the cantilever in the fourth mode. The largest attainable natural frequency of the triangular cantilever is higher than that of the rectangular one with the same dimensions (h, L) and of the same material (*see* section 5-4).

Internal and external damping.

It was found in the experimental study that Q-factor of silicon triangular cantilevers and their arrays with length from 500 nm to 100μm and thickness from 30 to 100 nm is as low as 5 in air and as high as 10^4 in vacuum. Typical mechanical characteristics of such cantilevers include bending spring constant within 8-15 N/m and natural frequency within 4 -10 MHz. Damping effects (*see* G.Y. Chen *et al.*, 1994) in a gas insignificantly (within 0.2-1.5 %) change the resonance frequency of rectangular and V-shaped cantilevers, but their quality factor changes up to 280% for rectangular cantilevers and up to 230% for V-shaped cantilevers in argon, air, and helium media. However, the natural frequency of V-shaped cantilevers can change within 180% in different liquids such as hexane, ethanol, water, and hexadecane.

The cantilever's resonant frequency depends on the *Q*-factor as follows

$$f_{0,Q}=f_0\frac{\sqrt{Q_{fac}^2-1/4}}{Q_{fac}}, \tag{8-7}$$

where $Q_{fac}=m\omega_0/C_y$, $f_0=\omega_0/(2\pi)$is the cantilever's fundamental frequency without damping [*see* expressions (5-87d)and (5)], *m* is an effective mass, and $C_y=2C_y m$ is damping coefficient of the cantilever (*see* equation 9 in Table 1-10). If $Q_{fac}>1$, then $f_{0,Q}\approx f_0[1-1/(8Q_{fac}^2)]$. D.M. Photiadis *et al.*, 2004, predicted that contribution to the loss factor for the fundamental mode of a typical cantilever beam attached to the substrate is less than $Q_{fac}^{-1}\leq 10^{-7}$. Therefore, the corresponding change in the fundamental resonant frequency of the cantilever from the attachment losses in this case is less than $(1/8)10^{-14}f_0$. More accurate predictions may be done with the formulae for Q_{fac}^{-1} presented in the referenced source, where is shown its dependence on the width, thickness, length of the oscillating cantilever, and the thickness of corresponding substrate in relation to the elastic oscillations wavenumber. In order to overcome the difficulties of resonant cantilever operation in liquids, cantilever transducers are used as a part of self-oscillating system with the positive feedback. Enhancement of the quality factor via driving force control is useful at frequency modulation and self-excitation techniques (*see* T.R. Albrecht, 1991). Polymer microcantilevers may have larger deformations without fracture. Their quality factor is in the range from 9 to 30 at the resonant frequencies from 2 to 10 kHz. Tensile test of polymer micro- and nanoscale fibers shows that their stress-strain relationships can be nonlinear (in some cases with strain in third power).

Ambient influence.

Thermal and ambient-induced (relative humidity, exposure to mercury vapor) deflections of *AFM* cantilevers may cause change of their resonance frequency as well (*see* T. Thundat *et al.*, 1994; S. Jeon *et al.*, 2004). Detailed analysis of the composite multilayer cantilever beams (silicon substrate with oxide, nitride, and polymer dielectric layers) is done by D. Lange *et al.* (2002) whose research of the temperature, pressure, and *Q*-factor influence on the cantilever's resonant frequency is informative. The combined inverse quality factor Q_{fac}^{-1} then is equal to the sum of inverse internal and external quality factors : $Q_{fac}^{-1}=\sum_i Q_i^{-1}$.

Temperature influence revisited. Accepted by D. Lange *et al.* (2002) models are helpful and satisfactory enough, except for the model of the temperature influence, in which the Poisson's ratio assumed to be constant with the changes of temperature. This is correct only for the isothermal conditions. The adiabatic moduli of elasticity differ from the isothermal one (L.D. Landau and E.M. Lishitz, 1970), and the Poisson's ratio in this case depends on the temperature (*see* Sect. 6-3). A deformation is adiabatic if there is no exchange of heat between the various parts of the body or between the body and the surrounding medium. Therefore, the resonant frequency model in more general temperature conditions should be represented as follows

$f(T) = \frac{\lambda_n^2}{\sqrt{12}}\sqrt{\frac{E(T)\, h(T)^3 b(T)}{[1-\nu_p(T)^2]\, M_g\, L(t)^3}}$, where λ_n is the constants of a frequency equation

for the rectangular cantilever (*see* note to Table 5-4) with $\lambda_{n=1}$= 0.1875, $\lambda_{n=2}$=4.694, $\lambda_{n=3}$= 7.855, $\lambda_{n=4}$= 10.996, $\lambda_{n=k>4}$ = π (2k-1)/2; M_g is the total mass of the cantilever, $h(T)$, $b(T)$ are the cantilever temperature dependent thickness and width;
$E(T)=E(T_n)[1+\alpha_E E(T-T_n)]$, $\nu_p(T)=\nu_p$ at isothermal conditions;
$E(T)=E_{ad}(T)=E(T_n)+E^2T\alpha_L^2/9C_p$, $\nu_p(T)=\nu_{pad}(T)=\nu_p+(1+\nu_p)ET\alpha_L^2/9C_p$
are the temperature dependent adiabatic Young's modulus and Poisson's ratio, respectively. C_p represents the specific heat of elastic material (*see* Table 7-6). However, the difference between isothermal and adiabatic Young's moduli for most metals is less than two percents at normal temperature.

Actuators and detectors. For cantilever probe to oscillate with sufficient amplitude (up to 100 nm), its resonant frequency should be supported by an actuator: bimorph ceramic/metal, piezotube, piezo-stacks.

Cantilever oscillations are conventionally monitored optically by focusing a laser ray onto the small and curved cantilever, which reflects the ray onto a photoreceptor. The optical feedback from light scattering can be reduced with high frequency laser current modulation. The advantage inherent in miniaturizing scanning probe microscopes has led to an AFM array probe with integrated deflection sensor and amplifiers for signal readout all combined on the same chip. In addition, the magnetovibrational coupling in small dielectric (such as silicon) cantilevers with a single-domain ferromagnetic film deposited at free end was studied. M.Vogel *et al.* (2003) demonstrated optically tunable mechanics of microcantilevers. Cantilever in the liquid should be isolated by parylene, silicon nitride film, etc.

"Minute size and mass of microfabricated cantilveres makes them susceptible to thermally induced noise…Therefore, cantilever sensor may operate in the resonant mode either with or without external excitation" (N.V. Lavrik *et al.*, 2004). The thermal frequency fluctuation of the oscillating cantilever is considered (T.R. Albrecht *et al.*, 1991) equal to:

$\frac{\delta f_{th}}{f_0} = \sqrt{\frac{k_B T B_f}{2\pi k_s \left\langle A_C^2\right\rangle f_0\, Q_{fac}}}$ and force noise is $\delta F_{th} = \frac{2k_s\sqrt{\left\langle A_C^2\right\rangle}}{f_0}\delta f_{th}$,

where k_s is the spring constant; B_f and $\left\langle A_C^2\right\rangle$ are the bandwidth and mean-square amplitude of thermal oscillations, respectively, and Q_{fac} is the cantilever's Q-factor. Elastic boundary conditions influence on the microfibers' oscillation at clamping by glue with different stiffness is discussed in X. Chen *et al.*, 2004, publication.

Calibration. A variety of methods have been proposed to determine the cantilever spring constant. The two most commonly employed methods are the thermal noise method and the added mass method, which determine the spring constant for a point load at the end of the cantilever. These methods are applicable for the calibration of bending and torsional spring constants. Some authors show that both constants can be calibrated simultaneously. It may be true from the practical standpoint, but it has some theoretical concerns because twisting of the cantilever changes its bending spring constant as in the pretwisted strips (*see* Chap. 2 and 5).

Recently a number of authors have proposed methods of calibration AFM cantilever spring constant involving the hydrodynamic drag of the cantilever or a particle attached to a cantilever. The force acting on the cantilever is equal to $F_{dr}=\frac{1}{2}c_d \rho_\gamma v^2 Lw$, where c_d is the drag coefficient, ρ_γ is the density of fluid, v is its velocity, L and w are the length and width of the cantilever.

A microfabricated array (MARS) of reference spiral-springs (P.J. Cumpson *et al.*, 2003) has limited options for the calibration of AFM cantilevers' spring constant, which is more than 0.05 N/m. Spiral spring-spoke (Fig. 8-9f) elastic behavior is a complicated one and similar to that in the curved beam, which is loaded perpendicular to its plane (*see* Table 3-8, schematics 4-6). The precision limitation of this calibration device for application in measurement of nano-newton and pico-newton forces may be imposed by vacuum fluctuations (Casimir force) and van der Waals molecular forces (attraction change in inverse ratio to the seventh power of distance between molecules) acting on the spiral springs as well.

Notch flexure hinges application.

They are especially effective in the connection with piezoelectric actuators, which are becoming increasingly popular. Because of their low displacement and high force outputs, piezoelectric actuators commonly coupled through flexure-hinged mechanical displacement amplifiers: bridge-type, lever-type (*see* Table 4-5), and others. Geometrical transmission ratio i_{brg} of a bridge-type amplifier with three notch hinges 1, 2,and 3 can be derived on the basis of simple trigonometric considerations (see Chap. 4 and Fig. 8-9a) and presented as follows: $i_{brg} \approx \Delta B_1 / \Delta L_1 \approx 1/\tan\psi_\alpha$. If the angle $\psi_\alpha \leq 0.1$ is small, we can assume that $i_{brg} \approx 1/2\,\psi_\alpha$ and therefore the transmission ratio is supposed to be approximately 10 in the bridge system with ψ_α =0.05. However, the experimental data show that the largest real amplification ratio in such a system is not more than 3.5. This result is due to the flexure hinge deformation in the parasitic direction that was also proved by the matrix calculation method. Let us recall that the planar cable mechanism PM-IV (*see* section 4-2) has a similar limitation. To achieve a larger amplification without increasing the length of the amplifier, a three-dimensional bridge-type hinge mechanism has been developed (J.H.Kim *et al.*, 2003) whose amplification is approximately equal to 10 as a result of motion sequential transmission through two bridge-type amplifiers (Fig. 8-9b) with notch hinges (depicted in accordance with J.H. Kim *et al.*, 2003). Notch hinge design can have motion side limiters 1(Fig. 8-9c) to prevent hinge overloading in rotation.

The dynamic properties of a monolithic mechanism with notch flexure hinges for precision control of orientation and position were discussed in detail by M. Tanaka as early as 1983.

Micromechanical arrays

The micromechanical oscillator array 2, 3 (Fig. 8-9d) that was fabricated with a 300 nm thick silicon-nitride film 4 on a silicon substrate 1 with up to 248 cantilevers of two alternative lengths 50/55 μm and pitch 40 μm was prepared and used for study of intrinsic localized vibrational modes. Similar cantilever arrays may be used for other scientific studies.

a) b)

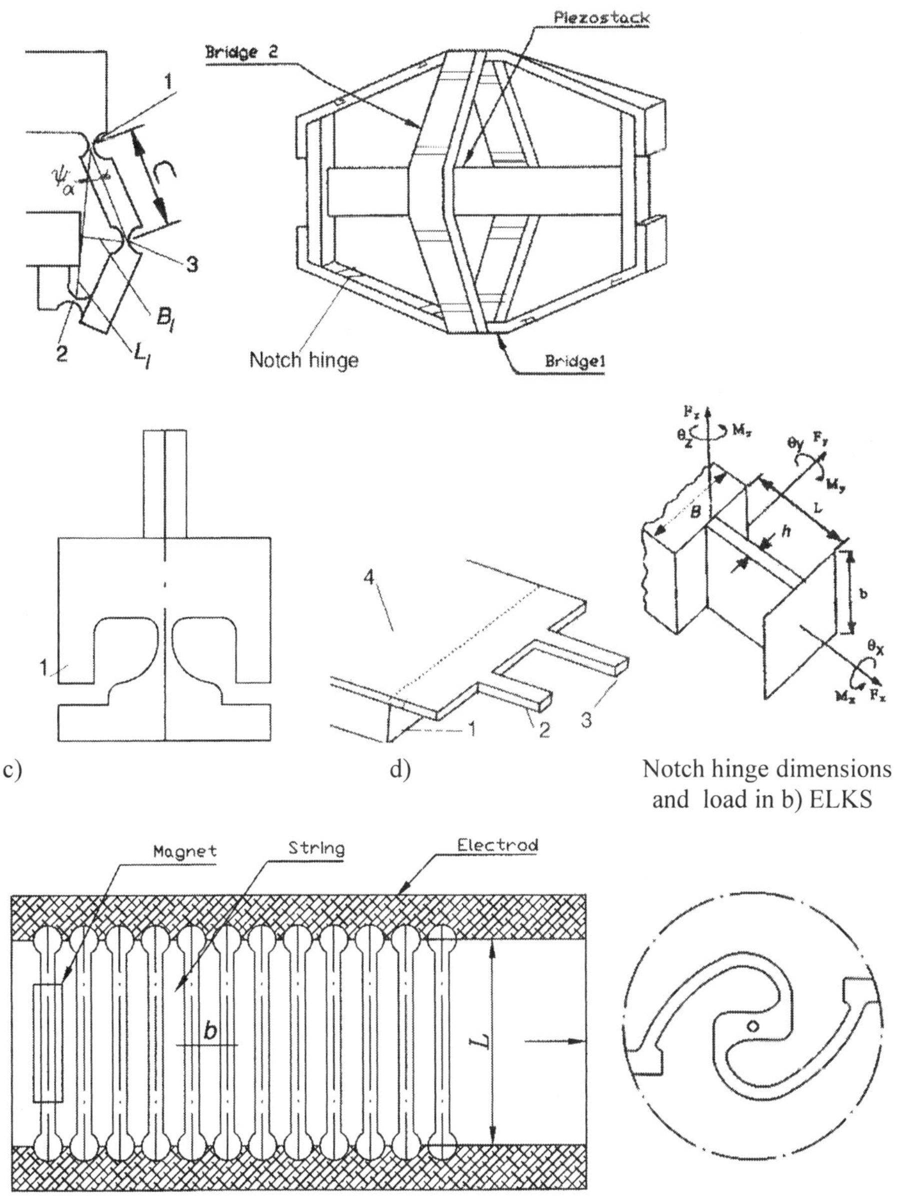

c) d) Notch hinge dimensions and load in b) ELKS

e) f)

Fig. 8-9. Microstructures with notch hinges, cantilevers, and spiral spokes.

The linear absolute and incremental encoder-transducers with the array of microstrings (Fig. 8-9e) were proposed by the author of this book in 1997 as a research development. Each string vibrates in the field of magnet which position corresponds to the comparator's dimension. The feedback current comes from self-excited oscillator, which is connected with strings through electrodes deposited on the substrate.

8-5. Effective Models for Precision Applications

In recent years many flexural mechanisms for ultraprecision manipulation and new precision measuring systems with elastic sensors are developed. Their resolution is evaluated by the analytical and experimental methods. Let us discuss the main principles of the structural synthesis in precision elasticity that will help further progress in this important field.

Principles.

The approximate order, developed methods of the structural synthesis, and main synthesized features are shown in Table 8-8 and Fig. 6-8b.

Table 8-8. Structural synthesis in application to precision elastic transmission system

Step #	Specified values	Synthesized features	Expressions, developed methods, limitations
1	Target, workout quantity &values, structure size limits, transmission ratio i_{ELKS}	Structural diagram, equivalent schemes, limits of elements, their dimensions and transmission ratios *of straight i_{st} and feedback i_{efbk} chains, ELS, EKS, ELKS scheme initial structure	Analogy and sequential approach, minimum rigid joints with structural damping, functional chain matrix equations * $a\sum n_{ij}$, Π-theorem ***satisfaction
2	Dynamic rate of the system	Elements' masses, damping, dynamic loads, effective mass factor	Transient $\theta_e(t)$, $h_e(t)$, transfer functions, differential-integral equations, unstable zones elimination**
3	Operating range of elastic elements	Dimensions, material^^, shape's stability of main scheme's elements with elimination of bifurcation, strain-stress evaluation	^Elastic members' functions approximation: springs, hinges, and other involved elastic links' strain and stress estimation by accurate and approximating methods, material strength features and endurance limit satisfaction
4	Environmental conditions, influence quantities	Condition's correctors for main influence quantities compensation, adjustment limits prediction, temperature influence minimization**** $\Delta L_T = \sum_{i=1}^{n} \Delta L_{Ti}(t) \to 0$	Influence functions topogram identification, matrix condition number K_c with $1+\eta_U$ upper limits yielding, functional bifurcation application ****

Table 8-8 Cont'd.

Step #	Specified values	Synthesized features	Expressions, developed methods, limitations
5	Driving force Input Output	Contact/non-contact force action, estimation of driver features and driven options	Force compensators ** installation, input force uncertainty evaluation, contact strains correlation
6	Accuracy (errors, uncertainty) limits	Trajectories, transmission ratios'uncertainty, and functional nonlinearity minimization *** $\eta_U = e^{H_{\ni}(x)} / \delta_p$ $\int P_n(x)\, dx = 0$; $\delta i_{ELKS} \leq \eta_U$ * $\xi_{sc} - i_{sc}\, i_{ifbe}\, \xi_{sifbe} = 0$	Correctors: number, position, range evaluation with criterion of negligible errors, dimensions with tolerances of precision elastic elements estimation, systems of functional equations with minimal series polynomial expansion, and the principle of energy similarity satisfaction^
7	Precision (repeatability limit)	Internal and structural friction compensation * $D_\xi = \dfrac{D_{\xi sc} + i_0^2 D_{\xi ifbe}}{(1+i_0)^2}$	Damping minimization , straight and equivalent feedback chains combination with criterion of negligible uncertainty limitation^

* Chap. 1; ** Chaps. 4, 5; *** Chap. 6; **** Chap.6, 7; ^ Chaps. 2-5; ^^Materials with larger ratios E/ρ_γ and σ_r/ρ_γ are preferable.

Reader can find the similar general steps of mechanism design analysis and synthesis in A. G. Erdman *et al.* (1984) publication, where is shown that the synthesis is the iteration process, and is not a one-way, single-pass effort.

Problems.

Let us discuss several methodical examples of problems, which are useful for the clear understanding of the developed methods' application and procedures of precision elastic systems' analysis and synthesis.

Fig 8-10 shows structural problems for: a) elastic strip motion with nonlinear relationships; b) axis and shear planes for piezoelectric material (after A.H. Slocum, 1992), c) vibro-contact system; and d) vibrating rod with a node-point a and the lug of concentrated mass m^*.

Problem 1. Let us discuss one of the methods for the spring parallelogram synthesis in nonlinear strain-stress zone based on the "exact" theory and polynomial expansion. The rigid carriage hanged on two initially flat steel strips translationally moves under the action of force $2Q_1$, which has an angular coefficient of similarity $\varsigma_0' = 90°$. It is necessary to estimate the lateral towards the strips s_1 and longitudinal in the direction of strips axes s_2 displacements of the carriage under force $2Q_1$ in the range, which corresponds to $Q_1 \in [Q_{10}; Q_{1m}]$. The calculation schematic for one of the strips is shown in Fig. 8-10a. We can obtain the approximate relationship $s_1 = f(\eta_\alpha; Q_1)$ on the basis of expression (2-8). In this case, with the approximating polynomial of the second degree $R_2(x)$, we have $Q_{1s} = 0.8284\ Q_{1m}$, $Q_{1r} = 0.4142\ Q_{1m}$, $Q_{1c} = Q_{1m}$, $Q_{1m} = 8$ N; $Q_{10} = 0$, $Q_2 = 0$. We calculate displacements s_1 of the strip end under forces Q_{1s}, Q_{1m}, and Q_{1c} by the accurate Popov's method for "inflection" fundamental class with the table of elastic parameters (*see* Table 8-9).

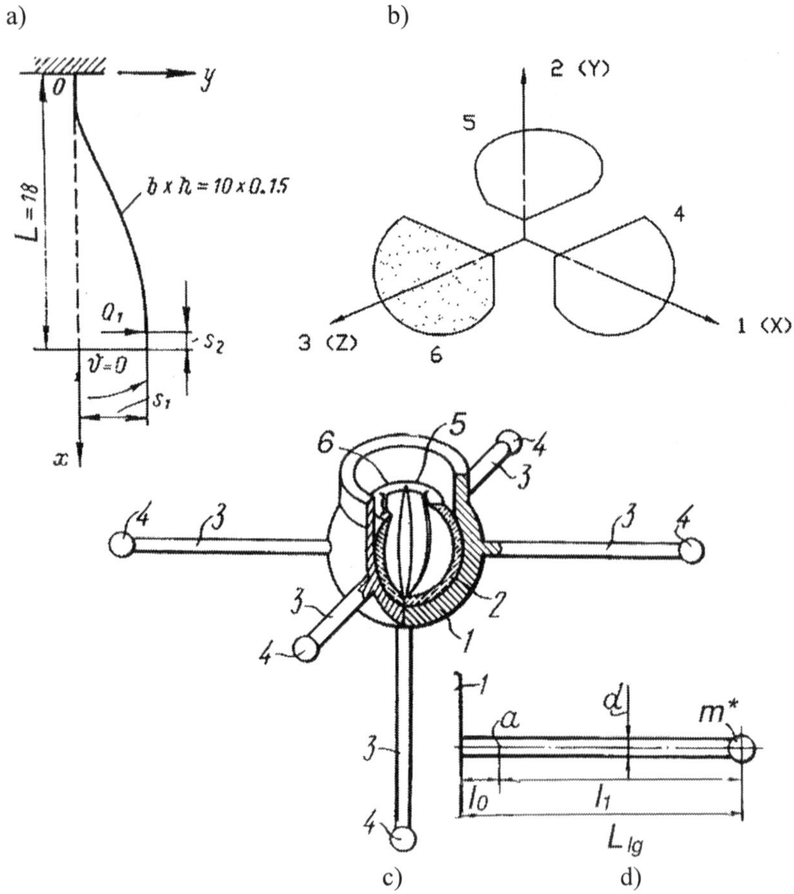

Fig. 8-10. Structural models for problems 1and 2.

Table 8-9. Calculation of the displacements $s_1$under $Q \in [0; 8\ \text{N}]$ with the accurate method

ς_0'	Q, N	$\varphi°$	α_k^o	λ_0''	s_2, mm	s_1, mm
90°	Q_{1s}=6.5	57.5	57.5	0.96	0.25	4.68
90°	Q_{1r}=3.25	65	50.8	0.68	0.09	2.32
90°	Q_{1c}=7.8	55	58.89	1.06	0.335	5.76

As a result, we obtain the equation for the estimation of the dimensionless parameter η_α as $4.68 - \eta_\alpha \dfrac{Q_{1s}\, L^3}{12\, H} = 0,$ (8-7a)

where H is the flexural rigidity of the strip's cross section. Substituting the values of the known quantities, we obtain $\eta_\alpha = \dfrac{4.68}{5.63} = 0.83\,.$ (8-8)

Thus, the sought nonlinear displacements s_1 under change of force Q_1 in the range from 0 to 8 N can be found (Y.M. Tseytlin, 1972) on the basis of the following linear relationship $s_1 = 0.83\dfrac{Q_1 L^3}{12\,H}$. (8-9)

Let us verify the condition (2-13), which proves the rightful choice of the approximating polynomial by inequality $\dfrac{|\gamma(Q_{1r})|-|\gamma(Q_{1c})|}{2}=\dfrac{0.1-0.06}{2}=0.02$ mm$\leq\{\Delta s\}$.

Therefore, the largest error $\{\Delta s\}$ of calculation with the proposed method can be equal 0.02 mm. If this error is permissible, then the condition (2-13) is satisfied and the choice of the polynomial is right. The relative error of the approximate s_1 calculation is not more than 1.73% when the largest driving force Q_{1m} =8 N.

Calculation of displacement s_1 on the basis of common linear bending strain theory in the corresponding region yields an error of 20% (*see* Y.M. Tseytlin, 1972). We see the same deflection from the common linear theory in the stiffness expression, which follows from (9)

$$Q_1/s_1 = 1.2\frac{12H}{L^3} \quad (8\text{-}10)$$

One can derive a similar formula for the stiffness of cantilever at large deflections from (13α) if to assume that the elastica in the figure (8-10a) consists of two cantilevers with length l=1/2L. In this case, we have

$Q/s = 1.2\dfrac{3H}{l^3}$, instead of $Q/s = 3H/l^3$ in accordance with common linear

mechanics for small cantilever's deflections. A similar 20%increase in stiffness (spring constant) with increase of bending forces and flexural displacement was found in the experiments on fabricated ultrashort nanocantilevers with length of 2μm, width of 150 nm, and thickness of 50 nm.

The more complicated approach is to yield the approximate relationship for longitudinal displacement s_2. We will use the equation (2-9) for this purpose because the absolute values of s_2 are small. Let us consider Q_{10} =0.4 N (displacements s_2 under smaller forces are negligible indeed), Q_{1m} = 4N, and Q_2 again is equal to zero. For the purpose of convenience, we can use the dimensionless force coefficient $(\lambda_0'')^2 = \dfrac{1}{4}\dfrac{Q_1 L^2}{H}$, instead of force Q_1, and then apply values $\dfrac{s_2}{(\lambda_0'')^2 L}$.

As a result, we have the sought expression as follows

$$0.0456 - \frac{s_2}{(\lambda_0'')^2 L}(Q_{1s}) = 0, \text{ which yields } \frac{s_2}{(\lambda_0'')^2 L}(Q_{1s}) = 0.0456. \quad (8\text{-}11)$$

The smallest force value Q_{10} yields the following equation $\dfrac{s_2^{(0)}}{(\lambda_0'')^2 L}(Q_{10}) = 0.0736$.

The approximate relationship for displacements s_2 can then be established (Table 8-10) on the basis of the similar triangles

$$s_2 = \left[0.027\frac{0.46-(\lambda_0'')^2}{0.4} + 0.0456\right](\lambda_0'')^2 L\,. \quad (8\text{-}12)$$

We assume that condition (2-13) for displacements s_2 is satisfied if an error in s_2 calculation is less than 10%. This error slightly increases if force Q_{1m} values are more than 4 N. In the discussed problem, the accurate solution by Popov's or another effective nonlinear method should be obtained only for the displacements corresponding to forces Q_{10} and Q_{1m}.

Table 8-10. Calculation of the displacements s_2 under $Q\in[0.4; 4\text{ N}]$ with the accurate method

ς_0'	Q, N	φ^o	α^o	λ_0''	$(\lambda_0'')^2$	η_0''	ξ_0''	$\frac{s_1}{L}$
90°	Q_{10}= 0.4	80	45	0.24	0.0576	0.239	-0.00265	0.011
90°	Q_{1s}= 3.3	65	51	0.68	0.4604	0.666	-0.1037	0.150
90°	Q_{1r}= 1.65	71	48	0.48	0.2304	0.474	-0.02	0.042
90°	Q_{1c}= 3.9	63	53	0.75	0.5550	0.730	-0.13	0.175

Table 8-10 Cont'd.

ς_0'	$\frac{x_1}{L}$	$\frac{s_2}{L}$	$\frac{s_2 L^{-1}}{(\lambda_0'')^2}$
90°	0.9958	0.0042	0.0736
90°	0.979	0.021	0.0456
90°	0.985	0.015	0.0650
90°	0.978	0.022	0.0395

Problem 2. The piezoelectric actuators, as well as the piezoelectric resonant sensors, have found wide use in many precision elastic systems, including contact sensors of coordinate measuring machines.

Materials. Let us recall that piezoelectric materials that are especially effective in motion are ferroelectric crystals. But they have large amounts of internal damping that make them difficult to control as actuators. We should consider them nonlinear in reciprocal motion and sensing systems. The latter requires a special approach to stabilization of their behavior as an actuator and a sensor. The acoustic contact between measuring lugs and the piezoelectric resonator are used in the sensor version. The principle of the piezoelectric sensors action in coordinate measuring machines is based on the change under static force or friction of the vibrating systems' dynamic features: Q-factor, resonant frequency, phase of ringing frequency, the factor of running wave, etc. When an applied frequency to the piezoelectric material is equal to its mechanical resonance frequency, then the excited material is said to be in piezoelectric resonance. Oscillator circuits, when suitably connected with the piezoelectric material, will give an output frequency at the mechanical resonance frequency of the excited material with the shifted phase if additional friction is involved in the mechanical chain. We recall that the resonance frequency of the resonating material is given by the equation

$$f_0 = \frac{1}{2h}\sqrt{K_j/\rho_\gamma}\,, \tag{8-13}$$

where h is the thickness of the piezoelectric material, f_0 is the natural mechanical resonance frequency of the first order, K_j is the stiffness constant of the material, and ρ_γ is the density of the material. Let us remind that the theory of the piezoelectric phenomenon is a complicated one (*see* N.N. Rogacheva, 1994). Piezoelectric materials usually behave anisotropically with respect to their coordinate axis (Fig. 8-10b). Therefore, their mechanical and electromechanical constants are described by tensor's notations with double subscripts *ij*. A subscript 1,2, or 3 corresponds to one of the material axes. A subscript 4 represents the plane parallel to the 2 and 3 axes, which is normal to axis 1; plane 5 is parallel to 1-3 axes; and plane 6 is parallel to 1-2 axes. When used as the second subscript in one of the constants, values greater than three indicate that shear strain or shear stress occurs in that plane (*see* A.H. Slocum, 1992). Despite large number of possible combination in *ij* subscripts, the real piezoelectric materials have small number of working piezoelectric directions. Some piezoelectric materials' features and typical relationships are shown in Table 8-11. It is clear from notes to the table that $1/(S^E_{ii}\times 10^9) \approx E$, in GPa. Note also that the difference between the modulus of elasticity for piezoelectric element in open E_{oc} and short E_{sc} circuits is proportional to square of the "transmission ratio" k^2.

Table 8-11. Physical properties of common piezoelectric materials

Name	t_T, °C limit, temp. Curie	Direction E_E	F	d_{ij}	d_{ij} pm/V	$S_{ij}^E\times 10^{-12}$ m²/N	E, GPa	K_{diel}	k	ρ_γ 10^3kg/m³
Quartz crystal	550	X	X	d_{11}	2.3	12.5	80-107	4.52	0.10	2.65
		X	//YZ	d_{14}	0.73			4.63		
		X	Y	d_{12}	2.3	12.5	80	4.52	0.10	
PZST	300	Z	Z	d_{33}	250	14.92	67	1200	0.64	7.5
		Z	X	d_{31}	105		81	1200	0.30	
PZT(45/55)	100	Z	Z	d_{33}	140	14.08	53-71	450	0.6	7.6
PZT-4 (hard)	325	Z	Z	d_{33}	289	15.5	77		0.69	7.5
		Z	X	d_{31}	-120	-5.31		1325	0.75	
PZT-5 (soft)	365	Z	X	d_{31}	-180	-7.22		1900		7.75
		Z	Z	d_{33}	374	18.8	66			
Lead metaniobate	300	Z	Z	d_{33}	80	16.7	60	225	0.42	6.0
		Z	X	d_{31}	11		60	225	0.04	

Notes: materials PZT are the piezo ceramics of lead, zirconate, titanate; PZST of lead, zirconate, strontium, titanate. The following notations and relationships are typical for piezoelectric materials E_E is the electrical field, F is the force, d_{ij} is the piezoelectric constant of strain ε_j in direction j in relation to the gradient E_{Ei} of electric field in i direction, $\varepsilon_j = d_{ij}E_{Ei}+S^E_{ij}\sigma_i$; $S^E_{ii}\approx 1/(E\times 10^9)$is the compliance at constant electrical field, in m²/N; $g_{ij}=d_{ij}/(K_{diel}\varepsilon_0)$ is the piezoelectric stress constant which reflects the relationship between the field gradient E_{Ei} (volts/thickness) in i direction and the stress σ_j in the j direction, $\varepsilon_0=8.85\times 10^{-12}$ F/m is the dielectric constant of vacuum, $E_{Ei}=g_{ij}\,\sigma_j$; E is the modulus of elasticity, k is a "transmission ratio" between modulus of elasticity in piezoelectric element open E_{oc} and short E_{sc} circuits, $E_{oc}(1-k^2)=E_{sc}$; $k_1^2=\dfrac{\textit{electrical energy created}}{\textit{mechanical energy applied}}$ and $k_2^2=\dfrac{\textit{mechanical energy created}}{\textit{electrical energy applied}}$ are two versions of k coefficient with $k_1=k_2$; K_{diel} is a dielectric constant of a material with possible free state K_{diel}(free) without mechanical stress and clamped state K_{diel}(clamped) which are related to k coefficient as follows K_{diel}(free) $(1-k^2)=K_{diel}$(clamped); $\sqrt{E/\rho_\gamma}\approx V_w$ is the velocity of sine waves in the material. (Data and relationships after A.H. Slocum, 1992; M. J. Madou , 2002; and N.N. Rogacheva, 1994).

Continued high-stress cycles can cause permanent changes of piezoelectric characteristics in soft piezoelectric materials, unlike hard ceramics, which stabilize to new levels. Let us recall that the temperature of a ferroelectric crystal at which its nonsymmetric structure returns to nonpiezoelectric symmetric structure is called the Curie temperature. However, precision piezoelectric actuator should not get that hot. *Resonant sensor*. The piezoelectric exciter with a lug forms an oscillatory system, which performs forced oscillations at the frequency of the signal coming from the generator. In a free stage, the current flowing through the sensor reaches the maximum. The contact lug touching the measured part sharply reduces the Q-factor of the oscillatory system and decreases the amplitude of the current flowing through the electrical circuit of the transducer, which is an indication of the contact between the lug and the measured part.

A progressive solution for the piezoelectric resonant contact sensor was proposed by R.J. Bansevichus *et al.* in US Patent 4, 397, 188 (1983). In this case, a common oscillatory system is formed by five contact lugs 4 (Fig. 8-10c) in conjunction with rods 3, cylindrical or spherical enclosure 1, and piezoelectric hollow exciter 2 in the enclosure with the acoustic contact. The ringing (self-oscillations) frequency of the enclosure with the rods and the contact lugs in the system is equal to the oscillation frequency of a sine-wave electrical signal. The piezoelectric vibration exciter contains exciting electrode 5 and one additional electrode 6 acting as a measuring electrode and furnishing an output signal whose phase variation, with respect to the phase of the sine-wave of the common oscillatory system, is indicative of the moment at which the contact lug contacts the measurand. The advantages of this sensor are substantially enhanced measuring precision, increased efficiency, and versatility of the coordinate measuring machine. The arrangement of the rods with their axes normal to the surface of the enclosure assures high sensitivity at various points on the surface of the contact lugs in different measuring directions. Furthermore, the rods may be arranged so that oscillations of the piezoelectric exciter occur in a perpendicular direction or along the rod axis, a factor assuring a high operational precision and stability of the sensor. It was recommended in the patent that each cantilever rod with lug should have length L_{lg} (Fig. 8-10d) determined by the relationship $L_{\mathrm{lg}} = l_0 + l_1$, (8-14)

where l_0 is the distance from the external surface of the enclosure to a nodal point "a" of the rod oscillations; $l_1 = \frac{C_{snd}}{4 f_0}(1+2n)$, (8-15)

C_{snd} is the speed of sound in the material of rod, f_0 is the natural frequency of the system formed by the enclosure with the rods and the contact lugs, and integer n = 0,1, 2... Point "a" with zero amplitude of vibration position (Fig. 8-10d) and the corresponding distance l_0 were found by the authors of patent experimentally. However, these parameters may be found from known theoretical relationships as well. If the concentrated exciting forces are applied during the forced vibrations to the clamped end of the cantilever rod, its differential equation for the forced bending vibrations become homogeneous and the exciting motion $a_\omega \sin \omega t$ is reflected in the boundary conditions. In this case, the vibration form function in the rod coordinates (A.N. Krylov, 1936, Y. G. Panovko, 1968) is equal to:

$$Y(x) = C_1\, S(\alpha x) + C_2\, T(\alpha x) + C_3\, U(\alpha x) + C_4\, V(\alpha x), \qquad (8\text{-}15a)$$

where $\alpha = \sqrt[4]{\frac{m_u \omega^2}{EJ}}$, (8-15b)

C_1 through C_4 are the constants which are defined from the boundary-value conditions, m_u is the mass of the rod's unit length; S, T, U, and V are A.N. Krylov's functions (*see* Table 4-11). The boundary-value conditions for the rod with the lug may be presented as follows

$$Y'(0) = 0;\ Y(0) = a_\omega;\ Y''(L) = 0;\ EJY'''(L) = m_* \omega^2 Y(L), \tag{8-16a}$$

where m_* is the concentrated mass of the lug on the free rod's end. The sequential derivatives of function $Y(x)$ with respect to variable x are equal to:

$$Y'(x) = \alpha[C_1V(\alpha x) + C_2S(\alpha x) + C_3T(\alpha x) + C_4U(\alpha x)];$$
$$Y''(x) = \alpha^2[C_1U(\alpha x) + C_2V(\alpha x) + C_3S(\alpha x) + C_4(\alpha x)]; \tag{8-16b}$$
$$Y'''(x) = \alpha^3[C_1T(\alpha x) + C_2U(\alpha x) + C_3V(\alpha x) + C_4S(\alpha x)].$$

By substituting equations (16b) into the corresponding expressions (16a) for the boundary-value conditions and recalling that Krylov's functions $S(0) = 1.00000$, $T(0)=0$; $U(0)=0$; $V(0)=0$, we yield the sought constants as follows

$C_1 \equiv a_\omega$, $C_2 \equiv 0$, $C_3S(\alpha L) + C_4 T(\alpha L) = 0$, and

$$C_3 = \frac{a_\omega[EJ\alpha^3T(\alpha L) - m_*\omega^2S(\alpha L)]}{m_*\omega^2[U(\alpha L) - S(\alpha L)V(\alpha L)/T(\alpha L)] - EJ\alpha^3[V(\alpha L)} - S^2(\alpha L)/T(\alpha L)]\,. \tag{8-17}$$

Substituting the found expressions for constants C_1 through C_4 and Krylov's functions $S(\alpha x)$ through $V(\alpha x)$ into (15a), we yield the following expression of rod bending vibration amplitude along of its axis x

$$Y(x) = a_\omega\{1/2(\cosh\alpha x - \cos\alpha x)[1 + \frac{EJ\alpha^3T(\alpha L) - m_*\omega^2S(\alpha L)}{m_*\omega^2[U(\alpha L) - S(\alpha L)V(\alpha L)/T(\alpha L)] - EJ\alpha^3[V(\alpha L) - S^2(\alpha L)/T(\alpha L)]}$$
$$- \frac{1}{2}(\sinh\alpha x - \sin\alpha x)\frac{S(\alpha L)}{T(\alpha L)}]\frac{EJ\alpha^3T(\alpha L) - m_*\omega^2S(\alpha L)}{m_*\omega^2[U(\alpha L) - S(\alpha L)V(\alpha L)/T(\alpha L)] - EJ\alpha^3[V(\alpha L) - S^2(\alpha L)/T(\alpha L)]}\}.$$

The position on the rod of the sought point $x = a$ with zero amplitude may be found from the latter expression because $Y(x) = 0$ if $\cosh \alpha x - \cos \alpha x = 0$ and $\sinh \alpha x - \sin \alpha x = 0$ simultaneously. Nonzero solution for the system of these simultaneous equations yields the following transcendent equation $\tanh \alpha x = \tan \alpha x$, which first root is equal to: $\alpha x = 3.927$.

Therefore, the position of sought point a on the vibrating rod has coordinate

$$x = l_0 = 3.927 / \alpha. \tag{8-18}$$

Let us calculate the value of α for the above described piezoelectric vibrating contact sensor system with the natural frequency f_0=105 kHz (*see* V.-A.A. Gapshis *et al.*, 1988) and with the corresponding angular natural frequency $\omega = 2\pi f_0 = 2\times\pi\times105000$, which steel contact rod has diameter d=1.5 mm and the corresponding sectional moment of inertia $J = \pi d^4/64 = 0.248$ mm^4. We will use the following parameters for the steel rod material: E = 200 GPa, density ρ_γ=7860 kg/m^3. Hence, the weight of 1 mm length of such a rod equals $w_u = m_u g = \rho_\gamma \pi d^2/4 = 1.36\times10^{-4}$ N/mm and the mass per unit length is $m_u = w_u/g$. Substituting all these parameters into (15b), one will obtain the value of α=0.591 mm^{-1} and the coordinate of a point $x = l_0$ =3.927/ 0.591 = 6.6 mm.

In accordance with the above presented expression (15) for the distance from found node point a with zero amplitude to the right end of the contacting rod, and assuming that speed of sound in steel with the bulk modulus of $B_E = \dfrac{E}{3(1-2\nu_p)}$ =160

GPa is equal to $C_{snd} = \sqrt{\frac{B_E}{\rho_\gamma}}$ =4512 m/s, we have the value for l_1 as follows

$l_1 = \frac{4512 \times 1000}{4 \times 105000}(1+2n) = 32.2$ mm, where n=1 for the first mode of oscillations' frequency. Therefore, the length of the corresponding contacting rod is approximately equal to $L_{lg} \approx 39$ mm.
The natural frequency equation for the longitudinal oscillations of the similar cantilever rod with the lug (mass m_*) on the free end can be presented (Y. G. Panovko, 1968) as follows $\frac{\omega L}{a_{E\rho}} \tan \frac{\omega L}{a_{E\rho}} = \frac{\rho_\gamma AL}{m*}$ and its oscillations' form is described by the formula $X_k(x) = C_k \sin \frac{\omega_k x}{a_{E\rho}}$, where $a_{E\rho}^2 = E / \rho_\gamma$ depends on the parameters of the material, ω_k is the natural angular frequency of oscillations, and C_k constants, which depend on the boundary-value conditions.
Let us recall that the lowest natural angular frequency of bending oscillations for the cantilever loaded with the distributed longitudinal force q is equal to:

$\omega_1^* = \omega_1 \sqrt{1 + \frac{qL^3}{8EJ}}$, where ω_1 is the lowest natural frequency of the rod oscillations without longitudinal load. This distributed longitudinal force may be from the density of the rod material and can vary at the different orientation ψ of the rod in the gravity field with cos ψ proportion.

Resonant acoustic oscillators in the shape of the tuning fork and piezoelectric quartz crystal bar are used in the contactless scanning microscopes for examination of an object surface in the short-range field with submicron resolution (*see* K. Dransfeld *et al.*, US Patent 5,212,987, 1993). The oscillations may be directed along the object surface, perpendicular to the latter, or oblique to the object. Note that the ambient air pressure has significant influence on the scanning microscope resolution.
Problem 3. Elastic kinematic systems are now widely used in nanotechnology because they provide very low friction, do not require grease, and can have miniature structures. A non-traditional miniature elastic kinematic system can be made on the basis of DNA molecule (*see* C. Bustamante *et al.*, 2003). Outside of traditional DNA biology, the easy of synthesis and well-characterized elasticity of DNA make it an ideal material for stiff molecular handles to manipulate other molecules. Such handles have been used to mechanically unfold molecules of RNA. The elasticity of the DNA molecule with double-stranded (dsDNA) and single-stranded (ssDNA) helical structure (Fig. 8-11a, b) is very carefully studied. We should mention that the DNA molecule's model in the shape of a thin cylindrical elastic rod (D. Swigon *et al.*, 1998) used by some authors seems to be not relevant to its physical structure which resembles the elastic helix, despite that some estimation results on this basis are correlated with experimental data.
Helix chain. From the theoretical stand point DNA molecule is presented as a worm-like helix chain (J.A.Schellman, 1974) with a simple relationship between the local distortion of helix regularity in the deviation of direction angles θ_i and a free energy $E_g(\theta)$ temperature dependent variation similar to the statistical fluctuation in the macro-model of precision elastic systems (*see* Chaps. 2, 4, 5, and 7) where the

standard deviation $\sigma_i(\theta_i)$ in accordance with the equipartition theorem is equal to: $\sigma_i(\theta_i) = \sqrt{\dfrac{K_B T \theta_i^2}{2E_g(\theta)}}$.

When tension in a nicked DNA molecule is increased to 65 pN, it displays a reversible, cooperative transition to an extended form that is approximately 70 % longer than normal *B*-DNA and with a substantially reduced twist. It was also found that the twist-stretch coupling in these molecules is equal to 50 nm. Evidence exists that overstreched DNA (polynucleotide) may present one of the following structures: (i) base-paired (dubbed *S* form, doubbed-helix) or (ii) two independent strands of ssDNA which are similar in structure (*see* B. Bhushan, ed., 2004) to polypeptide single helix (Fig. 8-11b).

a) b)

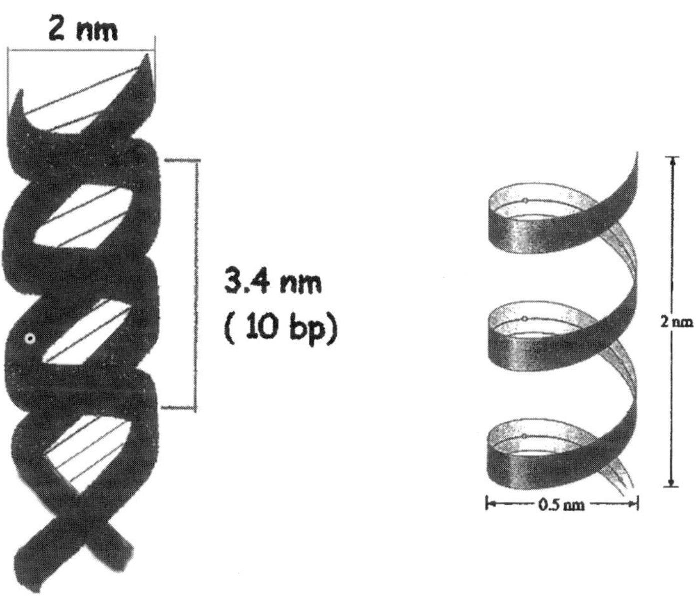

Fig. 8-11. Helico

Helicoidal model. We assume now that the stretched DNA model (i) in the mechanical aspects is similar to the elastic helicoid (Fig. 8-11a) made of a punched pretwisted strip and model (ii) is similar (Fig. 8-11b) to a coiled ribbon structure (*see* Chap. 4). Let us consider the features of base-paired structure (i). For twist-stretch coupling D_c estimation, we can use the conditional radius formula (4-86) of elastic pretwisted strip

$$D_c = \rho_c = \frac{j_{k\tau} - j_{\tau\tau} i_\theta}{j_{kk} - j_{\tau k} i_\theta}, \tag{8-19}$$

where $j_{\tau\tau} = GT_o(1 + \nu_{ipf} b_1^2 k/\lambda)$; $j_{kk} = EA$; $i_\theta = \dfrac{Ek_p}{4G\lambda}(1 + \nu_{ipf} b_1^2 k_p^2/\lambda)^{-1}$.

$j_{k\tau}=j_{\tau k}= EAb_1^2k_p/4;$

here $j_{k\tau}=j_{\tau k}$ is the tension stiffness of a pretwisted strip under its tension with a simultaneous twist or the stiffness of this strip under its twist with a simultaneous tension, $j_{\tau\tau}$ is the twist stiffness of a pretwisted strip under twist with an applied torque, but without tension, j_{kk} is the tension stiffness of the pretwisted strip under tension without twist, and $i_\theta = |\theta/s|$ is the ratio between the untwist angle θ and corresponding stretch s of the pretwisted strip in rad/nm. Let us also denote the ratio between elastic moduli $E/G = \nu_{E/G}$ and the ratio between the section torsional stiffness T_0 and area A magnitude as $T_0/A = \nu_{T_0/A}$. Substituting all of these expressions into (19), we obtain, after corresponding cancellations, the final formula for the twist-stretch DNA coupling calculation as follows

$$D_c=\frac{(b_1^2k_p/4)-\nu_{E/G}\,\nu_{T_0/A}\,\lambda^{-1}k_p/4}{1-(b_1^2k_p^2/4)\nu_{E/G}\,(\lambda+\nu_{ipf}b_1^2k_p^2)^{-1}/4}. \qquad (8\text{-}20)$$

If we now estimate this in accordance with known dimensions (Fig. 8-11a) of the B-DNA molecule: $b_1 = 1$ nm, $k_p = 2\pi/3.4 \approx 1.847$ nm^{-1} with a helix pitch of 3.4 nm or 10 base pairs, and assume the ratio of Young's and shear moduli $\nu_{E/G} = 2(1+\nu_p)$ $= 2.4$-3.0 as for any real elastic material with Poisson's ratio $\nu_p = 0.2$-0.5, the transmission form factor for a perforated pretwisted strip $\nu_{ifp}= 0.39$, $\lambda=0.02$ $= (h_{DNA}/2b_1)^2$, where $h_{DNA}/2b_1 \approx 0.14$ represents the minimal ratio between the thickness h_{DNA} and the width $2b_1$ of the DNA molecule's cross section, and the sectional twist stiffness to the cross-section area ratio $\nu_{T_0/A} = hl/\pi \geq 1.5$ nm^2 as for a thin shell with the section's contour length l and wall thickness h, then

$$D_c=\frac{(1^2\times1.847/4)-2.5\times1.5\times(0.02)^{-1}\times1.847/4}{1-(1^2\times1.847^2/4)(0.02+0.39\times1^2\times1.847^2)^{-1}/4}=-55.6 \text{ nm}.$$

This number corresponds to the data presented in contemporary literature on the DNA mechanics. The minus sign denotes the untwist motion in the structure under the stretch.

In accordance with Table 7-3, the limit of statistical fluctuations' standard deviation for the pretwisted strip structure is equal to:

$$\sigma_{\varepsilon M}=\sqrt{\frac{k_BTL(\lambda+\nu_{ipf}\,b_1^2k_p^2)}{EA(\lambda+0.18\,b_1^2k_p^2)}}. \qquad (8\text{-}21)$$

Note that (21) corresponds to the equipartition theorem.
Let us recall that $k_BT = 4$ pN·nm at normal reference temperature T=293 K. We also can assume that for double-stranded DNA the expression $EA\,(\lambda+0.18\,b_1^2k_p^2)/L(\lambda+\nu_{ifp}b_1^2k_p^2) \approx F/s$ is a stretch stiffness and may be equal to 65/(3.4×0.7)= 27 pN/nm for the molecule of 10 base pairs (bp) in length (*see* C. Bustamante *et al.*, 2003; T. Strunz *et al.*, 1999), where F is an acting force and F/s is the molecule's elastic rate. Therefore, from (21), we can obtain the limit of the statistical fluctuations' standard deviation for a DNA molecule with the above shown parameters within $\sigma_{\varepsilon M} = 0.4$ nm. If the flexural persistence length of the stretched DNA under the force of 10 pN is equal to A_{bp}=50 nm (as shown in many published

sources) and the elongation of the DNA molecule under the force up to 65 pN is linear and equal to s= 0.1 A_{bp}= 5 nm, when the force reaches 65 pN (*see* A. Sarcar *et al.*, 2001), then the elastic rate F/s = (65-10)/ 5 ≈ 0.14 pN/nm. In this case, the standard deviation of the fluctuations is equal to $\sigma_{\varepsilon M} = \sqrt{4/14} = 0.5$ nm. If we assume that Young's modulus of the DNA molecule is equal to (3.46±0.3) ×10^8 Pa (*see* S.B. Smith *et al.*, 1996), then our prediction for a standard deviation of the fluctuations in a molecule with the flexural persistence length L=A_{bp}= 50 nm on the basis of expression (21) is not less than $\sigma_{\varepsilon M} = \sqrt{\dfrac{4\times 50\times 1.35}{3.46\times 100\times 1.46\times 0.63}} = 0.9$ nm, where 1.46 nm^2=A is the estimation of the molecule's cross-section area, 1.35 and 0.63 are the numerical values of (λ+ 0.39$b_1^2k_p^2$) and (λ+ 0.18$b_1^2k_p^2$) members, respectively, with the above accepted parameters λ, b_1, and k_p of the DNA molecule. Let us recall also that on the basis of the worm-like DNA model (J.A. Schellman, 1976; C. Bustamante *et al.*, 2003) there is found the expression for Force (F_D)-Stretch S_D relationship as follows: $F_D = \dfrac{K_BT}{A_{bp}}\left[\dfrac{S_D}{L_{cont}} + \dfrac{1}{4(1-S_D/L_{cont})^2} - \dfrac{1}{4}\right]$,where A_{bp} is the persistence length and $L_{cont} = S_0N_t$ is the contour length of DNA molecule with N_t turns and helix pitch S_0= 10 bp or 3.4 nm. F_D-S_D expression has linear and nonlinear members.

The twisting of stretched DNA can lead to other structural transitions. For example, after a critical amount of overwinding has been introduced into a molecule, it may buckle. Complex force-extension curves with multiple force plateaus are seen when single DNA molecules are twisted in either direction and pulled to high forces. This allows us to assume possible jumps of hysteresis asymmetry at a fast and slow extension and the relaxation of the DNA molecules similar to those in the pretwisted helicoids (see Chap. 5).

Of course, all these assumptions and the described elastic helicoidal model's application to DNA mechanics should be experimentally verified and theoretically developed in more detail. This experimental verification is still very difficult as a result of the ultimately small sizes and acting forces. However, the closeness of biomechanics and precision nano-elastic mechanics is evident.

Further research in the structural synthesis of precision elastic chains may be directed to the more effective elastic correctors, study of mechanical motion in the quantum regime, simplification of enough accurate analytical methods for new effective elastic systems synthesis in the field of micro- and nano-scale, and improvement of their metrological traceability, especially in the area of nanotechnology, picometer, femto-Joule, subattogram, and subattonewton detection, including detection of tiny chemical, biological agents, and vapor molecules (*see* N.V. Lavrik *et al.*, 2004). Thorough research should be performed on new, more stable, non-corrosive elastic materials and films with negligible internal damping, increased range of a linear stress-strain relationships, larger endurance and tensile stress limits, lesser temperature expansion coefficients, lesser density, larger modulus of elasticity, larger *MYS*, and better "machinability". Special research is now intensified on new methods and materials for the fabrication of nonplanar spiral and helical microstructures, carbon nanocoils which may be the most sensitive mechanical transformers of micro-motion, joining elastic links by laser and electron ray welding without softening or cracking the elastic material, cold bonding setting with resins and adhesives [e.g., advanced Araldite, Yacca gum resin with E=5 GPa, Elmer's glue with E=0.6 GPa (*see* X. Chen *et al.*, 2004; S.W. Schediwy et al., 2005), and effective mechanical clamping. A new and promising option is the growing of carbon nanotube's tips and super-sharp (tip radius 1-2 nm) tungsten nanowire directly on the atomic force microscopy cantilever's movable end with probe and depositing fine-crystalline diamond particles grown on standard silicon tips that provide necessary stiffness and eliminate structural damping in this very responsible systems of highly sensitive scientific instruments. Effective growing attachment methods and synthesis of helical micro- and nano-springs include chemical-vapor deposition and electrospinning.

All these problems are related to the next generation of the structural synthesis in precision elasticity.

REFERENCES

H. Abramson (1947), The Mikrokator amplifying mechanism and its use in measuring lengths and loads, *Transactions: Instruments and Measurements Conference*, Stockholm, 177-181.

T.R. Albrecht, P. Gütter, D. Horne, D. Rugar (1991), Frequency modulation detection using high-*Q* cantilevers for enhanced force microscopy sensitivity, *J. Appl. Phys.*, **69**, 668-673.

I. S. Amosov, Y. M. Tseytlin (1969), Technique for evaluating the measuring effort, *Meas. Tech.* **12**, 1459-1461.

L.E. Andreeva (1962), *Elastic Elements of Instruments* (in Russian), Mashinostroenie, Moscow. English translation (1966), Israel Program for Scientific Translations.

L. E. Andreeva (1981), *Elastic Elements of Instruments* (in Russian), 2nd edn., Mashinostroenie, Moscow.

S. S. Antman (1995) *Nonlinear problems of Elasticity*, Springer-Verlag.

P.A. Arutyunov and A.L. Tolstikhina (2002), Atomic force microscopy and universal means of measuring physical quantities in the mesoscopic length range, *Meas. Tech.*, **45**, 714-723.

W.E.Ayrton, J.Perry (1884), The new form of Spring for Electric and other measuring, *Proc. Roy. Soc. A*,**36**, 297-319.

R.J. Bansevichus, V-A. A. Gapshis, A.J. Kasparaitis, V.V. Perfilov, A.G. Ragauskas (1983), Piezoelectric Resonant Contact Sensor, US Patent 4,397,188.

A.A. Becker (2004), *Finite Element Analysis*, The ASME, New York.

P.I. Begun (1975), Calculation of sensitivity of twisted tubular springs, *Izvestia Vysshikh Uchebnykh Zavedenii. Priborostroenie*, **6**, 91-97 (in Russian).

G. G. Berdzenishvili, Y. M. Tseytlin, M. G. Boguslavskii (1974), Computing and experimental method of determining temperature errors in linear measurements, *Meas. Tech.* **17**, 1170-1172.

G. Berndt (1929) *Grundlagen und Geräte technischer Längenmessungen*, Springer.

B. Bhushan, ed. (2004), *Springer Handbook of Nanotechnology*, Springer-Verlag, Berlin, New York.

V.L. Biderman (1963), Helical spring of rectangular cross section as a cylindrical shell, *Izv. AN SSSR*, Mechanika i Mashinostroenie, N4, 67-75 (in Russian).

V.L. Biderman and V.N. Shitikov (1970), Tension and torsion on cylindrical ribbon spring in the presence of large displacements, Izv. AN SSSR, *Mekhanika Tverdogo Tela*, **5**, 137-141 (121-125 in English translation).

I. A. Birger, N. A. Fedorov (1968), Calculation of rings and ring systems, in *Strength, Stability, Vibrations*, Handbook (1968) in Three volumes, I. A. Birger, Y. G. Panovko, eds., **1**, 287-416, Mashinostroenie, Moscow (in Russian).

X. Blase, Ph. Gillet, A.S. Miguel, and P. Melion (2004), Exceptional Ideal Strength of Carbon Clathrates, *Phys. Rev. Let.*, **92**, 215505-1-4.

F. Bleich (1952), *Buckling Strength of Metal Structures*, McGraw-Hill.

M.G. Boguslavskii, N. A. Pisareva, Y. M. Tseytlin (1973), Determination of the surface roughness of a small-diameter wire, *Meas. Tech.*, **16**, 1307-1309.

M.G. Boguslavskii, Y.M. Tseytlin (1976), *Instruments and Methods for Precision Measurement of Length and Angles*, Standartov, Moscow (in Russian).

B. Bollobás (1998), *Modern Graph Theory*, Springer-Verlag.

V. V. Bolotin (1956), The *Dynamic Stability of Elastic Systems* (in Russian), Gostekhteorisdat, Moscow. English transl. V.I. Weingarten *et.al.*, Holden-Day, 1964.

V.V. Bolotin, V.N. Moskalenko (1968), Vibration of plates, in *Strength, Stability,*

Vibrations, Handbook (1968), I. A. Birger, Y. G. Panovko, eds., **3**, 310-417, Mashinostroenie, Moscow (in Russian).

V. V. Bolotin (1984), *Random Vibrations of Elastic Systems*, English translation edited by H.H.E. Leipholz, Kluwer Academic Publisher.

L. N. Bol'shev, N. V. Smirnov (1983), *Tables of Mathematical Statistics*, Nauka, Moscow (in Russian).

P. Bondy (1961), Blattfeder-Parallelführungen, *Proceed.* 2nd *Internat. Meas. Conf.*, **I**, 271-287 (in German).

P. Bondy (1967), Räumliche Belastung von einfachen Blattfedergelenken, *Feingerätetechnik*, 115-116, (in German).

C. de Boor (2001), A Practical Guide to Splines, Springer-Verlag, New York.

M.Born (1956), *Physics in my Generation*, Pergamon Press.

V.B. Braginsky and A.B. Manukin (1977). *Meausrements of Weak Forces in Physics Experiments*, Chicago University Press.

V.B. Braginsky, V.P.Mitrofanov, V.I.Panov (1985), *Systems with Small Dissipation*, ed. K.S. Thorne and C.Eller, transl. in English by E. Cliner, Chicago University Press, Russian ed. Systemi s maloi dissipatsiei, 1981, Nauka, Moscow.

V.B. Braginsky, V.P. Mitrofanov, K.V. Tokmakov (1996), Energy dissipation in the pendulum mode of the test mass suspension of gravitational wave antenna, *Phys. Lett. A*, **218**, 164-166.

R. Breitinger (1969), Blattfederelemente für Messvorichfungen, *Werstattstechnik*, 78-83 (in German).

L. Brillouin (1962), *Science and Information Theory*, Academic Press.

I.N. Bronshtein, K.A. Semendiaev (1985), *Handbook of Mathematics*, English Transl., 3rd ed. by K.A. Hirsh, Van Nostrand Reinhold.

D.M. Buede (2000), *The Engineering Design of Systems: models and methods*, Wiley, New York.

C. Bustamante, Z. Bryant & S.B. Smith (2003), Ten years of tension: single-molecule DNA mechanics, *Nature*, **421**, 423-427.

H.-J. Butt, P. Siedle, K. Seirt, K. Fendler, T. Seeger *et al.* (1993), Scan speed limit in atomic force microscopy, *J. Microscopy*, **169**, 75-84.

R.D. Carnahan (1964), Microplasticity, its measurement and application to guidance and control components, *J. of Metals*, **16**, 990-994.

D.W. Carr, S. Evoy, L. Sekaric, H.G. Graighead, and J.M.Parpia (1999), Measurement of mechanical resonance and losses in nanometer scale silicon wire, *Appl. Phys. Lett.*, **75**, 920-922.

P.L. Chebyshev (1899), *Sochineniia*, *Oeuvres* de P.L.Chebyshev, **1**, **2**, M. A. Markoff et N. Sonin, S.Petersburg.

G.Y. Chen, R.J. Warmack, T. Thundat, D.P. Allison, and A. Huang (1994), Resonance response of scanning force microscopy cantilever, *Rev. Sci. Instrum.*, **65**, 2532-2537; idem. ibid., Friction effects in the deflection of atomic force microscope cantilevers, 394-398.

X. Chen, S. Zhang, G. J. Wagner, W. Ding, R. S. Ruff (2004), Mechanical resonance of quartz microfibers and boundary condition effects, *J. Appl. Phys.*, **95**, 4823-4828.

Chen Chu (1952a), The effect of initial twist on the torsional rigidity of thin prismatical and tubular members, *Proc. First US Nat.Congr. Appl. Mech.*, ASME, **4**, 265-269.

Chen Chu (1952b), A theory of twisted Bourdon tubes, *Proc. First US Nat.Congr. Appl. Mech.*, ASME, **4**, 271-280.

W. G. Cochran (1963), *Sampling Techniques*, Wiley.

Comptes Rendus des Séances (1961), XI Conférence Générale des Poids et Mesures.

H. Cramer (1971), *Mathematical Methods of Statistics*, Princeton University Press.

P.J. Cumpson, J. Hedley, P. Zhdan (2003), Accurate force measurement in the atomic force

microscope: a microfabricated array of reference springs for easy cantilever calibration, *Nanotechnology*, **14**, 918-924.
N.N. Davidenkov (1933), *String Method of Deformation Measurement*, Gostekhizdat (in Russian).
R.S. Decca, E. Fishbach, G.L. Klimchitskaya, D.E. Krause, D. Lopez, and V.M. Mostepanenko (2003), Improved tests of extra-dimensional physics and thermal quantum field theory from new Casimir force measurements, *Phys. Rev. D*, **68**, 116003(15).
B.V. Derjaguin, V.M. Muller, Y.P. Toporov (1975), Effect of contact deformations on the adhesion of particles, *J. Colloid Interface Sci.*, **53**, 314-326.
R. H. Dieck (1997), *Measurement Uncertainty*. Methods and Applications. 2nd edn.. Research Triangle Park, NC, ISA.
F.M. Dimentberg (1978), *Theory of Screws and its Applications*, Fizmatgiz (in Russian, Teoriya Vintov i ee Prilozheniia).
I.N. Dolitskii, T.A. Fedorenko (1972), Calculating the maximum attainable natural frequencies of elastic elements in electromechanical pressure transducers, *Meas. Tech.*, **15**, 1040-1042.
A.T. Draudin (1966), Calculation of the flat spring pivots' kinematics, *Trudy MIEM*, Énergia, 271-285 (in Russian).
A.T. Draudin (1967), Calculation of guides' accuracy, *NIIM Vusov*, Mashinostroenie, Moscow, **1**, 108-120 (in Russian).
F.S. Eastman (1935), Flexure pivots to replace knife edges and ball bearings, *Bulletin* No86, University of Washington Engineering Experiment Station, 3-47; idem. ibid., The design of Flexure Pivots, *J. Aeronaut. Sci.*,**5**, 16-21, 1937.
W. Eckerkunst (1966), *Automatisierung in der Längenmesstechnik*, Verlag Technik (in German).
B. Edlen (1953),The Dispersion of Standard Air, *JOSA*, **43**, 5, 339-344; idem. ibid. The refractive index of air, *Metrologia*, 1966, **2**, 71-80.
A.G. Erdman, G.N. Sander (1984), *Mechanism Design: Analysis and Synthesis*, Prentice-Hall, New Jersey.
L. Euler (1736), Solutio problematic ad geometriam situs pertinentis, Academiae Sci. I. Petropolitanae, **8**, 128-140.
L. Euler (1744), *Methodus invendi lineas Curves maximi minimive proprietate gaudents, sive Solutio problematic isoperimetrici latissimo sensu accepti*, Marcum-Michaelem Bousquet, Lousanne, English translation of Appendix I, *Leonard Euler's Elastic Curves*, transl. by W.A. Oldfather, C.A. Ellis, D.M. Brown, 1933, Saint Catherine, Bruges; Russian issue, GTTI, Moscow, 1934.
D. K. Faddeev (1959), About condition of matrices, *Proceedings of the Steklov Mathematical Institute*, **53**, 387-391, Kraus Reprint LTD VADUZ, 1963 (in Russian).
F. T. Farago (1968), *Handbook of Dimensional measurement*, Industrial Press, Inc.
V. I. Feodos'ev (1949), *Elastic Elements of Precision Instrumentation*, Oborongiz, Moscow.
R. S. Figliola, D. E. Beasley (2000), *Theory and Design for Mechanical Measurements*, Wiley, 3rd ed.
L.N.G. Filon (1900), On the resistance to Torsion of Certain Forms of Shafting, *Phil. Ttrans. Roy. Soc. of London*, Ser. A, **193,** 309-352.
L.A. Foygel, Z. Kh. Shekhter, S. V. Elukim (1959), Investigation of the relationship in elastic properties from geometric dimensions for the twisted tubular springs, *Priborostroenie*, **4**, 15-17, (in Russian).
A. Gaidarzhy, G. Zolfagharkhani, R.L. Badzey, P. Mohanty (2005), Evidence for Quantized Displacement in Macroscopic Nanomechanical Oscillators, *Phys. Rev. Lett.*, **94**, 030402-1-4.
V.-A.A. Gapshis, A.J. Kasparaitis, M.B. Modestov, V.-Z.A. Ramanauskas, N.A. Serkov, V.A. Chudov (1988), Coordinate Measuring Machines and their Application, Mashinostroenie, Moscow (in Russian).

M.F. Gardner, J.L. Barnes (1942), *Transients in Linear Systems studied by the Laplace Transformations*, Wiley.

P.J.Geary (1954), *Flexure Devices*, B.S.IR.A., Research report, M18.

P.J.Geary (1960), *Torsion Devices*, B.S.I.R.A., Research report, R249.

I.M. Gel'fand, G.E. Shilov (1964) *Generalized Functions*, v.**1**, Academic Press.

J.M. Gere (2001), *Mechanics of Materials*, 5th ed., Brooks/Cole.

G.I. González and P. R. Saulson (1994), Brownian motion of a mass suspended by an anelastic wire, *J. Acoust. Soc. Am.*, **96**, 207-212.

GOST8.050-73, *Reference Conditions for Linear and Angular Measurements* (Russian interstate standard/ theme leader Y. M. Tseytlin).

GOST8.051-81(73), Errors Permissible for Measurement of Linear Dimensions up to 500 mm (Russian interstate standard/ theme leader N. N. Markov).

A.I Gots, R.L. Polozhenskaya (1969), Calculations of beams with small rigidity, *Trudy, Vladimirskii Vechernii Polytech, Institute*, M., "Visshaya shkola", **6**, 49-52 (in Russian).

I.S. Gradshtein and I.M. Ryzhik (1994), *Tables of Integrals, Series, and Products*, ed. A. Jeffrey, transl. from Russian by Scripta Technica, Acad. Press, Boston.

A.N. Grishin (1970), Kinematic error of mechanisms using corrugated membrane and linear transfer member, *Meas. Tech.*, **12**, 1366-1368.

F. R. Hampel, E. M. Romchetti, P. J Rouseeuw, *et al.* (1986), *Robust Statistics*: the approach based on influence functions, Wiley.

C.-S. Han, S-H. Kim (2002), Three-axis lever actuator with flexure hinges for an optical disk system, *Rev. Sci. Instrum.*, **73**, 3678-3686.

J.A. Haringx (1949), The cross-spring pivot as a constructional element, *Appl. Sci. Research A*, **1**, 313-332.

H. Hasselmeier (1954), Allgemeine Untersuchungen uber das Kreuzfedergelenk, *Jenaer Jahrbuch,* Jena, 279-304 (in German).

A.Husain, J. Hone, H.W.Ch. Postma, X.M.H. Huang, T. Drake, M.Barbic, A.Scherer, and M.L. Roukes (2003), Nanowire-based very-high-frequency electromechanical resonator, *Appl. Phys. Lett.*, **83**, 1240-1242.

International Vocabulary of Basic and General Terms in Metrology (1993), BIPM, IEC, ISO, *et al.*

ISO (1995), *Guide to the Expression of Uncertainty in Measurement* (GUM).

J.N. Israelachvili (1992), *Intermolecular and Surface Forces*, Academic Press.

H. Jeffreys and B. Swirles (1956), *Methods of Mathematical Physics*, ed. 3, University Press, Cambridge, England.

S. Jeon, Y. Braiman, T. Thundat (2004), Cross talk between bending, twisting, and buckling modes of three types of microcantilever sensors, *Rev. Sci. Instrum.*, **75**, 4841-4844.

R. V. Jones (1988), *Instruments and Experience*, Wiley.

R. Kacker and A. Jones (2003), On use of Bayesian statistics to make the Guide to the Expression of Uncertainty in Measurement consistent, *Metrologia*, **40**, 235-248.

M. M. Kempinskii (1972), *Accuracy and Reliability of Measurement Instruments*, Mashinostroenie, Leningrad (in Russian).

A.Khan, J. Philip, and P. Hess (2004), Young's modulus of silicon nitride used in scanning force microscope cantilevers, *J. Appl. Phys.*, **95**, 1667-1672.

J. H. Kim, S. H. Kim, and Y. K. Kwak (2003), Development of a piezoelectric actuator using a three-dimensional bridge-type hinge mechanism, *Rev.Sci.Instrum.*, **74**, 2918-2924.

M. Z. Kolovskii (1976), *Automatic Control of Vibro-protective Systems*, Nauka, Moscow (in Russian).

M.Z. Kolovsky, A.N. Evgrafov, Yu.A. Semenov, A.V. Slousch (2000), *Advanced Theory of Mechanisms and Machines*, transl. by L. Lilov, Springer, Berlin.

G.A. Korn, T. M. Korn (1968), *Mathematical Handbook*, 2nd edn., McGraw-Hill, New York.

V. P. Korsunov (1966), Normalization of the twisted tubular manometric springs,

Priborostroenie, **5**, 6-8 (in Russian).

A.N. Krylov (1936), *Vibrations of ships*, Leningrad (in Russian); idem. ibid., Über die erzwungenen Schwingungen von gleichförmigen elastischen Stäben, Math. Ann., Leipzig, Bd. 61, 211-234 , 1905; idem. ibid., Izvestiia AN SSSR, N7, 963-1012(1931).

J.B. Kuperman (1971), *Approximate Linear Algebraic Equations*, Van Nostrand Reinhold, London.

L.D. Landau and E. M. Lifshitz (1958), *Statistical Physics*, Pergamon Press, Transl. from Russian by E.Pierls and R.F. Pierls.

L.D. Landau and E.M. Lifshitz (1969), *Mechanics*, Pergamon Press, Transl. from Russian by J.B. Sykes and J.S. Bell.

L.D. Landau and E.M. Lifshitz (1970), *Theory of Elasticity*, 2nd English ed., Pergamon Press, Transl. from Russian by J.B. Sykes and W.H. Reid.

D. Lange, O. Brand, H. Batles (2002), *CMOS cantilever Sensor Systems:Atomic force microscopy and Gas Sensing Applications*, Springer, Berlin.

N.V. Lavrik, M.J. Sepaniak, P.G. Datskos (2004), Cantilever transducers as a platform for chemical and biological sensors, *Rev. Sci. Instrum.*, **75**, 2229-2253.

Yu. N. Lyandon, N. N. Zyabreva (1941), Use of elastic elements in the instruments for linear quantities control, *Izmeritelnaya Tekhnika*, № 5 (in Russian).

Iu.V. Linnik (1961), *Method of Least Squares and principles of the Theory of Observations*, Transl. from Russian by R.C. Elandt, ed. N.L. Johnson, Pergamon Press, New York.

N. Lobontiu, J.S.N. Paine, E. Garcia, M. Goldfarb (2001), Corner-filleted flexure hinges, *Trans. ASME*, **123**, 346-352.

W. Lotze (1963), Grundzüge der Kinematik von Biegefederaufhängungen, *W. Zeitschrift*, T.U. Dresden, **12**, H.6, 1859-1866(in German).

A.E.H. Love (1944), *A Tretise on the Matematical Theory of Elasticity*, Dover Publication, New York.

B. Luan, M.O. Robbins (2004), Effect of Inertia and Elasticity on Stick-Slip Motion, *Phys. Rev. Lett.*, **93**, 036105-1/4.

A.I. Lurie (1947), *Static of Thin-walled Elastic Shells*, Gostekhizdat (in Russian).

A.I. Lurie (1970), *Theory of Elasticity*, Nauka, Moscow (in Russian).

A.V. Lykov (1968), *Analytical Heat Diffusion Theory*, ed. J. Hartnett, Academic Press, Transl. from Russian.

M. J. Madou (2002), *Fundamentals of Microfabrication*: The Science of Miniaturization, 2nd ed., CRC Press, Boca Raton, Fla.

E.B. Magrab (1979), *Vibration of Elastic Members*, Sijthoff & Noordhoff.

E. Majorana, Y. Ogawa (1997), Mechanical thermal noise in coupled oscillators, *Phys. Lett. A.*, **233**, 162-168.

A.N. Malakhov (1978), *Cumulant Analysis of Random non-Gaussian Processes and their Transformations*, Sov. Radio, Moscow (in Russian).

E.H. Mansfield (1989), *The Bending and Stretching of Plates*, Cambridge University Press (Inextensional theory, Quart. J. Mech. Appl. Math., 1955, 8, 338-352).

K.N. Manuilov (1959), Cylindrical tape-like spring as a transmission mechanism, *Trudy NIIGMP*, Moscow (in Russian).

V.M. Marchenko (1965), *The Temperature Fields and Stresses in the Design of Flying Apparatus*, Mashinostroenie, Moscow (in Russian).

D. Maugis (2000), Contact, Adhesion and Rupture of Elastic Solids, Springer, New York.

N.W. McLachlan (1964), *Theory and Application of Mathieu Functions*, Dover, New York.

J.H. Michell (1901), The Inversion of plane stress, *Proc. Lond. Math. Soc.*, **34**, 134-142.

S.G. Mikhlin and K.L. Smolitskiy (1967), *Approximate Methods for Solution of Differential and Integral Equations*, ed. R.E. Kalaba, American Elsevier.

S.G. Mikhlin (1988), *Some Questions of the Theory of Errors*, Leningrad University, Leningrad (in Russian), [similar English text with addition of two Appendices in *Errors Analysis in Numerical Processes*, Wiley, New York].

Ph. M. Morse (1948), *Vibration and Sound*, McGraw-Hill, New York.
D. Munz, T. Fett (1999), *Ceramics: mechanical properties, failure behavior, Material Science*, Springer.
L. S. Murashkin, S. L. Murashkin (1977), *Applied Nonlinear Mechanics of Machine Tools*, Mashinostroenie, Leningrad (in Russian).
NASA, Hexfoil Rotary Flexures, *TechBriefs*, www.nasatech.com/ Briefs/Jun02/NPO21154.html; by D. Moore and R.Calvet of Caltech for NASA's Jet Propulsion Laboratory, Pasadena, Calif., see also www.nasatech.com/tsp.
J.M.Neumeister, W.A.Ducker (1994), Lateral, normal, and longitudinal spring constants of atomic force microscopy cantilevers, *Rev. Sci. Instrum.*, **65**, 2527-2531.
L.W. Nickols (1955), An adjustable strut for use in measuring instruments, *J.Sci. Instrum.*, **32**, 149-150.
R. Noch, H. Huhn (1959), Untersuchungen uber den Einfluss von Temperaturanderungen auf Messwertanzeige von Feinmesszeigern mit glosser mechanisher Ubersetzung, *Zeitshrift fürInstrum. Kunde*, **67**, 11(in German).
P. V. Novitskii (1968), *Fundamentals of the Information Theory of Measuring Devices*, Énergia, Leningrad (in Russian).
P. V. Novitskii, V. G. Knorring, V. S. Gutnikov (1970), *Digital Instruments with Frequency Transducers*, Énergia, Leningrad (in Russian).
P. V. Novitskii (1978), Real and Imaginary Problems on the development of an information theory of measurement facilities, *Meas. Tech.*, **21**, 1079-1084.
P.V. Novitskii, I.A. Zograf, V.S. Labunets (1990), *Dynamics of Errors in Measurement Means*, Ěnegoatomizdat, Leningrad (in Russian).
N.P.L. (1956), *Application of Spring Strips to Instrument Design*, Notes on Applied Science, **15**, H.M.S.O., London.
J. T. Oden, E. A. Ripperger (1981), *Mechanics of Elastic Structures*, 2nd ed., McGraw-Hill, New York.
Ya.G. Panovko *et al.*(1960), *Construction Damping in Non-movable Joints*, Latvian AN, Riga (in Russian).
Ya. G. Panovko (1968), Natural and forced vibrations of beams and beam systems, in *Strength, Stability, Vibrations*, Handbook (1968), I. A. Birger, Y. G. Panovko, eds., **3**, 285-346, Mashinostroenie, Moscow (in Russian).
Ya. G. Panovko & I. I. Gubanova (1979), *Stability and Vibrations of Elastic Systems*, 3rd ed., Nauka, Moscow; English translation of 1st ed. Consultants Bureau, New York, 1965.
J.M. Paros and L. Weisbord (1965), How to design Flexure Hinges, *Machine Design*, **37**, Nov. 25, 151-156.
E. Parzen, K. Tanabe, G. Kitagawa (1998), ed., *Selected Papers of Hirotogu Akaike*, Springer.
Z. Peng and P.West (2005), Crystal sensor for microscopy applications, *Appl. Phys. Lett.*, **86**, 014107-1-3.
J. Perry (1890), Twisted strips, *Proc. Phys. Soc.of London*, **XLIV**, June, 343-349.
B.N.J. Persson (1998), *Sliding Friction: Physical principles and Applications*, Springer.
R.E. Peterson (1974), *Stress Concentration Factors*, Wiley, New York.
V.B. Petrov (1967), Stability of the twisted beams with unequal principal rigidity at bending, in *Strength Calculations*, **15**, Mashinostroenie, 195-216 (in Russian).
D.M. Photiadis and J.A. Judge (2004), Attachment losses of high Q oscillators, *Appl. Phys. Lett.*, **85**, 482-484.
O. Piétrement and M.Troyon (2000), General Equations Describing Elastic Indentation Depth and normal Contact stiffness versus Load, *J. Colloid Interface Sci.*, **226**, 166-171.
J.E. Plainevaux (1953), *Nuovo Cimento*, 10, 1451-1459; idem, ibid. Etude des déformation d'une lame de suspension élastique, *Nuovo Cimento*, 4, 922-928 , 1956a; idem, ibid Movement de tangage d'une suspension elementaire sur lames élastiques, *Nuovo Cimento*,5, 1133-1141, 1956b, (in French).

B.T. Poliak (1987) *Introduction to Optimization*, English, Optimization Software, New York.

P. Poncharal, Z. L. Wang, D. Ugarte, W.A. de Heer (1999), Electrostatic Deflection and Electromechanical Resonance of Carbon Nanotubes, *Science*, **283**, 1513-1516.

E.P. Popov (1948), *Nonlinear Problems of the Statics of Rods*, OGIZ, Moscow (in Russian).

V.N. Popov (2004), Carbon nanotubes: properties and application, *Materials Science and Engineering*, **R 43**, 61-102.

F.J. von Preissig (1989), Applicability of the classical curvature-stress relation for thin films on plate substrates, *J. Appl. Phys.*, **66**, 4262-4268.

T.J.Quinn, C.C.Speake, and R.S.Davis (1997), Novel torsion balance for the measurement of the Newtonian gravitational constant, *Metrologia*, **34**, 245-249; idem, ibid.(1992) Material problems, *Phil. Mag. A*, **65**, 261-276.

E. Ratajczyk (1968), Výpočet převodů mikrokátorů, *Merova Technika*, **7**, N6, 81-86 (in Czech).

J.W.S. Rayleigh (1945), *Theory of Sound*, 2nd ed., Dover Publications.

N.N. Rogacheva (1994), *The Theory of Piezoelectric Shells and Plates*, CRC Press, Boca Raton, Fla.

V.Y. Rosenberg (1978), A deterministic approach to the determining the amount of measurement information quantity, *Meas. Tech.*, **21**, 10, 1329-1334.

J. E. Sader (2003), Susceptibility of atomic force microscope cantilever to lateral forces, *Rev. Sci. Instrum.*, **74**, 2438-2443.

J.E. Sader and C.P. Green (2004), In-plain deformation of cantilever plates with applications to lateral force microscopy, *Rev. Sci. Instrum.*, **75**, 878-883.

M.C. Salvadori, I.G.Brown, A.R. Vaz, L.L. Melo, and M. Cattani (2003), Measurement of the elastic modulus of nanostructured gold and platinum thin films, *Phys. Rev. B*, **67**, 153404 (4).

D. Sarid (1997), *Exploring Scanning Probe Microscopy with Matematica*, Wiley.

A.Sarkar, J-F. Léger, D. Chatenay, and J.F.Marko (2001), Structural transition in DNA by external force and torque, *Phys. Rev. E*, **63**, 05193-1-9.

S. W. Schediwy, S. Gras, L. Ju, and D.G. Blair (2005), High Q factor bonding using natural resin for reduced thermal noise of test masses, *Rev. Sci. Instrum.*, **76**, 026117.

L. I. Sedov (1993), *Similarity and Dimensional Methods in Mechanics*, Transl. by A. G. Volkovets, 10-th ed., CRC Press.

Saint-Venant B. (1858), Torsion en général. Points dangereux et conditions de résistance, *Savants Étrangers*, **XIV**, 347-453, in Travaux et Titres Scientifiques, Imprimerie de Mallet-Bachelier.

A. I. Sergeeva, Y. M. Tseytlin (1981), Analyzing accuracy in the measurement of the thickness of dielectric films, *Meas. Tech.*, **24**, 557-559.

A.I. Sergeeva, Y. M. Tseytlin (1985), Effects of surface geometry in insulating films on the accuracy of thickness measurement with a contact instrument, *Meas. Tech.*, **28**, 951-953.

A.I. Sergeeva, Y. M. Tseytlin, N.I. Karmalitsyn (1990), Evaluation of corrections due to local deformations in contact measurements of the thickness of low-modulus articles, *Meas. Tech.*, **33**, 212-214.

J.A. Schellman (1974), Flexibility of DNA, *Biopolymers*, **13**, 217-226.

J.F. Shackelford, W. Alexander (2001), *CRC Materials Science and Engineering Handbook*, CRC Press, Boca Raton.

W.N. Sharpe (2002), Mechanical Properties of MEMS Materials, in *The MEMS Hadbook*, ed. Mahamed Gad-el-Hak, 3-1-3-33, CRC Press, Boca Raton.

V.A. Shemetillo (1975), Research of the temperature and vibration influence on an uncertainty of the set of the unified means for precision parts measurement, in *Experience of Development the Progressive Methods and Means for Quality Assurance*, LDNTP, Leningrad (in Russian).

A. Shimokohbe (2002), Development of nano-meter motion control in Japan, *J. Braz.*

Soc.Mech.Sci. <B>., **24**(4), 312-317.

B.F. Shorr (1961). Vibration of Twisted Beams, *Izv. AN SSSR, Mekhanika i Mashinostroenie*, **3**, 102-112.

B.F. Shorr (1968), Naturally twisted beams, in *Strength, Stability, Vibrations*, Handbook (1968), I. A. Birger, Y. G. Panovko, eds., **1**, 440-465, Mashinostroenie, Moscow (in Russian).

I.M. Sivokonenko and K.N. Yavlenskii (1964), Guides for rotational motion, *Handbook for the Designer of Precision Instrumentation*, ed. F.L. Litvin, 462-576, Mashinostroenie, Moscow (in Russian).

B.A. Skolozubov (1953), Twisted metallic strips as the transmission mechanisms of instruments, *Stanki i Instrument*, No9, 22 (in Russian).

A.H. Slocum (1992), *Precision Machine Design*, Prentice Hall, New Jersey.

S.B. Smith, Y. Cui, C. Bustamante (1996), Overstretching B-DNA: The elastic Response of Individual Double- Stranded and Single-stranded DNA Molecules, *Science*, **271**, 792-799.

S.T. Smith and D.G. Chetwynd (1992), *Foundations of Ultraprecision Mechanism Design*, Gordon and Breach.

S. T. Smith (2000), *Flexure Elements of Elastic Mechanisms*, Gordon and Breach.

C.J. Smithells (2004), *Smithells Metals Reference Book*, 8th ed. W.F.Gale and T.C. Totemeier, Elsevier, ASM.

G.G. Stoney (1909), The tension of metallic films deposited by electrolysis, *Proc. R. Soc. London*, A, **82**,172-175.

T.Strunz, K. Oroszlan, R. Schäfer, and H.-J. Gűntheodt (1999), Dynamic force spectroscopy of single DNA molecule, *PNAS*, **96**, 11277-11282.

M.J.O. Strutt (1932), *Laméshe-Mathieushe-und vermandte funktionen in physik und technik*, Verlag J. Springer, Berlin (in German).

D. Swigon, B.D. Coleman, and I. Tobias (1998), The Elastic Rod Model for DNA and its Application to the Tertiary Structure of DNA Minicircles in Mononucleosomes, *Biophys. J.*, **74,** 2515-2530.

R. Szilard (1974), *Theory and Analysis of Plates*, Prentice-Hall, New Jersey.

M. Tanaka (1983), The dynamic Properties of a Monolithic Mechanism with Notch Flexure Hinges for Precision Control of Orientation and Position, *Japanese J. Appl. Phys.*, **22**, 193-200.

S. B. Tarasov, Y. M. Tseytlin (1976), *The Increase of Vibro-stability of the Instruments for Linear Measurement*, LDNTP, Leningrad (in Russian).

Iu. V. Tarbeev, Y. M. Tseytlin *et al.* (1983a), Metrology of small length. *Investigations in the Field of Linear and Angular Measurement*, Ĕnergoatomisdat, Leningrad (in Russian).

Iu. V. Tarbeev, N. P. Gherasimov, Y. M. Tseytlin* (1983b), Metrecology-the problems and procedures, in *Theoretical and Practical Limits of measurement Accuracy*, Abstracts of 2nd Symposium of IMEKO, 55-56, Budapest, Hungary.

T. Thundat, R.J. Warmack, G.Y. Chen, and D.P.Allison (1994), Thermal and ambient-induced deflections of scanning force microscope cantilevers, *Appl. Phys. Lett.*, **64**, 2894-2896.

Z.A. Timofeeva and T.G.Petrova (1961), Physicomechanical properties of Torsion Suspension for Electrical measuring Instruments, *Meas. Tech.*, **4**, 127-130.

S.P. Timoshenko (1925), Analysis of Bi-Metal thermostats, *J. Opt. Soc. Amer. and Rev. Sci. Instrum.*, **11**, 233-255.

S.P. Timoshenko (1955), *Strength of materials*, Van Nostrand, New York.

S.P. Timoshenko, S. Woinowsky-Krieger (1959), *Theory of Plates and Shells*, 2nd ed., McGraw-Hill.

S.P. Timoshenko, D.H. Young (1964), *Vibration Problems in Engineering*, 3rd ed., Van Nostrand.

A. Timpe (1905), Probleme der Spannungsverteilung in ebenen Systemen, einfach gelöst mit Hilfe der Airyschen Funktion, *Math. Und Phys.*, **52**, 348-383 (in German).

J. Todd (1949), The condition of certain matrices, *Quart. J. Mech. Appl. Math.*, **2**,469-472.

H. Troeger (1962), Consideration in the application of flexural pivots, *Automatic control*, **17**, 41-46.

Y. M. Tseytlin* (1962a), Adjustment of contactless pneumatic systems for measuring Dimensions, *Meas. Tech.*, **5**, 367-371.

Y. M. Tseytlin*, I.V. L'vovich (1962b), New transducers for the automation of dimensional control, *Meas. Tech.*, **5**, 911-914.

Y. M. Tseytlin* (1963), Guides with flat springs for forward displacements, *Meas.Tech.*, **6**, 732-736.

Y. M. Tseytlin*, I.V. L'vovich, O.I. Yusov (1964 a), Photoelectric transducer with a spring mechanism, *Meas. Tech.*, **7**, 22-23.

Y. M. Tseytlin** (1964b), Nouveaux transducers de mesure pour l'automatization des opérations de vérification, *La Revue Polytechn.*, Switzerland, N°1202, 115-121(in French).

Y.M. Tseytlin (1964c), *Reliability of the spring mechanisms of measuring heads and transducers*, LDNTP, Leningrad (in Russian).

Y.M. Tseytlin (1964d), Articulated and spring transmission mechanisms, *Handbook for the Designer of Precision Instrumentation*, ed. F.L. Litvin, 633-674, Mashinostroenie, Moscow (in Russian).

Y.M. Tseytlin (1964e), Elastic-plastic twisting of the microstrips, Trudi of Leningrad Polytechnic Institute, **236**, 103-108, Mashinostroenie (in Russian).

Y. M. Tseytlin* (1965), Computation of spring-loaded instrument mechanisms, *Meas. Tech.*, **8**, 223-226.

Y.M. Tseytlin, B.M. Sorochkin, B.G. Larionov, I.M. Barkan (1968), *Photoelectric Automata for Dimensional Control*, Mashinostroenie, Leningrad (in Russian).

Y. M. Tseytlin* (1969), Spring-loaded instruments for testing dimension and displacements, and the prospect of their development, *Meas. Tech.*, **12**, 916-919.

Y. M. Tseytlin* (1971), Optimal adjustment of elastic mechanisms, *Meas. Tech.*, **14**,1120.

Y. M. Tseytlin (1972), *Elastic Kinematical Devices*, Mashinostroenie, Leningrad (in Russian).

Y. M. Tseytlin*, N. A. Pisareva (1973), Opticator having increased measurement limits, *Meas.Tech.*,**16**, 426-428.

Y. M. Tseytlin* (1974), Selection of normal conditions for linear and angular Measurements, *Meas. Tech.*, **17**, 342-345.

Y. M. Tseytlin*, G. B. Grechukhina (1976), Analysis of the conditions of execution of linear and angular measurements, *Meas. Tech.*, **19**, 79-81.

Y. M. Tseytlin* (1978 a), Automation of spring measuring heads, *Meas. Tech.*, **21**, 34-36.

Y. M. Tseytlin* (1978 b), Normal conditions for linear and angular measurements in feedback monitoring, *Meas. Tech.*, **21**, 615-616.

Y. M. Tseytlin* (1978c), Conditions for obtaining uniform statistical observation results in linear measurement, *Meas. Tech.*, **21**, 1218-1219.

Y.M. Tseytlin* (1979), Identification of the normal conditions of linear and angular measurements, in *Theoretical Problems of Metrology*, Trudy of the metrological institutions of USSR, **237**(297), 86-95 (in Russian).

Y. M. Tseytlin* (1980 a), Condition and prospects for metrological provisions for inspecting the thickness of especially thin coatings, *Soviet Journal of Nondestructive Test*, **16**, 476-478 (Plenum Publishing Corporation, 1981).

Y. M. Tseytlin* (1980 b), Normal conditions for synthesizing elastic measuring systems, *Meas. Tech.*, **23**, 304-306.

Y. M. Tseytlin* (1981a), Designing a displacement transducer of the helicoid multivibrator type, *Meas. Tech.*, **24**, 278-280.

Y.M. Tseytlin (1981b), *Normal Conditions of Measurement in the Mechanical Engineering*,

Mashinostroenie, Leningrad (in Russian).
Y.M. Tseytlin*, V. M. Brzhezinskii , G. B. Grechukhina, V.Y. Demchenko (1982 a), Simulation measures for especially thin coatings and means of checking them, *Meas. Tech.*, **25**, 570-573.
Y. M. Tseytlin* (1982 b), A helicoidal-multivibrator measurement transducer, *Mashinovedenie*, No1, 53-57,Acad. of Sciences USSR (in Russian).
Y, M, Tseytlin*, N. N. Konev, N.V. Lyubetskii (1984 a), Thickness checking for thin metal films, *Meas.Tech.*, **27**, 893-895.
Y. M. Tseytlin* *et al.* (1984 b), Flexible checking automation, *Meas. Tech.*, **27**, 1002-1004.
Y. M. Tseytlin*, N. P. Gherasimov (1984 c), Principles underlying the measurement assurance of thickness meters for very thin coatings, *Proc. Intern. Symp. Metrology for Quality Control in Production*, Tokyo, Japan, 260-265.
Y. M. Tseytlin* (1984d), Structural-functional model of measurements of geometric quantities, *Meas. Tech.*, **27**, 1011- 1014.
Y. M. Tseytlin* (1985), Stability of distribution laws of results under normal measurement conditions, *Meas. Tech.*, **28**, 964-967.
Y. M. Tseytlin* (1987), Extended normal conditions for measurements of the first and second kinds, *Meas. Tech.*, **30**, 419-422.
Y. M. Tseytlin* (1988), Internal friction in a helicoidal multivibrator, *Soviet Machine Science*, No**3**, 105-109, Allerton Press, New York..
Y.M. Tseytlin, Iu. V. Skachko, V.V. Kapirin (1989), *Modified String Transducers for Measuring of the Geometric Quantities*, Standartov, Moscow (in Russian).
Y.M. Tseytlin (1994), Special measures for thin coatings measurement, *Proc. 40th Intern. IS*, 181-189, ISA, Baltimore, USA.
Y. M. Tseytlin (1995), Helicoidal multivibrator as a bifurcating transducer, *Proc . 41st Intern. IS*, 499-508, ISA, Aurora, USA.
Y. M. Tseytlin (1997), Metrecology: Measurement Uncertainty and Environmental Margins, *Proc. 43rd Intern. IS*, 423-433, ISA, Orlando, USA.
Y. M. Tseytlin (1998), Quantum Metrology, non-linearity, and motion stability in MEMS, *Proc. 44th Intern. IS*, 664-673, ISA, Reno, Nevada, USA.
Y. M. Tseytlin (2002), Notch flexure hinges: an effective theory, *Rev. Sci. Instrum.*, **73**, 3363-3368.
Y.M. Tseytlin (2005), High resonant mass sensor evaluation, *Rev. Sci. Instrum*, 76.
L.M. Tsukernik (1976), Approximate calculation of the flat springs, in *Calculations on Strength*, **17**, 157-182 (in Russian).
Ya.Z. Tsypkin, B.T. Polyak (1992), Robust Absolute stability of Continuous Systems, in *Robustness of Dynamic Systems with Parameter Uncertainties*, Birkhauser Verlag, 113-124.
A. M. Turing (1948), Rounding-off errors in matrix processes, *Quart. Journ. Mech. Appl. Math.*, **1**, 3,287-308.
P.B. Usatin (1961), Computing Torsion Suspensions for Moving Parts of Electrical Meausring Instruments, *Meas. Tech.*, **4**, 130-135.
A.S. Valedinskii (1960), Spring hinges and mechanisms in measurement instruments, in *Calculations of Parts and Mechanisms of Precise Instruments*, 171-179, MVTU, Mashgiz, Moscow (in Russian).
V.Z. Vlasov (1961), *Thin-walled Elastic Beams*, transl. from *Tonkostennye uprugie Sterzhni*, 1959, Israel Program for Scientific translations, Jerusalem, published for National Science Foundation USA, Washington.
M. Vogel, C. Mooser, and K.Karrai (2003), Optically tunable mechanics of microlevers, *Appl. Physics Lett.*, **83**, 1337-1339.
A. M. Voutsas (1963), Twisted beam transducer, *AIAAJ*, **1**, 911-913.
J. E. Walsh (1949), Some significance tests for the median which are valid under very general conditions, *Ann. Math. Stat.*, **20**, 64-81.

L. Wang, H. Hu (2005), Flexural wave propagationin single-walled carbon nanotubes, *Phys. Rev. B*, **71**, 195412.

W. Weaver, S. P. Timoshenko, D. H. Young (1990), *Vibration Problems in Engineering*, 5th ed.,Wiley.

W.D. Weinstein (1965), Flexure-pivot bearings, *Machine Design*, **37**, (13), 150-157, (16), 136-145.

A. Wiemer (1962), *Pneumatic Measurement of Dimensions*, Mashgiz, Moscow (Russian translation from German: Pneumatische Längenmessung, VEB, Berlin, 1960).

N. Wiener (1961), *Cybernetics*, New York., M.I.T. Press

S. S. Wilks (1963), *Mathematical Statistics*, Wiley.

D.F. Wilkes (1968), Rolamite. A new mechanism, *Mechanical Engineering*, **94**, 11-29.

W.H. Wittrick (1948), The theory of symmetrical crossed flexure pivots, *Austral. J. Sci.Res.*, *A*, **1**, 121-134.

E.G. Wolff (2004), *Introduction to the Dimensional Stability of Composite Materials*, DEStech Publications, PA U.S.A.

E. R. Wong, P.E. Sheehan, C. M. Lieber (1997), Nanobeam mechanics: elasticity, strength, and toughness of nanorods and nanotubes, *Science*, **277**, 26 September, 1971-1975.

Y. Wu and Z. Zhou (2002), Design calculations for flexure hinges, *Rev. Sci. Instrum.*, **73**, 3101-3106.

R. Yang, G. Chen, G.J. Snyder, J-P. Fleurial (2004), Multistage thermoelectric microcoolers, *J. Appl. Phys.*, **95**, 8226-8232.

L.P.Yatsenko, M. Loeffler, B.W. Shore, K. Bergmann (2004), Interferometric analysis of nanostructured surface profiles: correcting material-dependent phase shifts, *Appl. Optics*, **43**, 3241-3250.

D. Young (1950), Vibration of Rectangular Plates by Ritz method, *J. Appl. Mech.*, **17**, 448-453.

W.C. Young and R.G. Budynas (2002), Roark's Formulas for Stress and Strain, 7th ed., McGraw-Hill, New York.

V. N. Zheludkov (1971), Kinematics of the leaf spring hinge at plane bending, *Izvestia Vuzov, Priborostroenie*, 79-83; idem, ibid. (1965), 143-151; (1963), 115-128 (in Russian).

P. Zenow (1970), Untersuchung der Einwirkung von kräften auf die dritten Ableitungen, Feingerätetechnick,2, 63-66(in German).

C. Zener (1948), Elasticity and Anelasticity of Metals, Chigago Univercity Press.

Note: Spell of the author's last name in the publications marked with * was Tseitlin; Spell of the author's name in the publication marked with **is Zeitlin Ja. M., in other publications and citations, the spell may be as Cejtlin J.M.; Цейтлин Я. М (in Russian),TSeitlin IA. M.

Patents SU referenced in the text are the Author's Certificates of the former Soviet Union.

INDEX

Mechanical Engineering Series *(continued from page ii)*

W.K. Gawronski, **Dynamics and Control of Structures: A Modal Approach**

G. Genta, **Dynamics of Rotating Systems**

D. Gross and T. Seelig, **Fracture Mechanics with Introduction to Micromechanics**

K.C. Gupta, **Mechanics and Control of Robots**

R. A. Howland, **Intermediate Dynamics: A Linear Algebraic Approach**

D. G. Hull, **Optimal Control Theory for Applications**

J. Ida and J.P.A. Bastos, **Electromagnetics and Calculations of Fields**

M. Kaviany, **Principles of Convective Heat Transfer, 2nd ed.**

M. Kaviany, **Principles of Heat Transfer in Porous Media, 2nd ed.**

E.N. Kuznetsov, **Underconstrained Structural Systems**

P. Ladevèze, **Nonlinear Computational Structural Mechanics: New Approaches and Non-Incremental Methods of Calculation**

P. Ladevèze and J.-P. Pelle, **Mastering Calculations in Linear and Nonlinear Mechanics**

A. Lawrence, **Modern Inertial Technology: Navigation, Guidance, and Control, 2nd ed.**

R.A. Layton, **Principles of Analytical System Dynamics**

F.F. Ling, W.M. Lai, D.A. Lucca, **Fundamentals of Surface Mechanics: With Applications, 2nd ed.**

C.V. Madhusudana, **Thermal Contact Conductance**

D.P. Miannay, **Fracture Mechanics**

D.P. Miannay, **Time-Dependent Fracture Mechanics**

D.K. Miu, **Mechatronics: Electromechanics and Contromechanics**

D. Post, B. Han, and P. Ifju, **High Sensitivity and Moiré: Experimental Analysis for Mechanics and Materials**

R. Rajamani, **Vehicle Dynamics and Control**

F.P. Rimrott, **Introductory Attitude Dynamics**

S.S. Sadhal, P.S. Ayyaswamy, and J.N. Chung, **Transport Phenomena with Drops and Bubbles**

A.A. Shabana, **Theory of Vibration: An Introduction, 2nd ed.**

A.A. Shabana, **Theory of Vibration: Discrete and Continuous Systems, 2nd ed.**